Gravity Interpretation

Wolfgang Jacoby · Peter L. Smilde

Gravity Interpretation

Fundamentals and Application of Gravity Inversion and Geological Interpretation

With CD-ROM

Prof. Dr. Wolfgang Jacoby
Johannes
Gutenberg-Universität Mainz
Institut für Geowissenschaften
Saarstr. 21
55099 Mainz
Germany
jacoby@uni-mainz.de

Dr. Peter L. Smilde
Finther Str. 6
55257 Budenheim
Germany
peter.smilde@smilde-becker.net

ISBN: 978-3-642-44850-8   ISBN: 978-3-540-85329-9 (eBook)

Cover design: deblik, Berlin

Printed on acid-free paper

9 8 7 6 5 4 3 2 1

springer.com

Preface

Gravity interpretation involves inversion of data into models, but it is more. Gravity *interpretation* is used in a "holistic" sense going beyond "inversion". Inversion is like optimization within certain a priori assumptions, i.e., all anticipated models lie in a limited domain of the a priori errors. No source should exist outside the anticipated model volume, but that is never literally true. Interpretation goes beyond by taking "outside" possibilities into account in the widest sense. Any neglected possibility carries the danger of seriously affecting the interpretation.

Gravity *interpretation* pertains to wider questions such as the shape of the Earth, the nature of the continental and oceanic crust, isostasy, forces and stresses, geological structure, finding useful resources, climate change, etc. Interpretation is often used synonymously with modelling and inversion of observations toward models. Interpretation places the inversion results into the wider geological or economic context and into the framework of science and humanity. Models play a central role in science. They are images of phenomena of the physical world, for example, scale images or metaphors, enabling the human mind to describe observations and relationships by abstract mathematical means. Models served orientation and survival in a complex, partly invisible physical and social environment.

Inversion of gravity anomalies is the mathematical derivation of density distributions and their confidence limits. This is a notoriously non-unique problem, while the so called forward problem of finding the gravity effects of given mass distributions is perfectly unique. The ambiguity of inversion simply results from the fact that knowledge of a sum does not imply knowledge of the addends. If you know $c = a + b$, but nothing about a and b, c reveals neither a nor b. There is always an infinite model space; the infinity of answers can be reduced only by invoking a priori information. It can be of any nature and depends on the problem at hand. If for example $b = 2a$, you get $a = c/3$ and $b = 2c/3$.

This treatise attempts to give a perspective of the problem and to prepare readers for finding their way to solutions. A priori information is central to gravity inversion. It ranges from "hard" geological and geophysical data, such as seismic results, to general ideas based on experience and to models of processes which would produce gravity signals. Generally the additional knowledge will be limited, but often very important aspects will be revealed. If the a priori information were complete, there would be no problem left to be solved.

This touches the question: what does gravity tell and what not? To endeavour along these lines is the exciting business of approaching the truth, but one can never be absolutely sure. Nature has built in too many obstacles. If gravity is measured on or above the Earth's surface, one cannot truly look inside. The signals come largely from within, though. There is a philosophical extension of these ideas about gravity interpretation: in our general intellectual human condition we are in a very similar situation concerning our world views including that of ourselves. We observe and receive signals from within and without, and we communicate. We build our virtual worlds that should be consistent. This aim feeds back into our approach to gravity interpretation.

Texts of Applied Geophysics generally have an "exploration outlook"; the present book has also a strong geodynamic "inclination". Gravity is active and passive: a *force* doing work toward equilibrium from a disturbed state and generating a *field* with observable signals to be interpreted. In many geological situations gravity has done work and we try to find out what happened. For example, a valley has been excavated and refilled by lower-density sediments, giving a negative gravity anomaly. Or hot, low-density mass has risen or is rising, and cold, high-density mass has sunk or is sinking and working against viscous forces and deflecting density surfaces from their equilibrium level. The density distributions generated give gravity signals which can be interpreted only in view of such model ideas. Without them, models of a totally different nature can "explain" the anomalies.

This situation causes confusion. Is it worth at all to interpret gravity? Some seem to think: not. This view is definitively wrong. Gravity plays two fundamentally useful roles in the earth sciences: it helps to inexpensively detect "anomalies" worth studying, and it falsifies and eliminates models by forward computations. The methodological side is the theory and practice of data gathering, forward modelling and of Bayesian inversion, including the various preliminary steps of measurement and data preparation. The practical side is the presentation of applications and case histories. The philosophical side is that it wants to teach general aspects of applying observations to science and to life.

Presentation of observational techniques is kept to a minimum, but some discussion is unavoidable. Gravity or geoid observations are affected by errors or confidence limits. Errors have an important effect on what can be learnt from gravity, so their discussion is carried through all chapters. With the development of new methods of terrestrial, marine, airborne and satellite-based observational methods, and with increasing accuracy of the observations the scope of interpretation widens. Many methods of forward calculation of gravity effects are well known and reference is given to other texts; but some aspects of the basic approach in this treatise are novel.

Much of a textbook is concerned with the reader learning to work in geophysical "practice". Many today, especially science administrators on all levels, suggest that teaching is the main function of universities, and efforts in "pure science" and conveying in-depth understanding is not so important. This attitude is short-sighted. Only deep understanding will produce reliable results, also in limited exploration projects. Good self-critical judgement, for example of the probability and possibility of errors in an interpretation, requires knowledge beyond technical skills. This our own experience we wish to share. Indeed, we endeavour to make readers wonder about problems.

Probably the best way of learning is from mistakes and from independent efforts in problem solving. We therefore include, as a CD, a collection of tasks or problems with some instructions for how to approach solutions. This will give readers the chance to make their own mistakes and to correct them. Answers are listed at the end, including discussion of the problems and solutions. Some of the tasks are applications of inversion (Chap. 7) to geological or theoretical modelling which add principal aspects discussed at some length.

One of our own examples serves as an illustration: when working out the solid-angle solution for a cube at one of its corners (see Sect. 2.9.6), the assumption that the gravitational vector effect points to the centre of mass led to the surprisingly beautiful result that the vertical gravity effect would be precisely 1/6 of that of the infinite Bouguer slab of the same thickness and density. But beauty is no proof, and the result did not stand the test. The mistake was that, contrary to widespread belief, the gravitational vector does *not* generally point to the centre of mass, except in certain cases of special symmetry (see Sect. 2.9.1.2) which should have been immediately evident, for example, from the Earth's ellipsoid or the geoid. The misconception arose from a mix-up with mechanics where the action of a force on a body is described by action on its barycentre, i.e. centre of mass or balance point.

The authors have consulted other texts covering the subject, especially the classical book (in German) by Karl Jung (1961), *Schwerkraftverfahren in der Angewandten Geophysik* (Gravity methods in Applied Geophysics; it will be referred to as KJ61), and the book *Interpretation Theory in Applied Geophysics* by F.S. Grant and G.F. West (1965) (referred to as GW65). Many useful ideas have been taken up and partly expanded. In those early days of computing machines their possibilities had been clearly seen and the foundations had been laid down for their application.

One of the authors (WJ) studied physics, geophysics and more and more geology and considers himself a geophysically guided geologist, interested in how the Earth works and concerned about how mankind treats its home planet. The other author (PS) studied geodesy and became more and more involved in geophysics and geology when working with WJ on his PhD thesis on gravity inversion, developing the program package INVERT, of which an executable copy is attached to this book on a CD. The thesis is also the basis of the most important last chapter of the book on optimization and inversion. The cooperation led to a synthesis of the geological-geophysical approach to the problems of interpretation and the geodetic, more mathematically inclined approach. It is the combination of geological imagination and experience, on the one hand, and abstract geophysical-mathematical reasoning, on the other, that is the basis of Earth science. Experience-based intuition must be checked by mathematical validation. Indeed, science is suspended between the two extremes of freedom of thinking and rigorous checking. Scientists surely endeavour to approach the truth in such suspense.

Many colleagues and friends in various institutions, not only from our own study field, have participated in teaching us this lesson, from our parents, families and some school teachers to our academic teachers, Karl Jung[†], Kiel, and Reiner Rummel, Delft, and to our later colleagues and students. Every one of them has chosen her/his own way and none is responsible for ours, but the – hopefully – mutual benefit has been immense. The intellectual challenges by colleagues and students are gratefully acknowledged. Geological teaching by Eugen Seybold, Kiel, and exchange with Richard Walcott, Richard Gibb, Alan Goodacre and Imre Nagy in Canada and with Gerhard Müller[†], Frankfurt (Main), were important. In Mainz, Georg Büchel, Evariste Sebazungu, Tanya Fedorova, Ina Müller, Chris Moos, Michaela Bock, Herbert Wallner, Hasan Çavşak, Tanya Smaglichenko and many others were influential on both of us.

Herbert Wallner helped intellectually by many discussions, with calculations and quite a number of figures. Tanya Fedorova provided some of the gravity inversion models. Evariste Sebazungu, in his own PhD thesis on potential field inversion, developed original ideas which entered into this treatise. Hasan Çavşak provided gravity calculations for various polyhedral bodies and helped discovering errors in some theoretical derivations. Pierre Keating provided information on some of the free modelling software. Discussions with Markus Krieger (Terrasys, Hamburg) led to several ideas and insights into the practical solution of interpretation problems.

All of them and many more contributed thought-provoking ideas and thus influenced the present treatise. Most importantly, the mutual discussions between the authors through the whole time of their cooperation were beneficial to both. Finally, lecturing on gravity (and magnetics) taught us more than anything else to endeavour to present the ideas clearly.

Petra Sigl was always helpful and did an excellent job in preparing most of the figures in this book. The book could hardly have been completed without the many forms of support by the *Institut für Geowissenschaften, Johannes Gutenberg-Universität Mainz*, various grants by *Deutsche Forschungsgemeinschaft*, Bonn, and by *Stiftung Rheinland-Pfalz für Innovation*, Mainz, the *Terrasys company, Hamburg*.

Bettie Higgs, Stefan Bürger, Herbert Wallner, Mark Pilkington, Pierre Keating critically proofread parts of the draft and partly checked the mathematics. The responsibility for any errors remains, however, exclusively with the authors. All help by persons and institutions is gratefully acknowledged, including the many that are not named.

Mainz, Germany   *Wolfgang Jacoby*
                                                                                         Peter L. Smilde

Contents

Symbols Used

Symbols as used in this book are grouped in the fields: ***general rules, mathematics (differentiation, coordinates, geometry), physics (properties), gravity***. Special expressions are explained in the text, particularly if exceptionally deviating from the general usage. The same applies to some constants used with different meanings in different connections.

General rules
Italics: variables x, y, a, ξ, etc.
Normal: physical properties, state, units etc. (exception density ρ, Vp, Vs)

Bold italics; vectors as: ***r***
Scalars shown as: r
Subscripts have different meanings:
(1) vector components, as in δg_x or δg_i, the x or i component of the vector $\boldsymbol{\delta g}(i = 1, 2, 3)$,
(2) counting, e.g., station numbers i, as in δg_i where the δg_i can be considered components of the different vector $\boldsymbol{\delta g}$ of all gravity values,
(3) partial derivatives (see below)
$i, j, k, l, m, n, p\ldots$ running indices, m, n, p often as upper limit. Written in italics in text, normal as subscripts or superscripts.
Δ usually indicates a finite part or segment or a finite difference of a quantity, as in Δx, Δy, Δz finite intervals of x, y, z or lengths
δ usually indicates an anomaly of a quantity
Generally SI units (Systeme International)

Mathematics (differentiation, coordinates, geometry)
General
$=!$ symbolizes a definition
$\times$, $\bullet$, $\cdot$ signify multiplication
$*$ signifies convolution
Underlining: arithmetic mean or average, as in $\underline{\rho}$ = mean density.
w weight or weighting, as in weighted mean; $\overline{W}$ Fourier transform of w

Differentials

df/dx or Df/Dx total differential of f after x etc.

$\partial f/\partial x$ partial differential, abbreviated as $\partial_x f$ or f_x, etc., e.g. in W_x, W_{zz} etc.
  Partial derivatives with respect to a coordinate are mostly self-explaining from the context; if not, as in the case of the x derivative of gravity as the norm of the vector g, specifically defined: $g_{(x)} = \partial W_z/\partial x = \partial^2 W/\partial z\partial x$.

Coordinates, geometry:

$O = (0, 0, 0)$ origin of coordinate system

$P = (x, y, z)$ point of observation or calculated effect; often $P \equiv O$

$Q = (x, y, z)$ point of a source, e.g. a point mass

$x,y,z \equiv x_1,x_2,x_3$ ordinary Cartesian coordinates, z vertical

$i = (1, 0, 0), j = (0, 1, 0), k = (0, 0, 1)$: unit vectors in x, y, z direction.

ξ, η, ζ (alternative) Cartesian coordinates in special cases

X, Y, Z centre of gravity (of an anomalous mass)

$r = (x^2 + y^2 + z^2)^{1/2}$ distance from O or P (see above) to an arbitrary point (x, y, z)
  [r also spherical coordinate, residual gravity anomaly or a number, e.g. of unregularized variables in 7.2.2.2.4].

a, b, c dimensions in x, y, z directions, respectively, e.g., of Cartesian prism [also Earth ellipsoid axes]

$r^* = (x^2 + z^2)^{1/2}$ sometimes used for "2D" distance from P to an arbitrary point (x, y, z)

R, Z, Λ: vertical cylinder coordinates [R also radius of Earth or of a sphere of mass]

r, φ, λ: spherical coordinates (especially: terrestrial), r radius, φ latitude, λ longitude [r also distance or number of unregularized variables in 7.2.2.2.4; φ also function or dip or plunge angle of r from P to Q; λ also $2^1/_2$D half length, Tikhonov regularization parameter]

$\phi = \text{colatitude} = \pi/2 - \varphi$, also geocentric latitude of Earth's normal ellipsoid (4.4.1)

Length, surface, volume, vector, tensor

r (if not otherwise stated) absolute distance in any direction [also spherical coordinate or number of unregularized variables in 7.2.2.2.4]

r_{ijk} triple index notation: distance from P $(0, 0, 0)$ to point (x, y, z) where $i, j, k = 0$ if $x, y, z = 0$ or 1 if $x, y, z \neq 0$ (2.8.3.3)

L general length

λ wavelength [or longitude or regularization parameter]

k wavenumber [also counting index]

κ radius of curvature [also susceptibility]

S, s, ds surface, surface element, respectively [also variance]

E or (E) sometimes used for finite surface element [also Eötvös unit]

V volume, dV infinitesimal volume element (e.g. $dV = dxdydz$)

n normal unit vector, e.g., surface-normal vector

s, $ds = sn$, dsn normal vector of planar surface, surface element, respectively

ds, ds infinitesimal surface element, its normal vector

s, ds also used for path of integration

p normal vector of surface or straight line (Appendix M1)

$\boldsymbol{a} = (a_1, a_2, a_3)$ vector by components

$\boldsymbol{a}^{\mathrm{T}}$ transposed vector

$\boldsymbol{a}^{-1}$ inverse vector

ab (or a·b, or a × b) product of scalars

Scalar product of vectors $\boldsymbol{ab}$ or $\boldsymbol{a}\,\boldsymbol{b}$ or $\boldsymbol{a}\cdot\boldsymbol{b}$ or $\boldsymbol{a}\bullet\boldsymbol{b}$

Vector product $\boldsymbol{a}\times\boldsymbol{b}$ or $\boldsymbol{a}\times\boldsymbol{b}$

$\{W_{ij}\}$ or $\underline{\boldsymbol{W}}$ matrix, tensor

$\underline{\boldsymbol{W}}^{-1}$ inverse matrix

$\underline{\boldsymbol{W}}^{\mathrm{T}}$ transposed matrix $\{W_{ji}\}$

Angles

α, β, γ, ϕ, φ, θ, ψ angles, if not otherwise specified (α, β, γ corresponding to x, y, z)

φ dip or plunge angle of r from P to Q [also least-squares function]

$\psi = \varphi - \pi/2$ complementary angle

$(\boldsymbol{a},\ \boldsymbol{b})$ may mean the angle between vectors $\boldsymbol{a}$ and $\boldsymbol{b}$, as in $\sin(\boldsymbol{a},\ \boldsymbol{b})$ or also in $\sin(x,\ X)$.

Ω, $\delta\Omega$ or $\Delta\Omega$ solid angle

$\underline{\boldsymbol{W}}^{\mathrm{T}}$ transposed matrix $\{W_{ji}\}$

Physics (properties)

f force [also gravimeter scale factor, earth flattening, number of degrees of freedom, function]

M, M mass of a body [m also index; M also "total mass" causing a gravity anomaly]

m general; M or $\Delta M = \Delta\rho \cdot V$ as determined by the gravity surface integral (2.6.6) [or index]

M_{earth} or $M_\oplus$ mass of Earth, may be called M, if meaning obvious.

dm infinitesimal mass element

$\rho = m/V$ density $(\mathrm{kg/m}^3)$.

ρ^* surface density $(\mathrm{kg/m}^2)$.

ρ^+ line density $(\mathrm{kg/m})$.

$\delta\rho^{\#}$ circular average

$\Delta\rho = \rho - \rho_0$: density contrast relative to a reference density ρ_0 (6.1.4, Fig. 6.1 – 1).

t time (units: s, a, Ma) [also thickness of wall, dyke, etc.]

T period of oscillation or...

T temperature

α thermal expansivity

ω angular velocity vector, especially that of Earth; ω its scalar value

Vp, Vs seismic compressional, shear velocities (usually in km/s)

K, μ elastic bulk, shear moduli, respectively

$\underline{m}$ mean atomic weight

$B_\oplus$ Earth's magnetic field strength

m specific magnetization (vector) = magnetic moment of volume element divided by volume

μ permeability, μ_o = p. of vacuum, μ_r = relative p.

κ susceptibility = $\mu_r - 1$, $B = \mu F = \mu_r \mu_o H$. ($B$ magnetic flux density, F mag. field strength) [also radius of curvature]

Q Königsberger ratio

Gravity

General

mGal = milliGal = 10^{-3} Gal

Gal = cm/s^2 = 10^{-2} m/s^2

E = 10^{-9}s^{-2} unit of gravity gradient

G universal gravitational constant:

$\quad$ G = $6.6742 \pm 0.001 \cdot 10^{-11}$ m^{3}kg^{-1} s^{-2} (Nm^{2}kg^{-2}) (or $6.6742\ 10^{-8}$cm^{3}g^{-1}s^{-2})

g gravity vector; g its scalar value

z centrifugal acceleration of point P, z its scalar value

g_o time-averaged gravity felt at point P

a_o time-averaged gravitational attraction felt at P

z_o time averaged centrifugal acceleration of point P

a_k gravitational attraction by cosmic masses felt at P

a_d gravitational attraction by Earth mass deformation felt at P (deformed minus average)

a_c acceleration of coordinate system in inertial system

z_o time-averaged gravitational centrifugal acceleration felt at point P

b_c acceleration of P in an inertial system, undetermined

a, b, c, f lengths of earth ellipsoid semi-axes and flattening, minor, intermediate, major, $f = (a-c)/a$ (see 4.4.1) [also x, y, z dimensions, e.g., of Cartesian prism]

m geodetic parameter (4.4 – 4) [also mass, index]

g_o or g_b reference gravity value (g_b at base station)

δg_i gravity value at point i, relative to some reference (mGal)

g_{oi} gravity value it station i versus g_b

f gravimeter scale factor [also force, earth flattening, number of degrees of freedom]

e_i error at station i, observed minus true value; sometimes synonymous to r_i [also eigen-vector].

ε_g closure error in gravity network

δg_{obs} specifically: observed gravity anomaly value (mGal); the term "anomaly" often refers to ensemble of observed gravity values that represent a distinct geometrical feature.

δg_o reference gravity value, sometimes also $\delta g_o = \delta g_{obs}$ if indicated

δg_m gravity effect calculated for a model.

δg_e = extreme value of a gravity anomaly or a gravity effect

$g_t = g_t(t)$ time-dependent gravity value; $\delta g_t = \partial g / \partial t$ rate of temporal gravity variation (mGal/a)

$\delta g^{(s)}$ gravity effect of a surface element s as superscript

$\boldsymbol{\delta g} = (\delta g_x, \delta g_y, \delta g_z)$ vector of gravitational effect or a gravity anomaly (mGal)

δg_p, δg_q gravity component in directions p and q parallel to s, δg_n normal to s (mGal)

$\delta g_c = \delta g_x + i\delta g_z$ complex gravity effect (mGal) (see 2.8.7.4.3)

δg^* non-dimensional gravity effect rate

$\delta g^{\#}$ circular average of δg

h height, elevation above geoid or just $h = -z$ upward

TOP topography, topographic height

BA Bouguer anomaly

FA free air anomaly

IA isostatic anomaly

rBA residual *BA*

w half width of gravity anomaly; δg_e extreme value, either along a profile or on a map; w distance between the points where $\delta g = \delta g_e/2$. Some authors call $w/2$ half width, i.e. distance from the extreme to half of it [also weighting in averaging]

w^* width of a model, e.g., of a 2D strip

2D, 3D short for two-dimensional, three-dimensional; in 2D usually y is very large $(\partial/\partial y = 0)$.

t thickness of wall, dyke, etc. [also time]

λ $2^1/_2$D model length [also wavelength, longitude, parametrization parameter]

$U \equiv U_g$, ΔU, δU gravitational potential at P (not including the centrifugal potential)

U_z, ΔU_z, δU_z potential of centrifugal force field at P

$W = U_g + U_z$, ΔW, δW potential of gravity field at P [W or w also weighting]

N geoid undulation [also index]

Γ gravitational flux

Optimization and inversion (Chap.7)

Simplified vector and matrix notation; if unambiguous, symbolized by italics (e.g. $Ax = y$)

$\tilde{x}, \tilde{y}$: observed, a priori, input, estimated values

$\hat{x}, \hat{y}$: adjusted, a posteriori, output, resulting values

A: matrix of linear observation equations, also meaning model or model relationship (7.1.2)

$\mathscr{A}$: matrix of non-linear observation equations.

x, x_k, $\boldsymbol{x}$; variables of models (general, numbered, vector) equivalent to parameters

p, p_k, $\boldsymbol{p}$: parameters of models (general, numbered, vector) equivalent to variable [also penalty function; normal vector of surface or straight line]

$\|\ldots\|_2 = a$, L_2 norm of vector $\boldsymbol{a}$

L_n: norm, statistics of distribution of residuals to the power of n.

$\phi(y|\tilde{y})$: probability of y if an actual observation $\tilde{y}$ is given with the covariance matrix C_y.

$x \propto N(\mu, C)$ normally distributed vector variables x with the expectations μ and the covariance matrix C

r residual gravity anomaly: $r = \delta g_o - \delta g_m$ [also distance vector from O, spherical coordinate]

r number of unregularized variables in 7.2.2.2.4

r_{ab} correlation coefficient between $a(x)$ and $b(x)$

f number of degrees of freedom [also force, earth flattening, scale factor]

e_i eigen-vector [also error at station i]

S variance [also surface]

φ function describing a least-squares solution [also angle]

σ_i: genuine standard deviation [also called s_i if specified, see below]

s_i: *criterion standard deviation*, computational quantity for the evaluation of inversion results (if specified, ordinary standard deviation, see above)

λ Tikhonov regularization parameter [or longitude or wavelength or $2^1/_2$D length]

Chapter 1
Introduction

1.1 The Subject and Scope

1.1.1 Gravity

Gravity interpretation is an important endeavour in the quest for understanding the Earth. This is so for several reasons. The shape and mass distribution of the Earth are governed by the central force of gravity counteracted mainly by the molecular and atomic forces against compression and deformation. Ongoing geodynamic processes are driven mainly by thermal disturbances of the equilibrium which gravity tries to achieve, to maintain or to restore. The processes generate density distributions which produce observable gravity signals which are the target of gravity study. Near the surface in the Earth's crust geological structures resulting from past geodynamic processes are "frozen in" and preserved over long periods, as the gravitationally driven forces inherent in the structures are too weak for the strength of the material to be overcome. Natural resources of all kinds are hidden in the structures. Gravity is an economic tool for exploring and discovering the resources.

The relationship between mass and effect is "asymmetric", the effect directly calculable but not vice versa. The resulting notorious ambiguity is the main drawback of gravity interpretation and gives it a "poor reputation". However, in many respects the reservation is not justified if gravity interpretation is seen in perspective. In the first place, information gained from gravity is not in every respect ambiguous; the ambiguity is in the geometrical density distribution, in particular the important parameter of depth. No ambiguity exists in the presence of an object of potential interest inferred from the presence of an observed gravity anomaly, although the opposite is not true: the lack of a gravity anomaly does not necessarily mean that no mass anomalies exist underneath. Equally, the horizontal location (coordinates) of the "centre of gravity" and the total amount of the anomalous mass are unambiguously obtained from the gravity anomaly. The depth of the anomalous mass and its shape or distribution *are* ambiguous. However, the ambiguity is reduced, even by only qualitative arguments, and some aspects of the interpretations can be highly probable. The ambiguity is further reduced by additional "a priori" information

W. Jacoby, P.L. Smilde, *Gravity Interpretation*,
© Springer-Verlag Berlin Heidelberg 2009

from many other sources such as other geological and geophysical data. If such data with uncertainty limits are well known, the problem of gravity interpretation can be "solved" by inversion, especially by "Bayesian inversion" which attempts to achieve the best compromise between all pieces of available information within their particular uncertainty limits, usually called "errors".

An important related aspect is that any data manipulation motivated by many different purposes is explicitly or implicitly an act of interpretation or at least affects the subsequent modelling, inversion and interpretation. Often data on a profile or a map are filtered or smoothed, perhaps to emphasize certain geologically interesting features or simply to render a clearer picture. It must be kept in mind that this does not go without effects on the final geological models. It is advisable to check the results, e.g. by filtering the inverted results with the same filter as applied to the original data. If the filtered results and the unfiltered ones are the same, then filtering is independent from the inversion results. Otherwise filtering did remove some effect present in the model itself which means that the residuals (unfiltered observations minus model effects) increase and the results are distorted.

1.1.2 Motivation

Passing on the experience gained with this type of gravity interpretation, and the insight into the teaching and learning processes involved, are the prime motivation for writing this book. An overview is attempted of the whole field as we know it at the beginning of the 21st century, both from the perspective of basic research and from the application to problems of exploration. Many new developments have taken place, partly in industry with its financial and technical capabilities not available to universities. It seems timely to write such a treatise, although it is difficult to obtain an overview of all the new developments.

The classical 1961 book by K. Jung: "*Schwerkraftverfahren in der Angewandten Geophysik*" (*Gravity methods in Applied Geophysics*, KJ61) is long out of print; it is still quoted even in the English language literature, because of its in-depth and far-sighted treatment of the subject, including topics which became really useful only with increasing computing power. We attempt to follow Karl Jung's footsteps and to present a concise treatise covering subjects from potential field theory, the observation techniques, reductions and data analysis to quantitative interpretation methods and inversion. KJ61 has, indeed, been an important guide in writing the present book. Naturally the authors (WJ and PS) have taken advantage of their knowledge, not only of Jung's book but also from personal study and advice. WJ had Karl Jung as his PhD advisor and PS did his PhD thesis on inversion with WJ as advisor.

There is a lack of recent texts in gravity interpretation. One exception is Blakely (1995), though his emphasis differs. Of course, the numerous books published on applied geophysics in general over the years, devote one or two chapters to gravity, and the other potential field of magnetics. Of these, the two classic books by Grant and West (1965), (GW65) and Telford et al. (1990) have been consulted extensively.

1.1.3 Aims

The aim of this treatise is to give students and professionals insights into potential field interpretation, based on the fundamental theory, especially the gravity field (Chap. 2). The interest in potential field theory goes back to astronomy and geodesy and was summarized in Newton's laws. Gravity is the main subject, but a brief introduction into the theory of magnetics (2.11) is added. While the emphasis is on geophysics and its geological applications, geodesy is intimately connected to gravity, and it is an important goal to bridge the gap between these different specializations.

It is the intention of the authors to give an overview of gravity observation (Chap. 3) as well as reduction and data analysis (Chap. 4). The principal aim of sharing our experience with, and insights into, gravity interpretation leads us to structure its discussion into three chapters: qualitative interpretation (Chap. 5), quantitative interpretation (Chap. 6) and optimization and inversion (Chap. 7). The notion of gravity *anomaly* is central to the whole subject and must be carefully reflected upon, and it usually means an ensemble of differing values in space, i.e. more than a single anomalous point value; it is an important aspect to which we shall return many times, especially in 1.4; 2.6 and 2.9; 3.4; 4.3, 4.6 and 4.77; and of course, in Chaps. 5, 6 and 7 as interpretation always concerns fields, not single points.

Interpretation involves, beside the determination of the source distributions, also their geological implications. "Geological" refers to any aspect of Earth structures and processes, irrespective of scale. In view of the double role of gravity in driving processes and generating useful signals, geodynamics is an important aspect. These aspects are closely related to each other and must be envisioned together when interpreting gravity anomalies, the more so, the bigger the volumes considered. They must be especially taken into account in problems of gravity inversion which attempts interpretation on a rigorous mathematical basis, by models within quantifiable error bounds.

It is our aim to show that in spite of the notorious ambiguity of the inverse problem in potential field theory, gravity is a very useful tool for studying the Earth's interior. True: only the forward problem has unique solutions, and the inverse problem is non-unique. If the source distribution is known, the field distribution is uniquely determined by mathematical relations involving integrals over the mass distributions or convolution of the mass with certain kernel functions. However, there are infinite numbers of mass (or magnetisation) distributions which generate the same gravity (or magnetic) "image". It is emphasized that the solution or model space (or domain) can be reduced with the aid of additional a priori information, and only with it, no matter whether it provides tight or loose constraints. It must thus be carefully evaluated. The task will be to find solutions that are reasonable, plausible or probable compromises between all pieces of information, even if they are in mutual conflict.

The task involves more than mathematical methods and requires more than knowledge of available methodologies. What is needed is something like feeling or intuition based on experience of successful solutions of the problem of gravity interpretation. Intuition may be defined (Wikipedia) as the gift of forming spontaneous, subconscious ideas, for example, insights into complex relationships or inventions – without explicit analytical deduction. Intuition involves an element of chance, but as Louis Pasteur is quoted: "chance in a prepared mind". It is also called "serendipity". It is hoped that the reader will gain some of this through studying this book. But intuition has also the other side of experience generally being guided by "current wisdom" and thus not without bias which may block the imagination. Imaginative minds are needed to transgress such limitations to open up new avenues of thinking about gravity interpretation. Quite often gravity anomalies do guide researchers to well constrained geological solutions or models.

1.1.4 Special Aspects

Some aspects are somewhat unconventional and novel. For example, in the integration of gravity effects of extended mass volumes (Sect. 2.8) a special approach is taken. The long known principle of integration along "rays" from the observation point to mass elements contained within a solid angle from the observation point P (KJ61, 148-155) is exploited systematically. Because this does not generally give the wanted vector effect, it has to be complemented by considering the gravitational components parallel to causative straight mass lines and planes. This way, many of the occurring forward problems are conceptually more easily treated than by schematic classical integration over customarily defined bodies. The widespread misconception that the gravity vector effect of an arbitrary body principally points to its centre of mass was mentioned in the Preface.

Although gravity is central to this treatise, observations of other field quantities (Sect. 2.7.2) are included and need not be transformed into gravity before modelling. As additional errors affect the transformations, it is principally better to model the observations directly; the expressions are provided in Sect. 2.7.2. For purposes of aiding creative imagination, transformations remain, of course, useful, as maps remain important when most data analysis and interpretation are done digitally, i.e. "invisibly".

1.1.5 The Book and the Reader

Experience is acquired by doing, not by theoretical or exclusively technical learning. Experience can be shared. Experience is gained by practicing problem solving.

Feeling and intuition grow with experience possibly guided by a book like this one. It provides the necessary theoretical and practical foundations and includes exercises. Practical problems of gravity inversion are posed and readers can do them. That will lead to surprises and failures as well as successes.

Students new to the field may follow the book and do the exercises along the way. Remember that some patience is needed when, in the beginning, a problem seems unfamiliar and complex. Early attempts to understand a section or to solve a problem may be fraught with mistakes, but with persistence the misunderstandings disappear and solutions fall into place. Or it may become clear that a problem posed cannot be solved. The authors themselves have gone, and are still going, through such a process, and the book can be a guide or a map helping the wanderer to find her/his own way with her/his own short-cuts and detours. For some of us it may be very helpful to always have a pencil and a piece of paper at hand and immediately sketch the situation described in the text. The human mind, while individually very variable, seems to strongly cooperate with the whole person and her/his body; maybe, it is the time and effort spent on sketching which gives the brain the time needed to grasp an idea fully.

Readers familiar with the basics can begin with the chapter on inversion and consult the earlier chapters when necessary; cross references are given frequently. Chapters 2–4 introduce the basic concepts of potential theory, of measuring gravity and of data treatment and analysis. Chapters 5 and 6 deal with qualitative and quantitative interpretation and Chap. 7 discusses gravity inversion extensively.

One last word about how to find more on gravity and new developments driven by new technologies of observation and computation or other earth science aspects. Today the internet is a source of useful science information, but can be difficult to use discriminately by those lacking basic knowledge and judgement in the fields of enquiry. The individual cannot hold all the wanted details in mind, but basic knowledge is the precondition for judging the available information and making use of it. This book wishes to provide the basic knowledge and understanding for further studying and applying the science of gravity interpretation.

1.2 Historical Review

1.2.1 Astronomy, Geodesy, Geophysics, 18th and 19th Centuries

Gravity is an everyday experience so that it is hardly noticed in daily life. Only if we have to lift a heavy weight or climb a steep mountain do we feel gravity, and we observe the notorious falling apple. Weight and gravity are not the same, gravity abstracts weight from mass; weight is the product of gravity and mass. Galileo Galilee may have been the first to clearly understand that gravity is the common acceleration all masses experience in free fall, i.e. "free" from any obstruction, air included. Isaac Newton, the first to postulate central forces acting through empty space, explained the observations, especially Kepler's three laws of planetary motion, and derived mathematically his two fundamental laws of mass gravitation and inertia. Gravitational attraction decreases with distance r as $1/r^2$. This can be understood as a quality of Euclidian space with the assumption of a constant gravitational flux (see Sect. 2.1 and 2.7.5; Eq. 2.1.1), emanating from any massive body, and evenly spreading over spherical surfaces which grow in area proportional to r^2. In contrast to magnetic flux, the notion of gravitational flux is not established, but nevertheless equally useful. Modern space and satellite geodesy must consider relativistic aspects; however, this subject does not affect today's down-to earth geophysics where, generally, classical Newtonian physics in space and time is fully adequate.

That gravity varies along the Earth's surface was experienced by the early explorers who took pendulum clocks along their voyages and noticed that, near the equator, the clocks ran late. With time, systematic variations of gravity were discovered all over and compared to the gravitational attraction of large masses calculated from Newton's law. Measuring the Earth, for example, the length of the meridian, in the 18th and 19th centuries clearly demonstrated the importance which gravity has for the figure of the Earth and for measuring it. Observation of deflections of the vertical by astronomical and geodetic means, gravity measurements with pendulums, since P. Bouguer (1698–1758), and measurement of gravity gradients with the torsion balance became the domain of geodesy. They also brought insights into the principles of mass layering, isostasy was suggested by G.B. Airy (1801–1892) and J.H. Pratt (1809–1871) as the Himalayan masses appeared to be compensated by a mass deficit at depth (a review, as an example of internet-based information is given in http://www.univie.ac.at/Wissenschaftstheorie/heat/heat-3/heat393f.htm).

1.2.2 20th Century

Measuring gravity directly, in the 20th century, became an important tool for mineral and hydrocarbon exploration. These achievements made it necessary to work out applications of Newton's law theoretically which is the essential basis for all gravity interpretation. Thus, gravity on Earth can be safely founded on classical Newtonian physics (although observations with the aid of satellites are bringing us into the age of Einsteinian relativity). Theory for gravity interpretation has been laid down in many classical works on geodesy and geophysics; the present treatise attempts a condensated presentation of the essential aspects in Chap. 2.

Progress in gravity research is closely linked to the observational precision, which is driven by the advance of experimental physics. Methods of gravity observation evolved slowly in the 18th and 19th centuries, when pendulums, telescopes and torsion balances were the exclusive tools. The 20th century brought an explosion of new instrumental developments, especially the gravity meters. Rather recently, borehole gradiometers of highest sensitivity have been constructed, tested and employed in exploration. Since artificial satellites have been launched, a new era of gravity observation from space has begun, and the best knowledge of the Earth's gravity field is now being gained by combinations of terrestrial and space observations of different nature. Beside gravity itself, the gravitational potential has become observable indirectly by radar satellites that measure the ocean surface topography which, to a first approximation, is the equipotential surface of the geoid. Measuring accuracies are reaching levels permitting the distinction of the sea surface topography from the geoid such that the effects of ocean currents, temperature and salinity variations can be isolated which is of high relevance for oceanography and climate research. A brief account of today's methods is given in Chap. 3.

Gravity measurements need to be reduced (Chap. 4) by numerically removing several calculable effects which obstruct their efficient interpretation. For the reductions, the geodetic coordinates or locations, including height or elevation are needed. Thus, the recent tools of satellite geodesy, especially GPS and GLONASS and future improved systems, as the European GALILEO, have an immense impact on gravity measurement, analysis and interpretation.

Besides, a contemporaneous technological development of great consequence to gravity research is the dramatic increase of computation power. The basic theory was worked out and many applications were formulated long before the advent of efficient numerical computation, but its possibilities have made it necessary to design new program tools that considerably enhance the usefulness of gravity studies. They involve handling of large digital data sets, their representation and analysis by spherical harmonics and Fourier series, modelling of complex structures (Chap. 6) and visualization. But that has not made simple modelling superfluous, because complex models can be applied reasonably only with the aid of good human imagination and intuition which are strongly aided by developing a feeling for the nature and size of model effects (Chap. 5). For this, i.e. the interaction of the human mind with computing power, visualisation is essential. It is not just nice, but is a tool for uncovering problems, suggesting solutions and also facilitating communication between scientists of different fields.

The reliability of gravity interpretation strongly depends on the accuracy of all data input. The more the errors are reduced and the more reliably they are estimated, the more successful can gravity inversion (Chap. 7) become, provided that also the additional or a priori information from other sources has a similarly improved quality.

1.2.3 Geodesy and Geophysics

Geodesy and geophysics have diverged in their development accommodating to their individual need and emphasis. Separate terminologies are now in existence which hamper communication and mutual learning from each other. Examples are "gravity anomaly", "gravity disturbance", "correlation", "nullspace", etc. The central theme of geodesy is measuring Earth and thus questions of errors or accuracy are of paramount concern; typically, a geodesist will ask a geophysicist, how accurately (s)he wants to have certain observations to be made. But the geophysicist is concerned with the more or less inaccessible interior of Earth and cannot generally anticipate what is needed, so may not be able to answer such a question precisely, except saying: "as accurate as possible". Moreover, the existence of uncontrollable effects, for example, of local density variations, limits the required precision. Only if both sides have a basic common understanding of each other's problems and thinking, will they be able to unite their efforts. It is time to attempt bridging the gap. As the present authors come from these two fields, geophysics and geodesy, they hope to be in a good position for such an attempt. It must include geology that is the object of much of geophysical research and has incorporated most of the geophysical insights into Earth's interior. The view of the triple geodesy – geophysics – geology is to be complemented by all other branches of the earth sciences which as a whole might best be called "geology" in the widest sense as the study of the Earth.

1.3 Purposes of Gravity Measurements

Today's aims of measuring gravity, as in the past, have a wide scope which extends with increased precision and in combination with other improved measurements, e.g., of distances and coordinates. Applications are inherent in *geodesy* – paramount for defining the Earth's shape, for example, in combination with levelling and other methods of surveying. This is especially evident with the Global Positioning System GPS which gives the radius of a point from the Earth's centre and requires knowledge of the geoid – the gravitational equipotential surface – to provide the point elevation above sea level. In *geophysics* and geology the aim is exploration of the Earth's interior and gravity has also important bearing on oceanography, archaeology, engineering and even on theoretical physics. In geodynamics temporal gravity change is becoming a topical subject as the space-time behaviour reflects processes as loading or unloading and flow inside the Earth. Precise recording of temporal gravity variation can reveal mechanical properties and even deep processes as Earth core oscillations.

A division is usually made between general geophysics and applied geophysics. It is rather artificial, since geophysical gravity observations are frequently applied to problems outside gravity. Nevertheless, a gap in outlook and terminology has developed also between these branches or communities of geophysics because there are differences in emphasis and aims of research in industry which must produce economic value, and university motivated by fundamental science. Both are equally important in human culture. Different motivations unavoidably influence thinking, but seeing this should also stimulate learning from each other. Thus, this book wants to serve both communities and to provide a basis for many kinds of application.

1.4 Gravity and Gravity Anomalies

"Gravity interpretation" means precisely "interpretation of gravity anomalies". As emphasized in Sect. 1.1.3, gravity anomalies are the very object of interest, although an anomaly always requires two things: an observation and a norm or reference or something expected to represent a normal field. We consider anomalous the deviation of the observation from the expected. The gross variation of gravity on Earth quite closely corresponds to what is expected from an idealized Earth with no lateral variations of structure and density, such as would be the case if a fluid would perfectly accommodate to the forces originating only from self-gravitation and rotation. One has come to call such an Earth the "normal earth". But, beyond the parameters of the normal earth and its gravity field, there are deviations, and it is these deviations from the norm that are here of interest, i.e. the gravity anomalies which are to be interpreted. The deviations from the ideal are, however, not large, indeed, if relative scales are considered.

The raw observations of gravity are not easily interpretable, if at all. They must first be reduced, i.e. referred to the reference normal gravity model, i.e., to the normal earth. The various ways of treating the normal earth and the visible deviations from it are subject of the various kinds of reductions and further data analysis (Chap. 4). In order to understand the gravity treatment better, a brief overview of the gravity variations or anomalies encountered on Earth will be given in the next Sect. 1.5.

Gravity anomalies are variations in space (and in time), and relative gravity meters can perfectly provide the wanted information; even absolute gravity observations are interesting in their variations for gravity interpretation. Variation implies an ensemble of points or a continuous field, and it has become customary to understand the term "anomaly" in the sense of "anomalous field". One isolated value of gravity is useless for interpretation as envisaged here (Sect. 1.1.3). An ensemble of discrete points of gravity values is not identical with a continuous anomaly field, indeed, generally field is an idea, and in this sense, defining an "anomaly" from discrete points is an act or part of interpretation. "Field" and interpretation thus mutually influence each other and the data points are but one part of this. Defining an anomaly from a limited set of points is thus not generally a trivial task. The theory chapter (2) deals with gravity effects usually in the form of continuous functions of coordinates, derivatives, relations in space and the possibilities of exploiting them for their interpretation. In the observation chapter (3) planning surveys (Sect. 3.4) is shown to be guided by expectations of anomalies. In the reduction chapter (4) the emphasis is on making anomalies "visible"; especially Sect. 4.7 on the analysis of anomalies, deals with the concrete construction of an anomaly from discrete points and with the notion of their errors (Sect. 4.7.1); and as a frequent task, Sect. 4.7.7 discusses the separation of regional and residual fields. The interpretation Chaps. 5, 6 and 7 are anyway always concerned with anomalies in space; in Sect. 5.1.5, in particular, the notions of anomalies and of gravity effects are confronted with each other and their mutual dependence is considered; they should be clearly kept apart.

1.5 Some Important Aspects of the Terrestrial Gravity Field and Internal Mass Distribution

1.5.1 General Considerations

Gravity interpretation does not happen in isolation, but in the world of shallow subsurface investigation, of mineral exploration, and of whole Earth geodynamics. Basic knowledge of the essential features of terrestrial gravity and mass or density distribution is therefore a precondition for a reasonable approach to the tasks at hand. Moreover, the fundamental ambiguity in gravity interpretation makes a priori knowledge mandatory for reducing this ambiguity to an acceptable level. However, a priori knowledge includes both basic ideas and high precision geological and geophysical data. Familiarity with, or a feeling for, the subject is essential for successful and efficient work, but thinking must go beyond the familiar limits. Questions, as to what kind of gravity and density variations are to be expected or what are their normal magnitudes, and hence what are the requirements of accuracy, will mutually influence measuring, modelling and interpreting different gravity effects.

In the 19th and 20th centuries knowledge about the Earth increased and recent progress has been fast. The dual role of gravity, as signalling density variations inside the earth, and generating them by driving dynamic processes becomes more and more relevant, and both are intimately interconnected. To interpret large-scale gravity anomalies one needs to know something about the processes and the material properties of the Earth's interior. To successfully apply gravity to the search for mineral resources, knowledge of the processes of mineral concentration and their geological associations is equally essential.

In thinking about gravity it is critical to distinguish between the different kinds of anomalies: the customary Bouguer anomaly (*BA*), the Free Air anomaly (*FA*) and the isostatic anomalies. The different reference models used in defining the various anomalies must be taken into account (see Chap. 4); otherwise gross misinterpretations are the result; the relations with topographical, geological and tectonic features is very different; for example, mountain ranges are usually accompanied by generally positive, but highly scattered *FA* values and at the same time by a strongly negative smooth *BA*; spreading ocean ridges have a similar gravity expression, except that the *BA* is positive, if referenced to sea level, but negative in comparison with the deep sea basins.

The following descriptions will necessarily be somewhat subjective. These will include: *the Earth's figure and constitution* (Sect. 1.5.2: ellipsoid, geoid, Earth's density and shells crust, mantle, and core), *continents and oceans* (Sect. 1.5.3: isostasy of large geological structures, fold mountain ranges, limits to lateral density variations), *plate tectonics and mantle flow* (Sect. 1.5.4: mantle dynamics: convection, ridges, subduction), *associated gravity anomalies* (Sect. 1.5.5: scale laws and kind and size of gravity variations or anomalies to be expected), *other large-scale gravity features* (Sect. 1.5.6: loading and unloading processes, postglacial rebound), *smaller-scale gravity anomalies relevant to exploration for economic minerals*

(Sect. 1.5.7: density anomalies inside the crust: objects of general geological research and mineral exploration), *harmonic spectrum of the gravity field* (Sect. 1.5.8: Kaula's rule, upward and downward continuation, mantle tomography).

This book is not an exhaustive treatise on geodynamics, geology, geochemistry and geophysics of the Earth. Only a brief outline is given here to emphasize the relations of gravity with Earth structures and processes. The reader is also referred to a number of historical and recent texts where many of these aspects are treated in one way or the other and help the reader to form an overall picture of the Earth (Wegener, 1915–1930, 1966, 1980; Holmes, 1944, 1993; Cox, 1973; Press & Siever, 1974; Press et al., 2003; Turcotte & Schubert, 1982, 2002; Skinner & Porter, 1989; Lowrie, 1997; Mussett and Khan, 2000; Schubert et al., 2001; Fowler, 2004).

1.5.2 The Earth's Figure and Constitution

As already mentioned, the Earth's gravity field and figure are intimately related to each other. Basically this follows from Newton's law of gravitation, in that a given arbitrary mass distribution generates a unique external gravity field which becomes smoother and more spherical with distance from the source. But the regularities of the Earth and its gravity field demonstrate that gravity played a major role in shaping and structuring the Earth. It is nearly a sphere, or more accurately, an ellipsoid of rotation and nearly exactly obeys Clairot's principle of the equilibrium figure of a rotating self-gravitating fluid body in space (A.C. Clairot, 1713–1765). Equilibrium and the deviations from it play an essential role in the whole field of gravity interpretation and geodynamics. However, the most voluminous part of the Earth, the mantle is not fluid; it is made of solid rock which transmits seismic transverse shear waves. Obviously this solid material has properties of a fluid if subjected to long-lasting forces such as gravitation and centrifugal acceleration.

The Earth's surface topography has a relief of about ± 10 km, that is about $\pm 1/600$ of Earth's radius. The ellipsoidal major and minor axes differ by the same amount, and the flattening is about 1/300 (the flattening is defined as $f = (a - c)/c$ with a = equator radius and c = polar radius). This leads to a difference between gravity at the poles and at the equator by about 0.5% or 1/200 ($\pm 1/400$). The gravity difference is relatively greater than the geometrical difference, because gravity reflects both the direct centrifugal force *and* the effect of the ellipsoidal shape. The exact value of flattening and gravity variation depends on the internal density structure, primarily the density depth distribution. Thus, by itself, gravity tells us about the density increase with depth supported by mechanics and astronomical observations concerning the Earth's angular momentum and hence momentum of inertia.

Speculation about the Earth's internal structure and material properties began from hard surface rock and proceeded to hot molten rock to explain volcanoes and religious ideas about a burning hell. Iron meteorites suggested to scientists that the interior might be molten iron. Astronomical and geodetic mathematical theory and improving physical understanding led to the derivation of the general densification with depth, which could be explained by pressure and chemical stratification.

A more detailed picture arose as a result of seismic studies of the propagation of waves through the Earth's interior, and by the early 20th century a fairly accurate knowledge of a three-layered Earth had been gained. It became natural to imagine crust, mantle and core as the basic "onion model". Crust is a thin veneer of quartz-rich rocks. Mantle, to 2900 km depth, consists of ultramafic silicate rocks, olivine rich peridotite, transformed by pressure and temperature at depth through phase transitions (olivine→spinel→perovskite and magnesiowüstite). The resulting layer boundaries roughly conform to equipotential surfaces, but significant lateral deviations are expected in response to dynamic processes. The core consists of iron with impurities, the outer core is molten, and the inner core solid frozen; this somewhat surprising situation is due to the pressure effect on melting.

Equilibrium would mean perfect density stratification, varying exclusively with depth according to the principle of minimum potential energy. Density boundaries would then perfectly conform to internal equipotential surfaces, and the variation with depth would affect the Earth's ellipticity and the normal gravity field which varies only with latitude (see Chap. 4). The Earth would have become static and "dead". Non-equilibrium boundary undulations with associated gravity anomalies, that is deviations from the normal gravitational field, can be maintained either by internal elastic strength and/or generated by dynamic processes disturbing the equilibrium. Thermal convection provides such a process. The gravity effects of undulating density boundaries are indistinguishable by spectral methods from effects generated by voluminous lateral density variations, which themselves are also caused by convective currents. This process is most relevant to the mantle; however, the large mantle viscosity makes the convection currents move slowly, probably not exceeding a few decimetres per year, under normal circumstances.

1.5.3 Continents and Oceans

The Earth's surface obviously deviates from fluid-like equilibration. Topography is governed by continents and oceans; about 30% of the Earth' surface are continent, including continental shelf area submerged under shallow seas, and 70% are deep ocean basins. Only small portions are occupied by continental slopes, deep sea trenches and active high mountain belts. The rest is mainly in two plateaus above and below sea level. It was early realized (e.g. by Alfred Wegener, 1912; Jacoby, 2001) that this kind of frequency distribution of elevation, called the "hypsographic curve", calls for a fundamental explanation with two kinds of Earth crust. Now it is known from a multitude of seismic studies that continental crust is 20–80 km thick, 20 km in exceptional low elevation regions as some shelf areas or plains as the Pannonian Basin, and perhaps up to 80 km under the highest mountain ranges and plateaus, as the Himalaya and Tibet or the Altiplano of the Andes. Continental crust contains cores of material that solidified 4 Ga ago or even earlier. Oceanic crust is about 7 km thick, on average, it is much younger, 180 Ma at most in the north-western Pacific, and it is made of rocks of basaltic composition as products of mantle melting. All ocean basins are traversed by ocean ridges, the active, i.e. spreading ridges gently sloping towards the abyssal plains, see below. Passive ridges, active or extinct volcanic islands and seamounts are partly organized in some order, along chains, which was not understood before the advent of continental drift and plate tectonics.

The notion of crust and crustal thickness has played an essential role in understanding the Earth and especially its gravity field. Before detailed seismic studies had been carried out in large style, it was clear from mechanical and gravity arguments that ocean crust and mantle are in approximate isostatic equilibrium with continental crust and mantle. It was understood that continents and oceans differed significantly in rock density in the upper 100 km depth range, or so. Seismology then provided proxy information on crustal densities on the basis of density-related seismic velocities (see Chap. 3). Thus, crust and mantle were basically geophysical notions (seismic, gravity, isostasy). But as geodynamic and petrologic knowledge increased, particularly after the advent of plate tectonics, the notions of crust and mantle have somewhat shifted from geophysical to petrologic definitions involving the processes of their origin. One now speaks about differences between the seismic (or gravity) and petrologic crust or mantle. Crustal basaltic/gabbroic material at the base of the continental crust may suffer a phase change to eclogite which in seismic and density properties resembles high-velocity and high-density mantle material, such that the seismic and gravity crust may appear thinner than the petrologic crust which includes mantle-like, but crustal eclogite at its bottom. Under the ocean basins an opposite process seems to occur when water percolates through the basaltic crust into the topmost mantle, peridotite reacts with the water and is transformed into low-density and low-velocity serpentinite. Thus the seismic (and gravity) oceanic crust may look geophysically thicker than the basaltic, i.e. petrologic crust.

1.5.4 Plate Tectonics and Mantle Flow

The discovery of the plate movements and plate tectonics has completely changed our view of the Earth, some 50 years after Alfred Wegener proposed continental drift in 1912. Understanding the Earth and its gravity field is intrinsically linked to the processes of plate interaction and the underlying mantle flow. Only these ongoing endogenic processes explain Earth structure, topography, gravity and many more features. However, large-scale geology is shaped also by exogenic processes as erosion, sediment transport and sedimentation in a complex interaction with the endogenic processes. High mountain ranges are actively eroded, so they must be geologically young and rising to be and remain high for some time. Erosion disturbs the isostatic equilibrium and the response is a rise; endogenic push-up or pull down may occur simultaneously. "Tectonically active" means that these processes are at work and this nearly always shows up in gravity anomalies. Eroded material is transported toward the ocean basins were it is ultimately deposited as sediments.

Continental margins, and especially deltas of big rivers, become sediment deposits which form loads on the crust or, better, on the lithosphere, and depress it. By this mechanism of isostatic adjustment, very thick sedimentary sequences may be laid down to form what earlier had been called "geosynclines".

When studied by surface geology and deep geophysical methods, old peneplaned regions turn out to be orogenic roots of former mountain ranges. In geology, such structural and mineralogical characteristics define an orogeny or mountain belt; the present-day topography is not an essential feature of an orogen. In the cold rigid upper crust, the old orogenic structures are frozen in, the strength of the rocks prevents flow and equilibration on the scales of the structures; this again is expressed in gravity anomalies. Most continental regions have experienced a series of orogenies, the younger overprinting the older ones and often younger mountain belts were attached to older solidified structures, themselves having formed in a similar fashion earlier; thus, continents look like a complex mosaic of orogenic belts with the younger one surrounding older cores. However, an old core may directly abut an ocean basin, and some continents have evidently been torn apart such that older geological structures are cut abruptly just to reappear on the other side of an ocean, e.g., the Atlantic.

Basically, convergence and continental collision form new continental crust. Convergence of lithospheric plates implies subduction of one plate under the other, or collision of continental plates, i.e. plates carrying continental crust at the convergent margin, which inhibits subduction because of the low density and large buoyancy of continental crustal material.

The opposite of convergence is plate divergence at active ocean ridges and in continental rift zones which may initiate large-scale divergence, continental separation and ocean floor spreading which may ultimately reverse and lead to convergence and orogeny completing a "Wilson cycle". Or rifts may fail and will then be preserved in continents at some stage of development. In the upper crust, such structures also freeze in and generate gravitational signals.

The third type plate interaction is that of transform faults where two plates move horizontally past each other, as in the case of the San Andreas Fault of western North America or the North Anatolian Fault in Turkey. The movement juxtaposes different structures and deforms them by shearing. Active transform faults connect other types of plate boundaries, especially offset ridge segments in the oceans, and beyond them, they become inactive fracture zones where lithosphere of different age is juxtaposed.

These processes are more complex than a simple description can convey. The relative movements are rarely exactly normal or parallel to the plate boundaries but usually oblique and encompass aspects of transpression or transtension: boundaries, at closer look, are not clear-cut faults but complex sets of faults and wide belts of deformation in which, more or less locally, the sense of deformation may even reverse, involving, e.g., grabens and rifts in collision zones.

Active volcanoes occur preferentially in tectonically active zones, mostly in plate-marginal regions. These regions are generally also seismically active, i.e., they are belts of seismicity or earthquake occurrence. That plate divergence is accompanied by volcanism is not surprising as hot material must rise into the gap. When rising mantle melts it produces basalt (see above) often called "MORB" (mid ocean ridge basalt), hence the result is ocean crust. But volcanism is also dominant in convergence zones, exactly speaking: on the upper plate above the subducted one where its surface reaches about 100 km depth. The products are different, typically andesitic (the term being derived from "Andes") which is a rock type intermediate between acid or silica-rich and basic or silica-poor. Melting of wet, i.e. H_2O-rich subducted material is a source of this type of volcanic rocks. They play an important role in orogeny and the forming of new or recycled or mixed continental material.

Volcanism of importance also occurs far from plate boundaries. Such an anomalous occurrence of volcanism is called a "hot spot", as e.g. Hawaii. Hot spots may also occur near or on plate boundaries as in Iceland, lying on a divergent plate boundary. The volcanic products, called OIB (ocean island basalt), are similar to MORB, but significant differences especially in trace elements suggest different sources. Hot spot volcanism is interpreted by the model of heat advection from possibly great depth. The generally accepted model is called "mantle plume" envisioned as concentrated upwelling which begins to be seen by seismic tomography (see below).

The relations of plate motion with mantle convection are also not simple, in contrast to early ideas that large steady convection cells encompass the mantle and that plate divergence occurs above the upwellings and subduction occurs over the downwellings. Actually little is known about the flow pattern in the mantle, but seismic mantle tomography does provide 3D P and S velocity perturbations from standard radial velocity models, as PREM (Preliminary Reference Earth Model; Dziewonski and Anderson, 1981; for tomography see e.g. Masters et al., 1996; Grand et al., 1997; Kennett & van der Hilst, 1998; Kennett et al., 1998, to mention just a few). If it is assumed, as often has been done, that velocity and density are related, for example, through temperature and/or chemical composition of mantle rocks, flow patterns may be implied or can be derived and related to the surface anomalies of gravity and/or the geoid. One difficulty is that the velocity-density relations may vary in the mantle. Theoretical relations between the geopotential and gravity with density (and thus indirectly with tomography) are given in Schubert et al., 2001 (pp. 279–280), and an application is mentioned (King & Masters, 1992).

Plumes which are part of convection may rise from different depths. Vertical movements imply work in the gravity field and the question of mass balance and thus also observable gravity anomalies. A very simple isostatic "Airy plume" model is discussed in Sect. 5.6.9.3. Mantle plumes are predicted, among others, by high-Rayleigh number convection experiments and numerical modelling and hence dynamic plume models need to be explored (see e.g. Schubert et al., 2001, pp. 537–543); however, this goes beyond the present treatise. Other types of melting anomalies or heterogeneities in the mantle, for example, inherited from former plate subduction, may also play a role.

Especially when relating the geodynamic aspects to gravity, the notion of rheology or the laws governing material strength, deformation, fracture and flow becomes of paramount importance. How much stress can be sustained elastically? Are there limits of force and mass imbalance (in the gravity field) which the Earth can sustain for some time or for how long time? This is immediately related to the question of maximum possible elevation differences and slopes or maximum possible density contrasts and dimensions of such contrasts. It is the same question, in different terms, as that of isostasy and its possible modes. What we observe today is a geological snapshot and geological history can tell us something about the relevant time constants of the processes involved. Some appear static (as in "isostatic") maintained for billions of years, some appear transient, as for example, glacial or postglacial isostasy or better: isostatic readjustment, lasting only thousands of years, in some hot regions as Iceland much shorter.

The above structures and processes are driven largely by gravity and thermal energy and they generate characteristic gravity anomalies or signals. Unfortunately for the study of Earth, the processes and signals are superimposed on each other in a complex fashion and the observed picture is very complicated. As emphasized in this treatise, however, gravity anomalies reflect horizontal or lateral density variations, such that the study of gravity is an essential aspect of regional geology. Below, a brief discussion follows of what may be expected.

1.5.5 *Associated Gravity Anomalies*

The reader is again reminded of the differences of Bouguer anomalies (*BA*) and Free Air anomalies (*FA*) as treated in Chap. 4. In the Bouguer model topography is but added to, or in the case of the oceans subtracted from, an otherwise idealized layered Earth. The Free Air earth model ignores the topography on the idealized earth which acknowledges its approximate isostatic compensation within the upper 100 km. The *BA* thus emphasizes such density variations inside the Earth while the *FA* is rather affected by mass excess or deficit along the vertical down to some depth.

Beside gravity as such, the gravitational potential or, as derived from it, the geoid undulations depict large-scale regional to global anomalies or the major features of the gravity field better than gravity itself does, because the potential as an integral quantity emphasizes the longer wavelengths. The geoid is the equipotential surface which coincides with the mean sea level and is, thus, the idealized shape of the Earth as defined by its gravity field. For geoid undulation we often briefly say "geoid". What is meant is the geoidal height reduced by the best fitting axial ellipsoid of rotation. The geoid resembles the *FA* anomaly, but is much smoother.

The longest-wavelength *FA* gravity variations (apart from those related to rotation and ellipticity) have dimensions of 10^{3-4} km and show little or no obvious relations with the continent-ocean distribution. Shorter wavelengths of order 10^2 km closely image the continental margins, especially where they are steep and marked. Passive, i.e. non-convergent, margins are usually accompanied by a dipolar band of distinct positive anomalies above the shelf break and less distinct negative anomalies above the foot of the continental slope. The amplitudes are tens of milligals and the width corresponds to that of the continental margin. The steeper the continental slope, the more pronounced the gravity anomalies. This medium-scale feature is mostly suppressed by the Bouguer reduction by which the change from positive to negative *BA* values (from ocean to continent) is brought out. The broad-scale *BA* is generally negative in continents, especially in highly elevated areas, and positive in oceans, and the regional mean *BA* values approximately correspond to the effects of the Bouguer reduction, i.e. the gravity effect of the topographic mass mathematically removed or subtracted on land and added in the oceans (see Chap. 4); on the large scale, it looks like the reduction being wrong or superfluous. The conclusion from this is that the homogeneously layered reference earth with topography only added or subtracted is wrong; to the contrary, elevated excess mass is compensated by low density deeper roots and vice versa, as mentioned above. This is the mass equilibrium principle of isostasy.

All major plate tectonic margins are accompanied by relatively short-wavelength gravity anomalies (*FA* and *BA*), especially where elevation has significant slopes. "Short" here means typically one to several hundred kilometres, to a large portion similar to the width of the geological and morphological structures. However, some of the characteristic gravity anomalies extend beyond the immediate structural limits. The relation with plate boundaries or margins is not simply one-to one but varies along the boundaries. Nevertheless, some general trends exist.

There is some obvious relation of the longest gravity wavelengths with plate tectonics. A belt of relatively positive *FA* anomalies accompanies the belt of plate convergence surrounding the Pacific and extending also to the Mediterranean-Himalayan-Sunda belt of convergence. In the Sunda region north of Australia exist the largest positive *FA* anomalies and the highest geoid undulation of order 100 m; it is the region with the largest concentration of plate convergence on Earth. The positive circum-Pacific belt is surrounded by a less clear belt of broad negative *FA* anomalies with the deepest minimum (order −100 m) on Earth south of India (near Sri Lanka). The gravity high accompanying convergent plate margins is a broad *FA*, up to several thousand kilometres wide accompanied by a narrow deep *FA* low over ocean trenches following approximately the crest of the broad high. In continental mountainous collision zones the deep *FA* low is missing. In the *BA* the relations with the deep sea trench or mountain belt are just reversed as the Bouguer mass reduction mostly outweighs the *FA* lows or highs. Topography and the Bouguer reductions vary with the broad *FA* gravity high, therefore strong variations occur along convergence zones and from zone to zone.

Active ocean ridges, i.e. the diverging plate boundaries, too, have a tendency of slightly positive *FA* values (order 10 mGal) broadly correlated with ridge topography. The ratio of the ridge related *FA* over topography, relative, say, to adjacent deep sea basins, is low for the broad ridges but much larger for short wavelength of, say, ridge crest rift valleys, sea mounts (and trenches). This reflects Newton's $1/r^2$ law: shallower, nearer masses have stronger gravity effects than the deeper compensating, more distant masses; moreover, different isostatic mechanisms may support this trend. The gravity over topography amplitude ratio, especially when systematically measured versus wavelength λ (or wave number $1/\lambda$ or $2\pi/\lambda$), is called admittance and its study constrains density and dynamical models more tightly than gravity can do alone, because the ambiguity of gravity modelling is reduced by including topography information.

The dominant gravity anomalies in the vast ocean areas, however, seem to bear little relation to the plate motions or ocean depth and are not those related to plate divergence. Geoid and *FA* highs occur also far from plate convergence as a large positive feature or broad anomaly belt extending from the north Atlantic, centred on Iceland, south-eastward across Africa to the southern Indian Ocean roughly marked by the Kerguelen hotspot. A somewhat similar, less distinct feature occupies the Southeast Central Pacific. Analysis reveals that these positive regions are correlated with a high spatial frequency (or density) of hotspots. It suggests some relation with dominant mantle upwelling.

That high *FA* and geoid anomalies accompany mantle upwelling, on the one hand, and plate convergence and downwelling, on the other hand, is surprising. Obviously, the relations of gravity with plate tectonics and mantle flow are not so simple and are probably related to the interaction between the rheological mantle structure and the convective flow.

Transform faults have not a very consistent gravity signature as they have also a varied morphological expression. Both depend on the local relative normal motion which varies in space and time because the faults are not straight, i.e. great circle lines. In zones under present convergence and compression with a transpressive character, ridges or transverse ranges are being pushed up; in zones under present divergence and tension with a transtensional character, pull-apart basins are opening which are filled with sediments in continents of form deep chasms in distal ocean areas (as the 7000 m deep Romanche Deep in the equatorial Atlantic). The associated gravity anomalies (*FA*) very much reflect the morphology with dimensions of the order of 10^{1-2} km (and more in length).

Hotspots have already been mentioned to preferably occur in high geoid and *FA* regions. There seems also to exist a typical shorter-wavelength signal consisting of a positive *FA* core surrounded by a negative rim, however, the anomalies are relatively weak in comparison to the regional anomaly variations so that they are not easily detected.

1.5.6 Other Large-Scale Gravity Features

Some broad gravity anomalies are associated and caused by transient external processes as build-up and disappearance of large ice sheets presently experienced 10 000 a after the latest ice age. Both northern North America and Fennoscandia are characterized by negative *FA* anomalies of up to several tens of milligals which roughly outline the former ice cover. Here it seems evident that isostatic equilibrium after melting of the ice has not yet been achieved. The observations can be used to estimate the viscosity of the mantle material flowing back towards the rising areas; the gross viscosity estimate is about 10^{21} Pa s, but the subject is more complex and not treated here. Although the basic interpretation has been debated, the general view is supported by many additional observations as, for example, the present uplift rates of the order of 1 cm/a in the central regions of former glaciation.

Transient features are also related to rapid erosion and sedimentation in limited regions as high standing young mountain belts or big river deltas. However, the rates of vertical change are generally much slower, typically one order of magnitude or more, than glacially driven changes, and so are the gravity effects.

1.5.7 Smaller-Scale Gravity Anomalies Relevant to Exploration for Economic Minerals

Gravity anomalies are often instrumental in guiding the exploration geophysicist and geologist to objects of interest. The tasks are varied, a complete account of all possibilities is not intended. Often gravity variations give the first overview of a previously little studied region, be it a whole country or an area, a kilometre or less in dimension. Correspondingly anomalies of interest range from the sub-milligal to tens of milligal amplitudes and wavelengths from meters to tens or hundreds of kilometres.

Generalisations of the targets of exploration are as difficult. They may be, from small to large scales: cavities, man-made construction features, faults, ore concentrations, synclines and anticlines to whole sedimentary basins and other large scale promising features. In very many cases, it is not the economic mineral as such which gives a significant gravity signal, but the geological structures which are the targets of study and which, from experience, are loci of oil or gas accumulation. Within a sedimentary basin, faults and anticlines or synclines may be prospective sites of oil, gas or mineral accumulations. Anticlines sometimes have dense cores or, in the case of salt structures, the opposite. Obviously it is essential that broad geologic and tectonic knowledge is coupled with geophysical understanding.

Technical applications of gravity in buildings are varied, and no general rule is evident, but modern gravity meters are accurate enough for detecting structural aspects as, for example, deformation when large weights are installed, or unknown inhomogeneities under the building.

As gravity interpretation is highly ambiguous, it is the combination with other data, for example, from seismic refraction or reflection, which will narrow down the spectrum of possible models. In some cases, gravity is used to fill in gaps of information; structural interpolation between seismic lines or between drill holes can gain considerably from gravity modelling, and in special cases "blind spots" of seismic models can be filled in this way.

1.5.8 Harmonic Spectrum of the Gravity Field

Beside the variation in space, spectral aspects of the gravity field are interesting for interpretation. In many cases the spectrum is more revealing and may disclose some facets of the sources. Furthermore, certain features of harmonic fields or field components are most attractive in their properties in space and since harmonic analysis is separation of individual harmonic components it has many advantages for gravity interpretation. Moreover, numerical treatment in the spectral domain has become very efficient. Global analysis is done by spherical harmonic expansion of the field. In small regions as in exploration geophysics, expansion in one or two-dimensional Fourier series serves the same purpose.

The global spatial spectrum is described in spherical harmonics, i.e. a series of Legendre polynomials to degree n in latitude φ multiplied by a Fourier series in longitude λ of orders $m = 0$ to n (see Sect. 2.10). The complete picture includes all terms of the series which describe superimposed waves in two coordinates, φ and λ, of minimum wavelength $2r\pi/n$ (r = earth's radius, $2\pi r = 40000$ km), and thus n and m play the role of the wave numbers in the Fourier spectrum. The complete set of coefficients (C_{nm}, S_{nm}) of all series terms corresponds to the amplitude and phase spectrum of the Fourier expansion. Often only the amplitude spectrum or the power spectrum, is considered; its dimensionless coefficients are $S_n^2 =_0 \Sigma^n (C_{nm}^2 + S_{nm}^2)$. The power spectrum has been defined in different ways in terms of the normalisation. The coefficients are transformed to dimensional values through multiplication with GM/r (disturbing potential in m^2/s^2), r (geoid heights in m), $GM(n+1)/r^2$ (gravity disturbance in m/s^2), $GM(n-1)/r^2$ (gravity anomaly in m/s^2). Kaula (1966) suggested a decrease of power in the form

$$S_n \approx 10^{-5}(2n+1)^{1/2}/n^2,$$

called "Kaula's rule of thumb".

Hipkin (2001), on the other hand, who corrected the normalisation of the (logarithmic) power spectrum found that it shows at least three straight-line sections versus n which can be related to three depth levels where mass concentrations occur preferentially (which, at those depths, would have the white spectra of delta functions or effective point masses). The interpretation is that anomalous masses (volume times density contrast) move buoyantly and thus interact with the rheological mantle structure (e.g. layering). The argument is based on the upward or downward continuation of the field (see Sect. 2.10) leading to two competitive effects: decay of gravity effect amplitude with depth and inhibition of sinking or rising. Since dominant wavelengths of anomalies increase with distance from the source or harmonic waves of defined wavelength decay in amplitude the more rapidly the smaller the wavelength, one may associate probable depths to the spectrum. However, one must not forget that long wavelength sources may exist at shallow depths.

Another aspect is the relations with the spectra of other geophysical or geological features. Relations exist, for example, between plate sizes and related large-scale features (expanded in harmonic series) and the low end of the gravity spectrum. Tomographic velocity variations at different depth levels in the mantle may have spectra similar to those of gravity and guide the associations of velocity and density. The approach is to compare the spectra, but whether or not relations are genetic is generally quite open; relations may have common causes or be but accidental.

In any case, ideas of possible or even probable interpretations may be suggested by spectral relationships between data sets, though it seems unlikely that under the circumstances probabilities can ever be reliably quantified. The principal ambiguity of the inverse problem of finding the source from the gravity anomalies is the same in the spectral domain as in the space domain.

On the small scales Fourier spectra of gravity, magnetics, geological structures and related data sets may reveal interesting relationships of exploration interest.

References

Blakely, R.J.: Potential Theory in Gravity and Magnetic Applications. *Cambridge Univ. Press*, 441pp., 1995

Cox, A.: Plate Tectonics and Geomagnetic Reversals. *Freeman, San Francisco*, 702pp., 1973

Dziewonski, A.M., Anderson, D.L.: Preliminary reference earth model. *Phys. Earth Planet. Int.*, 25, 297–365, 1981

Fowler, C.M.R.: The Solid Earth, 2nd ed. *Cambridge Univ. Press*, 685pp., 2004

Grand, S.P., van der Hilst, R.D., Widiyantoro, S.: Global seismic tomography: A snapshot of convection in the earth. *Geol. Soc. Amer. Today*, 7, 1–7, 1997

Grant, F.S., West, G.F.: Interpretation Theory in Applied Geophysics. *McGraw-Hill, New York, N.Y.*, 584pp., 1965 [**GW65**]

Hipkin, R.G.: The statistics of pink noise on a sphere: applications to mantle density anomalies. *Geophys. J. Internat.*, 144, 259–270, 2001

Holmes, A.H.: Principles of Physical Geology. Nelson & Sons, Edinburgh, 1288pp., 1944; 4th ed., edited by P.M.D. *Duff, Trans-Atlantic Publications, Philadelphia*, 807pp. 1993

Jacoby, W.R.: Origin of continents, English translation of A. Wegener, 1912: Die Entstehung der Kontinente. *J. Geodynamics* 32, 29–63, 2001

Jung, K.: Schwerkraftverfahren in der Angewandten Geophysik. *Akademischer Verlag Geest & Portig*, 348pp., 1961 [**KJ61**]

Kaula, W.M.: Theory of Satellite geodesy. *Blaisdell, Waltham*, 124pp., 1966

Kennett, B.L.N., van der Hilst, R.D.: Seismic structure of the mantle: From subduction zone to craton. In "The Earth's Mantle: Composition, Structure, and Evolution", ed. I. Jackson, 381–404,*Cambridge Univ. Press, Cambridge, England*, 1998

Kennett, B.L.N., Widiyantoro, S., van der Hilst, R.D.: Joint seismic tomography for bulk sound and shear wave speed in the Earth's mantle. *J. Geophys. Res.*, 103, 12469–124693, 1998

King, S.D., Masters, G.: An inversion for radial viscosity structure using seismic tomography. *Geophys. Res. Lett.*, 19, 1551–1554, 1992

Lowrie, W.: Fundamentals of Geophysics. *Cambridge Univ. Press*, 354pp., 1997

Masters, G., Johnson, S., Laske, G., Bolton, H.: A shear velocity model of the mantle. *Phil. Trans. Roy. Soc. London*, 354, 1385–1411, 1996

Mussett, A.E., Khan, M.A.: Looking into the Earth, An Introduction to Geological Geophysics. *Cambridge Univ. Press*, 470pp., 2000

Press, F., Siever, R.: Earth. *Freeman, San Francisco*, 945pp., 1974

Press, F., Siever, R., Grotzinger, J., Jordan, T.: Understanding Earth, 4th ed. *Freeman, New York*, 568pp., 2003

Schubert, G., Turcotte, D.L., Olsen, P.: Mantle Convection in the Earth and Planets. *Cambridge Univ. Press*, 940pp., 2001

Skinner, B.J., Porter, S.C.: The Dynamic Earth, an Introduction to Physical Geology [with Study Guide]. *Wiley, New York*, 541pp., 1989

Telford, W.M., Geldart, L.P., Sheriff, R.E.: Applied Geophysics, 2nd ed. *Cambridge Univ. Press, Cambridge, New York, N.Y.*, 770pp., 1990

Turcotte, D.L., Schubert, G.: Geodynamics: Applications of Continuum Physics to Geological Problems. *Wiley, New York*, 450pp., 1982
Turcotte, D.L., Schubert, G.: Geodynamics: Applications of Continuum Physics to Geological Problems, 2nd ed. *Cambridge Univ. Press, Cambridge, England*, 456pp., 2002
Wegener, A.: Die Entstehung der Kontinente. *Petermanns Geographische Mitteilungen, 581, 185–195, 253–256, 305–309, 1912*
Wegener, A.: Die Entstehung der Kontinente und Ozeane. *Vieweg, Braunschweig, 1915, 1920, 1922, 1930* (first and fourth edition reprinted, *Vieweg, 1980*)
Wegener, A.: The Origin of Continents and Oceans. *Methuen, London*, 1924; *Dover, New York*, 1966

List of some great classical scientists: mathematicians, astronomers, geodesists and geophysicists who advanced the theoretical basis of gravity analysis and interpretation

Airy, George Biddell, 1801–1892
Bessel, Friedrich Wilhelm, 1784–1846
Bouguer, Pierre, 1698–1758
Bruns, E. Heinrich, 1838–1919
Clairot, Alexis Claude, 1713–1765
Eötvös, Roland (Lorand), 1848–1919
Galileo Galilee, 1564–1642
Gauß (Gauss), Johann Carl Friedrich, 1777–1855
Hayford, John, 1868–1925
Heiskanen, Weikko, Aleksanteri, 1895–1971
Helmert, Friedrich Robert, 1843–1917
Jung, Karl, 1902–1972
Kepler, Johannes, 1571–1630
Laplace, Pierre Siméon, maquis de, 1749–1827
Legendre, Adrien-Marie, 1752–1833
Leonardo da Vinci, 1452–1519
Moritz, Helmut, 1923–
Newton, Sir Isaac, 1643–1727
Poisson, Simon Denis, 1781–1840
Pratt, John Henry, 1809–1871
Stokes, George Gabriel, 1819–1903

Chapter 2
Fundamentals of Gravity, Elements of Potential Theory

2.1 Introduction

Gravity is the vector $\mathbf{g}$ of gravity acceleration. Usually its norm g is called "gravity". It is composed of the vectorially added components of the Earth's gravitation and the rotational or centrifugal acceleration. Its value is roughly $9.81 \pm 0.03\,\mathrm{m/s^2}$ at the Earth's surface. A small part is time-varying. Gravity decreases with distance from the surface, both upward (Newton's law) and from some depth also downward (as only the masses below the observer exert gravitational attraction).

The gravitational flux is defined as

$$\Gamma = s\int \mathbf{g} \bullet \mathbf{ds} \qquad (2.1.1)$$

through a surface area S; $\mathbf{ds}$ is the vectorial surface element, i.e., its normal vector. Γ is constant for any surface S that completely encloses the same source of g, for example, the mass of the Earth M; it is a useful concept; among other aspects, it explains the $1/r^2$ relation in Newton's law of gravitation in three-dimensional Cartesian space (see Sect. 2.7). But gravitational flux is not as commonly used as magnetic flux.

2.2 Units

The MKS or SI unit of g is $\mathrm{m/s^2}$; for very precise values sometimes $10^{-9}\,\mathrm{m/s^2}$ or $\mathrm{nm/s^2}$ is used; traditionally geophysicists use $\mathrm{cm/s^2} = \mathrm{Gal}$ (from Galilee), $10^{-3}\,\mathrm{Gal}$ or mGal and $10^{-6}\,\mathrm{Gal}$ or $\mu\mathrm{Gal}(10\,\mathrm{nm/s^2})$. In older literature "gravity units", g.u., have been used where $1\,\mathrm{g.u.} = 0.1\,\mathrm{mGal} = \mu\mathrm{m/s^2}$.

For gravity gradients (see below) the SI unit is $\mathrm{m/s^2/m} = \mathrm{s^{-2}}$. Usually $10^{-9}\,\mathrm{s^{-2}} = 1\,\mathrm{E}$ (Eötvös). The standard vertical gradient of gravity near the surface is $\sim 3086\,\mathrm{E}$.

W. Jacoby, P.L. Smilde, *Gravity Interpretation*,
© Springer-Verlag Berlin Heidelberg 2009

2.3 Elements of g

Assume a point i on Earth (latitude φ_i, longitude λ_i, height h_i) with an associated value g_i, which for most purposes is the value of interest. Usually a small deviation δg_i, from a reference g_b is measured, not the total or absolute value of gravity acceleration:

$$\delta g_i = g_i - g_b \tag{2.3.1}$$

Most gravity observations are relative to a chosen base station b (Chap. 3). Besides, observed gravity is principally referred to theoretical reference fields by reductions (Sect. 4.5); δg_i will always refer to a standard or reference g_{bi} and can be converted to absolute g_i if g_{bi} is known. The symbol g may be used for δg where the difference is irrelevant.

Gravity g slightly varies with time (tides, changes in the hydrosphere and atmosphere as well as tectonics (Chaps. 3, 4)), so that:

$$g = g_0 + g_t(t) \tag{2.3.2}$$

The time-averaged value g_0 is composed by vector addition of the gravitational attraction $\boldsymbol{a}_0$ ($>99\%$) of the Earth mass and centrifugal acceleration $\boldsymbol{z}_0$ ($<1\%$) of the point, both time averaged:

$$\boldsymbol{g}_0 = \boldsymbol{a}_0 + \boldsymbol{z}_0 \tag{2.3.3}$$

This statement implies that the point of observation is fixed to the rotating Earth's body, which is more or less true also for the hydrosphere and atmosphere, but it is not true for satellites in space. For observation platforms moving relative to the solid Earth the centrifugal component is affected. The difference is taken into account in marine and airborne measurements in the form of the Eötvös reduction (see Sect. 3.2.10.3).

The time-dependent part $g_t(t)$ of gravity at point P consists of the gravitational attraction $\boldsymbol{a}_k$ of a unit mass at P by the cosmic masses (direct effect), the effect $\boldsymbol{a}_d$ of tidal deformation d of the Earth's body (indirect effect), and any acceleration $\boldsymbol{a}_c$ of the coordinate system in an inertial frame; $\boldsymbol{a}_c$ is not determined and thus neglected. The two tidal effects (direct and indirect) are of the order of $10^{-7} g$ and usually combined (see discussion in Chap. 3) and removed from the observations by the reduction procedures or by interpolation between repeated base station readings. Thus g_0 is the aim of most practical gravity measurements.

A very concise introduction (in German) into the basic definitions has been given by Jung (1961), referred to as KJ61, 84–89. Readers are referred to it or to more specialized literature. The main point is that every mass element in the Earth "feels" the gravitational attraction mainly from moon and sun and the accelerations due to the motion of the Moon-Earth system and the Sun-Earth system (analyzed separately), while the Earth rotation about its axis is primarily neglected. The vectorial summation of attraction and acceleration renders the system of tidal forces. Now the Earth rotation comes into play by taking any point through this system of tidal forces, which have two maxima and minima during each lunar day (24 h, 51 min); the superposition of the lunar and solar tides (about half as strong as the former) leads to spring tides during full and new moon and neap tides during half moon; the latter have about one third the strength of the former. The other neighbouring celestial bodies, especially Jupiter, have only small modifying effects.

2.4 Coordinate Systems

In practice three kinds of coordinate systems are in use: spherical, cylindrical and Cartesian. Note that, generally, O will denote the origin of the coordinates, P generally denotes an observation point, and Q is a source point or mass element. For many purposes it is convenient to place P at O; it simplifies the expressions and usually does not restrict the general applicability because the coordinate system can be shifted. Placing Q at O would be simple only when no integration over source volumes were required.

For local problems or small regions with dimensions small relative to the Earth's dimensions, it is generally convenient to take local coordinates, depicted in Fig. 2.4.1 with the respective mass elements. Cartesian x, y, z are shown as reference for the other systems. The z axis is taken positive downward, i.e. parallel to local $+\mathbf{g}$, as usual, upper left displays the general relation of P, Q and the distance r between them. Vertical cylinder coordinates, R, Λ, Z, and the corresponding volume element are presented on the right hand side. The volume element in spherical coordinates, least common for local problems, appears in the lower left. Local vertical cylinder coordinates are analytically simple, but only for axial points P. Most used are local right-handed Cartesian systems for machine-based calculations.

2.4.1 Spherical Coordinates

The natural coordinate system for global gravity are the spherical coordinates of a point: r outward from the Earth's centre O, latitude φ, or alternatively co-latitude $\phi = \pi/2 - \varphi$ and longitude λ. The centre O moves irregularly by several millimetres inside the solid Earth because of atmospheric and other mass shiftings, but in gravity modelling that is usually neglected.

Terrestrial or global Cartesian coordinates are centred at 0, x and y in the equatorial plane, x pointing toward longitude $\lambda = 0°$, y pointing toward $\lambda = 90°$, z pointing toward the north pole ($\phi = 0$ or $\varphi = 90°$); for points at the (spherical) Earth's surface:

$$x = r\cos\varphi\cos\lambda = r\sin\phi\cos\lambda$$
$$y = r\cos\varphi\sin\lambda = r\sin\phi\sin\lambda \qquad (2.4.1)$$
$$z = r\sin\varphi = r\cos\phi$$

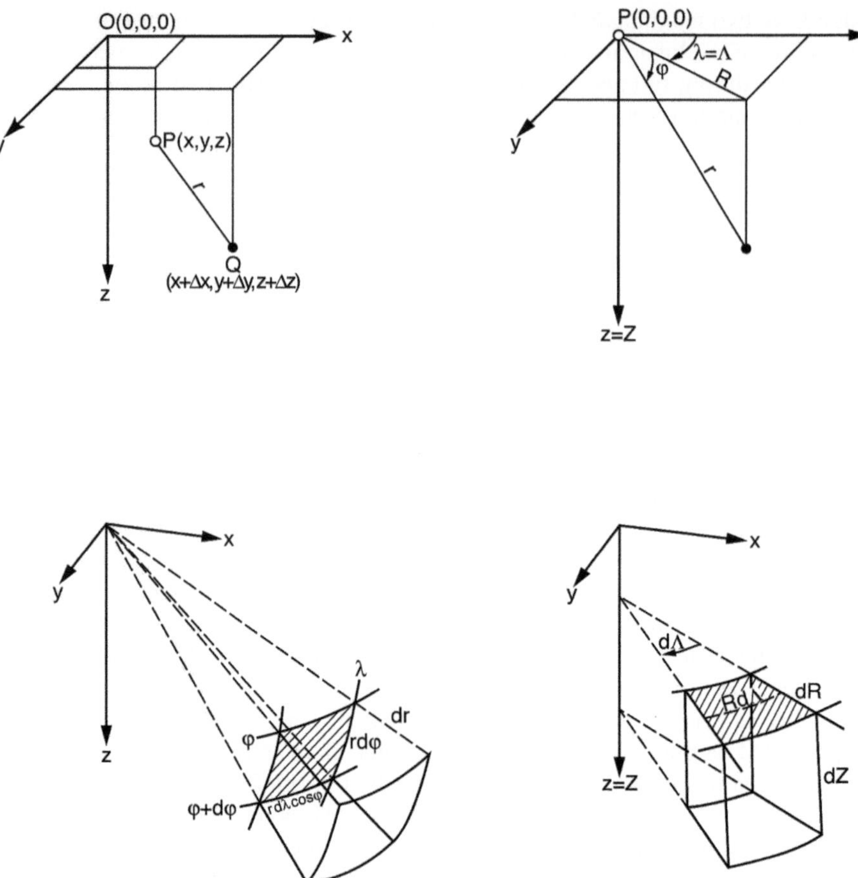

Fig. 2.4.1 Local coordinates with the respective mass elements; Cartesian x, y, z (*downward*) shown as reference. *Upper left*: general relation of P, Q and r. *Right-hand side*: vertical cylinder coordinates, R, Z, λ and volume element. *Lower left*: volume element in spherical coordinates

The natural volume element is $dV = r^2 dr \cos\varphi\, d\varphi\, d\lambda = r^2 dr \sin\phi\, d\phi\, d\lambda$, however, it is convenient only for calculations with reference to O at the centre of the sphere.

2.4.2 Vertical Cylinder Coordinates

Cylinder coordinates, R, Z, Λ (Z downward, parallel to $\boldsymbol{g}_o$, R horizontal distance and Λ an azimuth from some arbitrary reference direction) are handy for many purposes. While the gravitational attraction of a mass element onto the origin 0 depends on Λ, the Z component is independent from Λ, and the problem has cylinder geometry which is convenient. The volume element is $dV = RdR\, d\Lambda\, dZ$ and $r = (R^2 + Z^2)^{1/2}$.

2.4.3 Cartesian Coordinates

2.4.3.1 General Remarks

A local right-handed Cartesian system x, y, z, with an arbitrary (suitable) origin O has z pointing down (nadir), x pointing N (geographical) and y pointing E. The most flexibly applicable mass element is $dV = dx\,dy\,dz$. Down or nadir means exactly parallel to g. However, as the gravity vector varies spatially, influenced by the mass distribution, generally small x and y components will exist which are mostly neglected. With x, y, z, geometrical situations of geophysical interest can be adequately described in three-dimensional (3D) space if the objects are small relative to the Earth's size.

In many applications of gravity interpretation, one dimension, usually y, is neglected because it is actually dominant in the sense that geological structures are elongated in this direction; along y there is little or no variation, and the geometrical description reduces to two dimensions (2D), x, z. 2D modelling plays an important role (see e.g. Sect. 2.9.7).

2.4.3.2 Coordinate Transformations

This section refers only to Cartesian coordinates. In the task of integrating gravity effects over extended mass volumes, it is often necessary or at least useful to carry out coordinate transformations as gravity effects are invariant to coordinate transformations of translation and rotation. This is exploited by taking generally, in all expressions, the origin $(0, 0, 0)$ to represent the observation point P. For evaluation at a point P at x_i, y_i, z_i (or h_i) a translatory coordinate transformation of (x_k, y_k, z_k) is carried out: $\mathbf{x} = (x_k - x_i, y_k - y_i, z_k - z_i)$.

Forward gravity modelling is further greatly simplified for uniform bodies V in specific orientations which can be achieved by certain rotational coordinate transformations. Idealized bodies have planar surface elements, for instance, polyhedra in Cartesian coordinates limited by triangles bounded by straight lines. As shown below (Sect. 2.9.3 onward), integrating the gravitational effects by such bodies is simple for the components normal or parallel to the bounding planes or lines. For arbitrarily oriented linear and planar elements a rotational coordinate transformation is carried out, based on the coordinates of point couples or triples in x, y, z. The elementary analytical geometrical or vector operations are briefly summarized in view of their frequent application. When needed later, reference will be made to the required expressions.

Starting with a point P_i or vector $\mathbf{0} \rightarrow \mathbf{P_i} = \mathbf{r_i} = (x_i, y_i, z_i)$, define a straight line from $P_i \rightarrow P_j$ as $\mathbf{s_{ij}} = \mathbf{r_j} - \mathbf{r_i} = (\Delta_{ij}x, \Delta_{ij}y, \Delta_{ij}z)$ with the components $\Delta_{ij}x = x_j - x_i$, $\Delta_{ij}y = y_j - y_i$, $\Delta_{ij}z = z_j - z_i$, and $s_{ij} = (\Delta_{ij}x^2 + \Delta_{ij}y^2 + \Delta_{ij}z^2)^{1/2}$. The normal vector of a triangular plane $\mathbf{a_{123}}$ (a not to be confused with acceleration) may be defined by three non-identical and non-collinear points P_1, P_2, P_3 ($\mathbf{r_1}$, $\mathbf{r_2}$, $\mathbf{r_3}$), for instance, by the two vector sides s_{12} and s_{13} as the vector product: $a_{123} = s_{12} \times s_{13}$. Polygons and polyhedra (Sects. 2.9.4.2; 2.9.5.2; 2.9.6.2) are parameterized such that one side or edge with the corners P_1 and P_2 defines the local X axis with the unit vector $X = s_{12}/s_{12}$ in the (x, y, z)-system; the local Z unit vector is normal to the plane a_{123}, i.e. $Z = s_{12} \times s_{13}/(s_{12}\, s_{13})$ and $Y = Z \times X$. The angles between the X_j (or X, Y, Z) and x_i (x, y, z) coordinate axes are given by the components X_{ji} of the X_j unit vectors in the x_i-system; note that i and j, here, indicate the coordinates 1–3, not the points P_1, P_2 and P_3 defining the local coordinates; this must be kept in mind to avoid confusion.

A plane in x, y, z can be written in various forms, for example: $z = a + bx + cy$, in the form of axis intercepts x_0, y_0, z_0: $x/x_0 + y/y_0 + z/z_0 = 1$ or as Hessian normal form: $xp_x + yp_y + zp_z - p = 0$, where $p = (p_x, p_y, p_z)$ is the normal vector from the origin $O = (0, 0, 0)$ to the plane. The parameters of the different forms can be determined either by vector operations (above) combined with rotation (below), or they can be found by solving the linear equations represented by the point coordinates of P_1, P_2 and P_3 inserted (see Appendix M1). The normal distance p is useful for some of the integrations of this chapter. The line s_{ij} and its normal distance from O can be treated similarly (see Appendix M2).

When the local system $X_i = (X, Y, Z)$ is rotated into the global system $x_j = (x, y, z)$, any vector in one system can be transformed into the other system by multiplication of X_i with the orthonormal rotation matrix $\underline{R}$ which combines three successive rotations about the axes x, y and z with the Euler angles α, β, γ, individually (see Arfken, 2001):

$$\underline{R}_x = \left\{ \begin{array}{ccc} 1 & 0 & 0 \\ 0 & \cos\alpha & \sin\alpha \\ 0 & -\sin\alpha & \cos\alpha \end{array} \right\} \quad \underline{R}_y = \left\{ \begin{array}{ccc} \sin\beta & 0 & \cos\beta \\ 0 & 1 & 0 \\ \cos\beta & 0 & -\sin\beta \end{array} \right\}$$

$$\underline{R}_z = \left\{ \begin{array}{ccc} \cos\gamma & 0 & \sin\gamma \\ -\sin\gamma & \cos\gamma & 0 \\ 0 & 0 & 1 \end{array} \right\}$$

The general rotation matrix is derived, for example, by multiplication of the three matrices:

$$\underline{R} = \left\{ \begin{array}{ccc} \cos\alpha\cos\gamma - \sin\alpha\cos\beta\sin\gamma & -\sin\alpha\cos\gamma - \cos\alpha\cos\beta\sin\gamma & \sin\beta\sin\gamma \\ \cos\alpha\sin\gamma - \sin\alpha\cos\beta\cos\gamma & -\sin\alpha\sin\gamma - \cos\alpha\cos\beta\cos\gamma & -\sin\beta\cos\gamma \\ \cos\alpha\cos\beta & \cos\alpha\sin\beta & \cos\beta \end{array} \right\}$$

$$(2.4.2)$$

If the rotation is expressed by the single angle θ of rotation about an axis in the direction of a unit vector $v = (v_1, v_2, v_3)$:

$$\underline{R} = \left\{ \begin{array}{ccc} \cos\theta + v_1^2(1-\cos\theta) & v_1 v_2(1-\cos\theta) + v_3\sin\theta & v_1 v_3(1-\cos\theta) - v_2\sin\theta \\ v_1 v_2(1-\cos\theta) - v_3\sin\theta & \cos\theta + v_2^2(1-\cos\theta) & v_2 v_3(1-\cos\theta) + v_1\sin\theta \\ v_1 v_3(1-\cos\theta) - v_2\sin\theta & v_2 v_3(1-\cos\theta) - v_1\sin\theta & \cos\theta + v_3^2(1-\cos\theta) \end{array} \right\}$$

$$(2.4.3)$$

The above three matrices $\underline{\boldsymbol{R}}_x$, $\underline{\boldsymbol{R}}_y$, $\underline{\boldsymbol{R}}_z$ can be derived from (Eq. 2.4.3) by substituting the unit vectors $\boldsymbol{i}$, $\boldsymbol{j}$, $\boldsymbol{k}$ for v_i.

Back rotation into the global system is carried out after integration (see Sect. 2.9.3) and is achieved by applying the transposed rotation matrix $\boldsymbol{R}^T$ which is identical to the inverse $\underline{\boldsymbol{R}}^{-1}$, and the matrix product $\underline{\boldsymbol{R}}\boldsymbol{R}^{-1} = \underline{\boldsymbol{R}}\boldsymbol{R}^T = \boldsymbol{I}$, the unity matrix. Any points and vectors (geometry and gravity components) are treated this way. If the rules are carefully followed, no errors or confusion should occur. The execution will generally be by corresponding subroutines in the computer codes.

2D rotation is the special case where the axis of rotation is one of the global coordinate axes, say, $X_m \equiv x_m$. X_i ($i \neq m$) is correspondingly normal to x_m and, applying the same definition as above for the angle between the axes X_i and x_m: $\cos(X_i, x_m) = 0$ and $\cos(x_m, x_m) = 1$. If only two-dimensional coordinates are considered, $x_i = (x_1, x_2)$ and $X_i = (X_1, X_2)$, the rotation matrices are given as

$$_2\underline{\boldsymbol{R}} = \begin{pmatrix} \cos(X_1, x_1) & \cos(X_1, x_2) \\ \cos(X_2, x_1) & \cos(X_2, x_2) \end{pmatrix} \quad \text{or} \quad \begin{pmatrix} \cos(X_1, x_1) & \cos(X_1, x_2) \\ -\cos(X_1, x_2) & \cos(X_2, x_2) \end{pmatrix} \quad (2.4.4)$$

and

$$_2\underline{\boldsymbol{R}}^T = \begin{pmatrix} \cos(X_1, x_1) & -\cos(X_2, x_1) \\ \cos(X_1, x_2) & \cos(X_2, x_2) \end{pmatrix} \quad (2.4.5)$$

Determination of the normal distance p of an arbitrary straight line or an arbitrary plane from P(0,0) or P(0, 0, 0) [or from any other point] is a related task which has to be solved frequently; the expressions for p (Appendices M1.10, M2.3) are given in the Appendices M1 and M2.

2.5 Newton's Laws: Gravitation and Inertia Plus Centrifugal Acceleration = Gravity

From Kepler's laws of planetary orbits Newton derived his two fundamental laws of classical mechanics. Two point-like masses M and m, at positions Q and P, respectively, attract each other through space with a central force $\boldsymbol{f}$ along the radius vector $\boldsymbol{r}$ (pointing from Q to P, length r); the force acting on m is:

$$\boldsymbol{f} = -(GMm/r^2)\,\boldsymbol{r}/r \quad (2.5.1)$$

The minus sign is a matter of convention, depending of the definition of $\boldsymbol{r}$; in geophysics it is usually dropped, i.e. replaced by the plus sign, as generally done here. If uncertainties arise, the sign must be carefully considered. The notion of $\boldsymbol{r}$ may be sometimes confusing (see 2.9.1.2), if used for both the vector from a given source Q to receiver P (as in Eq. 2.5.1) and for the radius vector from the Earth's centre to a point P of observation. Gravity is directed against the latter $\boldsymbol{r}$, but for practical reasons, e.g. for calculating the gravitational effect of a buried mass, $\boldsymbol{g}$ is taken positive downward and $-\boldsymbol{r} = \boldsymbol{z}$ is the usual vertical coordinate. When we write

(2.5.1) as $\boldsymbol{f} = Gm/r^2\,\boldsymbol{r}/r$, $\boldsymbol{r}$ is the vector from P to Q, and its vertical component of usual practical interest is f_z/r (see Sect. 2.9.1.2).

The $1/r^2$ law, as mentioned above, expresses that a constant gravitational flux Γ (Eq. 2.1.1) flows from mass through Euklidian space. Other ways of expressing this are Gauss' integral theorem and the Laplace and Poisson's equations, see below (Sect. 2.7.6).

G is the universal gravitational constant which in the SI system is (National Institution of Standards and Technology, DODATA recommendation 2002):

$$G = 6.6742 \pm 0.001 \cdot 10^{-11}\,\mathrm{m^3 kg^{-1} s^{-2}}.$$

Mass has inertia; it moves with constant velocity $\boldsymbol{v}$ if no force acts on it and experiences an acceleration $\boldsymbol{a}$ if a force $\boldsymbol{f}_a$ acts on it:

$$\boldsymbol{f}_a = m\,\boldsymbol{a} \tag{2.5.2}$$

Combining the two equations ($\boldsymbol{f} = \boldsymbol{f}_a$) we get the gravitational acceleration as m cancels if both gravitation and inertia relate to the same mass, i.e., $\boldsymbol{a}$ is independent from the test mass m:

$$\boldsymbol{a} = -G\,M/r^2\boldsymbol{r}/r \tag{2.5.3}$$

We also introduce mass density ρ (or simply density) as total mass per total volume, V.

$$\rho = m/V \tag{2.5.4}$$

In geophysics volume and mass include pore space. Density is usually considered a continuum in space, i.e. atoms or elementary particles and voids are neglected. Density is assumed to be constant or vary continuously or discontinuously across surfaces that define step-like density changes of $\Delta\rho$. For the points lying exactly on discontinuous boundaries special care has to be taken.

Mostly in applied geophysics, only density anomalies $\Delta\rho$ count, i.e. density contrasts versus a reference density ρ_0, of the surrounding medium

$$\Delta\rho = \rho - \rho_0,$$

where ρ is the mean density in V. Thus, geophysicists often deal with "negative densities" which are not really negative. As a *gedanken experiment*, consider two hollow spheres within water and determine their mutual gravitational interaction (Task 2.4). The reference density ρ_0, is generally assumed either constant or varying only with depth z. A detailed discussion of ρ_0 follows in Sect. 6.1.5.1. Lateral variations themselves generate gravity variations and are therefore not generally useful as reference.

Note, at this point, that gravity is not concerned, in practical geophysics, with gravitational aspects of particle physics, nor are we here concerned with relativistic effects and non-Euclidian space. We are concerned with exploration and terrestrial as well as planetary scales.

As a basic rule, potentials of individual source masses do not mutually influence each other, i.e. they are independent from each other, and so are their derivatives: hence they can be added to each other or integrated over extended volumes V. Considering mass anomalies, their Newtonian gravity effect is:

$$\boldsymbol{\delta g} = \int_V \boldsymbol{dg} = G \int_V \Delta\rho\,(\boldsymbol{r}/r^3)\,\mathrm{d}V \qquad (2.5.5)$$

Although this may seem trivial, it need not be so, for in the case of magnetics mutual induction effects cause deviations from this rule (e.g. demagnetisation).

For earth gravity, to a first approximation, we often set $M = M_{\mathrm{earth}}$ although it is not point-like. For a homogeneous sphere (or for a radially symmetric density depending only on r) the integration over all mass elements leads to the expression with the total mass at the sphere centre; integration can be simplified by applying the solid-angle principle, see Sects. 2.9.1.1, 2.9.6.3 (Task 2.1).

Newton's laws are the basis for the whole field of geophysical gravity research and geological interpretation, aided, e.g., by the Laplace equation (2.7.1). Newton's $1/r^2$ space variation is fundamental in Cartesian geometry and physics and thus also the basis of the Laplace and Poisson equations (2.7.1 & 2.7.5).

At this point, the other – very small – contribution to gravity must be briefly described (although it is irrelevant for most tasks of interpretation of gravity anomalies). The centrifugal acceleration z acts normal to the instantaneous rotation axis or vector of angular velocity ω. It depends on the distance R from the axis of rotation (with $r =$ Earth's mean radius):

$$\boldsymbol{z} = \omega^2\,\boldsymbol{R} = \omega^2 r \cos\varphi\,(\boldsymbol{R}/R) = \omega^2 r \sin\phi\,(\boldsymbol{R}/R) \qquad (2.5.6)$$

and the vectorial sum (Eq. 2.3.3) renders, in view of the smallness of z relative to the gravitational acceleration a, $z_0 = z\cos\varphi = z\sin\phi$, such that:

$$z_0 = \omega^2 r \cos^2\varphi\,(\boldsymbol{R}/R) = \omega^2 r \sin^2\phi\,(\boldsymbol{R}/R) \qquad (2.5.7)$$

Finally it is worth mentioning here, that the gross properties of the Earth's gravity field are described by the idealized normal gravity formula which depends only on latitude

2.6 Gravity Potential and Equipotential Surfaces

The two components of time-averaged gravity are gravitational attraction $\boldsymbol{a_o}$ and centrifugal acceleration $\boldsymbol{z_o}$. Both components of the gravity field are conservative in the sense that the amount of work $\mathrm{d}W$ done along a path $\boldsymbol{ds}$ is $\mathrm{d}W = m\boldsymbol{g}\,\boldsymbol{ds}$, where $m\,\boldsymbol{dg}$ is the weight of the test mass m; the potential difference depends only on the locations of starting and end points. In many respects the mathematical treatment of a scalar field is simpler than that of a vector field. This can be exploited. It is also used to measure gravity variations in the oceans by radar satellites. If no other forces act on the fluid particles at the surface, they will form a level surface in equilibrium; a surface cannot maintain a slope, gravity would work and move the elements toward equilibrium. The observation that, apart from local topography, the Earth's shape is close to an equilibrium ellipsoid tells us that the long-term constitution of gross Earth material is fluid.

The gravitational potential U_g is implied in the $1/r^2$ law. It is most easily shown for a single mass element Δm or point mass at r_m attracting a unit mass at P (r_p) with:

$$\boldsymbol{\Delta g} = -\mathrm{G}\,\Delta m(\boldsymbol{r}_m - \boldsymbol{r}_p)/|\,\boldsymbol{r}_m - \boldsymbol{r}_p\,|^3 = -\mathrm{G}\,\Delta m\,\boldsymbol{\nabla}(1/|\,\boldsymbol{r}_m - \boldsymbol{r}_p\,|) \qquad (2.6.1)$$

with the gradient operator nabla $\boldsymbol{\nabla} = (\partial/\partial x, \partial/\partial y, \partial/\partial z)$. The minus signs are usually dropped (see below).

Gravity is a potential field and is itself the field strength or the negative gradient of the scalar potential U_g of Earth:

$$U_g = -\mathrm{G}\,M_{earth}/\,|\,\boldsymbol{r}_o - \boldsymbol{r}_p\,| \qquad (2.6.2)$$

where $\boldsymbol{r}_o$ refers to the geocentre; hence, $\boldsymbol{g} = -\boldsymbol{\nabla} U_g. = -\mathrm{G}\,M_{earth}\boldsymbol{\nabla}(1/\,|\,\boldsymbol{r}_o - \boldsymbol{r}_p\,|)$. Again, the minus sign is dropped in geophysics for practical reasons. This sign convention ignores the fact that the potential (to do work) increases with distance r from a source such that U_g must vary correctly as $-1/r$ and gravity points against $\boldsymbol{r}$. However, additional mass increases the amount of work or the potential to do work and, analogously, the positive gravitational force on P acts towards Q.

Equation (2.6.2) holds also for the integral over the whole Earth's body. This fact is just another way of saying that the work done or the potential is the scalar product of force $\boldsymbol{g}$ times path vector $\boldsymbol{ds}$ integrated across a path $\boldsymbol{s}$ from P1 to P2:

$$U_g^{1,2} = \int_{P1}^{P2} \boldsymbol{g} \bullet \boldsymbol{ds} = \mathrm{G}\,M_{earth}\,(1/r_2 - 1/r_1) \qquad (2.6.3)$$

or generally

$$U_g = \mathrm{G}\,M_{earth}/r + C, \qquad (2.6.4)$$

where C is an arbitrary constant of integration. This fact demonstrates that the potential is always determined only relative to some reference point or level.

The centrifugal acceleration $\boldsymbol{z}$ has also a potential, called here U_z, which uniquely depends on the distance R from the axis of rotation. With $\boldsymbol{R} = {}^1\!/_2\boldsymbol{\nabla}|\boldsymbol{R}|^2$, (Eq. 2.5.6) can be written:

$\boldsymbol{z} = \boldsymbol{\nabla}(1/2\omega^2|\boldsymbol{R}|^2)$ and hence the potential is

$$U_z = {}^1\!/_2\omega^2|\boldsymbol{R}|^2. \qquad (2.6.5)$$

Finally, the gravity potential is:

$$W = U_g + U_z = \mathrm{G}\int_{earth} dm/|\boldsymbol{r}_m - \boldsymbol{r}_p| + {}^1\!/_2\omega^2|\boldsymbol{R}|^2 \text{ and } \boldsymbol{g} = \boldsymbol{\nabla} W. \qquad (2.6.6)$$

Gravity anomalies δg are reduced for normal gravity which includes the centrifugal effect (Sect. 4.4.1). Anomalies and calculated gravity effects are both free from the centrifugal term and thus comparable. They represent only the z-components of the gravitational effects of mass anomalies. It is then proper to write δU_g, or short: δU, but δW is more usual.

W has the physical unit $m^2/s^2 = J/kg$ or $cm^2/s^2 = erg/g$ and is an abstract quantity not much used in practice. Instead of the potential W (or anomaly δW), an equivalent quantity which can be visualized, is generally chosen in the form of equipotential surfaces or briefly "W-surfaces": W or $\delta W = $ const. The work done by gravity is independent from the path (see above): one can thus construct unique surfaces with the potential difference between them constant and the vector $\boldsymbol{g}$ (or $\boldsymbol{\delta g}$) being normal; thus, along paths $\boldsymbol{ds}$ parallel to them, no work is done: $\boldsymbol{\delta g} \cdot \boldsymbol{ds} = 0$. The distance between surfaces separated by fixed ΔW varies and is inversely proportional to average $g = |\boldsymbol{g}|$ between them. Outside the source region W is differentiable and the W-surfaces are smooth. Equipotential surfaces are characterized by a scalar potential value, relative to some reference and are representations of the gravity field as valid as is gravity itself, but the potential emphasizes the longer wavelengths compared to gravity (Sects. 2.6; 2.10.5). It is worthwhile to make an effort to visualize the geometry of the W-surfaces as it offers several practical possibilities (see below). A remark may be added in view of the concept of anomaly (see Sect. 1.4): it is the irregularities or curvature variations (undulations) of the W-surface, which characterize the anomalies as a principal property of point ensembles, not of single points.

A special W-surface is the geoid which describes the potential that coincides with mean sea level if not affected by waves, ocean currents and temperature and salinity variations. Usually the geoid undulation N is given which represents the geoid referenced to the normal ellipsoid of rotation. From $g = \partial W/\partial z \approx \Delta W/N$ (see Chap. 1) follows Bruns' formula for the deviation N from the reference ellipsoid, e.g. given in the unit meter (m):

$$N \approx \Delta W/g \tag{2.6.7}$$

Where traditionally the value g of normal gravity is taken. In the precise relation between gravity anomalies and geoid undulations the integral of local gravity along its trajectory between the two surfaces has to be taken into account.

N may be regarded as the geoid surface or a point on it. In local gravity interpretation the fine structure of the W-surfaces is of interest, i.e. minute small-scale undulations. Their geometrical visualisation includes curvatures of vertical and horizontal sections and their mutual relationships is subject to geometrical analysis. They reveal relations between different components of the gravity field and quantities which cannot be directly measured. Curvatures, again, emphasize local structures, i.e. shallow mass anomalies. Smooth regional field components are suppressed, if not beforehand removed (see Sect. 4.7.7). Application, especially to local anomalies δg_z, is presented in the next Sect. (2.7).

Finally, in this context in spherical coordinates, the expressions at point $P = (r, \varphi, \lambda)$ for the gravitational potential and for the gravity effect at source point

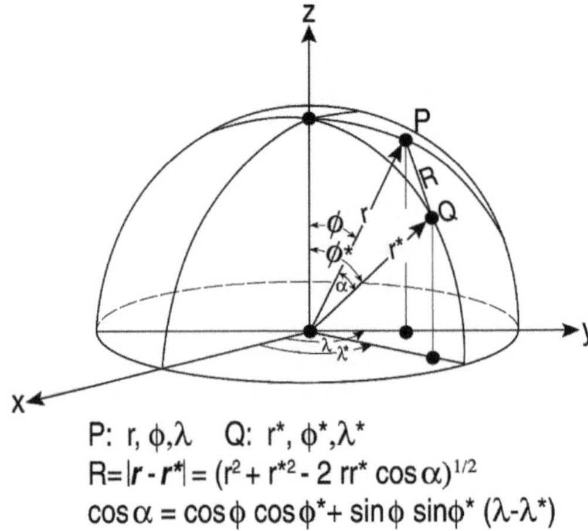

Fig. 2.6.1 Spherical geometry of the points P (observation) and Q (source dm) and distance R between them; notations of Eq. (2.6.10)

$Q = (r^*, \phi^*, \lambda^*)$ must be given (Fig. 2.6.1), written for the infinitesimal mass element $dm = \Delta\rho \, dV$:

$$d\delta W = G \, dm/R = G\Delta\rho \, dV/R \tag{2.6.8}$$

and

$$d\delta g = G \, d\delta W/r = G dm \partial/\partial r(1/R)\partial R/\partial r \tag{2.6.9}$$

where

$$R = (r^2 + r^{*2} - 2rr^*\cos\alpha)^{1/2} \quad \text{and} \quad \cos\alpha = \cos\phi\cos\phi^*$$
$$+ \sin\phi\sin\phi^*(\lambda - \lambda^*) \tag{2.6.10}$$

2.7 Laplace Equation, Field Quantities, Equivalent Stratum; Derivation of Some Field Quantities, Surface Integrals, Poisson Equations, Gravitational Flux Γ

The potential field variation in space $W(x,y,z)$ is governed by a simple partial differential equation of second order. It is especially simple in source-free space where the three second partial derivatives of W with respect to the coordinates add to zero, while within a continuously varying source volume, i.e. in density-filled space, the sum is proportional to density. In the first case the equation is called after Pierre Simon Laplace (1749–1827), in the second after Siméon Denis Poisson (1781–1840). Especially the Laplace equation has many direct and indirect applications in gravity analysis and interpretation. In this section, emphasis is on descriptions and relationships in space. The relationships are inherent in the idea of a field (see Sect. 1.4) and are generally based on the fact that it is given as a continuous and differentiable function of space coordinates, as Cartesian or other. In practice discrete observations are given, and they are taken to approximate functions which are then the basis for deriving other field quantities (Sect. 2.7.4) or field integrals (Sect. 2.7.6).

2.7.1 Source-Free Space: Laplace Equation

In empty space, the gravitational field $g = \nabla U_g = \nabla W$ is divergence-free – evidently: gravity has no interior source:

$$g = \text{div grad } W = \nabla\nabla W = W_{xx} + W_{yy} + W_{zz} = 0 \tag{2.7.1}$$

(Laplace equation). The operator $\nabla\nabla = \Delta$ is called Laplace operator, in Cartesian coordinates: $(\partial^2/\partial x_1^2 + \partial^2/\partial x_2^2 + \partial^2/\partial x_3^2)$. It is easily verified by differentiating $U_g = Gm/(x_1^2 + x_2^2 + x_3^2)^{1/2}$. It can also be demonstrated by applying Gauss' integral theorem $\int_v \nabla g \, dV = \int_s g \, dS$ for a point-like source at O and a sphere around it (volume V, surface S). Volume elements dV in spherical coordinates (Fig. 2.4.1, lower left) are bounded by six surface elements $dS = n_e dS$ (n_e = surface normal unit vector): four planar sides containing g (hence $g \, dSn = 0$) and two spherical caps separated by dr and normal to g, hence $g \, dS = g \, dS$. As S and g vary inversely to each other and gdS has opposite signs on both caps, both contributions to the surface integral cancel:

$$\int_s g dn = 0. \tag{2.7.2}$$

In other words, no net gravitational flux Γ (Eq. 2.1.1) passes through S if no mass is enclosed by the surface S. This can be generalized to the flux through limited solid angles for the integration over given mass bodies (see Sect. 2.9).

An empty volume, co-rotating with Earth, is not absolutely source-free, because the centrifugal acceleration field is not divergence-free. From $\nabla\nabla U_z = \nabla z$. with $z = \omega^2 R$, the derivation is $\nabla^2 U_z = \nabla z = \omega^2 \nabla R = 2\omega^2$ (from $R = (xe_x, ye_y, 0)$, $\nabla R = 2$).

Combining the two components gives

$$\Delta U_z = 2\omega^2 \tag{2.7.3}$$

The term $2\omega^2$ is constant and small. If gravity anomalies are considered after reductions for the normal gravity field, the term $2\omega^2$ is removed, and for the anomalous field, related to the mass anomalies, the Laplace equation holds. In the space around Earth the $2\omega^2$ term does not exist for free moving observers.

Note at this point that in so-called two-dimensional (2D) anomalous fields the Laplace equation degenerates to:

$$\nabla \delta g = \text{div grad } \delta W = \nabla \nabla \delta W = \delta W_{xx} + \delta W_{zz} = 0 \qquad (2.7.1.2D)$$

as $\delta W_{yy} = 0$. 2D approximations are important in gravity interpretation, when a gravity anomaly is characterized by one dominant elongation in which variations are then neglected and W is considered constant; usually in the y direction. In that case, the mass distribution is approximated by neglecting the dominant dimension, and described in the x, z plane (see Sect. 2.9.7).

2.7.2 The Field Quantities

Field quantities or elements of the field are the potential δW and its space derivatives. To reiterate: the Laplace equation completely describes the mutual relations between the field quantities in source-free space. Any complete representation as W, g or other spatial derivatives of W carries the same amount of information, though emphasizing different spectral windows; higher derivatives emphasize shorter wavelengths due to differentiation. Gravity g, as the norm of the gravity vector, alone, is an incomplete description of the gravity field, but its spatial variation often allows to recover aspects of the vector field.

Any of the field quantities can be defined for the point mass or infinitesimal mass element of integration. In the following, the expressions, in the three coordinate systems used in geophysics, are listed for the potential and its first, second and some third derivatives.

Starting with the **3D** *Cartesian coordinates*, assume an observer at $P = (0,0,0)$ and a point mass m at $Q = (x, y, z)$ separated by $r = (x^2 + y^2 + z^2)^{1/2}$. For integration the point mass is replaced by the infinitesimal mass: $dm = \rho dV = \rho dx\, dy\, dz$. Although mostly anomalies δW, δg etc. are considered here, "δ" is dropped for shortness, and the potential is called W, in contrast to U as in Eq. (2.6.3) used there to emphasize gravitation versus gravity with the centrifugal field included. As customary in geophysics, the quantities are taken to be positive. Subscripts, here, strictly mean partial differentiation of the scalar potential W, as $W_x \equiv \partial W/\partial x$. The other meaning in use for an index indicating components of the vector δg, as δg_x, will here not lead to confusion, but where the double meaning leads to ambiguity, differentiation will be explicitly expressed.

Only the geometrical aspects of the integrands, expressed in coordinates and distances r, denoted by italics, are listed below, while $G\rho\, dV$ is dropped. For example, for W we write f_W and instead of the complete $dW = G\rho\, dV/r$, we write $f_W = 1/r$, or for $dg = G\rho z dV/r^3$, we write $f_g = z/r^3$, etc. If density ρ is assumed constant, it is written before the integral symbols; otherwise it must be included in the integration. All the expressions, below, must be completed by $G\rho\, dV$, e.g., $W = G\rho \int f_W dV$ or $G \int \rho(x,y,z) f_W\, dV$ etc. The geometrical integrands are:

$$f_W = 1/r \quad r = (x^2 + y^2 + z^2)^{1/2} \qquad (2.7.4)$$

$$f_g = f_{gz} = f_{Wz} = z/r^3, f_{gx} = f_{Wx} = x/r^3, f_{gy} = f_{Wy} = y/r^3 \qquad (2.7.5)$$

$$f_{Wxx} = \partial f_{gx}/\partial x = 3x^2/r^5, f_{Wyy} = \partial f_{gy}/\partial x = 3y^2/r^5, \tag{2.7.6}$$

$$f_{Wzz} = \partial f_{gz}/\partial z = (2z^2 - (x^2 + y^2))/r^5, \tag{2.7.7}$$

$$f_{Wzx} = \partial f_{gz}/\partial x = 3xz/r^5,$$

$$f_{Wzy} = \partial f_{gz}/\partial y = 3yz/r^5, f_{Wxy} = \partial f_{gx}/\partial y = \partial f_{gy}/\partial x = 3xy/r^5 \tag{2.7.8}$$

$$f_{Wzzz} = \partial^2 f_{gz}/\partial z^2 = 3z(2z^2 - 3(x^2 + y^2))/r^7, \tag{2.7.9}$$

Note that the above differentiations pertain to the environment of the observation point P, not to the small volume source element $dV = dx\,dy\,dz$ or in spherical or cylinder coordinates where $dV = r^2 \sin\varphi\,dr\,d\varphi\,d\Lambda$ or $dV = R\,dR\,dZ\,d\Lambda$, respectively, vary with r and φ or with R and change size. Confusion is to be avoided when differentiating expressions contain the respective variables. When integrating the effects the situation changes to the source where any change in element size is relevant.

In the **2D case**, everything is constant for y from $-\infty$ to $+\infty$, and integration over y is carried out first. While generally $\rho = \rho(x,z)$, for simplicity, $\rho =$ const is assumed. With $r = (x^2 + z^2 + y^2)^{1/2}$ and $R = (x^2 + z^2)^{1/2}$, the 2D potential $W = G\rho_{-\infty}\int^{+\infty}[(1/r)\,dy]dx\,dz = 2G\rho_0\int^{\infty}[(1/r)\,dy]dx\,dz = 2G\rho \ln[(x^2 + z^2 + y^2)^{1/2} + y]\,|_0^\infty = 2G\rho[\infty - \ln R]$. This is the so-called *logarithmic potential*. The additive constant ∞ is irrelevant for practical applications, as any constant is, because the two-dimensionality is anyway not strictly realistic, and only the variations of the second term count, for example, when dW is differentiated with respect to x or z. Henceforth the minus sign of $\ln R$ is dropped (sign convention in geophysics), and "R" is written "r". Thus we write: $dW = 2G\rho \ln r = 2G\rho \ln(x^2 + z^2)^{1/2}$. The other field quantities can be derived by differentiating the potential with respect to z or by integrating the 3D quantities over y; the latter way avoids the infinite constant of the logarithmic potential.

As above, in the list, from the full expressions $G\rho dxdz$ has been dropped and only the integrands are shown. The factor 2, characteristic for 2D, has been kept. For integration, the expressions must be completed, e.g. for W by writing $G\rho \iint f_W\,dx\,dz$.

$$f_W = 2\ln r \quad r = (x^2 + z^2)^{1/2} \tag{2.7.4.2D}$$

$$f_g = f_{gz} = f_{Wz} = 2z/r^2, f_{gx} = f_{Wx} = 2x/r^2, f_{gy} = 0 \tag{2.7.5.2D}$$

$$f_{Wxx} = \partial f_{gx}/\partial x = 2(x^2 - z^2)/r^4, f_{Wyy} = \partial f_{gy}/\partial x = 0 \tag{2.7.6.2D}$$

$$f_{Wzz} = -f_{Wxx} = \partial f_{gz}/\partial z = 2(z^2 - x^2)/r^4, \tag{2.7.7.2D}$$

$$f_{Wzx} = \partial f_{gz}/\partial x = 4xz/r^4, f_{Wzy} = \partial f_{gz}/\partial y \equiv 0, f_{Wxy} = \partial f_{gx}/\partial y = \partial f_{gy}/\partial x = 0 \tag{2.7.8.2D}$$

$$f_{Wzzz} = \partial^2 f_{gz}/\partial z^2 = 4z(z^2 - 3x^2)/r^6, \tag{2.7.9.2D}$$

Useful variants of the above expressions are based on *horizontal cylinder coordinates* with y as axis through P, $r = (x^2 + z^2)^{1/2}$ as defined above, $\sin\varphi = z/r$, the 2D volume element $r\,dr\,d\varphi$ and the density constant or $\rho = \rho(r,\varphi)$ over which has to be integrated. All the expressions, except f_W, contain terms with $z/r = \sin\varphi$ and/or $x/r = \cos\varphi$, such that the 2D field quantities will be listed with only $G\rho\,dr\,d\varphi$

dropped (with the consequence that dW contains a factor r and that the power of r is decreased by 1 per differentiation). Again, as an example for a full expression: $W = G\rho \iint f_W\, r\, dr\, d\varphi$.

$$f_W = 2r\ln r \quad r = (x^2 + z^2)^{1/2} \tag{2.7.4.2Da}$$

$$f_g = f_{gz} = f_{Wz} = 2\sin\varphi, f_{gx} = f_{Wx} = 2\cos\varphi \tag{2.7.5.2Da}$$

$$f_{Wxx} = \partial f_{gx}/\partial x = 2\cos 2\varphi/r \tag{2.7.6.2Da}$$

$$f_{Wzz} = -f_{Wxx} = \partial f_{gz}/\partial z = -2\cos(2\varphi)/r \tag{2.7.7.2Da}$$

$$f_{Wzx} = \partial dg_z/\partial x = 2\sin 2\varphi/r, \tag{2.7.8.2Da}$$

$$f_{Wzzz} = \partial^2 dg_z/\partial z^2 = -4\sin 3\varphi/r^2 \tag{2.7.9.2Da}$$

In *spherical coordinates*, P at the centre of the sphere ($r = 0$) is the reference point and the mass element at $Q = (r, \varphi, \lambda)$ is $dm = \rho\, dV = \rho\, r^2 \cos\varphi\, dr\, d\varphi\, d\lambda$; φ corresponds to latitude and λ corresponds to longitude or azimuth and for symmetry no azimuthal components exist. The quantities of interest are the same as above, as the x, y, z coordinates continue to be used. The relations are (Eq. 2.4.1):

$$x = r\cos\varphi\cos\lambda; \; y = r\cos\varphi\sin\lambda; \; z = r\sin\varphi.$$

This set-up is not designed to calculate effects of mass at Q on P, where both points are anywhere in a sphere, for example, the globe. Again, the expressions in italics are normalized by $G\rho\, dV = G\rho r^2 \cos\varphi dr d\varphi d\lambda$, such that, for example: $W = G\rho \iiint f_W\, r^2 \cos\varphi\, dr d\varphi\, d\lambda$:

$$f_W = 1/r \quad r = r \tag{2.7.4a}$$

$$f_{gz} = f_{Wz} = \sin\varphi/r^2, f_{gx} = f_{Wx} = \cos\varphi\cos\lambda/r^2, f_{gy} = f_{Wy} = \cos\varphi\sin\lambda/r^2 \tag{2.7.5a}$$

$$f_{Wxx} = 3\cos^2\varphi\cos^2\lambda/r^3, f_{Wyy} = 3\cos^2\varphi\sin^2\lambda/r^3, \tag{2.7.6a}$$

$$f_{Wzz} = (3\sin\varphi - 1)/r^3, \tag{2.7.7a}$$

$$f_{Wzx} = 3\sin\varphi\cos\varphi\cos\lambda/r^3, f_{Wzy} = 3\sin\varphi\cos\varphi\sin\lambda/r^3,$$

$$f_{Wxy} = 3/2\cos^2\varphi\sin 2\lambda/r^3 \tag{2.7.8a}$$

$$f_{Wzzz} = (5\sin^3\varphi - 3\sin\varphi)/r^4, \tag{2.7.9a}$$

Briefly, the radial effects are, of course, $f_{gr} = \partial f_W/\partial r = (f_{gx}^2 + f_{gy}^2 + f_{gz}^2)^{1/2} = 1/r^2$, $f_{Wrr} = 2/r^3$ and $f_{Wrrr} = 6/r^4$; differentiation with respect to φ and λ renders, of course, zero. Remember that differentiations is done at the observation point P coordinates, not at the source element dV; confusing these leads to totally different and inconsistent results.

The following expressions are for the *vertical cylinder mass element* $dm = \rho R dR dZ d\Lambda$. Again, the volume element at Q grows in size with distance R, but differentiation is carried out for P on the axis at $R = 0$, $Z = 0$, with $r = (R^2 + Z^2)^{1/2}$ ($x = R\cos\Lambda, y = R\sin\Lambda, z = Z$). The example for a full expression is, again, $W = G\rho \iint f_W R dR dZ d\Lambda$. The normalized integrands are:

$$f_W = 1/r \quad r = (R^2 + Z^2)^{1/2} \tag{2.7.4b}$$

$$f_{gz} = Z/r^3, f_{gx} = R\cos\Lambda/r^3, f_{gy} = R\sin\Lambda/r^3 \tag{2.7.5b}$$

$$f_{Wxx} = 3R^2\cos^2\Lambda/r^5, f_{Wyy} = 3R^2\sin^2\Lambda/r^5, \tag{2.7.6b}$$

$$f_{Wzz} = (2Z^2 - R)/r^5, \tag{2.7.7b}$$

$$f_{Wzx} = 3ZR\cos\Lambda/r^5, f_{Wzy} = 3ZR\sin\Lambda/r^5, f_{Wxy} = 3/2\sin 2\Lambda/r^5 \tag{2.7.8b}$$

$$f_{Wzzz} = 3Z(2Z^2 - 3R^2)/r^7, \tag{2.7.9b}$$

The horizontal radial components are $f_{gR} = R/r^3$; $f_{WRR} = 3R^2/r^5$.

2.7.3 The Equivalent Stratum

The Laplace equation implies, that within empty space the variation of $W(x,y,z)$ is sufficiently defined by the values at the boundaries of the source space. Thus, the solution of the Laplace equation is a boundary value problem which implies that the real mass distribution behind the surface can be replaced by any equivalent mass generating the same gravitation at the boundary. This is the ambiguity problem of gravity inversion.

An example is the equivalent stratum, i.e. an infinitely thin surface mass layer (Sects. 5.6.1, 5.6.5; Task 5.2) with surface density ρ^* (kg/m^2) normal to gravity, at depth $z = $ const below P, where, as in most following arguments and figures, P is placed at O $= (0,0,0)$; $\rho^* = \rho^*(x,y)$. Generally, as P $\to$ Q on the stratum, i.e. $z \to 0$, there appears a formal problem, as $r \to 0$ and the integrand in the δg integral over the surface mass becomes indefinite, i.e. zero divided by zero, 0/0, at $r = 0$. If ρ^* is continuous and differentiable, then always a small radius ε exists about Q such that $\rho^*(Q)$ can be replaced by its constant mean value $\rho^*(\varepsilon)$, and the integral over the whole infinite plane $\int \rho^*(\Omega)\mathrm{d}\Omega \to \rho^*(x_Q, y_Q)\,\Omega$ with $\Omega \to 2\pi$, i.e., the solid angle Ω under which the stratum is seen from P is dominated by the local value of ρ^*. In other words, with ρ^* before the integral $\delta g^{(P)} \to \delta g^{(Q)} = \delta g^{(P\to Q)} \to 2\pi\,G\,\rho^{*(Q)} = 2\pi\,G\,\rho^*$, from which follows:

$$\rho^* = \delta g(x,y)/(2\pi G). \tag{2.7.10}$$

without causing an error greater than any pre-set small limit; if $|\rho^* - \rho^{*(Q)}| < \varepsilon/6\pi$, $|\delta g^{(P)} - 2\pi G\rho^{*(Q)}| < \varepsilon G$ (see KJ61, 112–113).

By inserting (Eq. 2.7.10) into the integrals, expressions (KJ61, 113–117) follow for the upward continuation (see also Sect. 2.10.5.3) of the field quantities whose normalized integrands are listed in Sect. 2.7.2 as (Eqs. 2.7.4, 2.7.5, 2.7.7 & 2.7.9) and their variants. Some integrals are listed below in Cartesian (x, y, z) and vertical cylinder coordinates (R, Z, Λ), numbered as in (Eqs. 2.7.4, 2.7.5, 2.7.7 & 2.7.9). In cylinder coordinates, the azimuthal integration $({}_0\!\int^{2\pi} \delta g\,\mathrm{d}\Lambda)$ or averaging for horizontal distances $R = $ const is carried out first: $\delta g = \delta g(R) \equiv 1/(2\pi){}_0\!\int^{2\pi} \delta g(R,\Lambda)\mathrm{d}\Lambda,$

by definition, with the vertical distance Z between P and the equivalent stratum and $r = (x^2 + y^2 + z^2)^{1/2} = (R^2 + Z^2)^{1/2}$:

$$\delta W = 1/(2\pi) \int_x \int_y \delta g(x,y)/r \, dx \, dy = \int_0^\infty \underline{\delta g} \, R/r \, dR \qquad (2.7.4c)$$

$$\delta g = z/(2\pi) \int_x \int_y \delta g(x,y)/r^3 \, dx \, dy = Z \int_0^\infty \underline{\delta g} \, R/r^3 \, dR \qquad (2.7.5c)$$

$$\delta g_z = 1/(2\pi) \int_x \int_y \delta g(x,y)[(2z^2 - (x^2 + y^2))]/r^5 \, dx \, dy = \int_0^\infty \underline{\delta g}(2Z^2 - R^2)R/r^5 \, dR$$
$$(2.7.7c)$$

$$\delta g_{zz} = 3z/(2\pi) \int_x \int_y \delta g(x,y)[(2z^2 - 3(x^2 + y^2))]/r^7 \, dx \, dy$$
$$= 3Z \int_0^\infty \underline{\delta g}(2Z^2 - 3R^2)R/r^7 \, dR \qquad (2.7.9c)$$

The corresponding 2D expressions, by combining (Eq. 2.7.4.2D) etc. with (Eq. 2.7.10), are, with $r = (x^2 + z^2)^{1/2}$ (no integration over dz):

$$\delta W = 1/\pi \int_0^\infty \delta g(x) \, \ln(z/r) \, dx \qquad (2.7.4.2Dc)$$

$$\delta g = z/\pi \int_0^\infty \delta g(x)/r^2 \, dx \qquad (2.7.5.2Dc)$$

$$\delta g_z = 1/\pi \int_0^\infty \delta g(x) \, (z^2 - x^2)/r^4 \, dx \qquad (2.7.7.2Dc)$$

$$\delta g_{zz} = 2z/\pi \int_0^\infty \delta g(x)(z^2 - 3x^2)/r^6 \, dx \qquad (2.7.9.2Dc)$$

2.7.4 Applications: Estimation of Field Quantities as δW_x, δW_y, δW_{zzz}

The task of estimating unobservable field quantities trains the interpreter's imagination of geometrical relationships and constraints when analyzing observations. It serves interpretation purposes:

(1) geometrical relations among the field quantities allow unobservable quantities to be derived from observed ones (for a more thorough discussion (see KJ61, 5–17);

(2) equivalent mass distributions as the equivalent stratum can be used to calculate the field quantities without knowledge of the true mass distribution and

(3) differentiable representations as expansions of the observations, for example, by Fourier and other series, can serve the same purpose.

Directly observable are:

- gravity: $\mathbf{g} = \nabla \delta W = \partial \delta W / \partial z$ (or δg), where z points along $\mathbf{g}$, measured by gravimeters;
- deflection of the plumb line: $\delta g_x / g$ and $\delta g_y / g$ with the horizontal components of $\mathbf{g}$, $\delta g_x = \partial \delta W / \partial x = \delta W_x$ and $\delta g_y = \partial \delta W / \partial y = \delta W_y$; deflections of the plumb line are defined relative to a global coordinate system based on the normal ellipsoid (see Sects. 4.3, 4.41), deflections of the local vertical, plumb line or nadir (opposite to zenith) are measured against the normal vertical which is established astronomically from latitude and longitude determinations;
- water surface topography as approximation of equipotential surfaces, observed by satellite altimeters;
- δW_{zz} only approximately by gravity measurements at points separated vertically by Δh;
- δW_{zx}, δW_{zy}, $\delta W_{yy} - \delta W_{xx}$, δW_{xy} measured with the torsion balance and all elements of the gravity tensor (Sect. 2.8) observed with appropriate gradiometers.

Elements that are unreliably observed include

- δW or W;
- δW_{zz} and δW_{xx} and δW_{yy}, separately (with the torsion balance only $\delta W_{yy} - \delta W_{xx}$);
- Although listed above as observable, δW_{zx} and δW_y are mentioned here, since the observation is cumbersome. Note that these are the so-called horizontal gradients of the gravity anomalies δW_z which are sometimes also written δg_x and δg_y which is a case of confusion to be avoided. Some of these quantities are used in geodesy and in gravity field analysis and interpretation.

Elements that are not directly observable include

- δW_{zxx}, δW_{zyy}, δW_{zzz}, the so-called second derivatives, and similar elements.

The Laplace equation and Newton's law offer the possibility to estimate these latter quantities from spatial variation $\delta g(x, y)$. The spatial variation is traditionally visualized by contour lines and by computer graphics including colours and shadings (see Chap. 3).

A special case is the vertical derivative of gravity, according to Ackermmann & Dix (1955): $\delta W_{zz}(Z) = {}_o\int^\infty \delta W_{zzz}(R) R / r dR$, where capital Z stands for the vertical cylinder coordinates used. In this case, with $Z \rightarrow 0, \delta g_z(0) = \delta W_{zz}(0) = {}_o\int^\infty \delta W_{zzz}(R) \, dR$. If the vertical gradient of gravity, as sometimes, is called δg_z, it must be clearly said not to mean the usual vertical component of gravity.

Equipotential surfaces are also useful if visualized since their curvature implies the existence of horizontal components of $\boldsymbol{\delta g}$, say, $\boldsymbol{\delta g}_u$ in any direction (coordinate u). Referred to point P_o, where $\mathbf{g}_o = (g_0, 0, 0)$, the surface may have a local maximum or minimum, a saddle point, or may be a ridge or a trench. According to differential geometry of second order surfaces, horizontal sections through them are conical sections, i.e. elliptical, hyperbolic or parabolic, in the latter case consisting of two parallel straight lines. A vertical section in an arbitrary u direction through

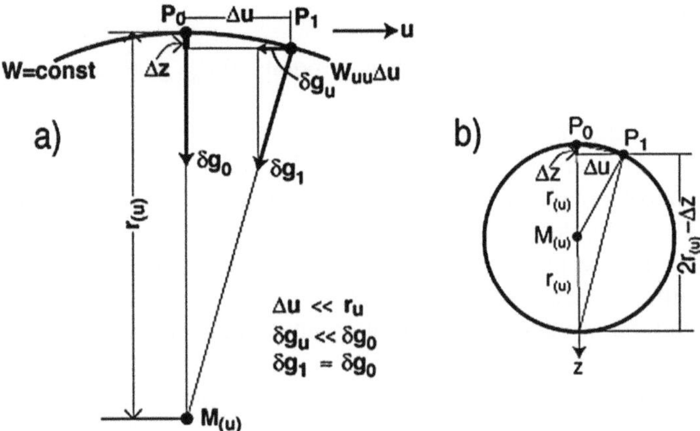

Fig. 2.7.1 (**a**) Curvature of a vertical section of the equipotential surface $W = $ const through P_0 and P_1 separated by Δu (u direction arbitrary) and relationship of $\boldsymbol{\delta g_o}$ and $\boldsymbol{\delta g_1}$. For small Δu the section is approximated by a circular arc with centre $M_{(u)}$. (**b**) Demonstration of theorem of height (Δu) in the *right triangle* P_0P_1Z

points P_0 and P_1, a short distance $u_1 - u_0 = \Delta u$ apart, contains $\boldsymbol{g_0}$ and approximately $\boldsymbol{g_1}$ ($g_1 \approx g_0 = g$, see Fig. 2.7.1). The section is a short, nearly circular arc (convex or concave from above) with the curvature $\kappa_u = 1/r_u$ ($r_u = $ radius of curvature, sign convention: convex positive, concave negative). The similarity (Fig. 2.7.1):

$$\Delta u/r_u \approx g_u/g \approx \Delta u W_{uu}/g$$

renders

$$r_u = 1/\kappa_u \approx g/W_{uu}$$

or

$$g_u \approx g\Delta u/r_u.$$

Furthermore, if the circular arc of Fig. 2.7.1 is completed to a full circle, the theorem of height in right triangles renders $\Delta u^2 = \Delta z(2r_u - \Delta z) \approx 2r_u\Delta z$.

The horizontal component of gravity g_u and the corresponding plumb line deflection g_u/g can be estimated from g and W_{uu}, measured, e.g., with the torsion balance. If only gravity observations exist, curvature must be estimated otherwise (below).

Conical sections have two principal axes, or in the case of hyperbolae, axial sections: say, in y direction (Δ_1: vertical section with maximum curvature $\kappa_1 = 1/r_1$) and x direction (Δ_2: minimum curvature $\kappa_2 = 1/r_2$); in the parabolic case $\Delta_1 \to \infty$. In the example of the ellipse, from its definition $x^2/\Delta_2{}^2 + y^2/\Delta_1{}^2 = 1$, follows that $\Delta u^2 = (\Delta_1{}^2 + \Delta_2{}^2)/(\Delta_1{}^2 \sin^2 \psi + \Delta_2{}^2 \cos^2 \psi)$, where ψ is the angle between the principal section I and the direction of u. Generalized to include the hyperbolic and parabolic cases, each Δ is multiplied by $+1$, if curvature is positive (convex from above), and -1, if curvature is negative (concave from above). The above relation $\Delta u^2 \approx 2r_u\Delta z$ is true also for $\Delta_1{}^2 \approx 2r_1\Delta z$ and $\Delta_2{}^2 \approx 2r_u\Delta z$, and combining these leads to Euler's theorem describing the relations between $\kappa_u = \kappa_1 \cos^2 \psi + \kappa_2 \sin^2 \psi$ or $\kappa_u = (\kappa_1 + \kappa_2)/2 + \cos 2\psi(\kappa_1 - \kappa_2)/2$. This gives a complete description of the curvature of the W-surface around P_o in its approximation by second-order surfaces. For a more detailed derivation and illustrations, see KJ61, 9–13.

The shape of the equipotential surfaces or W-surfaces is reflected in the curvature and density of the customary contour lines of observed gravity anomalies. The Laplace equation (2.7.1) implies: $W_{zzz} = -\partial W_{zx}/\partial x - \partial W_{zy}/\partial z$ (demonstrating, by the way, that for very smooth linear fields $W_{zzz} \approx 0$). The relation between the curvature κ of the δg contour lines and W_{zzz} can be derived analytically from the implicit or parameter form (parameter s) $x = v(s)$ and $y = w(s)$ with (see KJ61, 16–17):

$$dy/dx = w_s(s)/v_s(s) \text{ and } \kappa = (v_s w_{ss} - w_s v_{ss})/(v_s{}^2 + w_s{}^2)^{3/2} :$$
$$\kappa = (2g_x g_y g_{xy} - g_y{}^2 g_{xx} - g_x{}^2 g_{yy})/(g_x{}^2 + g_y{}^2)^{3/2}.$$

With special coordinates t and n, locally tangential (t) and normal (n) to a contour line, we obtain:

$$W_{zzz} = -g_{tt} - g_{nn} = \kappa \partial g/\partial n - \partial^2 g/\partial n^2,$$

demonstrating that curvature and variation of distance between contour lines relate to the second derivative of δg; or only the latter term: $W_{zzz} = -\partial/\partial n(\partial g/\partial n)$, if the contour lines are straight.

2.7.5 Source Space: Poisson Equation and Gravitational Flux Γ

Within space filled with a continuum of mass with density ρ, the gravitational vector field is, of course, not divergence-free. The formal mathematical problem is that the source point Q (at r^*) and the observation point P (at r) approach each other ($|r - r^*| \to 0$) and the potential function $\delta U_g = G \int_V \rho(r^*) dV/|r - r^*|$ becomes singular. If we consider the mass volume as divided into two parts, (1) a small sphere around Q and (2) an outer volume (in which no problem arises with $|r - r^*| \to 0$) and if the inner volume has a continuous or smooth density, such that it can be assumed constant if r is small enough, then its mass $\rho dV(r)$ approaches zero steadily, the expression can be evaluated for continuous functions $\rho(r^*)$ (e.g. Grant and West, 1965; henceforth referred to by GW65):

$$\nabla^2 \delta U_g(r) = \text{div grad } \delta U_g = 4\pi G \rho(r). \tag{2.7.11}$$

This is Poisson's equation stating that the source strength or divergence of gravitation, i.e. the gradient of the gravitational potential U_g in mass-filled space is the product of density, G and the full solid angle 4π (= surface area of a unit sphere of radius 1).

In terms of Gauss' integral theorem $\int_v \nabla \bullet \delta g \, dV = \int_s \delta g \, dS = \Gamma$; for a point mass m within the spherical surface $S : \Gamma = 4\pi m$ G; for constant density ρ, $m = (4\pi\rho/3)\rho r^3$, with $r =$ radius of the sphere S, the relation is identical. For δU_g or δg

with limited solid angles the fundamental relationships are valid accordingly with useful applications to estimating or computing gravity effects (s. Chap. 5).

If Cartesian coordinates are used in limited regions, the earth is replaced by a semi-infinite mass with radius $r \rightarrow \infty$, but such that g is as observed. In this case, the plane $z = 0$ divides the source volume below ($z > 0$) from source-free space ($z < 0$, upward). The plane $z = 0$ is only half of the full containing surface and the gravitational flux Γ through $z = 0$ is $\Gamma = 2\pi Gm$, where m is the total mass. As explained, m can be replaced by a surface mass of identical magnitude.

2.7.6 Surface Integrals: Total Mass, Centre of Mass

An application of Poisson's equation coupled with Gauss' theorem is the calculation of the total mass, here called M, generating an anomaly by integrating the anomaly over the whole (infinite) surface. In the Cartesian idealisation of the world, in x, y, z space, with P at O and all density anomalies below the surface, $\Delta\rho = \rho(x,y,z) - \rho_0$ ($\rho_0 =$ reference density, constant or only depth-dependent), the surface area integral over a gravity anomaly is

$$I = \int_{x,y} \delta g(x,y)\mathrm{d}x\mathrm{d}y = 2\pi GM \qquad (2.7.12)$$

where $M = \int_V \Delta\rho\,\mathrm{d}\underline{V}$. The solid angle is 2π, half of the full 4π. If gravity is known on a closed surface, as a sphere or ellipsoid, the integral leads to the total gravimetric flux $\Gamma = 4\pi GM$ of the anomalous mass M to be determined. The 2π value on the plane can be understood intuitively when the x, y plane at $z = 0$ is considered half of a closed surface which fully confines M; the other half surface may have any shape, it may be another infinite parallel x, y plane at $z = z_b$ where M is limited between these planes which intersect at infinity. For proof, application of Gauss' integral theorem $\int_v \nabla\delta g\mathrm{d}V = \int_s \delta g\mathrm{d}S$ is discussed, e.g. by GW65, 228.

Simpler proof is obtained by writing $\delta g(x,y) = \mathrm{G}\iiint_V \Delta\rho(x',y',z')\,\mathrm{d}x'\,\mathrm{d}y'\,\mathrm{d}z'$ in (Eq. 2.7.12), where the prime indicates integration over the source: $I = \iint_{x,y}(\mathrm{G}\iiint_{V'} \Delta\rho z'\mathrm{d}V'/r'^3)\mathrm{d}x\mathrm{d}y$, and reversing the sequence of integration by first carrying out the integration over the infinite x, y surface for an infinitesimal, i.e. point-like mass element $\mathrm{d}m = \Delta\rho(\mathrm{x'}\,\mathrm{y'}, \mathrm{z'})$; it must then be shown that $_\infty\int^\infty\,_\infty\int^\infty\,_{x,y}z'r'^{-3}\mathrm{d}x\mathrm{d}y = 2\pi$ (without proof). We then obtain again (Eq. 2.7.12): $I = 2\pi\mathrm{G}\iiint_{V'} \Delta\rho\mathrm{d}V'$. Another simple proof can be written for the equivalent stratum (Sect. 2.7.3) and Eq. (2.7.10) for surface density ρ^*. It leads to the result that the total surface mass $M^* = M$, the equivalent volume mass.

Without repeating the deduction which is quite similar, the surface anomaly integral can be generalized to horizontal anomaly moments to arrive at similar expressions for the horizontal centre of gravity X_c, Y_c of the gravity anomaly, and X_c, Y_c are identical for the mass anomaly itself; the reader may work this out for her/himself. The depth coordinate Z_c is, however; not resolved this way. With coordinate indices $i = 1, 2$:

$$X_{ci} = I^{-1} \int_{x,y} \delta g(x,y) x_i \, dx \, dy \tag{2.7.13}$$

This is extendable to higher moments, i.e. products of δg with x^n and y^m, n, m arbitrary, with some useful applications to gravity interpretation (GW65).

Application to the 2D case refers only to 2D density ρ^* or mass M^* per section area, but in this case ρ^* refers to the vertical x, z plane, and the 2D equivalent stratum density (kg/m) is ρ^+ per length unit x. While the total mass below the infinite x, y plane is infinite, the 2D mass is $M^* = {}_{x,z}\int \rho^*(x,z)\, dx\, dz = {}_{-\infty}\int^{\infty} \rho^+(x)\, dx$. With these definitions, (2.7.10) is valid also in 2D:

$$I^* = \int_{-\infty}^{\infty} \delta g(x)\, dx = 2\pi G M^* \tag{2.7.12.2D}$$

formally identical to (Eq. 2.7.12). Similarly, the centre of gravity of the 2D anomaly and that of the anomalous mass M^* follow from

$$X_c = I^{*-1} \int_x \delta g(x)\, x\, dx \tag{2.7.13.2D}$$

The higher moments can be treated accordingly.

Note that M is only the anomalous mass of a geological body of volume V, defined by the density anomaly $\Delta\rho : M = \Delta\rho V$. The total mass is $M_{tot} = (\rho_0 + \Delta\rho)V$, which is the quantity relevant, for example, to estimate the total of minable ore. V can be estimated from M if $\Delta\rho$ is known or assumed: $V = M/\Delta\rho$. Note also that the integral I, and hence M, is a quantity in gravity analysis not subject to the principal ambiguity of interpretation. However, it is subject to other uncertainties as that of defining anomalous gravity which involves the separation of fields and the definition of "regional" and "local" or residual anomalies (see Sects. 2.10.2 & 2.10.3, 4.7.7).

2.8 The Gravity Tensor (Eötvös Tensor)

The second derivatives of the gravity potential (also called gravity gradients, see below) are used in many ways in geodesy and geophysics.

In a local orthogonal coordinate system (see above) we have 9 elements of which only 6 are independent. With Cartesian coordinates $x_i (i = 1, 2, 3)$ corresponding to (x, y, z) we write:

$\partial^2 W / \partial x_i \partial x_j = W_{ij}$ where $W_{ij} = W_{ji}$ as the sequence of differentiation is interchangeable.

The 9 elements are traditionally ordered as

$$\{W_{ij}\} = \left\{ \begin{array}{ccc} W_{11} & W_{12} & W_{13} \\ W_{21} & W_{22} & W_{23} \\ W_{31} & W_{32} & W_{33} \end{array} \right\} \tag{2.8.1}$$

In two-dimensional fields with $\partial/\partial y \equiv 0$, the Eötvös tensor reduced to 4 (3 independent) components which simplifies the operations:

$$\{W_{ij}\} = \left\{ \begin{array}{cc} W_{11} & W_{13} \\ W_{31} & W_{33} \end{array} \right\} \qquad (2.8.1.2D)$$

The elements or combinations of some of them have physical meaning and important applications. Elements of the gravity tensor can be measured with the torsion balance and even in a satellite moving in free fall, i.e. with gravity and centrifugal acceleration cancelling each other at its centre of gravity. This principle is realized in recent gravity missions as CHAMP, GRACE and GOCE; the latter, especially, will carry accelerometers at 0.5 m distance in three coordinate directions: radial, i.e. pointing away from the Earth's centre, forward and to the side, i.e. pointing in the direction of the local vector of angular velocity (Sect. 3.2.8). Recent developments of terrestrial measurements measuring the full gravity tensor have found intensive application in the exploration industry (Sect. 3.2.9). Differential geometry of equipotential surfaces has been applied extensively in connection with the torsion balance and is again needed with the new gravity tensor devices. Since at present measurements are nearly always done with gravity meters readers are referred to the older literature, e.g. KJ61.

Generally the second derivatives describe curvature which is related to the closeness to compact (point-like) sources. A vertical section of an equipotential surface is approximated, near a point P, by a circle with the radius of curvature $r_k = -g/W_{kk}$ where k denotes an arbitrary horizontal direction (not necessarily x or y).

The vertical variation of gravity $g = |W_3|$ is described by the vertical potential gradient W_{33}. The horizontal gradients W_{31} and W_{32} (observable with the torsion balance) in map view point normal to gravity contour lines along (locally) the direction of maximum change with a magnitude inverse to distance between the contours. The term "gravity gradient" is not strictly correct because of the vector nature of gravity, but it is justified in view of g being treated as the magnitude of $\boldsymbol{g}$.

2.9 Gravity Effects and Anomalies – Summation and Integration

2.9.1 General Considerations

As has been emphasized, effects are distinguished from anomalies, the former being usually calculated for given masses and the latter being derived from observations. The two are referred to each other in the course of gravity interpretation; remember, however, that anomalies are not uniquely determined by limited discrete observations (Sect. 1.4; for a more thorough discussion, see Sect. 5.15).

Geological mass anomalies are three-dimensional (3D) volumes or bodies of anomalous density $\Delta\rho$, generally irregular in shape, i.e. not easily describable by mathematical expressions. To compute gravity effects of geological bodies they must be made tractable by analytical description, i.e. as idealized forms, in most cases with uniform density ρ or simple space relationships, for example, linear dependencies on the coordinates. In a way, the situation is aggravated by the fact, that what counts is the density contrasts $\Delta\rho = \rho - \rho_0$ where the reference density ρ_0 of the surrounding medium is generally also heterogeneous so that assuming ρ_0 constant or otherwise too simple is problematic. Models are usually idealized bodies and the most appropriate kind of idealization must be guessed. A popular idealisation is two-dimensionality, 2D, where only the cross section of a structure is considered; this is appropriate, i.e. fairly accurate, if one dimension is dominant in the sense of strong elongation of the structures and related gravity anomalies.

Coordinate systems in which integrations are to be carried out affect the strategies chosen. For geological gravity effects embedded into the Earth's normal gravity field, the azimuthal independence of the vertical component δg_z sometimes makes vertical cylinder coordinates natural and appropriate, but the emphasis here is on Cartesian coordinates appropriate for sufficiently small scales relative to the Earth's size and more easily applied to calculations at many observation points P. On the other hand, large scales require that the shape of the globe must be taken into account, and spherical coordinates are adequate or mandatory.

This section is divided into subsections on general aspects (Sects. 2.9.1, 2.9.1.1, 2.9.1.2, 2.9.2) and handling of 3D (Sects. 2.9.3–2.9.6) and 2D (Sect. 2.8.7) as well as $2\frac{1}{2}$D cases (Sects. 2.8.7–8). Integration in terrestrial spherical coordinates is deferred to subsections on some general aspects (Sect. 2.10.7.2) and on Gauss-Legendre quadrature (Sect. 2.10.7.3).

2.9.1.1 Integration of Effects of Finite Dense Masses

The main aspect of this section is the general aspect of integration of model gravity effects. Detailed applications are treated in Chaps. 5 and 6. The classical schemes to idealize, subdivide and/or approximate natural density distributions are universal, but the emphasis here differs from other texts. It is on a number of principal rules of simplification and outlook, which are not new as such but were not treated systematically elsewhere. The rules are listed below.

1. Most importantly, the solid angle $\Delta\Omega$ under which planar mass elements are seen from P offers conceptual and practical advantages to gravity effect integration. The solid angle is defined as the area of the projection onto the unit sphere around P. It directly gives the gravitational component normal to the plane considered, for example, the vertical effect Δg_z if the plane is horizontal. The approach is identical to building the kernel z/r^3 into the volume element of integration (Fig. 2.9.1).

2. If any other component of the gravitational vector effect is needed to define δg_z, it is necessary to calculate also the components parallel to the planar mass elements. As in the case of the solid angle the planar elements can also be limited or defined by rays from P so that the $1/r^2$ relation can still be exploited. The basic mass element of integration is the mass line parallel to the component considered,

e. g., for δg_z it is the vertical mass line as an element of a vertical plane. Two orthogonal plane-parallel components combined with the plane-normal component (above) define the vector effect. Vertical orientation or simply "vertical" (in quotes) is used to nickname the orientation parallel to the vector component sought, i.e. $\boldsymbol{g}$-parallel, which need not be vertical.

3. Vector calculus is often easier for defining trigonometric functions, e.g. $\cos \alpha = \boldsymbol{a}\,\boldsymbol{b}/ab$ than applying trigonometry itself.

4. It is often profitable to calculate the effect on the gravitational potential δW and then find δg by differentiation.

5. Generally the best basic body for composing more arbitrary model geometries is the simplest model body of its kind. For example, the rectangular prism expanded from P at (0,0,0) to arbitrary Cartesian coordinates x, y, z is the simpler than the commonly used prism ($\Delta x \cdot \Delta y \cdot \Delta z$) somewhere in Cartesian space. Any such prism can be constructed of several x, y, z-prisms with P at one corner, appropriately subtracted or added to each other. In vertical cylinder coordinates the corresponding basic body is the cylinder with radius R and depth Z, instead of a cylinder ring sector between $R = R_i{}^*$, $R_{i+1}{}^*$, $Z = Z_j{}^*$, $Z_{j+1}{}^*$, and $\Delta \Lambda^* = \Lambda_{k+1} - \Lambda_k$. Or a dyke can be composed of two steps.

6. The gravity anomaly δg is proportional to the length scale L, as will be evident below. In most expressions for δg the length scale, L, is explicitly L^4/L^3, as in the basic expression below (Eq. 2.9.2.4). However, L or L^2 are hidden in the definitions of surface density (kg/m^2) or line density (kg/m). The linear L-dependence of the δg expressions has the useful advantage that $\delta g = L\delta g^*(L)$, where the latter means that the geometrical variables are normalized by L, as x/L, r/L, R/L etc., for example,

$$\delta g = L\delta g^*(x/L, y/L, z/L) \tag{2.9.1}$$

i.e., dependent on shape and scale L. The effect δW of the gravity potential is given by integration of gravity over distance which leads to and augments the power of length by 1: δW scales with L^2. Conversely, differentiation decreases the power of length by 1: the first derivatives of δg scale with $L^4/L^4 = 1$ and are independent from L, the second derivatives scale with L^{-1}, etc.

7. Integrals of analytical expressions are taken from the tables of Gröbner & Hofreiter (1949, 1950), referred to as GH49 and GH50 with the number of the integral type, e.g. "GH49, 342,3c".

The aim of the following is to provide procedures and expressions for the task of calculating the gravitational effects of voluminous geometrical or geological bodies. Various parametrizations offer themselves. Horizontal disks lend themselves directly to approximate geological masses described by contours. Often bodies are divided into other types of uniform massive elements, as prisms and polyhedra. The elements of integration are points, lines and planes. Curved surfaces may be approximated by finite or infinitesimal planar elements, in the latter case analytical integration may be possible for mathematically defined surfaces. The class of parametrization in two dimensions (2D) is treated separately (Sect. 2.9.7). The most

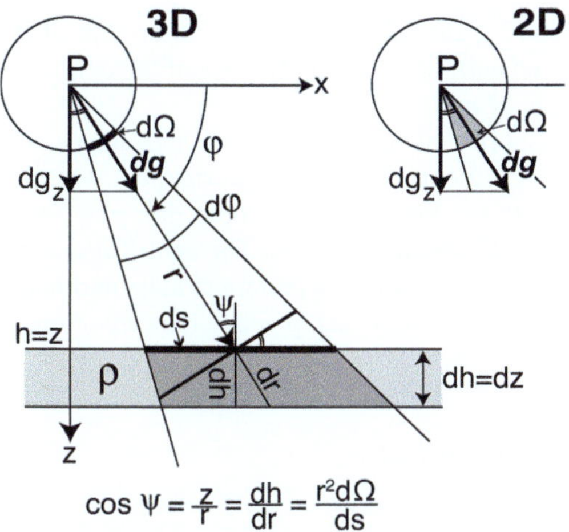

$$\cos \psi = \frac{z}{r} = \frac{dh}{dr} = \frac{r^2 d\Omega}{ds}$$

Fig. 2.9.1 The solid angle-based volume element $r^2 d\Omega dh$. *Left*: three-dimensional (3D) case; the solid angle $d\Omega$ is depicted as a thick line on the unit sphere. *Right*: two-dimensional (2D) case; the solid angle $d\Omega = 2\varphi$ is a bi-angle on the unit sphere seen from the side

profitable approximations must be chosen from case to case. The above rules 1–6 are applied where possible.

2.9.1.2 Direction of the Vector of the Gravitational Effect

The gravitational vector effect $\boldsymbol{\delta g}$ at P of a mass anomaly Δm at point Q is directed from P towards Q. If the mass Δm has a finite volume, the vector $\boldsymbol{\delta g}$ does not generally point toward its centre of gravity, in the sense of centre of mass (defined by $X_i \int dm = \int x_i dm$; compare Task 2.2), only in cases of special symmetry it does. Note that this is contrary to customary thinking. Special symmetry is, for example, that of a homogeneous sphere, as will be shown below. That the horizontal centre of gravity of the gravity anomaly, δg_z, is equal to that of the mass anomaly (see Sect. 2.7.6), is not to be confused with the vector direction.

In practice gravity interpretation is usually concerned with scalar values δg as deviations from a reference $g_o \approx 10 \, \text{m/s}^2$ of the terrestrial gravity vector $\boldsymbol{g}$. Generally the anomalous vector points in a direction different from normal $\boldsymbol{g_o}$ and the gravity effect is not the anomalous vector but only the $\boldsymbol{\Delta g}$ component in $\boldsymbol{g}$-direction. Here we admit that the direction of $\boldsymbol{g}$ is affected by the anomalous vector, but the directional variation is very small indeed, as $|\boldsymbol{\delta g}| << g$, certainly $<10^{-3} g$, and mostly much smaller. Consequently in practice, the directional variation of $\boldsymbol{g}$ is neglected. If in a fixed Cartesian coordinate system x, y, z (e.g. x to north, y to east and z vertical), assumed for a given limited region, earth curvature is neglected and z is assumed to be parallel to reference $\boldsymbol{g_o}$ at the origin; the gravity effect is then defined by δg_z, where $\delta g_z = \delta g \cos \vartheta = \delta g z / r$, where ϑ is the angle enclosed by $\boldsymbol{z}$ and $\boldsymbol{r}$

from P to Q. In large scales the standard $\boldsymbol{g}$ direction is assumed to be that of the (negative) radius vector of a standardized earth model, $-\boldsymbol{r}$. Beware of the confusion which may arise from the two meanings of $\boldsymbol{r}$.

The plumb line deflections are proportional to δg_x and δg_y and are usually neglected in the interpretation. This represents a loss of information which is insignificant in many terrestrial applications. Vector effects come into play, if arbitrary components are to be calculated for special purposes. The centrifugal component of observed gravity is also neglected, as it does not pertain to mass anomalies and has been essentially removed in the reductions which precede interpretation.

2.9.2 Coordinate Systems and Integration

The coordinate systems and the respective volume elements were shown above in Fig. 2.4.1. Since the emphasis is on tasks in geological modelling Cartesian coordinates are used here mostly. Spherical and cylinder coordinates, as mentioned in Sect. 2.4, are treated more briefly.

2.9.2.1 Spherical Coordinates

Spherical coordinates (Fig. 2.4.1a) are, of course, best suited to problems in which the sphericity of earth must be taken into account. The mass element is $dV = r^2 dr \cos\varphi \, d\varphi \, d\lambda = r^2 dr \sin\phi \, d\phi \, d\lambda$. The local vertical z coordinate at a point P corresponds to $-\boldsymbol{r}$ in global spherical coordinates. Several possibilities to integrate gravity effects of extended density anomalies are briefly outlined in Sects. 2.10.7.2 & 2.10.7.3.

2.9.2.2 Vertical Cylinder Coordinates: Cylinder Ring Sectors

Cylinder coordinates (Fig. 2.4.1, rhs).: Z parallel to $\boldsymbol{g}_0$, R horizontal, longitude Λ are most appropriate to the symmetry of the problem about the vertical $\boldsymbol{g}$ axis because the gravity effect is independent from azimuth. At points P at the cylinder axis the integration of $d\delta g$ is easy because on the boundaries of the volume elements, which are cylinder ring segments, $dV = RdRd\Lambda dZ$, $r = (R^2 + Z^2)^{1/2}$, the coordinates are constant:

$$\delta g = G\rho \, \Delta \Lambda^* \int_R \int_Z [1/(R^2 + Z^2)^{3/2}] \, R \, dR \, Z \, dZ \qquad (2.9.2)$$

If rule 5 is followed, the simplest mass element (and formula) is the cylinder (radius R^*, bottom at Z^*, density ρ, sector angle 2π or $\Delta\lambda^*$) with axial P at its surface ($Z = 0$). Integration renders:

Fig. 2.9.2 Scheme of calculating the gravity effect at $P = (0, 0)$ of the cylinder ring between $R = R_i{}^*, R_{i+1}{}^*$, $Z = Z_j{}^*, Z_{j+1}{}^*$ from basic elements ($R = 0 - R_k$, $Z = 0 - Z_j$); hachures rising to the right: elements added (positive density), rising to the left: elements subtracted ("negative density")

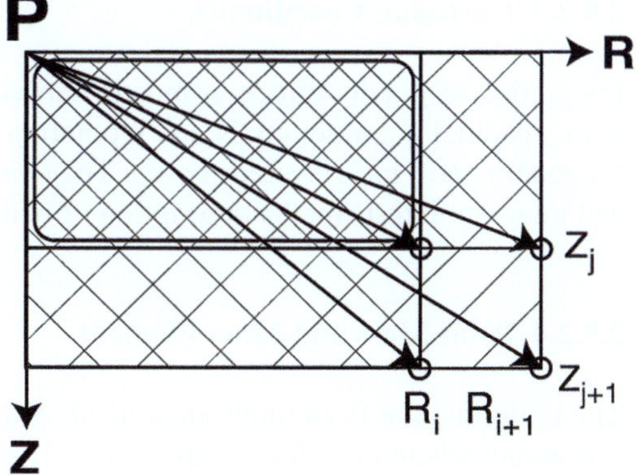

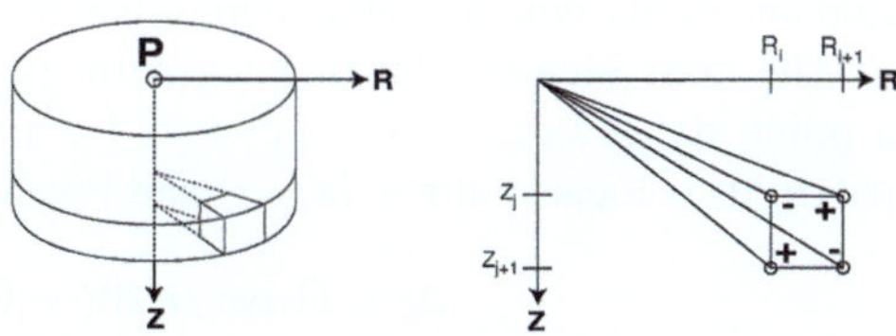

$$\delta g = G\rho\Delta\lambda^* \int_R \int_Z [1/(R^2 + Z^2)^{3/2}]\, R\,dR\,Z\,dZ = G\rho\,\Delta\,\lambda^*(Z^* + R^* - r^*)$$

$$\text{with } r^* = (R^{*2} + Z^{*2})^{1/2}$$

(2.9.3)

A cylinder ring sector element between $R = R_i{}^*$, $R_{i+1}{}^*$, $Z = Z_j{}^*$, $Z_{j+1}{}^*$, and $\Delta\lambda^* = \Lambda_{k+1} - \Lambda_k$ follows by adding elements with appropriate parameters and signs according to the scheme sketched in Fig. 2.9.2. (In this case, rule 5 has only little advantage over integration directly of a cylinder ring between R and $R + \Delta R$ and between Z and $Z + \Delta Z$). The expression

$$\delta g_{ijk} = G\rho\Delta\Lambda_{k,k+1}{}^*(-r_{ij}{}^* + r_{i+1,j}{}^* - r_{i+1,j+1}{}^* + r_{i,j+1})$$

(2.9.4)

is but the sum of the radii from P to the corners of the ring section with alternating signs (Fig 2.9.2). Finally; it is also easy to integrate the effect for the infinite Bouguer plate, with $(r^2 = R^2 + Z^2)$, $\Delta h = d$: $\delta g = G\,\rho^*{}_{R=0}\int^\infty {}_{A=0}\int^{2\pi} Z R\,dR\,d\Lambda/r^3 = 2\pi\,G$ $\rho^* \int_0^\infty Z\,R\,dR/r^3 = 2\pi G\rho^* z/r_0|^\infty = 2\pi\,G\,\rho^* = 2\pi\,G\,\rho\,d$ (see below Sect. 2.9.3.3).

In practice, dividing the source volume into point P-based cylinder coordinates must be repeated individually for each observation point P. This leads to tedious labour and is not very suitable for rapid gravity effect calculations, but was much used before the times of powerful computing, e.g., for estimating terrain effects. Alternatively, for non-axial P, elliptical integrals must be evaluated (Nagy, 1965) which is also difficult.

2.9.2.3 Cartesian Coordinates

Integration of gravity effects is mostly carried out in Cartesian coordinates x, y, z, with z parallel $\mathbf{g}$, x pointing N and y pointing E, but arbitrary orientation of the orthogonal x and y axes is often chosen. The mass element was depicted in Fig. 2.4.1c, and integration rests on the description of bodies in x, y, z.

2.9.2.4 Point Mass and Mass Element

The point mass m is an unphysical idealisation. It is readily justified for planetary astronomy where the planets appear from Earth nearly as points in the sky. A finite mass distribution of perfect radial symmetry, as a homogeneous sphere, has exactly the same external effect as all the mass condensed at its centre at P (this is always said and rarely demonstrated; here it has been posed as a problem; Task 2.1).

The mass element dm of integration is infinitesimal or point-like, treated like a point: dx, dy, $dz << r = (x^2 + y^2 + z^2)^{1/2}$. If the mass element $dm = \rho\, dV = \rho\, dx\, dy\, dz$ is located at $\mathbf{r} = (x, y, z)$ and P is located at $(0,0,0)$, its gravity effect is:

$$dg = G\, dm z / r^3 dV = G\, \rho\, dx\, dy\, dz\, z / r^3 \qquad (2.9.5)$$

$dg = dg_z$ is the vertical or z component of the infinitesimal gravitational vector effect $\mathbf{dg}$. Placing P at the coordinate origin does not limit generality because translational coordinate transformation is always possible.

2.9.2.5 Integration

Integration of Eq. (2.9.3) is based on the mutual independence of gravitating masses such that their effects can be summed:

$$\delta g = \iiint_V dg = G \iiint_V \rho(\mathbf{r}) z / r^{3\cdot} dV \qquad (2.9.6)$$

or, if $\rho = \mathrm{const}$,

$$\delta g = G\rho \iiint_V z / r^{3\cdot} dV \qquad (2.9.6a)$$

Integration can be approximated by summation of the effects of finite mass elements. A body may be discretized by a grid of small finite mass elements $\rho \Delta V (\Delta V = \Delta x \Delta y \Delta z)$ which are represented by equivalent mass points or spheres with radius, $R = (3\Delta x \Delta y \Delta z / 4\pi)^{1/3}$ of equal volume at their centres of mass, if effects of non-spherical shape and partial overlap are negligible. Even simpler is it to approximate sufficiently sphere-like or compact geological masses by spheres of equivalent location, radius R and density $\Delta \rho$; tolerable non-spherical deviations depend on the distance from the observer. Note, that observed effects are not those of the total mass, but only of the mass excess: $\Delta m = \Delta \rho V$ and $V = 4\pi R^3/3$ (see Sect. 2.7.6).

Explicit volume integration is limited by boundaries, i.e. surfaces, leading to some kind of surface integration. In Cartesian coordinates, the distance r in (Eq. 2.9.6) is generally not constant along the surfaces x, y, or $z = $ const. More general surface geometries pose additional analytical difficulties. Special definitions of the volume elements (besides those treated in Sects. 2.9.2.1, 2.9.2.2) facilitate the integration considerably, but they usually require also suitable orientation of the volume boundaries which cannot generally be achieved. This then will require co-ordinate transformations, before integration is carried out, and back transformation into gravity orientation. Since this task will be frequently encountered, it has been treated above in connection with the Cartesian coordinates (Sect. 2.4.3) and the expressions derived will be referred to in the subsequent text.

In the following subsections Eq. 2.9.12, the more general three-dimensional (3D) case is treated, while the more specialised and simpler 2D case is presented in Sect. 2.9.7. A compromise between 2D and 3D is the so-called $2\frac{1}{2}$D case of Sect. 2.9.8.

2.9.3 Special Mass Elements: Integration in One and Two Dimensions, Mass Lines and Mass Planes

Integration can be carried out for volume elements that are not infinitesimal in all co-ordinates, but may be finite, even infinite, in one or two directions, in which density does not vary, at least not significantly. Such mass elements are semi-infinitesimal, e.g., mass lines and surfaces. Finite or semi-infinite mass line elements may be written: $\rho \mathrm{d}V = \rho\,\Delta z\,\mathrm{d}x\,\mathrm{d}y$ and surface masses are $\rho\,x\,y\,\mathrm{d}z$, where x and y stand for any desired finite values. A mass line carries a constant line-density in kg/m: $\rho^+ = \rho\,\mathrm{d}x\,\mathrm{d}z$, and surface density is given, e.g. as $\rho^* = \rho\,\mathrm{d}z$ in kg/m^2. Generally, the introduction of semi-infinitesimal mass elements means that part of the integration is carried out at a preliminary step. For lines the preliminary integration covers one coordinate, for example, in case of two-dimensional models; for planar elements integration is carried out for two directions, and here we encounter the solid angle (rule 1).

2.9.3.1 Horizontal Uniform Mass Line

Two-dimensional models (see below) consist of infinite horizontal mass lines along one dimension, say y. The line density ρ^+ is constant. "Horizontal" implies that, as usual, the vertical gravity effect δg_z is considered. More generally, it is the case of mass lines normal to the component of gravitation under consideration; in some applications it may be any oblique direction.

To get from the standard infinitesimal volume element $\mathrm{d}x\,\mathrm{d}y\,\mathrm{d}z$ to the element $\mathrm{d}x\,\mathrm{d}z$, the integration is carried out over y from $-\infty$ to $+\infty$ beforehand such that we obtain $\mathrm{d}V = \mathrm{d}x\,\mathrm{d}z y_{(\to\infty)}$. The standard approach is that the effect of the infinite mass line at $(x, y_{-\infty}{}^\infty, z)$ on $\mathrm{P} = (0,0,0)$ is calculated by integrating the effect of the point mass $\rho^+ dy$ along the line from $-\infty$ to $+\infty$. With $r^* = (x^2 + z^2)^{1/2}$

$$\delta g = G\rho^+ \int_{-\infty}^{+\infty} z dy/r^3 = 2G\rho^+ \int_{o}^{+\infty} z dy/r^3 = 2G\rho^+ z/r^{*2} \qquad (2.9.7)$$

It differs from the point mass effect by the characteristic factor 2 and the $1/r^2$ dependence on distance from the line, instead of $1/r^3$ from the point.

The horizontal *line mass of finite length* 2λ (λ to each side from $y = 0$) has the effect:

$$\delta g^{(2\frac{1}{2}\cdot)} = 2G\rho^+ z/r^{*2}a \quad \text{with} \quad a = \lambda/(x^2 + \lambda^2 + z^2)^{1/2} \qquad (2.9.8)$$

where $0 < a < 1$. The superscript $(2\frac{1}{2})$ refers to the so-called $2\frac{1}{2}$ dimensional modelling or $(2\frac{1}{2}D)$ which takes the finite length of elongated structures into account (see Sect. 2.9.8). Any generalization of these definitions to other orientations and coordinate systems is possible.

The gravitational potential of the infinite mass line, as defined above in Sect. 2.7.2, is

$$\delta U = 2G\,\rho^+ \int_{o}^{+\infty} dy/r = -2G\,\rho^+ \ln(y + (x^2 + y^2 + z^2)^{1/2})|_0^\infty$$

$$= 2G\rho^+ \ln(z/(x^2 + y^2)^{1/2}) - \infty \qquad (2.9.9)$$

The constant, $-\infty$, is irrelevant, as any potential is relative to some reference, only the variations in a limited region are of interest, and so are they for differentiation $\delta g = \partial \delta U/\partial z$ which again leads to (Eq. 2.9.7).

2.9.3.2 Vertical Mass Line

Another important linear mass element is called the "vertical mass line" or "vertical rod" (in view of the usually needed vertical component δg_z) or the "parallel rod" (in view of the δg-component parallel to the line). The line-parallel component of the gravitational effect of a uniform mass line is required, for example, as an integration element for parallel walls (Sect. 2.9.3.4). Located at x, y and extending from z_o (top) to bottom at infinity and with $\rho^+ = \rho dx dy$ and $r_o = (x^2 + y^2 + z_o^2)^{1/2}$, the rod exerts on P at (0,0) the effect:

$$\delta g^{(l)} = G\rho^+ \int_{z_o}^{\infty} z dz/r^3 = G\rho^+/r_o \qquad (2.9.10)$$

By combining the effects of lines at x, y with tops at z_1 and z_2, the line extending from z_1 to z_2 has the effect:

$$\delta g^{(l)} = G\rho^+[1/r_1 - 1/r_2] = G\rho^+(r_2 - r_2)/r_1 r_2 = G\rho[1/r_1 - 1/r_2]dx dy \quad (2.9.11)$$

These equations break down when $r_1 \to 0$, because $G\rho^+/r_1 \to \infty$. With $\rho dx dy$ this is uncritical, and assuming, for the start, vertical cylinder geometry, Eqs. (2.9.2)

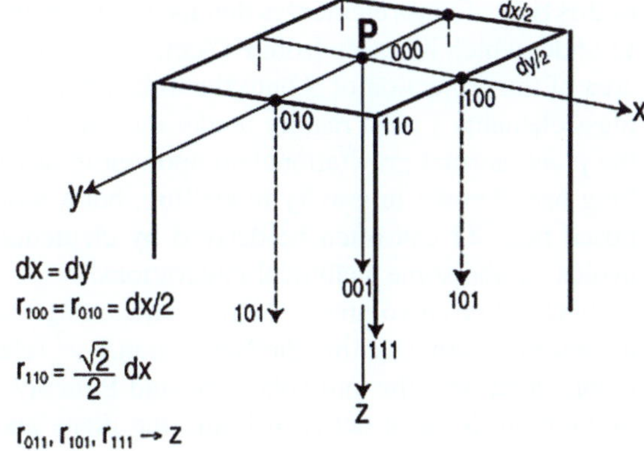

Fig. 2.9.3 Infinitesimal vertical rod of cross section $dx\,dy$ and finite length z with P at one end ($z = 0$). The triple indices are explained in Sect. 2.9.3.3 and Fig. 2.9.6

and (2.9.3), with top $Z = 0$, bottom $Z \gg dR$ (infinitesimal) renders $d\delta g = G\,\rho\,2\pi\,dR = G\rho C$ where $C = 2\pi\,dR$ is the circumference of the infinitesimally thin vertical cylinder.

Another example is the infinitesimally thin Cartesian rod or prism with the finite vertical extent z and the horizontal cross section $dx\,dy$, where $dy = dx$. At the centre of its top surface, the effect $\delta g_z^{(\text{rod xy})}$ is:

$$\delta g_z^{(\text{rod xy})} \to G\rho\,dx/2\ln[1 + 1/\sqrt{2}] = 0.881 G\rho \times 4dx = 0.881\,G\rho C$$

where $C = 4dx$ is, again, the circumference of the rod.

For the derivation, the expressions (2.9.6.1) in Sect. 2.7.6 are applied which describe the components of the gravitational effect of a finite massive cuboid on one of its corner points P. The present rod is therefor divided into 4 equal rods or prisms with cross section $dx/2\ dy/2$ (Fig. 2.9.3) for the limit $a \to dx$, $b \to dy$ and $dx = dy \ll z$. In the limit

$$\delta g_z^{(x)} = \delta g_z^{(y)} \to G\rho\,dx/2\ln[1 + 1/\sqrt{2}] = 0.881\,G\rho\,dx/2$$
$$\delta g_z^{(z)} \to G\rho\,c(\arcsin[1/\sqrt{2}] + \arcsin[1/\sqrt{2}] - \pi/2) \to 0.$$

The total rod effect is the sum of the effects of the 4 equal quarter rods with two contributions of each. Details may be worked out in Task 2.5.

This expression is, however, valid only for the special location of P on a thin rod and not applicable for integration over finite x or y (see below Sects. 2.9.3.4 & 2.9.5.1).

2.9.3.3 Horizontal Mass Plane or Floor: The Solid Angle $\Delta\Omega$

Horizontal planes S of surface density $\rho\,dz = \rho*$ in kg/m^2 form an important versatile class of infinitesimally thin surface mass elements which can be exploited for the integration of the gravity effect δg_z of arbitrary geological bodies. The advantage of this type of mass element is that its gravity effect is proportional to the solid angle Ω under which it is seen from P (Sect. 2.9.1.1: rule 1). The solid angle is the surface area of the projection of S onto the unit sphere around P. Horizontal planar surface mass elements S most readily render the vertical gravity component; this is true for the plane-normal gravitational component in any orientation of the plane. This has long been known in gravity modelling, but a more generalized application is proposed here. Ω can often be derived by elementary geometrical arguments which avoids cumbersome analytical integrations.

If Ω defines a volume V by rays emanating from P, the surface area grows with distance r^2, compensating the Newtonian $1/r^2$ relationship. The projection of a horizontal area onto the unit sphere around P involves a $\cos \psi = h/r$ term, where h is the normal distance between P and the plane and $\cos \psi$ is the angle between the radius vector $\boldsymbol{r}$ and the plane normal vector $\boldsymbol{s}$: $\cos \psi = \boldsymbol{s} \cdot \boldsymbol{r}/sr$.

A mass element dV defined by $d\Omega$ cutting out a horizontal surface element dS of a layer of thickness dz and density ρ (Fig. 2.9.1), automatically has built in the kernel z/r^3 of the integral in (Eq. 2.9.5). This type of dV is, hence, tailored for gravity integration: $d\Omega = \cos \psi \, dS/r^2 = dS h/r^3$ and $dg_n^{(s)} = G\rho \, dh \, ds \, h/r^3 = G\rho \, dh \, d\Omega$; $dg_n^{(s)}$ is the $\boldsymbol{dg}^{(s)}$ component normal to dS or parallel to $\boldsymbol{ds}$. In other words, dV grows with $r^2 r/h$ when ds (and dV) are limited by rays from P:

$$dm^* = \rho^* \, d\Omega \, r^3/h = \rho \, dh \, d\Omega \, r^3/h \qquad (2.9.12)$$

The basic integral (Eq. 2.9.5) for the component Δg_n normal to the surface takes the form:

$$\delta g_n = G\rho \int_\Omega \int_h dh \, d\Omega \qquad (2.9.13)$$

With the projection $\Delta \Omega$ of a finite planar element:

$$\int_\Omega \int_h dh \, d\Omega = \Delta \Omega \Delta h. \qquad (2.9.14)$$

Thus integration is reduced to the determination of $\Delta \Omega$ and to measure the distances Δh within, and normal to, the mass volume $\rho \Delta V$. For simple shapes as triangles and rectangles this is easy; remember the infinite Bouguer slab: $\Delta \Omega = 2\pi$; we exploit the fact that the quadrature of the circle or sphere has long been solved by invoking the transcendental number π. The situation can be used to design mass approximations accordingly.

The gravitational component normal to the surface considered cannot be generally used to infer the vector modulus $\delta g^{(E)} = \delta g_n^{(E)}/\cos \psi^{(E)}$, only if the direction $\psi^{(E)}$ of the vector is known its size is found by back projection (Fig. 2.9.4). The idea that $\boldsymbol{\delta g}^{(E)}$ always points to the centre of gravity C of E, is incorrect, though often implicitly assumed. If $\psi^{(E)}$ is not known, three components of the gravitational attraction must be calculated, $\boldsymbol{\Delta g}^{(s)} = (\Delta g_p^{(s)}, \Delta g_q^{(s)}, \Delta g_n^{(s)})$, for example, to project it onto another coordinate system; two plane-parallel components orthogonal to each other are treated in Sect. 2.9.3.4.

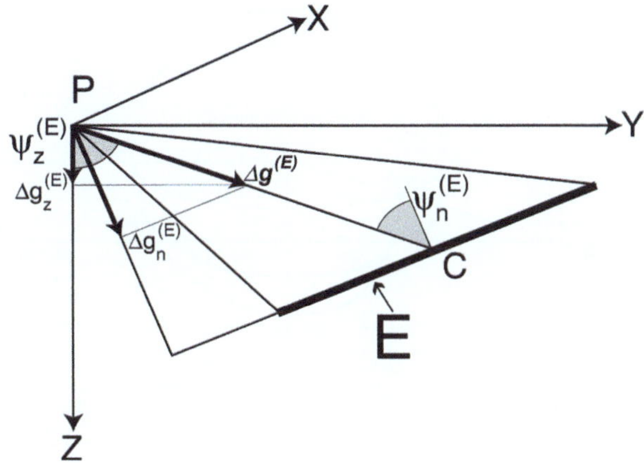

Fig. 2.9.4 Back projection of the plane-normal component δg_n onto the vector $\boldsymbol{\delta g}$ of known *a priori* direction

Horizontal Rectangular Plane Element

The special case of a rectangular horizontal plane mass element of thickness dz is presented as an example and in preparation for treating the rectangular prism or cuboid. For that also the rectangular planes normal to the other coordinate directions x, y are included. The planes are assembled into a box or hollow cuboid $a \times b \times c$ with $P = (0,0,0)$ at one corner (Fig. 2.9.5). Thus the box is made up of 6 planes: $x = 0$ and $x = a$, $y = 0$ and $y = b$, $z = 0$ and $z = c$, i.e. the infinitesimally thin rectangles $dx \times b \times c$, $dy \times a \times c$ and $dz \times a \times b$. The prism dimensions may be normalized with the vertical c (rule 6), i.e. we let $c = 1$, and the edges are then $a, b, 1$. For these elements the solid angles from P: $\Delta\Omega^{(x)}$, $\Delta\Omega^{(y)}$ and $\Delta\Omega^{(z)}$ are determined (Fig. 2.9.5a,b), and the plane-normal gravitational components (Eq. 2.9.13) are

$$\left.\begin{aligned}
\delta g^{(x)}{}_x &= G\rho\Delta\Omega^{(x)}dx \\
\delta g^{(y)}{}_y &= G\rho\Delta\Omega^{(y)}dy \\
\delta g^{(z)}{}_z &= G\rho\Delta\Omega^{(z)}dz
\end{aligned}\right\} \tag{2.9.15}$$

where the bracketed superscripts indicate the respective planes. $\Delta\Omega$ is derived by basic geometrical arguments. The opposite planes, i.e. those through P intersect the unit sphere as great circles, hence $\Delta\Omega^{(o)} = 0$.

The projection of a corner point x, y, z of the cuboid from P onto the unit sphere is called x', y', z'. The following notation is introduced: corners and their projections from $P = (0,0,0)$ are characterized by index triples of zeroes and ones, depending whether x, y, or $z = 0$ or $\neq 0$; for example, 100 means a point $(x,0,0)$, especially $(a, 0, 0)$; or 111: (x, y, z) or (a, b, c) – diagonally opposite P. This notation is also shown on Fig. 2.9.6. Lengths are called:

$$\begin{aligned}
r_{000} &= 0, \\
r_{100} &= a, r_{010} = b, r_{001} = c \text{ or } 1, \\
r_{110} &= (a^2 + b^2)^{1/2}, r_{101} = (a^2 + c^2)^{1/2} \text{ or } (a^2 + 1)^{1/2} \\
r_{111} &(a^2 + b^2 + c^2)^{1/2} \text{ or } (a^2 + b^2 + 1)^{1/2}, \\
r_{011} &= (b^2 + c^2)^{1/2} \text{ or } (b^2 + 1)^{1/2},
\end{aligned}$$

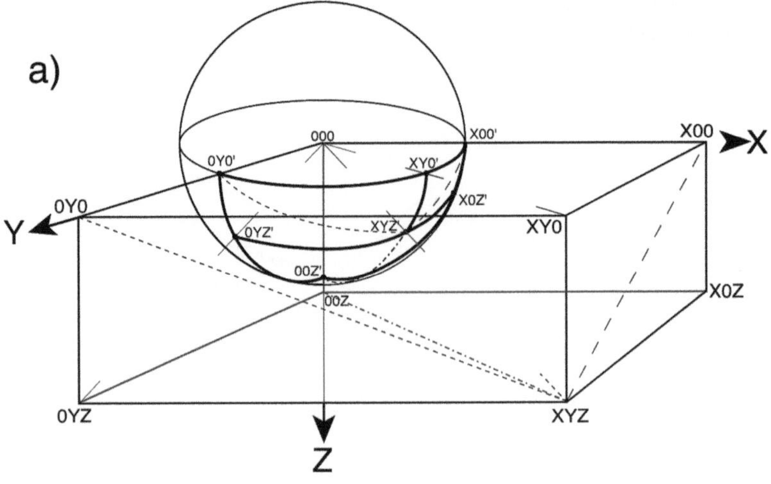

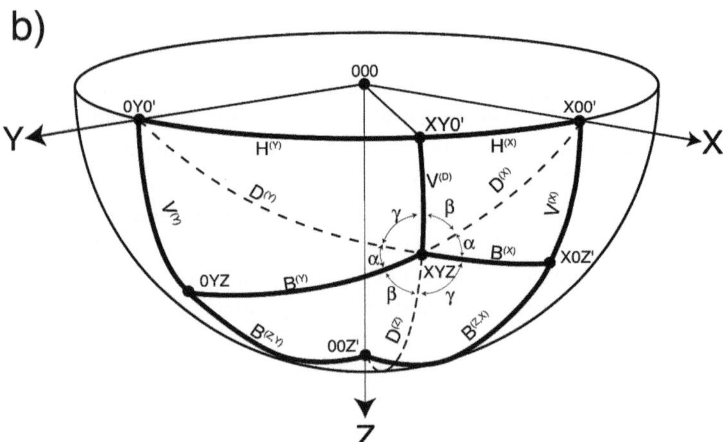

Fig. 2.9.5 The rectangular prism or cuboid with P at one corner. (**a**) Prism and unit sphere; sketched projection of prism edges onto unit sphere; (**b**) more detailed projection of prism edges onto unit sphere with symbols shown

The projection of a face onto the P unit sphere is a spherical quadrangle. The three spherical quadrangles have each three rectangular corners, only at x', y', z' the full circle is divided into three, usually unequal, corner angles, as x', y', z' is the intersection of three great circles (Fig. 2.9.5b). $\Delta\Omega$ is equal to the spherical excess ε, i.e. the excess over 180° of the sum of the inner corner angles in a spherical triangle: $\varepsilon = \Delta\Omega$ for each quadrangle. Consider the triangles made of the diagonal and two quadrangle sides and determine the angles α, β, γ, where $\alpha + \beta + \gamma = \pi$:

$$\Delta\Omega^{(x)} = \beta + \gamma - \pi/2, \ \Delta\Omega^{(y)} = \alpha + \gamma - \pi/2, \ \Delta\Omega^{(z)} = \alpha + \beta - \pi/2 \qquad (2.9.16)$$

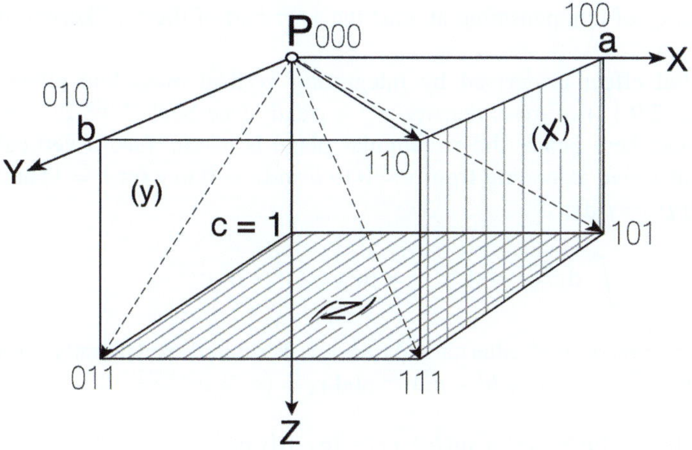

Fig. 2.9.6 Three digit notation for cuboid with P at one corner

For each spherical triangle all sides are defined from prism geometry
(Fig. 2.9.5b), e. g.

$$H^{(x)} = \arcsin(r_{010}/r_{110}), \; H^{(y)} = \arcsin(r_{100}/r_{110}r_{111}), \text{etc.},$$

and α, β, γ are found by the law of sines and:

$$\alpha = \arcsin([r_{100}r_{111}]/[r_{110}r_{101}]), \; \beta = \arcsin([r_{010}r_{111}]/[r_{011}r_{110}]),$$
$$\gamma = \arcsin([r_{001}r_{111}]/[r_{101}r_{011}]). \tag{2.9.17}$$

Hence, from (Eqs. 2.9.15, 2.9.16, 2.9.17), the plane-normal gravitational components are:

$$\left.\begin{aligned}
\delta g_x^{(x)} &= G \, \rho \, dx(\arcsin[r_{010}r_{111}/r_{011}r_{110}] + \arcsin[r_{001}r_{111}/r_{101}r_{011}] - \pi/2) \\
\delta g_y^{(y)} &= G\rho dy(\arcsin[r_{100}r_{111}/r_{110}r_{101}] + \arcsin[r_{001}r_{111}/r_{101}r_{011}] - \pi/2) \\
\delta g_z^{(z)} &= G\rho dz(\arcsin[r_{100}r_{111}/r_{110}r_{101}] + \arcsin[r_{010}r_{111}/r_{011}r_{110}] - \pi/2)
\end{aligned}\right\}$$
$$\tag{2.9.18}$$

2.9.3.4 Vertical Plane or Wall: The Logarithmic Effect

Walls are vertical planar surface mass elements exerting a gravity effect, i.e. the vertical component of their gravitational effect. "Vertical" can be generalized to imply
"parallel to gravity", and the wall element is especially designed here to calculate
the plane-parallel effect in any orientation. Together with the plane-normal effect
(Sect. 2.9.3.3) two plane-parallel effects describe the vector effect $\boldsymbol{\delta g}$ which, for example, if the plane is oblique, is needed, to calculate δg_z in an Earth-bound system.
Parallel wall elements, which are defined by rays from P, grow in area as r^2 and thus

preserve the advantage of compensating at least the $1/r^2$ part of the z/r^3 kernel of integration.

The plane-parallel effect is derived by integrating vertical mass line effects (Eq. 2.9.10) or (Eq. 2.9.11) of mass density $\rho^+ = \rho\,dxdy$ (see Sect. 2.9.3.2) over the appropriate plane; here again, the rectangular plane is taken. Take a vertical plane normal to x, at $x = a$, extending from $y = 0$ to b and $z = 0$ to c (or $c = 1$) and let $\rho^+ = \rho^*dy$ with $\rho^* = \rho\,dx$:

$$\delta g_z^{*(x)} = \int_0^c dy/(y^2 + A^2)^{1/2} = \ln(y + (y^2 + A^2)^{1/2})|_0^b$$

where A^2 stands for a^2 or $a^2 + c^2$, alternatively $a^2 + 1$. Written dimensionally with $r_1 = a$, $r_2 = (a^2 + b^2)^{1/2}$, $r_3 = (a^2 + b^2 + c^2)^{1/2}$ and $r_4 = (a^2 + c^2)^{1/2}$:

$$\delta g_z^{(x)} = -G\rho[\ln r_1 - \ln(b + r_2) + \ln(b + r_3) - \ln r_4]\,dx \text{ or}$$

$$\delta g_z^{(x)} = -G\rho\,dx\ln[r_1(b + r_3)/(b + r_2)r_4] = G\rho\,dx\ln[r_4(b + r_2)/(b + r_3)r_1]$$
$$(2.9.19)$$

With the notation, introduced above (Fig. 2.9.6) for the corners of a cuboid that is expanded from $P = (0, 0, 0)$:

$$\delta g_z^{(x)} = G\rho\,dx\ln[r_{101}(r_{010} + r_{110})/(r_{010} + r_{111})r_{100}] \qquad (2.9.20)$$

A remark on the vertical wall is inserted here: it is a model of a geological dyke. With $b \to \infty$, $(b + r_2)/(b + r_3) \to 1$, and from $-\infty$ to ∞, i.e. the effect of the two-dimensional dyke is (see Sect. 2.9.7.1.2):

$$\delta g_\infty = 2G\rho\,dx\ln[r_4/r_1] = 2G\rho\,dx\ln[r_{101}/r_{100}] \qquad (2.9.21)$$

If b finite, but $c \to \infty$, $r_4/(b + r_3) \to 1$, and from $-b$ to $+b$, the double effect is

$$\delta g = 2G\rho\,dx\ln[(b + r_2)/r_1] = 2G\rho\,dx\ln[b + (a^2 + b^2)^{1/2}/a]$$
$$= 2G\rho\,dx\ln[(r_{010} + r_{110})/r_{100}] \qquad (2.9.22)$$

If b and $c \to \infty$, $r_4/(b + r_3) \to 1$, division of the numerator and denominator by bc leaves the former finite while the latter goes to zero, thus the 2D bottomless vertical dyke generates an infinite gravity anomaly.

Also plane-parallel is the y component $\delta g_y^{(x)}$ of gravitational attraction of the same rectangle; the expression is derived analogously:

$$\delta g_y^{*(x)} = \int_0^c dz/(z^2 + B^2)^{1/2} = \ln(z + (z^2 + B^2)^{1/2})|_0^c$$

where $B^2 = a^2$ or $a^2 + b^2$, and dimensionally, as above

$$\delta g_y^{(x)} = G\rho\,dx\ln([(r_{001} + r_{101})r_{110}]/[r_{100}(r_{001} + r_{111})]) \qquad (2.9.23)$$

The planes $y = b$ and $z = 1$ (or $z = c$) are treated identically for their plane-parallel effects (y: $\delta g_x^{(y)}$ and $\delta g_z^{(y)}$; z: $\delta g_x^{(z)}$ and $\delta g_y^{(z)}$). Beginning with the plane face normal to x, at $x = a$, from $y = 0$ to b and $z = 0$ to 1 or c (Fig. 2.9.5). $\delta g_z^{(x)} = cG\rho\,dx\,\delta g_z^{(x)*}$ can be written:

$$\delta g_z^{(x)} = G\rho\,dx[\ln(b+(a^2+b^2)^{1/2}) - \ln a + \ln(a^2+1)^{1/2} - \ln(b+(a^2+b^2+1)^{1/2})].$$

All expressions for the rectangular prism, with, of course, $\ln 1 = 0$, contain for each rectangular plane, the arguments of the logarithms of the 4 radii from P to the 4 corner points:

$$\left.\begin{aligned}
\delta g_y^{(x)} &= G\rho[\ln(1+(a^2+1)^{1/2}) - \ln a + \ln(a^2+b^2)^{1/2} - \ln(1+(a^2+b^2+1)^{1/2})]dx \\
\delta g_z^{(x)} &= G\rho[\ln(b+(a^2+b^2)^{1/2}) - \ln a + \ln(a^2+1)^{1/2} - \ln(b+(a^2+b^2+1)^{1/2})]dx \\
\delta g_x^{(y)} &= G\rho[\ln(1+(b^2+1)^{1/2}) - \ln b + \ln(a^2+b^2)^{1/2} - \ln(1+(a^2+b^2+1)^{1/2})]dy \\
\delta g_z^{(y)} &= G\rho[\ln(a+(a^2+b^2)^{1/2}) - \ln b + ln(b^2+1)^{1/2} - \ln(a+(a^2+b^2+1)^{1/2})]dy \\
\delta g_x^{(z)} &= G\rho[\ln(b+(b^2+1)^{1/2}) - \ln 1 + \ln(a^2+1)^{1/2} - \ln(b+(a^2+b^2+1)^{1/2})]dz \\
\delta g_y^{(z)} &= G\rho[\ln(a+(a^2+1)^{1/2}) - \ln 1 + \ln(b^2+1)^{1/2} - \ln(a+(a^2+b^2+1)^{1/2})]dz
\end{aligned}\right\}$$

$$(2.9.24)$$

or, in the triple-index notation (Fig. 2.9.6):

$$\left.\begin{aligned}
\delta g_y^{(x)} &= G\rho\,dx\ln([(r_{001}+r_{101})r_{110}]/[(r_{001}+r_{111})r_{100}]) \\
\delta g_z^{(x)} &= G\rho\,dx\ln([(r_{010}+r_{110})r_{101}]/[(r_{010}+r_{111})r_{100}]) \\
\delta g_x^{(y)} &= G\rho\,dy\ln([(r_{001}+r_{011})r_{110}]/[(r_{001}+r_{111})r_{010}]) \\
\delta g_z^{(y)} &= G\rho\,dy\ln([(r_{100}+r_{110})r_{011}]/[(r_{100}+r_{111})r_{010}]) \\
\delta g_x^{(z)} &= G\rho\,dz\ln([(r_{010}+r_{011})r_{101}]/[(r_{010}+r_{111})r_{001}]) \\
\delta g_y^{(z)} &= G\rho\,dz\ln([(r_{100}+r_{101})r_{011}]/[(r_{100}+r_{111})r_{001}])
\end{aligned}\right\}$$

$$(2.9.25)$$

For P $\to$ the corner of the wall, the distances in the above denominators, r_{100}, r_{010}, $r_{001} \to 0$ and the respective logarithms, for example, in the case of $\delta g_z^{(x)}$, approach $\ln[2c/(1+(1+(c/b)^{1/2})dx/2)]$. However, $\lim_{dx\to 0}(dx\ln(1/dx)) \to 0$, and, for example:

$$\delta g_z^{(x)} \to G\rho\,dx/2(\ln[2c/(1+(1+(c/b)^{1/2}))] - \ln[dx/2])$$

where the second terms are small. These expressions are used below for rectangular shells (Sect. 2.9.5.1) and extended to massive rectangular prisms (Sect. 2.9.6.1). Horizontal planar mass elements have, however, more diverse applications.

2.9.4 Disks

Horizontal disks are useful as parametrization of geological bodies given by contour lines, e.g., in case for topography. Arbitrary orientations do occur and then disk-parallel components are required. Based on the above (Sects. 2.9.3.3 floors and 2.9.3.4 walls) where $\rho dh = \rho^*(kg/m^2)$, small finite thicknesses Δh are handled likewise – with caution. Three cases are distinguished: (1) rectangular disks, (2) disks bounded by polygonal contours and (3) disks with continuously curved boundaries or contours. Case (1) has essentially been treated above; case (2) is the most common approximation of arbitrary contours and case (3) is presented as a general approach which allows integration in closed form if the curve is mathematically defined.

2.9.4.1 Horizontal Rectangular Disk

The horizontal rectangle of density ρ and thickness Δz at depth z below $\mathrm{P} = (0, 0, 0)$ is treated as an example. The vertically oriented rectangles of Sects. 2.9.3.3 and 4 are analogous; for obliquely oriented disks coordinate transformations are required; as discussed for the case of oblique polygons (Sects. 2.9.4.2) and polyhedra, 2.9.5.2). Applying rule 5, we regard the rectangle with one corner at $\mathrm{P}'' = (0,0,z)$ and calculate general rectangles from 4 rectangles of the basic type by adding and subtracting their effects appropriately. Based on (Eqs. 2.9.16, 2.9.17, 2.9.18), the vertical component is given as

$$\delta g_z{}^{(z)} = \mathrm{G}\rho\,\Delta z(\arcsin[r_{100}r_{111}/r_{110}r_{101}] + \arcsin[r_{010}r_{111}/r_{011}r_{110}] - \pi/2)$$

$$(2.9.26)$$

Analogous relations for the other rectangles or other components can be taken from the above sections.

2.9.4.2 Horizontal Polygon

The polygon is a special case of a contour c made of straight line segments. It represents the classical approximation of an arbitrary empirical curve. The gravity effect δg_z can be integrated along the straight lines and summed for the whole polygon with surface mass $\rho^* = \rho dz$ (Talwani & Ewing, 1960).

The deduction of δg_z with the solid angle from the observation point $\mathrm{P} = (0, 0, 0)$ at a distance z or h above (or below) the plane is illustrated in Fig. 2.9.7. A local coordinate system X, Y, Z is chosen with X parallel to the straight line $\underline{\mathrm{AB}}$ from A to B. The normal distance p or n can be derived from (Appendices M2.3); in two coordinates X, Y, or global x, y (n is invariant to the coordinate transformation) it is simpler:

$$n = \underline{\mathrm{P}''\mathrm{N}} = (\Delta x y_\mathrm{A} - \Delta y x_\mathrm{A})/(\Delta x^2 + \Delta y^2)^{1/2} \qquad (2.9.27)$$

where $\underline{\mathrm{P}''\mathrm{N}}$ is the straight line from P'' to N and $\Delta x = x_\mathrm{A} - x_\mathrm{B}$, $\Delta y = y_\mathrm{A} - y_\mathrm{B}$. Local Y is parallel to $\underline{\mathrm{P}''\mathrm{N}}$, and $Z \equiv z$; X, Y, Z form a right-handed system. The coordinate transformation, according to (Eq. 2.4.4) is a rotation about z through an anticlockwise angle α of X versus x with the aid of the rotation matrix:

$$\underline{R}_{\mathrm{X,x}} := \begin{pmatrix} \cos\alpha & -\sin\alpha \\ \sin\alpha & \cos\alpha \end{pmatrix} \qquad (2.9.28)$$

with $\cos\alpha = \Delta x/(\Delta x^2 + \Delta y^2)^{1/2}$, $\sin\alpha = \Delta y/(\Delta x^2 + \Delta y^2)^{1/2}$. Any point $\boldsymbol{X} = (X, Y)$ is found by $\boldsymbol{X} = \underline{\boldsymbol{R}}_{X,x}\boldsymbol{x}$. The back transformation is $\boldsymbol{x} = \underline{\boldsymbol{R}}_{x,X}\boldsymbol{X}$ where $\underline{\boldsymbol{R}}_{x,X} = \underline{\boldsymbol{R}}_{X,x}^{T}$., i.e. $\underline{\boldsymbol{R}}_{x,X}$ transposed.

Each straight line segment of the polygon at $Z = $ const expands with P″, the projection of P onto the plane, a triangle AP″B. The right triangle AP″N (or BP″N) is the basic element, as shown as the triangle projection onto the unit sphere in perspective by Fig. 2.9.7. The effect of the triangle AP″B, $\delta g_{z(AB)}$, is obtained as the difference between the effects of BP″N and AP″N, where only the value of X must be changed:

$$\delta g_{z(AB)} = \delta g_{z(A)} - \delta g_{z(B)}.$$

With the above triple index notation (Fig. 2.9.6), N $\leftrightarrow r_{011}$, A $\leftrightarrow r_{111}$, $\delta g_{z(A)}$ proceeds in analogy to that of the rectangular prism. On the unit sphere around P the projected triangle corners are 111′, 001′, 011′ (Fig. 2.9.7, sphere); the spherical sides are a, b, c and the corner angles α, β, $\gamma = \pi/2$; b, c and α are defined directly by the given geometry (Fig. 2.9.7):

$$b = \arcsin(r_{010}/r_{011}); c = \arcsin(r_{110}/r_{111}); \alpha = \arcsin(r_{100}/r_{110}).$$

The law of sines gives

$$\sin\beta = \sin b/\sin c = r_{010}r_{111}/(r_{011}r_{110}).$$

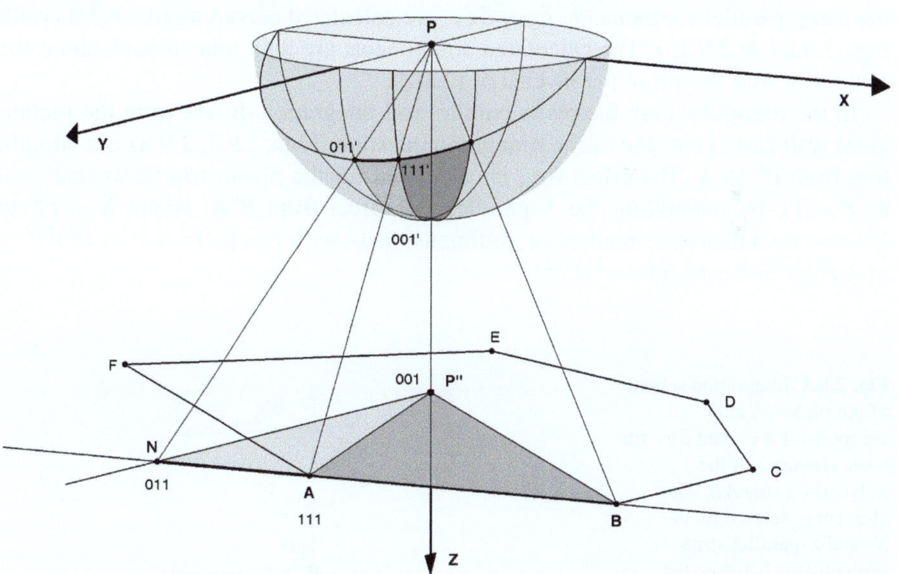

Fig. 2.9.7 Integration scheme for the normal gravity effect of an arbitrary n-cornered horizontal polygon ABCDEF perspective overview; $X, Y, Z \equiv z$ represent the local system with X parallel to the side $\boldsymbol{AB}$. Projection of the polygon side $\boldsymbol{AB}$ onto the unit sphere around the origin P = (0, 0, 0) for the calculation of the plane-normal effect. The basic triangles for the δg_z calculation are AP″N and BP″N, and the effect of the polygon triangle ABP″ is the difference between the two

The solid angle is equal to the spherical excess $\varepsilon = \Delta\Omega = \alpha + \beta + \pi/2 - \pi = \alpha + \beta - \pi/2$;

$$\Delta\Omega = \arcsin(r_{100}/r_{110}) + \arcsin(r_{010}r_{111}/(r_{011}r_{110})) - \pi/2. \qquad (2.9.29)$$

and, as before,

$$\delta g_Z = \Delta\Omega\, G\rho\, dZ = G\rho\, dZ(\arcsin(r_{100}/r_{110}) + \arcsin(r_{010}r_{111}/(r_{011}r_{110})) - \pi/2).$$
$$(2.9.30)$$

The sign of (Eq. 2.9.29) takes care of any possible location of A and B.

The sides of a polygon must then be treated consecutively. Each corner point is treated twice, once for each adjoining side. Encircling the polygon will automatically give its effect only. – A different, though somewhat similar, derivation of $\Delta\Omega$ for a polygon is given by Goguel (1961); see KJ61, 330–334.

2.9.4.3 Oblique Polygon

The complete vector effect $\boldsymbol{\delta g} = (\delta g_x, \delta g_y, \delta g_z)$ is needed if the polygon has an oblique orientation for rotation into the global x, y, z system. The coordinates must first be rotated according to (Eq. 2.4.2) into X, Y, Z, such that, again, the side $\underline{AB}$ (vector $\boldsymbol{AB}$) is parallel to the X axis $\boldsymbol{X}$; the rotation axis through $P = (0, 0, 0)$ is given by $\boldsymbol{x} \times \boldsymbol{AB}$. The plane-normal component δg_Z is calculated as in Sect. 2.9.3.3, and the plane-parallel components, δg_X, δg_Y, are calculated according to 2.8.3.4 (wall: Eqs. 2.9.24 & 2.9.25). The calculated components are then rotated back about the same axis with the aid of the inverse of (2.4.2).

In the triangular case the plane-parallel rod integration differs from the rectangular wall case. One side of the triangle in question (Figs. 2.9.7, 2.9.8) is a straight line from P'' to A. The effect δg_X is calculated for the plane strip between $Y = 0$ to $Y = Y_A$ by integrating the X-parallel rod effect from $P''A$, where $X = cY$ to $X \rightarrow \infty$, i.e. integration renders (non-dimensional), with $r = (c^2Y^2 + Y^2 + h^2)^{1/2} = (1+c^2)^{1/2}(Y^2 + h^2/(1+c^2))^{1/2}$:

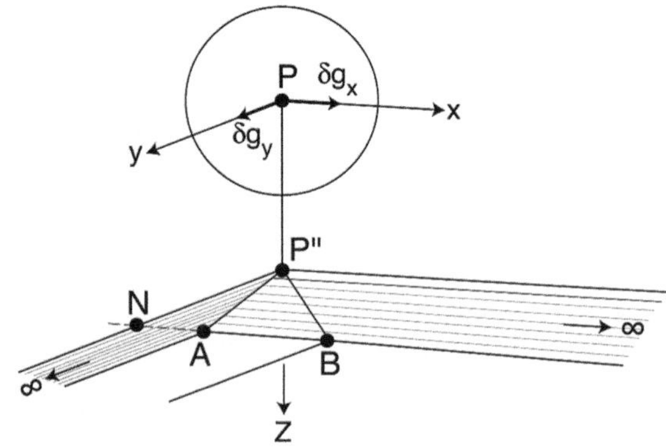

Fig. 2.9.8 Integration scheme of the plane-parallel components δg_X and δg_Y for basic elements of the polyhedron side $\boldsymbol{AB}$. The element is defined by the X- and Y-parallel strips, respectively, between the straight line $\boldsymbol{P''B}$ (or $\boldsymbol{P''A}$) and infinity. The effect of the triangular element $AP''B$ is the difference of the effects of $BP''N$ and $AP''N$

$$\delta g_x{}^* = \int_o^{Y_A} dY/r = (1+c^2)^{-1/2}[\ln(Y_A + (Y_A^2 + h^2/(1+c^2))^{1/2}) - \ln(h/(1+c^2)^{1/2})]$$

$$(2.9.31)$$

The Y-parallel effect δg_Y* is obtained in the same way, but the straight line is $y = x/c$:

$$\delta g_y{}^* = c(1+c^2)^{-1/2}[\ln(y_A + (y_A^2 + c^2h^2/(1+c^2))^{1/2}) - \ln(ch/(1+c^2)^{1/2})]$$

$$(2.9.32)$$

Dimensionally the gravitational components are $\delta g_i = \delta g_i{}^* G\rho^* = \delta g_i{}^* G\rho\,dh$. Evaluation for the point B (or Y_B) is analogous. Subtraction of the effect from that for A leads to the effects of the triangle $AP''N$. The sides of a polygon are treated consecutively; each time an individual rotation of the x, y, z components into a local X, Y, Z system and a back rotation of the results into the global system are carried and the sum of all contributions is formed.

2.9.4.4 Plane Horizontal Disc with Smoothly Curved Boundary

Treatment of a disc with a smooth, continuous boundary c and of thickness dh and density ρ (or surface density ρ^*) starts from (Eq. 2.9.13): $\delta g_n = G\rho \int_\Omega \int_h dh\,d\Omega$, the infinitesimal integration element is taken as the triangular surface (Fig. 2.9.9a) extended from P'' to the boundary element dc. P'' is the projection of P onto the

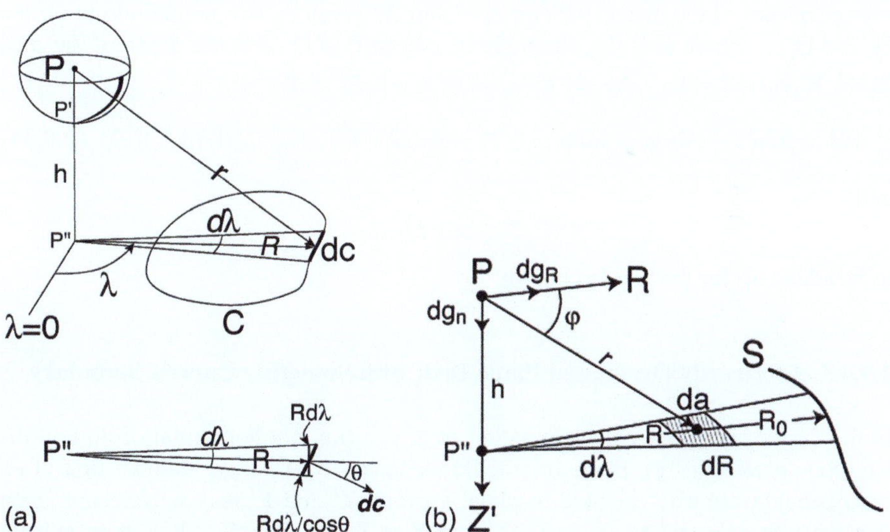

Fig. 2.9.9 Integration of the plane-normal (**a**) and plane-parallel (**b**) effects of infinitesimal triangular surface element from P'' to the element dc of a smooth boundary. P'' is the vertical projection of P onto the plane. The infinitesimal plane-normal effect $d\delta g_n$ (**a**) is derived directly from $d\Omega$, while the plane-parallel effect $d\delta g_p$ (**b**) of the infinitesimally narrow triangle of angle $d\lambda$ at P must be integrated with the surface element $da = Rdrd\lambda$

plane. The radii R from P$''$ to ds limit the incremental azimuth $d\lambda = dc/(R\cos\theta)$. The projection onto the unit sphere around P is the infinitesimal cap sector $d\Omega = d\lambda(1 - h/r)$. Hence $d\delta g_n = G\rho^* d\Omega = G\rho^*(1 - h/r)d\lambda$, and

$$\delta g_n = G\rho^* \int_c (1 - h/r(\lambda))d\lambda = G\rho^* \left(\int_c d\lambda - \int_c h/r(\lambda)d\lambda \right)$$

$$= G\rho^* \left(C - \int_c h/r(\lambda)d\lambda \right) \qquad (2.9.33)$$

where

$$C = \begin{cases} 0 & \text{if P outside contoured planar disc} \\ \pi & \text{if P exactly on the contour (if smooth or continuous)} \\ 2\pi & \text{if P inside the contour.} \end{cases}$$

Integration along a closed contour line takes automatically care of the finite disc by the sign of the $d\lambda$ increments along the contour. Analytical integration of (Eq. 2.9.33) requires $h/r(\lambda)$ to be expressed in mathematical terms (see below).

Alternatively, contours c can be defined by sequential points Q'_i, i.e. approximated by small finite segments Δc and the corresponding angular increments $\Delta\lambda$ (Jacoby, 1967). Effects of contour shape and of the finite thickness d are negligible, and if P lies close to a contour line, special measures have to be taken, as for example, filling in a gap along c with points Q'. The solid angle $\Delta\Omega$ is estimated for the triangular part of the disc expanded by P and Δc given by two neighbouring points Q'_i and Q'_{i+1}, where φ is the dip angle of r from P to Q', h is the depth of the disc below P, and $\Delta x = x_{Q'} - x_P, \Delta y = y_{Q'} - y_P, r_i = (\Delta x^2 + \Delta y^2 + h^2)^{1/2}$:

$$\Delta\Omega \approx \Delta\lambda(1 - (\sin\varphi_i + \sin\varphi_{i+1})/2) = \Delta\lambda[2 - h(r_i + r_{i+1})/(r_i r_{i+1})]/2 \quad (2.9.34)$$

and

$$\Delta g_z \approx \Delta\Omega\rho d$$

to be added up for the whole contour.

2.9.4.5 Arbitrarily Orientated Plane Disk with Smoothly Curved Boundary

In the case of arbitrary disc orientation in x, y, z (i.e. not horizontal or no coordinate axis parallel to g), plane-oriented coordinates X, Y, Z are defined first. Two orthogonal, arbitrarily oriented in-plane vectors, X' and Y' may be selected, such that the unit axis vectors $X = X'/X'$ and $Y = Y'/Y'$; $Z = X \times Y$ is then normal to the plane. The global x, y, z coordinates are transformed into X, Y, Z by establishing and evaluating the rotation matrix (Eq. 2.4.2). The calculation of the disc-normal component δg_n is then done as above with Eq. (2.9.33), but does not suffice for obtaining the vector components needed for back transformation into the Earth-bound x, y, z system; it requires also the plane-parallel components δg_X and δg_Y.

The plane-parallel component δg_p exerted by the dλ planar element can be integrated along the radius R, but the surface element, da, is only the small quadrangle of incremental R and λ : $da = dRR d\lambda$; with ρ^* as before and $r = (R^2 + h^2)^{1/2}$ (Fig. 2.9.9b), the infinitesimal gravity contribution is $d\delta g_p = G\rho^* da R/r^3 = G\rho^* d\lambda R^2 dR/(R^2 + h^2)^{3/2}$. If the variables R and λ are independent, the integration over R gives:

$$\int R^2 dR/(R^2 + h^2)^{3/2} = \ln(R + (R^2 + h^2)^{1/2}) - R/(R^2 + h^2)^{1/2} = \ln(R + r) - R/r$$

$$= \ln R(1 + \cos\varphi) - \cos\varphi$$

where $\varphi = \arctan(h/R)$ is the dip angle under which dc is seen from P (Fig. 2.9.9b). If $R \to \infty, \cos\varphi \to 1$ and even the infinitesimal element with the opening angle dλ, generates d$\delta g \to \infty$. This is not surprising, as the mass element $RdRdy$ grows linearly with R while the gravitational effect decreases as $1/R^2$. Hence the effect of da decreases only as $1/R$, and the integral of the type $\int dR/R$ renders the logarithm of R.

If R is given as an analytical function of λ on a curve c, the integration can be carried out and Δg_p can be evaluated in closed form:

$$\delta g_p = G\rho^* \int (\ln(R(\lambda) + r(\lambda)) - R(\lambda)/r(\lambda))d\lambda \qquad (2.9.35)$$

If the curves c are circles, the solution leads to elliptic integrals. Alternatively the integration may be carried out numerically or approximated in some appropriate way for any special case.

Finally the three components δg_X, δg_Y and δg_Z are back-transformed into the x, y, z system by applying the inverse of (Eq. 2.4.2).

2.9.5 Shells

Shells are hollow bodies treated as a preliminary step to the massive bodies.

2.9.5.1 The Rectangular Shell or Box

The box with P at its upper corner (0, 0, 0) is composed of two rectangular horizontal floors and 4 rectangular vertical walls (Eqs. 2.9.18 & 2.9.25). The three near sides have P at one corner such that the expressions for the $\delta g^{(j)}{}_i$ effects must be re-evaluated for the case of one of the distances $r \to 0$. The expression for the far side effects exactly correspond to (Eqs. 2.9.18 & 2.9.25) with the infinitesimal thicknesses replaced by the small but finite thickness t, i.e. $t \ll a, b, c$ (or 1), respectively. Remember that the superscripts (x), (y) or (z) denote the effects associated with box walls or surfaces whose exterior normal vectors point in x, y or z directions, respectively; in the text these surfaces are named "facing (x)" or "(x) facing", etc.

The far side effects are in the ijk notation (Fig. 2.9.6):

$$
\begin{aligned}
\delta g_{\mathrm{x}}{}^{(x)} &= G\rho\, t(\arcsin[r_{010}r_{111}/r_{011}r_{110}] + \arcsin[r_{001}r_{111}/r_{101}r_{011}] - \pi/2) \\
\delta g_{\mathrm{y}}{}^{(x)} &= G\rho\, t\ln([(r_{001}+r_{101})r_{110}]/[(r_{001}+r_{111})r_{100}]) \\
\delta g_{\mathrm{z}}{}^{(x)} &= G\rho\, t\ln([(r_{010}+r_{110})r_{101}]/[(r_{010}+r_{111})r_{100}]) \\
\delta g_{\mathrm{x}}{}^{(y)} &= G\rho\, t\ln([(r_{001}+r_{011})r_{110}]/[(r_{001}+r_{111})r_{010}]) \\
\delta g_{\mathrm{y}}{}^{(y)} &= G\rho\, t(\arcsin[r_{100}r_{111}/r_{110}r_{101}] + \arcsin[r_{001}r_{111}/r_{101}r_{011}] - \pi/2) \\
\delta g_{\mathrm{z}}{}^{(y)} &= G\rho\, t\ln([(r_{100}+r_{110})r_{011}]/[(r_{100}+r_{111})r_{010}]) \\
\delta g_{\mathrm{x}}{}^{(z)} &= G\rho\, t\ln([(r_{010}+r_{011})r_{101}]/[(r_{010}+r_{111})r_{001}]) \\
\delta g_{\mathrm{y}}{}^{(z)} &= G\rho\, t\ln([(r_{100}+r_{101})r_{011}]/[(r_{100}+r_{111})r_{001}]) \\
\delta g_{\mathrm{z}}{}^{(z)} &= G\rho\, t(\arcsin[r_{100}r_{111}/r_{110}r_{101}] + \arcsin[r_{010}r_{111}/r_{011}r_{110}] - \pi/2)
\end{aligned}
$$

$$(2.9.36)$$

In evaluating the near side effects applying (Eq. 2.9.36), each side is considered a rectangular plate of thickness $t \ll a, b, c$. For example, the plate facing x is defined by $x = 0$ and $x = t$ with the surface area $b \times c$. In the above case, the plate has the dimension $t \times b \times c$. Full evaluation requires for each component consideration of three elements, one plane-normal and two plane-parallel, i.e. 9 elements for each side. It turns out that the large faces, in the above case, $b \times c$, expand the dominant mass to exert significant effects relative to those of the narrow faces $b \times t$ and $c \times t$, in the above case. The detailed evaluation is based on the effect of the vertical rod as presented in Sect. 2.9.3.2 which, in turn, is best treated as a special case of the massive cuboid considered in Sect. 2.7.6 below; Task 2.5 is meant to substantiate the derivation.

Take for the above plate, e.g., the vertical component ${}^{\mathrm{xnear}}\delta g_{\mathrm{z}}{}^{(x)} = G\rho\, t(\ln[2bc/(b+ <b,c>)] - \ln t)$ where $<b,c> = (b^2 + c^2)^{1/2}$ and terms as $<t,b> \approx b$; with $t \to 0$, $\ln t \to -\infty$, while the product $(t\ln t) \to 0$, such that this term is uncritical; nevertheless, the expression must be fully evaluated. The two narrow faces of the plate (y) and (z) contribute ${}^{\mathrm{xnear}}\delta g_{\mathrm{z}}{}^{(y)} = G\rho b(\ln[b<b,c>/(b<b,c>)] \to \ln 1 = 0$, and ${}^{\mathrm{xnear}}\delta g_{\mathrm{z}}{}^{(z)} = G\rho c(\arcsin[t<b,c>)/(bc)] + \arcsin[b<b,c>/(<b,c>b] - \pi/2) \to \pi/2 - \pi/2 = 0$. Summarizing all the significant effects, the (x) facing near plate contributes:

$$
\left.
\begin{aligned}
{}^{\mathrm{xnear}}\delta g_{\mathrm{x}}{}^{(x)} &= G\rho\, t\pi/2 \\
{}^{\mathrm{xnear}}\delta g_{\mathrm{y}}{}^{(x)} &= G\rho\, t(\ln[2bc/(c+ <b,c>)] - \ln t) \\
{}^{\mathrm{xnear}}\delta g_{\mathrm{z}}{}^{(x)} &= G\rho\, t(\ln[2bc/(b+ <b,c>)] - \ln t)
\end{aligned}
\right\}
\qquad (2.9.37a)
$$

the (y) facing near plate contributes:

$$
\left.
\begin{aligned}
{}^{\mathrm{ynear}}\delta g_{\mathrm{x}}{}^{(y)} &= G\rho\, t(\ln[2ac/(c+ <a,c>)] - \ln t) \\
{}^{\mathrm{ynear}}\delta g_{\mathrm{y}}{}^{(y)} &= G\rho\, t\pi/2 \\
{}^{\mathrm{ynear}}\delta g_{\mathrm{z}}{}^{(y)} &= G\rho\, t(\ln[2ac/(a+ <a,c>)] - \ln t)
\end{aligned}
\right\}
\qquad (2.9.37b)
$$

and the (z) facing near plate contributes:

$$\left.\begin{array}{l} {}^{\text{znear}}\delta g_{\text{x}}{}^{(z)} = \mathrm{G}\rho t(\ln[2ab/(b+<a,b>)] - \ln t) \\[4pt] {}^{\text{znear}}\delta g_{\text{y}}{}^{(z)} = \mathrm{G}\rho t(\ln[2ab/(a+<a,b>)] - \ln t) \\[4pt] {}^{\text{znear}}\delta g_{\text{z}}{}^{(z)} = \mathrm{G}\rho t\pi/2 \end{array}\right\} \qquad (2.9.37c)$$

The above expressions (Eqs. 2.9.36 & 2.9.37) are ordered by the vector effects of each side (x), (y) and (z). The effect of the whole box with thin walls of mass density $\rho^* = \rho t$ is found by summing component wise.

2.9.5.2   Polyhedral Shell

The polyhedron is composed of contagious plane polygons, usually triangles. At edges and corners the surface orientation changes discontinuously. As in the case of the horizontal polygon (Sect. 2.9.4.2), the basic surface elements are the right triangles (Fig. 2.9.7), whose solid angle contributions $\Delta\Omega$ are to be calculated with Eq. (2.9.30).

Since the planes have generally oblique orientations in the x, y, z system, application of the above procedure requires first a coordinate rotation about a generally oblique axis and furthermore, for the determination of the gravitational vector contribution, also the plane-parallel gravitational components must be known as discussed above for the polygon (Sect. 2.9.4.2). The three dimensionless components $\delta g_{\text{X}}{}^*$, $\delta g_{\text{Y}}{}^*$, $\delta g_{\text{Z}}{}^*$ (Eqs. 2.9.4–2.9.7) in the local coordinates X, Y, Z for each polyhedral plane surface must be back rotated into the global system and added up for the whole polyhedron. The dimensional values of δg_{x}, δg_{y}, δg_{z} are obtained by multiplying each of the components $\delta g_{\text{x}}{}^*$, $\delta g_{\text{y}}{}^*$, $\delta g_{\text{z}}{}^*$ values by $\mathrm{G}\rho\Delta h$ with Δh being the finite thickness of each of the planes. – The details of the calculation for the hollow polygonal shell are not given here, since the case has little practical value; coordinate transformations and the resulting specific formulae can be worked out from Sects. 2.9.3 to 2.9.4.3.

2.9.5.3   Generalized Smoothly Curved Surface or Shell

For arbitrary curved surfaces the element is the infinitesimal surface ds which is continuously changing direction (Fig. 2.9.10a,b). The infinitesimal vector contribution is dg, with $d\Omega = ds\cos\psi/r^2, \cos\psi = ds\cdot r/(ds\cdot r)$, and dh or Δh as in Fig. 2.9.10b is:

$$dg_{(r)} = \mathrm{G}\rho\Delta h\, d\Omega/\cos\psi \qquad (2.9.38)$$

pointing towards $-r$; dh is the thickness of the curved shell. Integration over $d\Omega$, i.e. ds, is then done by components $(r = (x, y, z))$ with ϕ_{x}, etc. the angles of $-r$ with x and $\cos\phi_{\text{x}} = x/r$, etc.:

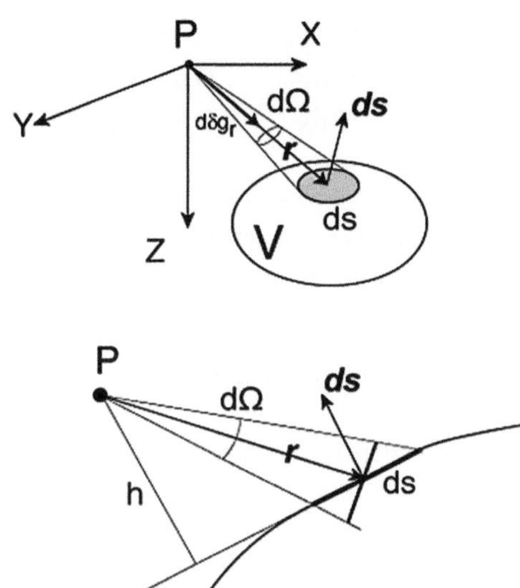

Fig. 2.9.10 Surface element of smooth shell, solid angle $d\Omega$ and gravitational attraction δg_r. *Top*: 3D view; *bottom*: geometrical relations

$$\left. \begin{aligned} \delta g_x &= G\rho \int_S \Delta h \cos\phi_x / \cos\psi \, d\Omega = -G\rho\Delta h \int_S x/(r\cos\psi)\,d\Omega \\ \delta g_y &= -G\rho\Delta h \int_S y/(r\cos\psi)\,d\Omega \\ \delta g_z &= -G\rho\Delta h \int_S z/(r\cos\psi)\,d\Omega \end{aligned} \right\} \qquad (2.9.39)$$

If the direction of δg is known beforehand (from P, e.g., to centre of a uniform sphere), the corresponding $\boldsymbol{dg}$ component can be integrated directly:

$$\delta g = G\rho\,\Delta h \int_S \cos\phi / \cos\vartheta \, d\Omega \qquad (2.9.40)$$

where ϕ is the angle of $-\boldsymbol{r}$ relative to $\boldsymbol{\Delta g}$.

2.9.6 Uniform Massive Volumes

Massive bodies are the main subject of gravity forward calculations. They are assumed to be uniform in density $\Delta\rho$. If a geological body is non-uniform, it has to be modelled by several uniform bodies, either superimposed, i.e. fully, partly or not overlapping, contagious or separated. Variations of the density contrast $\Delta\rho$ within a geological body may originate from external variations of the reference density ρ_0. Again, the principal approach is that which uses the solid angle. The expressions are, however, designed to apply to parametrizations of the model volume independent from the individual observation points such that effective modelling is possible. The solid angle $\Delta\Omega$ must be easily calculable from the given global coordinates. Remember the arbitrarily oriented surface element with $dV = dr\ ds = dr\ d\Omega r^3/h$, for which the normal component Δg_n takes the form

(Eq. 2.9.13): $\int_{\Omega}\int_{h}\mathrm{d}h\,\mathrm{d}\Omega = \Delta\Omega\,\Delta h$. If the solid angle $\Delta\Omega$ has been found, integration of massive bodies means only measuring the distances Δh within, and normal to, the mass volume $\rho\Delta V$. For example, a two-dimensional horizontal strip of surface density $\rho\,\mathrm{d}z$ seen from $P = (0, 0, 0)$ between two planes intersecting each other at P at an angle of $\Delta\psi$ has the effect at P of $\delta g_z = 2G\rho\,\Delta\psi\,\mathrm{d}z$, while the effect of a thick (Δz) body of trapezoid cross section is $\delta g_z = 2G\rho\,\Delta z\,\Delta\psi$.

As above, three types of bodies are presented here: the rectangular prism, the polyhedron and the lump of mass with a smoothly curved surface.

2.9.6.1 Massive Rectangular Prism or Cuboid

The calculation of the effect of a massive rectangular prism on one of its corners $P = (0, 0, 0)$ is now easy, i.e. it has already been prepared in Sects. 2.9.3.3 (floor), 2.9.3.4 (wall) and 2.9.5.1 (rectangular box). The last missing step is the integration over the thickness of the mass elements now extended from $x = 0$, $y = 0$, $z = 0$ to a, b, c (or l).

The massive rectangular prism is thus a special case of the box where in Eq. (2.9.36) for the floor and the walls the thickness t is replaced by the dimensions, for x, y, z: $t = a, b, c$, respectively.

$$
\begin{aligned}
\delta g_x^{(x)} &= G\rho\,a\,(\arcsin[r_{010}r_{111}/r_{011}r_{110}] + \arcsin[r_{001}r_{111}/r_{101}r_{011}] - \pi/2)\\
\delta g_y^{(x)} &= G\rho\,a\ln([(r_{001}+r_{101})r_{110}]/[(r_{001}+r_{111})r_{100}])\\
\delta g_z^{(x)} &= G\rho\,a\ln([(r_{010}+r_{110})r_{101}]/[(r_{010}+r_{111})r_{100}])\\
\delta g_x^{(y)} &= G\rho\,b\ln([(r_{001}+r_{011})r_{110}]/[(r_{001}+r_{111})r_{010}])\\
\delta g_y^{(y)} &= G\rho\,b(\arcsin[r_{100}r_{111}/r_{110}r_{101}] + \arcsin[r_{001}r_{111}/r_{101}r_{011}] - \pi/2)\\
\delta g_z^{(y)} &= G\rho\,b\ln([(r_{100}+r_{110})r_{011}]/[(r_{100}+r_{111})r_{010}])\\
\delta g_x^{(z)} &= G\rho\,c\ln([(r_{010}+r_{011})r_{101}]/[(r_{010}+r_{111})r_{001}])\\
\delta g_y^{(z)} &= G\rho\,c\ln([(r_{100}+r_{101})r_{011}]/[(r_{100}+r_{111})r_{001}])\\
\delta g_z^{(z)} &= G\rho\,c(\arcsin[r_{100}r_{111}/r_{110}r_{101}] + \arcsin[r_{010}r_{111}/r_{011}r_{110}] - \pi/2)
\end{aligned}
$$
$$(2.9.41)$$

As above: $\delta g_i = G\rho\sum_{(k)}\delta g_i^{(k)}$, where i, j, k correspond to x, y, z:

According to rule 5, an arbitrary prism with $P \neq (0, 0, 0)$ somewhere outside or inside, is composed of prisms of the same type but different dimensions by subtraction and/or addition. For this it is more convenient to number the corner points, or better: the discrete coordinate intervals, e.g. x_l, y_m, z_n of the arbitrary prism and to re-order the above expressions (Eq. 2.9.41) by functions of individual corner points or radii from P to these corners. The component effect $\Delta g_i^{(p)}{}_{lmn}$ of an arbitrary prism ($\Delta x_{l,l+1} \cdot \Delta y_{m,m+1} \cdot \Delta z_{n,n+1}$) is then expressed by cyclic summation with alternating signs of $f_{lmn} = f(x_l, y_m, z_n)$, for all corner with $r = (x^2 + y^2 + z^2)^{1/2}$ and:

$$f(x, y, z) = y\ln(x+r) + x\ln(y+r) - z\arctan(xy/zr) \tag{2.9.42}$$

$$\begin{aligned}
\delta g_{lmn} = G\rho(&f_{l,m,n} - f_{l+1,m,n} + f_{l+1,m+1,n}, -f_{l,m+1,n} + f_{l,m+1,n+1} - f_{l,m,n+1}\\
&+ f_{l+1,m,n+1} - f_{l+1,m+1,n+1})
\end{aligned} \tag{2.9.43}$$

Fig. 2.9.11 Scheme of summing the corner functions (Eq. 2.9.42) constituting the gravity effect δg_z at $P = (0,0,0)$ for an arbitrary rectangular prism with alternating signs; the corners (x, y, z) are counted as x, y, $z = 1$ or 2, 1 indicating the smaller numerical value

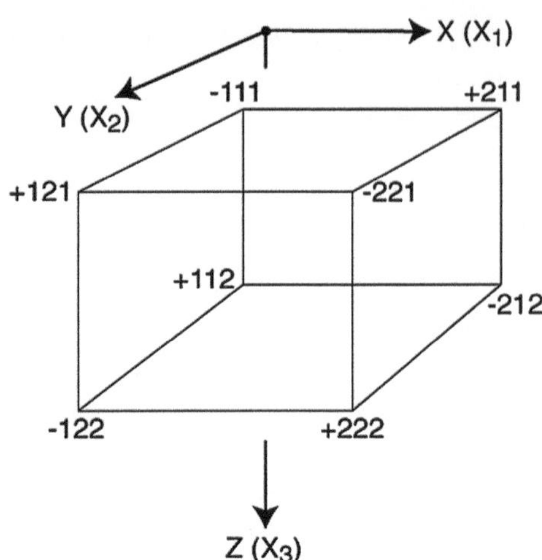

This is also obtained by conventional integration for $(\Delta x_{l,l+1} \cdot \Delta y_{m,m+1} \cdot \Delta z_{n,n+1})$. The scheme of summation is shown in (Fig. 2.9.11), where the corners are numbered as 111, 112, 211, etc. instead of showing the indices $l, m, n, l+1, m+1, n+1$.

Rounding errors may become large, if r greatly exceeds the side lengths, because Δg_{lmn} (see below) is expressed as a very small difference between large numbers f_{lmn} and the number of volume elements becomes large. Tests are necessary to investigate from where on the point approximation renders better results. Alternatively small elements may be combined to bigger ones.

2.9.6.2 Massive Polyhedron

Massive polyhedra are flexible approximations to arbitrarily shaped geological bodies. Their treatment somewhat differs from that of polyhedral shells (Sect. 2.9.5.2). The difference is in the infinitesimal mass elements. The size of the shell mass element $\rho^* ds$ depends on its orientation relative to the radius vector r, see (Eq. 2.9.12), and the arguments leading to it. In contrast, the mass element of the massive polyhedron is the oblique cone or pyramid expanded from ds and P (0, 0, 0); its volume is generally large relative to the infinitesimal dimensions of its base ds which, hence, is negligible and independent from the orientation of ds. Therefore the integration proceeds with mass elements ρdV, where dV is the volume of the infinitesimally thin cone defined by the solid angle $d\Omega$ whose volume grows with dr as $dV = r^2 dr d\Omega$; $d\Omega$ depends on the orientation as $d\Omega = \cos \psi \, ds / r^2$. If ds is at a given arbitrary r_0, $dV = \cos \psi \, ds \, dr$, and the volume of the cone ΔV depends on r as $\cos \psi \, ds_0 \int^{r_0} dr = \cos \psi \, ds \, r_0$.

The basic finite polyhedral elements Δs are the variably oriented right triangles which, from P, expand massive oblique tetrahedra. For the plane-normal components, the planar elements are projected, as above, onto the unit sphere at P to directly render $\Delta \Omega$.

In the special case that the direction $\psi^{(E)}$ of the vector effect $\delta\boldsymbol{g}^{(E)}$ is known, the components in x, y, z are found by back projection (Fig. 2.9.4): $\delta g_n^{(E)} = \delta g^{(E)}\cos\psi_n$, $\delta g_n^{(E)}\cos\psi_h/\cos\psi^{(E)}$. Generally, the vector or its direction is not known, and thus also the plane-parallel components must be calculated. The procedure is described by that for the polygon (Sects. 2.9.4.2 & 2.9.4.3; Figs. 2.9.7 & 2.9.8; Eqs. 2.9.29, 2.9.30, 2.9.31); integration of the massive oblique pyramid is reduced to $\int dh = h$, because the planar dh elements grow as h^2, compensating $1/r^2$. For each polyhedron triangle the vector is calculated in the local coordinates (X, Y, Z): $\delta\boldsymbol{g} = (\delta g_X, \delta g_Y, \delta g_Z)$ and rotated into the global coordinates: δg_x, δg_y, δg_z and added up component wise for the whole polyhedron, with $i = 1, 2, 3$ for x, y, z:

$$\delta g i = \sum\nolimits_{Sk} \delta g_{ik}$$

Another approach to calculating the gravity effect of a polyhedron(Çavşak, 1992) is first to integrate the disturbing potential effect ΔU of an arbitrarily oriented pyramid from similar volume elements as used here and then calculating the vertical derivative $\delta g_z = \partial\Delta U/\partial z$. It requires coordinate transformations. The approach is facilitated by using vector calculus. Several solutions and algorithms of gravity integration over uniform polyhedra have been published, at least since the 1960s. Pohánka (1988) and Holstein and co-workers in a series of papers (Holstein, 2002a,b; Holstein et al., 1999) summarized and compared them with each other, especially in view of computational precision. Polyhedra are treated with the aim to unify the calculations of what is called the "gravimagnetic effects" and to make optimal use of similarities common to all these related potential field problems. The methods may be distinguished as vertex, line and surface methods. The formulations are essentially all alike, but the approach is different: abstract, mathematical, based on the application of Gauss' and Stokes' integral theorems. In contrast, it is here attempted to design tailored mass elements (solid angle and vertical mass line, both growing with r^2) in a more visual approach. It encompasses special cases where mass elements degenerate to zero (on a polyhedron facet, an edge or a vertex) where analytical treatment has problems. Computational aspects are discussed in Chap. 6.

2.9.6.3 Lumps of Mass with Smoothly Curved Surfaces

If a uniform body is defined by a continuously curved closed surface S (Fig. 2.9.10) the mass parametrization is similar to that for the polyhedron (Sect. 2.9.6.2). The infinitesimal surface element ds is the same, it expands a dense cone of infinitesimal solid angle $d\Omega$ opening from P. In contrast to the polyhedron, the gravitational vector effect $\boldsymbol{dg}$ points in the direction of the axial radius vector $\boldsymbol{r}$ from P to ds; since the deviation from this at the generally oblique surface $\boldsymbol{ds}$ is negligible relative to the mass of the finite-length cone. (The approach is basically that of Poisson's equation $\Delta U = 4\pi G\rho = \mathrm{div\,grad\,}U = \mathrm{div}\boldsymbol{g}$, combined with Gauss' theorem $\int_V \mathrm{div}\boldsymbol{g}\,dV = \int_S \boldsymbol{g}\bullet\boldsymbol{ds}$, where $\boldsymbol{ds}$ is the surface vector, but the direction is reversed to that from the source Q to the observation point P). Letting $\psi = 0$, i.e. ds normal to $\boldsymbol{r}$:

$$|\mathbf{\Delta g}| = G \iiint_V dm/r^2 = G\rho \int_\Omega \int_r r^2\, dr\, d\Omega / r^2 = G\rho \int_\Omega \int_r dr\, d\Omega = G\rho\, \Delta\Omega\, \Delta r$$

$$(2.9.44)$$

This is exact in the case of spherical surfaces for which $r = $ const, but for different shapes the difference is negligible. However, the orientation of $\mathbf{ds}$ does enter the solid angle $d\Omega$ calculation. The infinitesimal vector contribution $\mathbf{\delta g}$, from an element $dr\, r^2\, d\Omega$ can be written, with $d\Omega = ds\cos\psi/r^2$ and $\cos\psi = \mathbf{ds}\cdot\mathbf{r}/(ds\, r)$:

$$\mathbf{\delta g} = G\rho dr \cos\psi\, ds / r^2 (\mathbf{r}/r) = G\rho dr d\Omega (\mathbf{r}/r) \qquad (2.9.45)$$

Integration of δg over r leads to the, yet infinitesimal, vector contribution $\mathbf{dg}_{(r)}$ from the $d\Omega$ cone or ray (r) expanded by P and ds:

$$\mathbf{dg}_{(r)} = G\rho\, r\, d\Omega (\mathbf{r}/r) \qquad (2.9.46)$$

No problem arises from the zero distance of mass to P. Integration over S or Ω is done by components ($\mathbf{r} = (x, y, z)$ with φ_x, the angle of $\mathbf{r}$ with x, and $\cos\varphi_x = x/r$, and φ_y, φ_z accordingly):

$$\left.\begin{array}{l} \delta g_x = G\rho \int_S r\cos\varphi_x d\Omega = G\rho \int_S r\,(x/r)\,d\Omega = G\rho \int_S x\,d\Omega \\ \delta g_y = G\rho \int_S y\,d\Omega \\ \delta g_z = G\rho \int_S z\,d\Omega \end{array}\right\} \qquad (2.9.47)$$

If the direction of δg is known beforehand (from P, e.g., to the centre of a homogeneous sphere), the corresponding δg component can be integrated directly:

$$\delta g = G\rho \int_S r\cos\varphi d\Omega \qquad (2.9.48)$$

where φ is the angle of $\mathbf{r}$ relative to $\mathbf{\Delta g}$.

Analytical integration is possible if the quantities $r = (x, y, z)$ and ψ can be described mathematically. Integrating over the whole closed surface S takes care of the mass inside s. The equation expresses that the lengths of rays r associated with mass (ρ) are to be intergrated multiplied with $\cos\phi$. In Task 2.1 this is applied to a homogeneous sphere and r means the secant lengths of the rays. It proves the well known fact that the gravitational attraction of a homogeneous sphere is identical to that of its mass concentrated at its centre. Even easier is it to show that nowhere inside a homogeneous spherical shell a test mass feels its attraction (Task 2.3).

2.9.7 Two-Dimensional Bodies

Two-dimensional (2D) mass distributions vary in two dimensions (usually x, z) and are uniform in the third dimension (y); 2D models play an important role in gravity interpretation (Chap. 6), mostly because they are much easier visualized and handled than three dimensions, imagination of 2D sections is easier than that of three-dimensional (3D) bodies, and 2D density variations $\rho(x, z)$ are conveniently drawn on two dimensional paper. If a gravity anomaly is characterized by one dominant elongation the mass distribution is approximated by neglecting the dominant dimension. Principally unrealistic, as it extends beyond even the limited universe and has an infinite gravitational potential, it nevertheless generates a realistic gravity effect, essentially because the effects of distant parts are negligible. The term "two-dimensional" has become popular and is generally understood correctly by the users although it is somewhat misleading: the structure is really three-dimensional but the one infinite dimension with no variation is uninteresting.

2.9.7.1 Two-Dimensional Mass Elements

2.9.7.1.1 Mass Line Elements

What is called "two-dimensional geometry" is described in Cartesian coordinates. The solid angle and rod approach correspond to horizontal cylinder coordinates. In analogy to the 3D mass elements, mass lines and mass planes, both horizontal and vertical are needed for the calculation of the gravitational components which are line-normal (or plane-normal) and parallel.

The infinite uniform horizontal mass line with density $\rho^+ = \rho\,dx\,dz$ (kg/m) is the fundamental mass element of 2D modelling. Its effect (Eq. 2.9.3.1) has been calculated in Cartesian coordinates in 2.8.3.1. The correction factor (Eq. 2.9.8) $a = \lambda/(x^2 + \lambda^2 + z^2)^{1/2}$ for finite length λ, instead of ∞ (see Sect. 2.9.3.1), depends only on the ratio $\eta = \lambda/z$ and approaches 1 quickly: for $\eta = 0.5, 1, 2, 5$: $a = 0.4$, 0.7, 0.9, 0.98, respectively.

The gravitational potential of the infinite mass line, δU (Eq. 2.9.9) is of interest in interpreting satellite radar measurements of ocean surface topography which, to a first approximation, follows the equipotential surface of the geoid, i.e. $\delta N \approx \delta U/g_0$, where g_0 is normal gravity at the Earth's surface.

For 2D geometry the solid-angle approach is simple. The mass line corresponds to an infinitesimally thin horizontal strip, infinite in y, seen from P under the angle $d\varphi$; it projects onto the unit sphere as a great circle bi-angle whose area is $d\Omega = 2d\varphi$ (Fig. 2.9.1). The 2D volume element is thus chosen $dV = dz\,ds = dz\,d\Omega r^3/h = 2dz\,d\varphi r^3/z$, hence the vertical line-normal effects is:

$$d\delta g_n = d\delta g_z = 2G\rho\,dz\,d\varphi \qquad (2.9.49)$$

If the y axis is taken as a horizontal cylinder axis, φ as the azimuth corresponds to horizontal cylinder coordinates, where $r = R$, the axial distance of a point or line; Eqs. (2.9.9) and (2.9.49) thus describe axis-normal effects. Equation (2.9.49) is integrated for planes, arbitrarily shaped pipes (e.g., of wall thickness t) and to massive beams of density ρ (Fig. 2.9.12), for example, defined by closed 2D contours c (i.e. infinite pipe-like surfaces) or polygons in cross section. As in 3D, the discussion is ordered from disks to hollow pipes and massive beams.

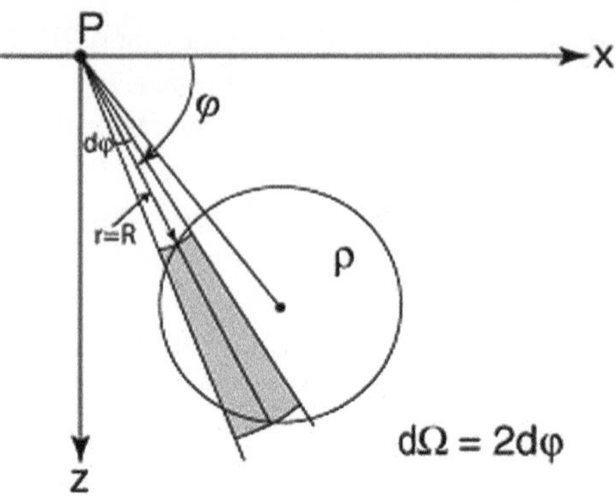

Fig. 2.9.12 Definition of symbols for $\Delta\Omega$ integration of δg_z of horizontal 2D cylindrical bodies of smooth cross section (not necessarily circular cylinders)

The rod or line-parallel effect, δg_y, along the infinite dimension has been calculated in Sect. (2.9.3.1) as (Eqs. 2.9.10, 2.9.11), however, for symmetry from $-\infty$ to $+\infty$, both sides cancel each other. The semi-infinite case, as in (Eqs. 2.9.10, 2.9.11), is not 2D and does not belong here.

The other line-parallel component is $d\delta g_x$ which can be formulated in analogy to (Eq. 2.9.49) with $dV = dx\,ds = dx\,d\Omega\, r^3/x = 2dx\,d\varphi\, r^3/x$, hence the horizontal line-normal effect is $d\delta g_x = 2G\rho\,dx\,d\varphi$. The line extends vertically along z and has been formulated in Sect. 2.9.3.1 – Vertical mass line: (Eq. 2.9.10) and integrated as (Eq. 2.9.11): $d\delta g_p = d\delta g_z = G\rho\,[1/r_1 - 1/r_2]\,dx\,dy$. The plane-parallel component is needed for defining the vector $\boldsymbol{\delta g}$.

2.9.7.1.2 Planar Mass Elements

Planar mass elements normal or parallel to gravity, as in 3D modelling, render the effects δg_n and δg_p, or in x, y, z: δg_z and δg_x, while $\delta g_y = 0$. The derivation is based on the solid angle and the rod integration. The plane normal effect is the integral (Eq. 2.9.49) over the finite solid angle $\int d\varphi = \Delta\varphi$ where $\Delta\varphi = \varphi_2 - \varphi_1$, $\varphi_1 = \arctan(x_1/z) = \arcsin(x_1/r_1)$, $\varphi_2 = \arctan(x_2/z) = \arcsin(x_2/r_2)$, $r_1 = (x_1^2 + z^2)^{1/2}$, $r_2 = (x_2^2 + z^2)^{1/2}$:

$$\delta g_n = \delta g_z = 2G\rho\,dz\Delta\varphi \qquad (2.9.50)$$

The plane-parallel effect is obtained by integration of $G\rho\,[1/r_1 - 1/r_2]\,dx\,dy$ over y from $-\infty$ to $+\infty$, with distance to top $r_1 = (x^2 + z_1^2)^{1/2}$ and distance to bottom $r_2 = (x^2 + z_2^2)^{1/2}$:

$$\delta g_p = \delta g_z = 2G\rho\,\ln(r_2/r_1)\,dx. \qquad (2.9.51)$$

see also (Sect. 2.9.3.4) for the effect of a finite rectangular wall (Eq. 2.9.21) extended to infinity in one horizontal direction.

2.9.7.2   Two-Dimensional Horizontal Disks

The two-dimensional horizontal disk is identical to the above uniform planar strip, except for a finite thickness. The strip is a trapezoid defined by two horizontal planes, separated by $\Delta z = t$ and two planes through P and the $\underline{y}$ axis intersecting at an angle of $\Delta\varphi$. Then $\Delta\Omega = 2\Delta\varphi$, and:

$$\delta g = 2G\rho\,\Delta\varphi\Delta z. \tag{2.9.52}$$

For the Bouguer slab $\Delta\varphi = \pi$. Strips are simple means for quick estimates of gravity effects (see Chap. 5 & Sect. 6.2.1.1 (3)).

2.9.7.3   Hollow Structures: Pipes

Hollow structures or pipes are built of contagious finite or infinitesimal narrow strips enclosing an area or two-dimensional volume. The formulations for rectangular, polygonal and smoothly curved cross sections have mostly been derived above.

2.9.7.3.1   2D Rectangular Cross Sections

Rectangular pipes consist of two element types: horizontal and vertical strips for which the plane-normal and plane-parallel components, respectively, are given by (Eqs. 2.9.50 & 2.9.51); the thickness of the pipe wall may be assumed finite t, where t is much smaller than any of the dimensions. Rule 6 could be applied by beginning with a pipe with P at its upper left-hand corner and subtracting and adding appropriate pipes of the same type but of different dimensions, however, it is of no advantage in this case. Instead, the pipe is here defined by x_1, x_2 and z_1, z_2 and the corner radii are $r_{11} = (x_1{}^2 + z_1{}^2)^{1/2}$, $r_{12} = (x_1{}^2 + z_2{}^2)^{1/2}$, $r_{21} = (x_2{}^2 + z_1{}^2)^{1/2}$, $r_{22} = (x_2{}^2 + z_2{}^2)^{1/2}$, such that Eqs. (2.9.50) & (2.9.51) give for the top, bottom and two side strips with the indices $i, j = 1, 2$ of the corners x_i, z_j:

$$\delta g_{z(pipe)} = 2G\rho\,t[(\varphi_{11} - \varphi_{21} + \varphi_{12} - \varphi_{22}) + \ln[r_{12}r_{22}/r_{11}r_{21})] \tag{2.9.53}$$

Such pipes are artificial but may be used to model structures built of shells with systematically varying density.

2.9.7.3.2   2D Polygonal Cross Sections

Polygonal cross sections describe 2D polyhedra, and the treatment is analogous to that of 3D polyhedra. The faces are strips of arbitrary orientation. Therefore plane-normal and plane-parallel components are derived for each face, in order to enable the vector projection onto the Earth-bound x, y, z system to be calculated.

As the section, i.e. its corner points x_k, z_k are originally given in this system, for each face the coordinates are first rotated about the y axis such that X is parallel to the face and Z is normal to it (see Sect. 2.9.4.2, where rotation about the z axis is described).

The plane-normal and parallel components (δg_Z, δg_X) are calculated in the rotated system with Eqs. (2.9.50) & (2.9.51) and then back-rotated into the Earth-bound x, y, z system. It is important to simply reverse the original rotation to avoid ambiguity. Finally the contributions (δg_x, δg_z) are summed for the whole polygonal section.

2.9.7.3.3 Arbitrarily Shaped Pipes of Continuous Smooth Cross Section

For pipes of surface density $\rho\, t$ and smooth cross section the procedure is the same as in 3D shells. The infinitesimal 2D surface strip mass element $\rho\, t\, \boldsymbol{ds}$ continuously changes direction along S (Fig. 2.9.13, compare Figs. 2.9.12 & 2.9.10b, where Δh corresponds to t). Because of the smallness of ds, the infinitesimal gravitational vector effect $\boldsymbol{d\delta g}$ pointing toward P, has the direction $-\boldsymbol{r}$, given by the centre of gravity of ds, such that $d\Omega = ds\cos\psi/r^2$:

$$\boldsymbol{d\delta g}^{(\mathrm{r})} = \mathrm{G}\rho t\, ds/r^2 = \mathrm{G}\rho t d\Omega/\cos\psi = 2\mathrm{G}\rho t d\varphi/\cos\psi \qquad (2.9.54)$$

where φ is the angle of $\boldsymbol{r}$ with x. Integration by components with $\boldsymbol{r} = (x,0,z)$ and $\cos\varphi = x/r$ leads to:

$$\left.\begin{aligned}
\delta g_x &= -2\mathrm{G}\rho t \int_S (\cos\varphi/\cos\psi)d\varphi = -2\mathrm{G}\rho t \int_S x/(r\cos\psi)d\varphi \\
\delta g_y &= 0 \\
\delta g_z &= -2\mathrm{G}\rho t \int_S z/(r\cos\psi)d\varphi
\end{aligned}\right\} \qquad (2.9.55)$$

or, if φ_0 is known *a priori* as the angle of $\boldsymbol{\delta g}$ relative to $-\boldsymbol{r}$, as is the case of a circular cylinder, for example, where δg points to its axis:

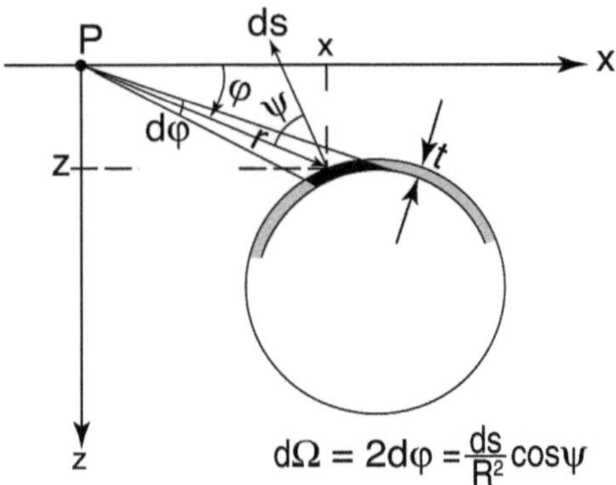

Fig. 2.9.13 Definition of symbols for $\Delta\Omega$ integration of δg_z of horizontal 2D cylindrical pipes of smooth cross section (not necessarily circular cylinders)

$$\delta g = G\rho\, t \int_S \cos\varphi_0 / \cos\psi\, d\Omega \qquad (2.9.56)$$

For other integrable mathematical expressions, the integration must be carried out component-wise.

2.9.7.4 Massive Beams and Steps

Massive beams are the most generally applicable 2D approximations of elongated geological bodies. The integration can be carried out in several ways. One is based on stacking planar mass elements (see Eqs. 2.9.50 & 2.9.51) defined by steps, vertical or oblique (Fig. 2.9.14). Integration over the full distance d from P to the plane under consideration now simply renders $\rho\, d$ (instead of $\rho\, t$ in the case of pipes).

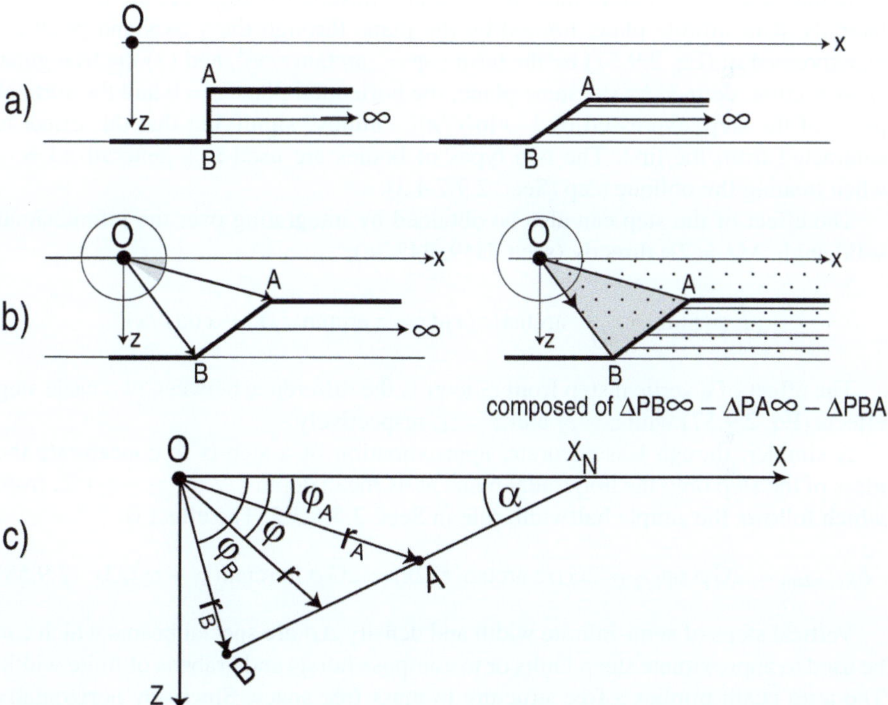

Fig. 2.9.14 Steps: (**a**) *left*: vertical; *right*: oblique; parametrization is by horizontal layers shown by horizontal hachures. (**b**) *Left*: the oblique step is defined by the inclined *straight line* (2D plane) between B and A as well as the horizontal planes $z = z_B$ and $z = z_A$; it is an element of a 2D polygonal cross section. *Right*: the step extending upward to $z = 0$ is the basic body of the solid angle approach between 0 and z_A or z_B; it is composed of: (1) the semi-infinite plate bounded to the *left* by the oblique radii from O to A and B, [OA∞] and [OB∞] (horizontal hachures, marking the degenerated triangles with one corner at infinity, and (2) the triangular beams [OAN] and [OBN] marked in *grey*. The effect of body (2) is subtracted from that of body (1). (**c**) Definition of symbols used in text

No singularity exists at P where $r \to 0$ as shown for the Poisson equation (2.7.11). Another way is to use complex integrands (see Sect. 2.9.7.4.3). Below, vertical steps, rectangular sections, oblique steps, polygonal sections and smoothly curved cross sections are presented in turn.

2.9.7.4.1 Vertical Steps

According to rule 6 the most basic body is the vertical step at position x from depth z to 0 which, alternatively can be considered a semi-infinite horizontal plate (Fig. 2.9.14a, left). Its effect, from (Eqs. 2.9.50 & 2.9.51), is:

$$\delta g^{(\text{step})} = 2G\rho[z\varphi - x\ln(r/x)] = 2G\rho[z\arctan(z/x) - x\ln(r/x)] \qquad (2.9.57)$$

with $r = (x^2 + z^2)^{1/2}$ and the distance to the step top at $(x, 0)$ is $r_1 = x$ in (Eq. 2.9.51). Note that (Eq. 2.9.57) combines two basically different bodies: (1) the horizontal, laterally semi-infinite plate, limited by the plane through the y axis and point (x, z), expressed in (Eq. 2.9.57) by the term $[z\varphi = z\arctan(z/x)]$, and (2) the triangular cross section, defined by the same plane, the horizontal plane $z = 0$ and the vertical plane of the step, expressed by $[-x\ln(r/x)]$, "minus" signifying that this effect is subtracted from the first. The two types of bodies are used in a generalized way when treating the oblique step (Sect. 2.9.7.4.3).

The effect of the step can also be obtained by integrating over the infinitesimal solid angle $\Delta\Omega = 2\varphi$ directly (see GH49, 342,3c):

$$\int \varphi(z)\mathrm{d}z = {}_0\!\!\int^z \arctan(z/x)\mathrm{d}x = z\arctan(z/x) - x\ln(r/x)$$

The effect of a vertical step from z_2 to z_1 is the difference between two basic step effects (Eq. 2.9.57) with $z = z_2$ and $z = z_1$, respectively.

A simpler, though less accurate, approximation of a step is to concentrate the mass of the step onto the horizontal plane at its mean depth $z/2$ or $(z_1 + z_2)/2$, from which follows the simple halfwidth rule in Sect. 2.5.6.3.2. The effect is:

$$\delta g_{\text{condens}} = 2G\rho\, z\varphi_{1/2} = 2G\rho\, z\arctan(z/2x) = 2G\rho\, z\arctan(z_1 + z_2)/2x \quad (2.9.58)$$

Vertical steps of semi-infinite width and density $\Delta\rho$ are special beams which can be used to approximate steep faults or to compose horsts and grabens of finite width. The term beam implies a free structure in mass-free space. Since any horizontally layered density distribution may be superimposed, the half space below may have the same density as the beam which thus becomes a horst, and if the density below is that of the surrounding mass, the beam becomes a graben. Furthermore, reversing the sign of the density contrast $\Delta\rho$ also reverses the tectonic nature of the beam. These considerations do not affect the calculations and equally apply to oblique steps and polygonal Sects. (see 2.9.7.4 & 2.9.7.4).

2.9.7.4.2 Rectangular Cross Sections

The effect of a rectangular beam is the difference between two basic steps (either from z to 0 or from z_2 to z_1), but can be derived also directly from the rectangular hollow pipe (Sect. 2.9.7.3).

The normal gravitational component generated by the beam is obtained from (Eq. 2.9.53) by replacing the constant wall thickness t by the appropriate thicknesses in depth z or horizontal distance x between P and the 4 faces and (instead of adding the effects of all 4 walls) by adding or subtracting the contributions related to each face such that the effective mass is only that enclosed by the rectangle:

$$\delta g_z = 2G\rho \left[z_2(\varphi_{12} - \varphi_{22}) - z_1(\varphi_{11} - \varphi_{21}) + x_2 \ln(r_{22}/r_{21}) - x_1 \ln(r_{12}/r_{11}) \right]$$
$$(2.9.59)$$

This can be rewritten as the sum of functions of the 4 corners $x_i z_j$, i, $j = 1$, 2, with

$$f_{ij} = (-1)^{i+j}(x_i \ln r_{ij} - z_j \varphi_{ij}) \, \delta g_{z \, \text{beam}} = 2G\rho \sum_i \sum_j f_{ij} \qquad (2.9.60)$$

2.9.7.4.3 Oblique Steps

The basic oblique step (Fig. 2.9.14b, left) is defined by an inclined straight plane face (line) between the points (x_B, z_B) and $(0, 0)$ and the horizontal planes $z = 0$ and $z = z_B$, extending to infinity. The effect of the step has been obtained by Talwani et al. (1959) by integrating over infinitesimally thin (mass density $\rho \, dz$) horizontal planes extending from x of the step face to ∞, expanding with P the solid angle $\Omega = 2\varphi = 2\arcsin(z/(x^2 + z^2)^{1/2})$: $_A\int^B \varphi \, dz$; φ is defined by the straight line (face) from $z = z_A$ to z_B. The step is the basic element of a 2D polygonal cross Sect. (2.9.7.4) and the corresponding method of forward gravity modelling is known as the "Talwani method".

The gravity effect of the step can be obtained directly from the solid angle (Fig. 2.9.14b, right). First, the most basic body is chosen as the step extending to $z = 0$. The notation [ABC] indicates the body of triangular cross section or simply 2D triangle defined by the corners A, B and C, including degenerated triangles with one corner at infinity. Physically and computationally, the step between depths z_B and z_A, [AB∞], can be considered as the difference between the two equivalent basic steps extending to $z = 0$, i.e. [NB∞] and [NA∞]. For each, (in the figure: [NB∞]) two bodies are considered: (1) the semi-infinite plate [OB∞] limited by the plane through O and B, and (2) the triangular beam [OBN], expanded by the step face and O. The effect of body (2) has to be subtracted from the effect of (1). The effect of (1) follows directly from the solid angle under which the body is seen from O, with $\Omega = 2\varphi$ and $\varphi = \arcsin(z_B/(x_B^2 + z_B^2)^{1/2})$: $\Omega = 2\varphi$, $\delta g_z^{(\text{plate})} = 2G\rho \, \varphi z_B$.

Body (2), i.e., the triangular beam [OBN] has been treated by analytical integration (KJ61, 201, Eq. (146)). Its effect is at $O = (0,0)$, with $\alpha = \arctan(z_B/(x_N - x_B))$ and $r_B = (x_B^2 + z_B^2)^{1/2}$, (for the symbols see Fig. 2.9.14c):

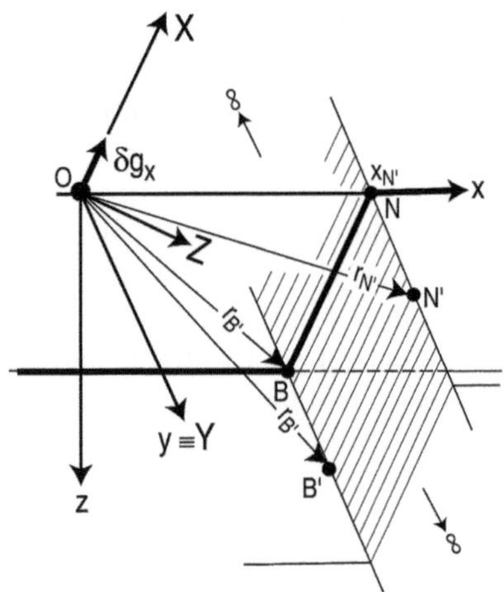

Fig. 2.9.15 Perspective view of the oblique step and coordinate systems: global x, y, z (z along $\boldsymbol{g}$) and X, Y, Z (X along step slope, $Y \equiv y$). The radii $r_{N'}$ and r_B appear only in the integration of the step along Y, not in the result of the calculation of the step slope-parallel component δg_X

$$\delta g_z^{(\text{tri})} = 2G\rho\, x_N \sin\alpha \left[\ln(r_B/x_N)\sin\alpha + \varphi_B \cos\alpha\right] \qquad (2.9.61)$$

Application of the Ω approach to [OBN] renders only the gravitational component normal to the step face such that also the face-parallel component must be determined by rod integration. This requires rotation of x, z about $y \equiv Y$ into X, Z: X is parallel to $\boldsymbol{BA}$ or $\boldsymbol{BN}$. The geometry of the integration scheme is illustrated in Fig. 2.9.15. The step face between B at z_B and N at $z = 0$, which lie in the x, z section ($y = 0$), is composed of rods along slope lines, generally expanded between B' and N' at arbitrary y values. Integration of the X-parallel rod effects from $y = 0$ to ∞ leads to the face-parallel effect as given by (Eq. 2.9.51):

$\delta g_X^{(\text{tri})} = 2G\rho\, Z_0 \ln(x_N/r_B)$ where $Z_0 = x_N \sin\alpha$ (Fig. 2.9.14c). The face-normal effect follows from (Eq. 2.9.50): $\delta g_Z^{(\text{tri})} = 2G\rho\, Z_0 \varphi_B$. Back projection or rotation renders:

$$\delta g_x^{(\text{tri})} = \delta g_X^{(\text{tri})} \cos\alpha + \delta g_Z^{(\text{tri})} \sin\alpha, \; \delta g_z^{(\text{tri})} = -\delta g_X^{(\text{tri})} \sin\alpha + \delta g_Z^{(\text{tri})} \cos\alpha, \; \text{hence}:$$

$$\delta g_z^{(\text{tri})} = 2G\rho\, Z_0 \left[\ln(r_B/x_N)\sin\alpha + \varphi_B \cos\alpha\right]$$

which is, with Z_0 inserted, identical to Eq. (2.9.61), and:

$$\delta g_z^{(\text{step})} = \delta g_z^{(\text{plate})} - \delta g_Z^{(\text{tri})} = 2G\rho\left[\varphi_B z_B - x_N \sin\alpha(\ln(x_N/r_B)\sin\alpha + \varphi_B\cos\alpha)\right]$$

$$= 2G\rho\, x_B \tan\alpha\left[\varphi_B(\sin^2\alpha - x_B/x_N) - \ln(x_N/r_B)\cos\alpha)\right].$$

Conversion to expressions of the corner point coordinates x_B, z_B, x_N is achieved by simple arithmetic operations. The expressions for steps [NB∞] and [NA∞] contain the identical quantities α and x_N.

Complex integrands. An alternative approach is to use complex integrands (KJ61, 199-204). The complex gravity effect $\delta g_c = \delta g_x - i\delta g_z$ ($i = \sqrt{-1}$) is a representation of the 2D vector:

$$\delta g_c = 2G\rho \int_x \int_z dxdz/(x+iz) = 2G\rho \int_\varphi \int_r r(\varphi)dr\,d\varphi/re^{i\varphi} = 2G\rho \int_\varphi r(\varphi)e^{-i\varphi}d\varphi.$$

$$(2.9.62)$$

(multiply enumerator and denominator with $x - iz$!). From the sine law (Fig. 2.9.14c):

$r = x_N \sin\alpha / \sin(\alpha + \varphi) = x_N \sin\alpha 2i/[\exp(i(\alpha + \varphi)) - \exp(-i(\alpha + \varphi))].$

Hence

$$\delta g_c = 2G\rho\, x_N \sin\alpha_{\varphi A} \int_{\varphi A}^{\varphi B} (2ie^{-i\varphi}d\varphi)/(e^{i(\alpha+\varphi)} - e^{-i(\alpha+\varphi)})$$

$$= 2G\rho\, x_N e^{-i\alpha} \int d(2i\varphi)/(e^{2i\varphi} - e^{-2i\alpha}) = e^{-i\alpha}I.$$

The integral I is of the type $\int dx/(e^x - a) = a^{-1}(\ln(e^x - a)x)$, thus, after some elementary steps:

$$I = \ln[(\sin(\alpha + \varphi_B)e^{2i\varphi B})/(\sin(\alpha + \varphi_A)e^{2i\varphi A})].$$

Invoking the sine law again:

$$\delta g_c = 2G\rho\, x_N \sin\alpha e^{i\alpha} \ln[r_B e^{i\varphi B}/(r_A e^{i\varphi A})] \qquad (2.9.63)$$

Separating the real and imaginary parts, we get

$$\delta g_z = 2G\rho\, x_N \sin\alpha[\sin\alpha \ln[r_B/r_A] + \cos\alpha(\varphi_B - \varphi_A)] \qquad (2.9.64)$$

This expression is identical to (Eq. 2.9.61) applied to the difference between the triangular beams: [OBN] – [OAN], where the identical quantities α and x_N cancel. For numerical calculations, r_B, r_A, φ_B, φ_A have to be expressed as functions of the coordinates x_A, z_A, x_B, z_B.

Which approach to the integration is preferred depends on personal taste. The solid angle approach has close affinity to complex analysis, as KJ61 remarks, which can be applied to calculating the gravitational vector effect $\delta\mathbf{g}$ and to projecting it on any component direction which, however, is not followed up here.

2.9.7.4.4 Polygonal Beams

Infinite 2D beams of polygonal cross section (degenerated 3D polyhedra) can be composed either of steps (1) or triangular beams (2) (Fig. 2.9.13b). In encircling the polygon, the external masses are removed. Which expression for the individual step is chosen is irrelevant, but while $\Delta\delta g_x^{(step)} \to \infty$, $\Delta\delta g_x^{(tri)}$ can be calculated. The faces, i.e. the polygon sides, are defined consecutively by x_k, z_k and

x_{k+1}, z_{k+1}. Each side is composed of two basic steps where $z_N = 0$: $\delta g_z^{(face,k)} = \delta g_z^{(step,k+1)} - \delta g_z^{(step,k)}$. The parameters in (Eq. 2.9.61) are $\alpha \to \alpha_k$ with $\tan \alpha_k = (z_{k+1} - z_k)/(x_{k+1} - x_k)$ and $x_N \to x_{Nk} = z_k(\tan \alpha + x_k/z_k)$. Hence:

$$\delta g_z^{(face,k)} = 2G\rho\, x_{Nk} \sin \alpha_k [\ln(r_{k+1}/r_k) \sin \alpha + (\varphi_{k+1} - \varphi_k) \cos \alpha] \qquad (2.9.65)$$

For each face k, i.e. $k \to k+1$, $\Delta\delta g_{Xk}$ and $\Delta\delta g_{Zk}$, are calculated, as above in face-oriented X, Z coordinates, transferred back to Earth-oriented x, z coordinates and added to the total sum. The polygonal cross section is encircled, for example, clockwise: $\delta g_z^{(polygon)} = \Sigma_k \delta g_z^{(face,k)}$. The procedure is repeated for all m stations at x_i, $i = 1$, n, after in each case the x coordinates are shifted, such that P is at 0, 0: $x_k = x_{k,global} - x_{i,global}$. – For detailed applications see Chap. 6.

2.9.7.4.5 Arbitrarily Shaped Beams with Smooth Cross Section

For the massive beam with arbitrary smooth boundaries (Fig. 2.9.12), the general equations are equivalent to those of the 3D case of a volume enclosed by an arbitrary closed surface S (Eqs. 2.9.47 & 2.9.48):

$$\left. \begin{array}{l} \delta g_x = 2G\rho \int_S r\cos \varphi_x d\varphi = -2G\rho \int_S x d\varphi \\ \delta g_y = 0 \\ \delta g_z = 2G\rho \int_S z d\varphi \end{array} \right\} \qquad (2.9.66)$$

If the vector direction is known (e.g., from P to the axis of a homogeneous horizontal spherical cylinder, with the angle α of $\boldsymbol{r}$ relative to $\boldsymbol{\Delta g}$:

$$\delta g = G\rho \int_S r\cos \alpha d\Omega \qquad (2.9.67)$$

Evaluation is possible for curves c for which $z = z(\varphi)$ can be expressed analytically or numerically. They can, of course, be generalized to any wanted oblique component of the attraction vector at P by rotating the coordinates accordingly and replacing $z(\varphi)$ in (Eq. 2.9.66) by the distance $h(\varphi)$ in the rotated coordinates.

2.9.8 Two-and-a-half Dimensional Models ($2\frac{1}{2}$D)

Two-dimensionality leads to significant errors of gravity calculation, if the length over depth ratio is not large. It is therefore advisable to estimate the errors. For this purpose, one can calculate the end corrections for finite length λ, i.e. by the factor a (Eq. 2.9.8) or construct simple 3D models for comparison.

It is tempting to save the simplicity of the 2D modelling by formulations generally applying end corrections of the 2D models and performing what has been called $21/2$ dimensional modelling. However, while infinite length is never realistic, the limited lengths are usually difficult to define. The lateral extent of geological structures is often roughly known or can be inferred from the gravity maps, but it is generally uncertain what lies beyond the limited body and usually varies in three dimensions with significant gravity effects. In many cases it is best to immediately construct 3D models, but these aspects are the topic of Chap. 6.

Remark 1

It should be noted that the Bouguer slab as any infinite mass plane generates a finite gravity effect proportional to its density ρ and thickness Δz, i.e. constant and independent from h above the plane. Hence, for such a slab the potential is simply $\delta U =_z \int^{z+h} \delta g \mathrm{d}z = 2\pi G \rho\, h \Delta z$. A laterally homogeneous mass, for example, one-dimensional $\rho(z)$, varying only with depth z, generates no gravity *variations* and is not seen and, hence, neglected. Since this is true only for lateral infinity, one must be cautious in the real world. The situation is the same on a perfect sphere with only radial density variation $\rho(r)$; even the ellipticity of the earth leads to deviations from the rule (Clairot's theorem), but for local to regional studies, the effects are negligible.

Note, however, that the gravitational invisibility of $\rho(z)$ or $\rho(r)$ is true only for observations above the masses. It is not true for stations inside the masses, for example, in mines or basements of buildings below the nearby mean earth's surface. And, of course, it is never true for lateral variations as, for example, the topographic relief.

Remark 2

Çavşak's (1992) integration of δg_z for *polyhedra* is based on the basic tetrahedra expanded from P to the arbitrarily oriented plane triangles (corners A, B, C, equivalent to vectors $\boldsymbol{A}, \boldsymbol{B}, \boldsymbol{C}$) taken as the basic mass elements $\rho \Delta V$. First the potential δU of the mass element is calculated in a suitable Cartesian coordinate system (X, Y, Z) before $\delta g = \partial \delta U / \partial z$ is derived. X is chosen parallel to the side $\boldsymbol{AB}$, Z parallel to $\boldsymbol{AB} \times \boldsymbol{BC}$ and Y normal to X and Z, i.e. parallel to the plane ABC. Integration is then fairly simple, being similar to the solid angle approach. To derive δg_z requires a rotational coordinate transformation (2.3.3.1) from (X, Y, Z) back to (x, y, z), for which we need the matrix of the components of the vector $\boldsymbol{x} = (x, y, z)$ or $x_i (i = 1, 2, 3)$ in the $\boldsymbol{X} = (X, Y, Z)$ or X_k ($k = 1, 2, 3$) system; the matrix elements are $\cos(\boldsymbol{x}_i, \boldsymbol{X}_k)$ of the angles between all $\boldsymbol{x}_i, \boldsymbol{X}_k$. Since the $\boldsymbol{X}_k$ are defined in (x, y, z), their x, y, z components $\cos(\boldsymbol{x}_i, \boldsymbol{X}_k) = \cos(\boldsymbol{X}_k, \boldsymbol{x}_i)$ are known. Numerical routines for elementary vector and tensor (or matrix) operations facilitate the calculations. The potential and gravity effects (δU, δg) of a polyhedron of triangles are derived by summing the contributions of all tetrahedra with a proper sign convention. Each edge separates two triangles and occurs thus twice. The final expression is principally the sum of functions of all corner points, i.e. their x, y, z coordinates, with the sign depending on the orientation of each triangle or the sign of the scalar product of $\boldsymbol{r} \cdot \boldsymbol{n}$, where $\boldsymbol{n}$ is the outward surface normal vector. Details are in the dissertation by Çavşak (1992).

2.10 Some Theoretical Aspects of Anomaly Analysis

2.10.1 Goals of Post-reduction Data Analysis

The anomalies need to be analyzed further in view of the ultimate goal of interpretation. Again, it is mentioned that the subject are continuous fields which in practice are derived from discrete point values and that this is not trivial (see Sects. 1.4, discussion in 5.15) and must be kept in mind throughout this whole section which assumes the existence of fields.

Some of the important theoretical relationships are presented here that are to be exploited along the path from observations to models. Humans are guided by ideas and images. Images are continua in colour or contoured. Visualization on the basis of discrete data requires transformation of the discrete data to tractable representations (see Chap. 5). For the start graphical methods are valuable, but analytical methods are often more efficient. The latter usually require regular data grids or spatial series (like time series), but observations are only rarely obtained at regularly spaced stations. Conversion to equidistant data values, either along profiles (one-dimensional) or on grids (two-dimensional) by interpolation, however, leads to some loss of authenticity and accuracy.

Mathematical and numerical tractability is facilitated by expansion into continuous functions of the coordinates or series of functions. Fitting functions to the data by minimizing the residuals is a common procedure where the residuals are considered random errors. Continuous functions permit various mathematical and numerical treatments, for instance, filtering and transformation into other field quantities. Alternatively a stochastic approach may be taken by which the most likely values are estimated for a continuum of points.

There are many suitable functions. In some cases single functions may fit desired features of the data. Functional families or series of functions, systematically ordered, are more flexible in describing arbitrary spatial variations. Series expansion is a formal procedure which requires convergence. Examples are polynomials, the orthogonal harmonic series of sine and cosine functions; they are of infinite extent, while wavelets are defined only in finite domains and vanish outside; other more specialized series may be chosen. Orthogonality guarantees that the included terms are independent from each other.

Series expansion of data serves more purposes. *Smoothing* of inaccurate (scattered) observations with errors of no geophysical interest is a useful application. An important application is the *exploitation of the properties of the functions*, for instance, of their known space derivatives. This is, for instance, true for the harmonic functions $\sin kx$, $\cos kx$ which permit easy upward and downward continuation (see below).

Functions can be chosen especially to express gravimetric effects of certain types of geological mass distributions and can be applied to gravity analysis. Then fitting becomes *inversion* or modelling (Chap. 7). Such methods of interpretation have been termed "direct" because a single mathematical operation renders a source model, in contrast to the "indirect" methods of trial and error (see Chap. 5). Actually, all functions chosen for gravity analysis have the aspect of representing idealized source distributions which are more or less capable of describing geological structures (see remarks below). A caveat about all interpretative efforts is that the ambiguity is, and remains, principal and unsolvable.

Functional series representation of the anomalies opens, in addition, new totally different *functional domains*, for example, as frequency spectra of amplitude and phase, instead of spatial variations. The anomaly description in the spectral domain is as valid as that in the space domain if complete transformation between the domains is possible both ways. Spatial spectra are not so familiar as acoustic spectra of time functions which we directly hear. Spectral qualities in geometrical space are, e.g., smoothness, roughness, waviness etc., Moreover, one can learn, and spatial spectra can be used in data analysis and interpretation, for example, for arbitrary spectral *filtering* (by wavelengths or scales) and frequency analysis, as customary in acoustics or seismology.

This section begins with the usual concepts of smoothing and of deterministic functions as polynomials, Fourier series, spherical harmonics, wavelets as well as a few geophysically motivated special functions. The theory is applied to practical tasks of data reduction and analysis in Chap. 4 as preparation for interpretation and inversion (Chaps. 5, 6, 7). For an exhaustive treatment the reader is referred to texts quoted below. The approach involves probabilistic and stochastic aspects as common averaging and smoothing, but a more general stochastic treatment of field representation and interpretation is briefly discussed at the end (Sect. 2.10.9).

2.10.2 Smoothing of Spatial Series

Dividing a data set into a smooth component and irregular residuals generally depends on subjective judgement on what is considered a "random" error or "noise" and what a meaningful signal. Errors are considered irrelevant to the study but must be judged critically. They may be of observational nature and/or stem from uncontrollable local density variations. What is called "local anomalies" in contrast to "regional anomalies" (which one may want to isolate from them) does not belong into the realm of random errors. The related operations of smoothing and filtering do serve the separation of different components of the data or the field, but as such these operations can usually not distinguish noise from signals.

Simple smoothing in the above sense can be achieved by several methods (Sects. 4.8.3 & 4.8.4) including various spatial averaging schemes or by functional fitting. They are usually chosen by intuition and experience, and the behaviour of the residuals should be analysed *a posteriori*. Overlapping averaging schemes usually apply some spatial weighting function. In spectral filter theory (Sect. 2.10.5) the procedure is, as any filtering, "convolution" of the anomaly $\delta g(x)$ or $\delta g(x,y)$ with the space-shifted weighting function $w(x)$ or $w(x,y)$. The filtered value δg^* is given by the cross covariance function, *KOV*, of $\delta g(x)$ and $w(-x)$:

$$\delta g^*(x) = \int_{-\infty}^{\infty} w(x)\delta g(x-\xi)\mathrm{d}\xi = \int_{-\infty}^{\infty} w(x-\xi)\delta g(x)\mathrm{d}\xi = w(x)_*\delta g(x)$$

$$= KOV(w,\delta g) \qquad\qquad (2.10.1)$$

In the spectral domain (see Sect. 2.10.5), convolution is transformed into a multiplication of the (complete) spectra of the weighting function and the anomaly; "complete" means amplitude *and* phase, for example, in complex form, with $k = 2\pi/L$:

$$\underline{\delta g}^*(k) = W(k) \bullet \delta g(k) \qquad\qquad (2.10.2)$$

A famous continuous filter function $w(x,y)$ is the Gaussian $\exp[-(x^2+y^2)/4L^2]$, the Fourier transform (see Sect. 2.10.5) of which is also a Gaussian: $W(k_x,k_y,L) = \exp[-k^2L^2]$, where $k^2 = k_x^2 + k_y^2$, and L characterizes the width of the filter (in the sense of the standard deviation). By choosing k_x and k_y differently, one can design anisotropic filters. The discrete case is treated in Sect. 4.7.4.2.

Triangular filter functions have been used for simplicity: $f(x) = x/B^2 + 1/B(-B \le x \le 0)$ and $-x/B^2 + 1/B(0 \le x \le B)$; everywhere outside $f(x) = 0$. The convolution integral leads to $W(k) = [\sin(kB/2)/(kB/2)]^2$, i.e. the squared $\sin k/k$ function with zero crossings at $k = 2n\pi/B$ or $\lambda = B/n(n \ne 0)$.

Thus, filter theory permits a better insight into the behaviour of overlapping averaging, and (Eqs. 2.10.1) and especially (2.10.2) reveal the essential spectral properties of the averaging schemes.

2.10.3 Polynomials

Traditional polynomial fitting renders coefficients with mostly no physical meaning, because ordinary polynomials are non-orthogonal. For example, while in a linear polynomial as $y = a + bx$, b is the physical mean gradient of the data y, b looses this meaning in $y = a + bx + cx^2$. Therefore the theoretical basis is only formal. Nevertheless, gravity anomaly profiles (δg_i values at locations x_i) can be expressed as: $\delta g(x) = a_0 + a_1 x + a_2 x^2 + a_3 x^3 + \ldots + r(x)$, where the a_k are constant coefficients and r is the residual, but as stated, the coefficients change when the polynomial degree is changed. If the r_i values can be assumed to behave as random errors e_i, one may determine the a_k from the least-squares condition $\Phi^2 = \sum_i e_i^2 = \min$ by solving the linear normal equations (see Chap. 7). If a low maximum degree is assumed, the polynomial may be arbitrarily taken as a regional field with residuals representing the local anomalies. But the latter will generally not be random as error theory demands and because of the non-orthogonality a high-degree polynomial expansion cannot be separated into a residual high-degree part and a regional low-degree part. In any case, note the many caveats concerning regional-residual separation in later chapters.

Two-dimensional polynomials are equivalent and can be written as:

$$\delta g(\mathrm{x},\mathrm{y}) = a_{00} + a_{10}x + a_{20}x^2 + a_{30}x^3 + \dots$$
$$+ a_{01}y + a_{11}xy + a_{21}x^2y + \dots$$
$$+ a_{02}y^2 + a_{12}xy^2 + a_{22}x^2y^2 +$$
$$+ a_{03}y^3 + \dots\dots\dots\dots\dots + r(x,y) \qquad (2.10.3)$$

fully analogous to the one-dimensional case.

Polynomials can be *orthogonalized* by enforcing the condition of orthogonality: integrals over the study region (or sums over all occupied coordinates) of mixed products of any of the polynomial terms (or any functions applied) must vanish which defines a set of equations for orthogonalizing coefficients. It is facilitated if the points are on a regular grid. The condition that the interval integral or sum over the squared orthogonalized functions be 1 normalizes the coefficients. The normal equations (Chap. 7) resolve into independent equations for each unknown for the fitting of $\delta g(x)$ or $\delta g(x,y)$. The extra work is not justified, if effective computing facilities offer more general functions for fitting gravity anomalies (see below). Special cases are the Legendre polynomials and the Hermite polynomials treated in Sect. 2.10.6, below.

2.10.4 The Field Quantities: Differentiation and Integration

The gravity field, usually observed as δg (not the vector), can be expressed in the form of other field quantities, for example, the potential δW and its higher derivatives (e.g. gravity tensor observed by gradiometers or torsion balances). As outlined in Sect. 2.7.2, on the basis of the Laplace equation, $\nabla^2 W = 0$, in source-free space, the other quantities can be derived from boundary values. Three approaches presented in 2.6.4 are (1) a geometrical one related to the equipotential surface, (2) application of equivalent mass distributions, (3) expanding the observations, for example, by Fourier and other series, and exploiting the known derivatives or integrals of these functions. The *operations* involve numerical differentiations and integrations of the given data sets (irregular or gridded).

The anomalous potential at the origin $(0,0,0)$, z_0 above an *equivalent stratum* in Cartesian coordinates, with $r = (x^2 + y^2 + z_0{}^2)^{1/2}$ and $\Delta\rho^*(x,y,z_0) = \delta g/(2\pi G)$ (Eq. 2.10.3), is:

$$\delta W(0,0,0) = G \int_{-\infty}^{\infty} \int_{-\infty}^{\infty} (\Delta\rho^*/r)\,\mathrm{d}x\,\mathrm{d}y, \qquad (2.10.4)$$

or in vertical cylinder coordinates and dropping the asterisk it is, with $r = (R^2 + Z_0{}^2)^{1/2}$:

$$\delta W(0,0,0) = G \int_{0}^{2\pi} \int_{0}^{\infty} (\Delta\rho(R,\lambda,Z_0)/r)\,\mathrm{d}R\,R\,\mathrm{d}\lambda \qquad (2.10.5)$$

and the circular average $\Delta\rho^{\#} = (2\pi R)^{-1}\int_0^{2\pi}\Delta\rho^*(R,\lambda,Z_o)d\lambda$:

$$\delta W(0,0,0) = G\int_o^{\infty}\Delta\rho^{\#}(R,Z_o)dR = (2\pi)^{-1}\int_o^{\infty}\delta g^{\#}(R,Z_o)dR. \qquad (2.10.6)$$

This operation is an upward continuation of the field, away from the source into mass-free space, here in terms of the anomalous potential which can be converted to gravity δW_z by differentiation. But δW_z can also be derived directly in an analogous manner. However, in the inverse direction approaching the source or into the source volume is not possible in this general form; it is solved below by invoking the harmonic functions (Sect. 2.10.5).

There is no general analytical solution for the integrals over arbitrary $\delta g^{\#}$. The potential increment $d\delta W = 2\pi G\Delta\rho^* dz = \delta g dz$ for continuous δg, immediately above the equivalent stratum is locally proportional to δg; the second derivatives δW_{xx}, δW_{yy} and δW_{zz} describe the curvature (or shape) of the equipotential surfaces (Sect. 2.7.4). In the close neighbourhood to the equivalent stratum δW and δW_z are correlated with $\delta\rho^*$, i.e. not phase shifted, and $\delta W_{zz} = 0$, but as one moves away from the stratum, δW_{zz} grows and its value is also correlated to $\delta\rho^*$. This permits the relative curvature of δW to be estimated numerically from second differences (or more sophisticated schemes). Only for demonstration, the one-dimensional case of $\delta g = \delta W_z$ leads to $\delta W_{xx} \propto (\delta W_{i-1} + \delta W_{i+1} - 2\delta W_i)/\Delta x^2$. Curvature can also be estimated by functional fitting; one possibility is the function describing the effect of a point mass which is discussed for illustration purposes: assume that locally the linear trend of $\delta g(x,y)$ is removed and that the effect of a point source it fitted (depth z and mass m determined), then all the local field quantities are $\delta W_z = Gmz/r^3$, $\delta W = Gm/r$, $\delta W_{zz} = Gm(r^2 - 3z^2)/r^5$, and analysis could go on from here. The most popular kind, however, are the harmonic functions (see below).

2.10.5 Harmonic Functions

2.10.5.1 Introduction to Plane Harmonic Functions

In many ways, series of harmonic functions are especially suited to describe variations in space. As is the case with the point mass effect (this is always true if the source is known), all field quantities are automatically defined in mass-free space (through the Laplace equation; but again the equivalent stratum is the known source). Through the Fourier theorem, any space variation, for example, at the surface, can be described as a Fourier series. Moreover, harmonic functions are orthogonal, the elements of the function family are independent, so that inclusion or deletion of any element (of a certain frequency) does not alter the amplitudes and phases of the other included elements. Along profiles with one variable x, anomalies $\delta g(x)$ are analogues of time series and can be treated the same way by employing series of $\sin(jkx)$ and $\cos(jkx)$. The corresponding 2D density variation at depth would consist of distributions in x and z. Extension to more than one variable, e.g. x, y, is easy, and two-dimensional harmonic descriptions of $\delta g(x,y)$ would correspond to 3D density distributions $\rho(x,y,z)$. Time series analysis has become a standard procedure (Morse & Feshbach, 1953; Byerly, 1959; Moon & Spencer, 1988; Klingen, 2001).

The Fourier theorem states that a physical function varying periodically with wavelength λ or angular frequency $k = 2\pi/\lambda$ can be expressed as a sum or series of sinusoidal components of frequencies k, $2k$, $3k$, $4k$, etc.: $\delta g(x) \approx {}_j\Sigma_0^\infty A_j \cos(jkx + \varphi_j) = {}_j\Sigma_0^\infty (a_j \cos(jkx) + b_j \sin(jkx))$, each wave of the series having its own amplitude A_j and phase φ_j. The two sets of parameters or coefficients are related by:

$$A = (a^2 + b^2)^{1/2} \text{ and } \tan\varphi = b/a.$$

The functional fit lies in the calculation of the coefficients a_j, b_j; the procedure is also called "Fourier analysis" or "frequency analysis":

$$a_j = (2/\lambda) \int_0^\lambda \delta g(x) \cos(jkx)\mathrm{d}x$$

$$b_j = (2/\lambda) \int_0^\lambda \delta g(x) \sin(jkx)\mathrm{d}x \qquad (2.10.7)$$

The harmonic functions can be expressed through Euler's formula by complex exponential functions where phase and amplitude are coded by the two "coordinates", real and imaginary:

$$e^{ix} = \cos x + i\sin x, \text{ and } e^{-ix} = \cos x - i\sin x.$$

The reverse is Moivre's formula:

$$\cos x = (e^{ix} + e^{-ix})/2 \text{ and } \sin x = (e^{ix} - e^{-ix})/(2i).$$

The Fourier series is written with the complex coefficients C_j, as:

$$\delta g(x) \approx {}_j\sum_0^\infty C_j e^{ijkx} = {}_j\sum_0^\infty C_j \exp(ik_j x) \qquad (2.10.8)$$

with $jk \equiv k_j$. Again, the coefficients:

$$C_j = (1/\lambda) \int_0^\lambda \delta g(x) \exp(-ik_j x)\mathrm{d}x \qquad (2.10.9)$$

represent the spectral domain. To be useful, the fit of arbitrary gravity variations is, of course, required to converge which is no problem for wavy fields but is for edgy features as abrupt steps. The above expressions are valid for continuous functions $\delta g(x)$. In practice the integrals are numerically evaluated as sums over equidistant discrete points (see Sect. 4.7.4) The description is valid in Cartesian space. For planetary-scale problems, in spherical coordinates, the analogue are the spherical harmonics (see below).

The fact that gravity anomalies $\delta g(x)$ are generally non-periodic and defined in limited intervals, while the harmonic functions are unlimited or infinite, has no influence on the series expression within the given interval of fundamental wavelength

λ_o outside which the anomalies are neglected, but problems arise in modelling or upward continuation (see below). One can also extend the fundamental wavelength λ_o to infinity (but that does not change the fact that the data are limited). Then the Fourier series becomes the Fourier integral:

$$\delta g(x) = \int_{-\infty}^{\infty} F(k)e^{ikx}dk \qquad (2.10.10)$$

where the coefficients C_j become a continuous function $F(k)$ and $\delta g(x)$ is expressed as the convolution of $F(k)$ with e^{ikx}.

As mentioned, the anomaly spectrum $F(k)$ in the frequency domain is an alternative view of an anomaly. The relation between the space domain and the frequency domain is expressed by the above Fourier integral. The inverse, i.e. the definition of the complex spectrum $F(k)$ from the given spatial function $\delta g(x)$ is expressed by the inverse Fourier integral which is called the Fourier-transform, i.e. the convolution of $\delta g(x)$ with e^{-ikx}:

$$F(k) = (2\pi)^{-1} \int_{-\infty}^{\infty} \delta g(x)e^{-ikx}dk \qquad (2.10.11)$$

Negative wave numbers refer to waves in negative x direction. In the case of gravity, negative wave numbers are superfluous or redundant (in the above sine and cosine series: $a_{-j} = a_j$, $b_{-j} = b_j$, $b_o = 0$), and it is sufficient to consider only waves in x direction. The integration is then performed only from 0 to ∞.

Of course, infinity is an unrealistic idealisation, and neither is $\delta g(x)$ defined between $-\infty$ and $+\infty$, nor is it known as an analytically integrable function $f(x)$. A fundamental wavelength λ_o with the wave number $k_o = 2\pi/\lambda_o$ and its integer multiples k_1, k_2, limits the spatial resolution. Integration is approximated by finite sums and efficient numerical schemes exist (Chap. 4).

In upward continuation the spatially infinite periodicity of the harmonic functions implies an unrealistic existence of gravity variations and their sources outside the interval of definition, and that affects the results. It is sometimes considered as the main problem posed by the use of harmonic functions. However, any other unrealistic assumption about the anomalies and their sources outside the interval of definition creates corresponding problems. In this respect, the limited wavelets do not really solve the fundamental problem of truncation in space, time or frequency since they imply zero gravity outside the region of study which is equally unrealistic. But generally, it is possible and advisable to test the effects of the truncations in space and spectrum of given gravity (or other) data.

2.10.5.2 Differentiation and Integration of Harmonic Components

In source-free space the order of integration and differentiation can be changed, and the mathematical operations on the given anomaly as a function in space can be carried out on the harmonic components before subsequent synthesis of the function.

The practical advantages are enormous because the solutions are known for the harmonic functions. Essentially it means that the outer field generated by a harmonic source is completely defined analytically and this is exploited. Caution must be exercised, however, with respect to convergence and resolution.

Differentiation after the explicit coordinates x (and y) is straightforward and leads only to a change in amplitude and phase. The argument kx (or $k_x x$ and $k_y y$) results in factors of k (k_x and/or k_y) for the first derivative and $k^2(= k_x{}^2 + k_y{}^2)$ for the second derivative. Integration has the opposite effect, i.e. the factor is $1/k$. Factors of i, $-i$ or -1 and exchange of sine and cosine describe phase shifts of $\pm\pi/2$ and $\pm\pi$.

δW at the equivalent stratum (depth z), for simplicity, is $\delta W(x,0,z) = -1/k\,\cos(kx)$,

$$\delta W_x(x,0,z) = \sin(kx),\ \delta W_{xx}(x,y,z) = k\cos(kx) \tag{2.10.12}$$

Differentiation after z cannot be obtained directly from $\delta g(x,y)$ with no explicit z dependency at the surface $z = $ const, but the Laplace equation in free space, $\delta W_{xx} + \delta W_{yy} + \delta W_{zz} = 0$, implies the vertical derivative through the horizontal derivatives. At level $z = 0$, i.e. elevated by z above the surface at z, with (2.10.15):

$$\delta W_z(x,0,0) = \mathrm{e}^{-kz}\cos kx,\ \delta W_{zz}(x,0,0) = -k\mathrm{e}^{-kz}\cos kx \tag{2.10.13}$$

where, for $z = 0$, $\mathrm{e}^{-kz} = 1$, and the equivalent stratum at depth z has a surface density $\rho_0^* = \cos(kx)/(2\pi G)$. The Laplace equation at the surface is obviously satisfied as $\nabla\delta W = \delta W_{xx} + \delta W_{yy} + \delta W_{zz} = k\cos(kx) + 0 - k\cos(kx) = 0$. It is nearly as easy to show this for two-dimensional harmonic fields.

2.10.5.3 Upward and Downward Continuation

Upward continuation into mass-free space has been mentioned above in connection with the Laplace equation and the harmonic functions. It is a frequent task in gravity field analysis and interpretation, and it is an unambiguous forward problem. The inversion, i.e. the downward continuation is the real problem, most naturally solved by expanding gravity anomalies into harmonic series (Fourier transform).

The basic operations are demonstrated for an individual component $F(k)\mathrm{e}^{ikx}$ of a one-dimensional series (in two dimensions x, y: $F(k)\exp(i(k_x x + k_y y)) = F(k)\exp(ik_x x)\exp(ik_y y)$, with $k = (k_x{}^2 + k_y{}^2)^{1/2}$).

Starting from the convolution integrals (similar to the above) over the equivalent stratum:

$\delta W_z(0,0,0) = (z_0/2\pi)\int_{-\infty}^{\infty}\int_{-\infty}^{\infty}(\delta g(x,y,z_0)/r^3)\mathrm{d}x\,\mathrm{d}y$ with $r = (x^2 + y^2 + z_0{}^2)^{1/2}$, and assuming $\delta g(x,y,z_0) = \mathrm{e}^{ikx}$, we write in vertical cylinder coordinates ($x = r\cos\varphi$, $y = r\sin\varphi$):

$$\delta W_z(0,0,0) = (z_0/2\pi)\int_0^{2\pi}\int_0^{\infty}(\mathrm{e}^{ik\cos\varphi r}/(r^2 + z_0^2)^{3/2})\mathrm{d}r\,r\,\mathrm{d}\varphi \text{ with } r = (R^2 + z_0^2)^{1/2},$$

$$\tag{2.10.14}$$

and the φ integral has the solution $2\pi J_0(k,r)$ with the Bessel function of order zero $J_0(k,r)$. This can be shown by expanding the exponential function into series and integrating by terms. The r integration from $-\infty$ to $+\infty$ is carried out similarly and leads to the final and general solution (where the index 0 of z_0 is dropped, and remember: P is elevated by z above the source):

$$\delta g(\mathrm{x},0,0) = z\mathrm{e}^{\mathrm{ikx}}\mathrm{e}^{-kz}/z = \mathrm{e}^{\mathrm{ikx}}\mathrm{e}^{-kz}. \tag{2.10.15}$$

It means that the amplitude of the sine wave decays exponentially with distance z from the equivalent stratum and equally with the wave number $k = 2\pi/\lambda$, and naturally no phase shift occurs. This result is obvious for harmonic functions which – intuitively – can change only in amplitude.

Inversion of the above expression is trivial by inverting the sign of the exponential in the decay function, and it is correct outside the source volume. It is still in order within a homogeneous body generating no gravity variations except vertical. In practice (see Chaps. 4 & 5), inhomogeneities (and data errors) are ubiquitous and lead to convergence problems. Downward continuation is an act of interpretation or inversion in terms of source distributions as surface mass layers, possibly converted to undulations with a density contrast, and this must be used with caution (Task 5.2). Practical frequency analysis and applications are treated in application sections of Chaps. 4–6 in connection with data analysis and modelling.

2.10.6 Special Functions

Special cases of invoking the harmonic functions are based on analytical functions f(x) or f(x,y) that have certain properties suitable for gravity problems and to which the Fourier transform is applied analytically instead of treating the given discrete anomaly values numerically. The first step is to expand a given gravity anomaly $\delta g(\mathrm{x})$ or $\delta g(\mathrm{x},\mathrm{y})$ in terms of, e.g.:

$$\delta g(x) = \sum_{\mathrm{l}=0}^{\infty} \mathrm{a_l}\, \mathrm{f_l}(x).$$

The coefficients $\mathrm{a_l}$ carry the information on the given gravity anomaly. The rest then follows from the Fourier transform or the harmonic functions, i.e. their capabilities of field continuation.

2.10.6.1 Legendre Polynomials

A brief description of the Legendre polynomials is given in view of their importance for gravity description, for example, with spherical harmonics (Sect. 2.10.7) and interpretation and in large dimensions (see Sect. 2.10.7.3). Legendre polynomials P_n (order n) are orthogonal power series. A function $f(x)$ is orthogonal in the interval A, B if $_\mathrm{A}\!\int^\mathrm{B} f_\mathrm{n}(x)f_\mathrm{m}(x)\mathrm{d}x = 0$ for $n \neq m$; the integral is $\neq 0$, if $n = m$: The term "orthogonal" has been taken from the three independent coordinates in Euclidian space;

orthogonality of other useful functions, as sine and cosine, has been mentioned and exploited before. As power series, Legendre polynomials are orthogonalized in the limits $A = -1$, $B = +1$ from the basis functions $x^o, x^1, x^2, \ldots x^n, \ldots$, with $P_0 = 1$ by successive application of

$$\int_{-1}^{+1} P_n P_m dx = 0; \text{ this renders}: P_1 = x, \ P_2 = 1/2(3x^2 - 1), \ P_3 = 1/2(5x^3 - 3x),$$

$$P_4 = 1/8(35x^4 - 30x^2 + 3), \text{etc.}$$

with the generally useful recurrence formula

$$P_n = (2 - 1/n) \, x \, P_{n-1} - (1 - 1/n)P_{n-2}, \text{ or derivable from}:$$
$$P_n(x) = (1/2^n n!) d^n/dx^n (x^2 - 1)^n.$$

In this form the P_n can be considered normalized for all n by the condition $P_n(1) = P_n(-1) = 1$ (Weisstein, 1999). Another normalisation on the basis of the area under $P_n^2(x)$, i.e.:

$$\int_{-1}^{+1} P_n^2(x) dx = 1 \text{ (KJ61, 275), leads to } P_n(x) = K_n P_n(x), \text{ with the factor}$$

$$K_n = (n + 1/2)^{-1/2}$$

which ensures that computed coefficients (as in Eq. 2.10.25) are comparable among each other and physically meaningful. This normalisation is achieved directly if P_n is defined by letting

$$\int_{-1}^{+1} P_n P_m dx = (n + 1/2)^{-1} \delta_{nm},$$

where δ_{nm} is the Kronecker delta (Weisstein, 1999).

An important class are the orthogonal *associated Legendre functions* $P_{nm}(x)$, which are solutions of a differential equation with two integer parameters, n and m and are part of the spherical harmonics which describe longitudinal variations beside the colatitudinal ones. The $P_{nm}(x)$ also follow from power series expansion (powers of $\cos \alpha$) of $1/R$ where R is the distance between point P and source point Q (Fig. 2.6.1 and Eq. 2.6.10):

$$R = (r^2 + r^{*2} - 2rr^* \cos \alpha))^{1/2} \text{ and } \cos \alpha = \cos \varphi \cos \varphi^* + \sin \varphi \sin \varphi^* (\lambda - \lambda^*);$$
$$\text{if P or Q are at the Earth's north pole}, \alpha = \varphi.$$

The P_{nm} can also be defined by differentiation of the ordinary or unassociated Legendre polynomials $P_{no}(x) = P_n(x)$, but attention must be paid to different sign conventions:

$$P_{nm}(x) = (-1)^m (1 - x^2)^{m/2} d^m/dx^m (P_n(x)) = (-1)^m (2^n n!)^{-1} (1 - x^2)^{m/2} d^{m+n}/dx^{m+n} (x^2 - 1)^n.$$

If $(-1)^m$ is dropped, one often writes P_n^m. A recurrence relation is:

$$(n-m)P_\text{n}^\text{m}(x) = x(2n-1)P_{\text{n}-1}^\text{m}(x) - (n+m-1)P_{\text{n}-2}^\text{m}.$$

Normalisation is accomplished by defining $_{-1}\int^{+1} P_\text{n}^\text{m} P_1^\text{m} \mathrm{d}x = (n+1/2)^{-1}[(n+m)!/(n-m)!]\delta_{\text{n}1}$ (Weisstein, 1999), but again, caution is necessary in view of different normalisations. If, in the case of spherical harmonics, $x = \cos\phi$, with co-latitude ϕ (see Eqs. 2.10.23–2.10.25), $\mathrm{d}P_\text{n}^\text{m}(\cos\phi)/\mathrm{d}\phi = (1 - \cos^2\phi)^{-1/2}[n\cos\phi\, P_\text{n}^\text{m}(\cos\phi) - (n+m)P_{\text{n}-1}^\text{m}(\cos\phi)]$ and the recurrence relation follows:

$$(2n+1)\cos\phi\, P_\text{n}^\text{m}(\cos\phi) = (n+m)P_{\text{n}-1}^\text{m}(\cos\phi) + (n+m+1)P_{\text{n}+1}^\text{m}(\cos\phi).$$

Legendre polynomials have wide applications in physics and geophysics, and an extensive mathematical literature exists, e.g. Abramowitz & Stegun (1972); Arfken (2001); Sansone (1991); Weisstein (1999): http://mathworld.wolfram.com/LegendrePolynomial.html.

2.10.6.2 Hermite Polynomials

A classical method (Tsuboi & Fuchida, 1937) is called after the orthogonal Hermite polynomials $H_1(x)$ which are part of a family of orthogonal functions based on the Gaussian and products with x, x^2, etc. The method is useful for downward continuation and interpretation of locally limited anomalies. The functions f_1 derive from the orthogonalizing integrals ($-\infty$ to $+\infty$) of $\exp(-x^2/2)$, $x\,\exp(-x^2/2)$, $x^2\exp(-x^2/2)$, etc. In these expressions the variable x is normalized with a length s such that $x' = sx$, but the prime has been dropped here. All the functions are of bell shape (f_0 is the Gaussian) or oscillate within such an envelope approaching zero with growing x:

$$f_\ell(x) = (\pi a^{-1/2} 2^\ell \ell!)^{-1/2} H_\ell(x)\exp(-x^2/2).$$

The Hermite polynomials are:

$$H_0 = 1, \ H_1 = 2x, \ H_2 = 4x^2 - 2, \ H_3 = 8x^2 - 12x, \ldots$$
$$H_\ell = (2x)^\ell - \ell(\ell-1)/\ell!(2x)^{\ell-2} + \ell(\ell-1)(\ell-2)(\ell-3)/2!(2x)^{\ell-4} - + \ldots.$$

$$(2.10.16)$$

or with the recursion formula: $H_\ell = 2xH_{\ell-1} - 2\ell H_{\ell-2}$; it is derived from

$$H_\ell = (-1)^\ell \exp(x^2)\partial^\ell/\partial x^\ell[\exp(-x^2)]. \tag{2.10.17}$$

The Fourier transform in the cosine and sine forms, $_cF(k)$ and $_sF(k)$, respectively, is

for even j:

$$_cF(k) = (2/\pi)_{j=0} \sum^{\infty} a_j \int_0^{\infty} f_j(x)\cos(kx)dx \text{ and } {}_sF(k) = 0, \qquad (2.10.18)$$

and for odd j:

$$_cF(k) = 0 \text{ and } {}_sF(k) = (2/\pi)_{j=0} \sum^{\infty} a_j \int_0^{\infty} f_j(x)\sin(kx)dx. \qquad (2.10.19)$$

The integrals can be calculated and render the final results:

$$_{c,s}F(k) = (2/\pi)^{1/2}{}_{j=0} \sum^{\infty} (-1)^j a_j f_j(k) \text{ for even and odd } j.$$

Here $k' = sk = 2\pi s/\lambda$ is the non-dimensional wave number, but again the prime has been dropped. Hence:

$$\delta g(x,0) = (2/\pi)^{1/2}{}_{j=0} \sum^{\infty} (-1)^j a_j \int_0^{\infty} f_j(\cos(kx)) \mid (\sin(kx))dk \quad (\text{even} \mid \text{odd} j)$$

$$\delta g(x,z) = (2/\pi)^{1/2}{}_{j=0} \sum^{\infty} (-1)^j a_j \int_0^{\infty} f_j(\cos(kx)) \mid (\sin(kx))e^{kz}dk \quad (\text{even} \mid \text{odd} j)$$

$$[kz = kz]$$
$$\rho^*(x,z) = \delta g(x,z)/(2\pi G) \qquad (2.10.20)$$

The integrals must be evaluated numerically.

Briefly, consider $f_o(x)$, i.e. a Gaussian gravity anomaly $\delta g(x,0) = a_o f_o(x) = a\exp(-x^2/2)$:

$$\delta g(x,z) = (2/\pi)^{1/2}a \int_0^{\infty} \exp(-k^2/2)\exp(kz)\cos(kx)dk = (2/\pi)^{1/2}a$$
$$\int_0^{\infty} \exp(-k^2/2 + kz)\cos(kx)dk$$

The function resembles a Gaussian, but it oscillates with growing x, and the amplitudes grow exponentially with z and k.

Again, the result follows from the fact that the field is fully determined and calculable in source-free space above a boundary at which δg is described by the above functions. The downward continuation has the aspect of a direct interpretation method, but the results must be – cautiously – translated into realistic geological models. Anomalies varying in two spatial variables x and y can be treated in an analogous manner; the formalism is the same, only some of the constants are modified.

2.10.6.3 The $\sin x/x$ Function

A similar method based on the $\sin x/x$ function can also be used for describing localized gravity anomalies. The function has a similar shape to that of the above f_o, in that it is maximum ($= 1$) at $x = 0$ and approaches zero with growing $|x|$; it also oscillates about zero with decaying amplitudes. Again, lengths x and z are normalized with s: $x' = x/s$; $k' = ks$ (primes dropped); then

$$\delta g = C \sin(kx)/(kx)$$

Application of the Fourier transform in cosine, sine form is convolution with $\cos(px)$ and $\sin(px)$, where $p = 2\pi/\Delta$:

$$_cG(p) = 2C/\pi \int_0^\infty [\sin(kx(/(kx)] \cos(px)dx = Cs/\pi\{1\,|\,1/2\,|\,0\} \text{ for}$$
$$\{k > p\,|\,k = p\,|\,k < p\};_s\, G(p) = 0.$$

Then

$$\delta g(x,z) = C\, s/\pi \int_0^k \cos(px)\, e^{pz}dx = C\, s/\pi\, e^{pz}/(x^2 + z^2)(z\cos(px) + x\sin(px))|_0^k$$
$$= C\, e^{kz}/(k^2(x^2 + z^2))(kz\, \cos(kx) - ke^{-kz} + kx\, \sin(kx))$$

If all lengths and wave numbers are normalized with s ($z' = z/s$, $x' = x/s$, primes dropped),

$$\delta g(x,z) = C\, e^{kz}/(\pi(x^2 + z^2))(z\cos(kx) - e^{-kz}/s + x\sin(kx))$$

Again, the method can be extended to two coordinates x, y:

$$\delta g(x,y,0) = C\, \sin(k_x x)\sin(k_y y)/(k_x x k_y y).$$

2.10.6.4 The $z/(x^2 + z_0{}^2)$ Function

The function $f(x) = z/(x^2 + z_0{}^2)$ has a similar bell shape but no oscillations. It can be treated the same way. Since $f(x)$ is proportional to the effect of a horizontal mass line at $x = 0$, $z = z_0$, the result of the treatment will not be surprising. From $_sG(k) = 0$, $_cG(k) = 2/\pi z_0$ it follows that

$$\int_0^\infty \cos(kx)/(x^2 + z_0{}^2)\, dx = 2/\pi \underline{z}_0\, \pi/(2z_0)\exp(-k\, z_0) = \exp(-kz_0),$$

and:

$$g^*(x,\, z) = \int_0^\infty \cos(kx)\exp(-kz_0 - z) = \exp(-kz_0)\, dk = (z_0 - z)/(x^2 + (z_0 - z)^2)$$

$$(2.10.21)$$

which describes the effect at depth z of the same mass line. At $z = z_o$ the result degenerates to the delta function. The example shows that the method works. However, it is not really necessary, because fitting data to the function $f(x)$ gives $\delta g(x) = G\rho^+ z/(x^2 + z_0^2)$, with the line density ρ^+ (kg/m), from which the parameters ρ^+ and z_0 and any wanted field quantities can be worked out directly.

2.10.6.5 Summary of Special Functions

Especially the last example demonstrates that the above analytical functions applied to gravity anomaly analysis, represent idealized mass distributions (mass line, special equivalent strata). This means that their application is equivalent to anomaly inversion and interpretation (direct methods). The success lies in eliminating the principal inversion ambiguity by drastic reduction of the number of unknown parameters. Any mass distribution whose effect can be suitably expressed analytically may, indeed, be chosen, and this guides the way to inversion generally, where – in the later chapters – the problems of ambiguity will be central to the whole discussion.

2.10.7 Spherical Harmonics

2.10.7.1 Derivation and Expansions

The scope of gravity interpretation includes scales up to that of global mantle dynamics, and large regional and global scales must be treated in spherical geometry. While Cartesian geometry often suffices and the emphasis in geophysical applications lies on smaller scales, large scales must be included, requiring spherical coordinates and the corresponding spatial frequencies described by spherical harmonics. However, a brief overview of some important relations is presented here; for thorough discussions the reader is referred to texts as (Byerly, 1959 [p. 244]; Moon & Spencer, 1988 [p. 26]; Morse & Feshbach, 1953 [pp. 514, 658]) and also into mathematical treatments in the internet as (Weisstein, 1999) http://mathworld.wolfram.com/HelmholtzDifferentialEquationSphericalCoordinates.html

> http://mathworld.wolfram.com/SphericalHarmonic.html
> http://hyperphysics.phy-astr.gsu.edu/hbase/math/sphhar.html
> http://www.google.de/search?q=spherical+harmonics+geodesy&ie=ISO-8859
> -1&hl=en&btnG=Google+Search

The spherical harmonics correspond to Cartesian harmonic functions and follow directly from the solution of the Laplace equation written in spherical coordinates (radius r, latitude φ and longitude λ), named after A.M. Legendre (1752–1833) or H. v. Helmholtz (1821–1894) (Byerly, 1959; p. 244):

$$\nabla^2 W = r^{-2}\partial/\partial r(r^2 \partial W/\partial r) + (r^2 \sin^2 \varphi)^{-1}\partial^2 W/\partial \lambda^2$$
$$+ (r^2 \sin \varphi)^{-1}\partial/\partial \varphi(\sin \varphi \partial W/\partial \varphi) \tag{2.10.22}$$

The solution proceeds by separation of the variables into $W(r,\varphi,\lambda) = R(r)\Phi(\varphi)\Lambda(\lambda)$ (http://mathworld.wolfram.com/HelmholtzDifferentialEquationSpherical Coordinates.html).

In spherical geometry longitudinal relations are described by harmonic functions, i.e., normal Fourier series with the parameter λ and the base wavelength $\lambda_o = 2\pi$ (or $2\pi a$ where a = major Earth's radius). Space restrictions on the sphere require a different description of the latitudinal relations achieved by associated Legendre functions $P_{nm}(x)$ where $x = \sin(\varphi)$ or more commonly $\cos(\phi)$ with colatitude $\phi = \varphi - \pi/2$ (see Sect. 2.10.6.1). The radial (r) relation appears in powers of r:

$$\delta W(\phi,\lambda) = (GM/r)\,_{n=2}\sum^{N}(a/r)^n\,_{m=0}\sum^{n}(C_{nm}\cos(m\lambda) + S_{nm}\sin(m\lambda))$$
$$P_{nm}(\cos\phi) \tag{2.10.23}$$

with M = Earth's mass. By differentiating with respect to $-r$

$$\delta g(\phi,\lambda) = (GM/r^2)\,_{n=2}\sum^{N}(n-1)(a/r)^n\,_{m=0}\sum^{n}(C_{nm}\cos(m\lambda)$$
$$+ S_{nm}\sin(m\lambda))P_{nm}(\cos\phi) \tag{2.10.24}$$

The coefficients C_{nm} and S_{nm} are defined by convolution integrals over the Earth's surface (δW^* observed):

$$C_{nm} = 1/4\pi \int_{\varphi}\int_{\lambda} \delta W^*(\cos(m\lambda))P_{nm}(\cos\phi)\cos\phi\,d\phi\,d\lambda$$
$$S_{nm} = 1/4\pi \int_{\varphi}\int_{\lambda} \delta W^*(\sin(m\lambda))P_{nm}(\cos\phi)\cos\phi\,d\phi\,d\lambda \tag{2.10.25}$$

where, as above, sums over finite compartments replace the integrals; n is called order or rank (German: Grad) and m is called degree or component (German: Ordnung). Eqs. (2.10.23–2.10.25) express the fact that transformation works both ways between the domains of space (ϕ, λ) and harmonic spectrum (C_{nm}, S_{nm}).

The associated Legendre functions $P_{nm}(\cos\phi)$ describe generally, in conjunction with the longitudinal Fourier series of degree m, wavy tesseral patterns on the sphere; the special case of $m = 0$ describes zonal patterns, as the longitudinal terms reduce to $\sin 0 = 0$ and $\cos 0 = 1$ and $P_{no} = P_n$ (see Sect. 2.10.6.1); finally the P_{om} terms define sectorial patterns, as $P_{om} = P_o$, where $P_{oo} = 1$ and the only variations are longitudinal waves along the latitudinal parallels.

Spherical harmonics are global orthogonal functions (just as are the Cartesian harmonic functions). In practice, when describing the external gravity field the expansion will always be truncated at some feasible limit N. Convergence is no problem in free space, but care must be taken where the geoid (mean sea level) is below the physical Earth's surface.

Areal spherical harmonics for constant r or, roughly, the Earth's surface, can be used to describe any distribution on a sphere, but a complete description requires degree and order to extend to infinity. Convergence may then be a problem. Practical difficulties can arise from cut-off spatial frequencies. If a truncated series is transformed back into the space domain and plotted, for example, by colouring or contour lines, the character will be wavy which is a feature of the functions used rather than one of the original data. This must not be misinterpreted.

The spherical harmonics are defined over the whole sphere, and uneven distribution of data points will lead to field descriptions unconstrained in data holes with no relevance to the real field. It is strictly wrong to interpret such anomalies unconstrained by observations. Therefore local instead of global functions have been proposed to be used.

2.10.7.2 Some Remarks About Integration of Gravity Effects

For the integration of effects of mass anomalies described in spherical coordinates, several options exist.

(1) A trivial option is to transform the given coordinates into Cartesian coordinates with the observation point P at O $(0, 0, 0)$ and z pointing toward the Earth's centre and to perform the integration with a suitable method from Sect. 2.9.

(2) Any mass anomaly can be adequately expanded into spherical harmonics. For small bodies or abrupt variations the expansion must be extended to high degrees and orders; the point mass which is a delta function with an infinite "white" spatial spectrum. A thin mass layer (thickness b) is especially suited to expansion, and its effects can be integrated in spherical harmonics and upward continued such that they can be calculated at any point of interest. Such cases are the undulated density contrast surfaces as, e.g., the 660 km discontinuity in the mantle or the core mantle boundary where the contrast is called $\Delta \rho_b$. Combined with the boundary undulation $\delta r_b(r_b, \phi, \lambda)$ it can be described by an equivalent surface density variation $\rho^* = \Delta \rho_b \delta r_b$ or the density variation, $\delta \rho = \rho^*/b$, in a thin layer of constant thickness b at the mean radius of the undulated contrast surface, r_b, hence: $\delta \rho \approx \Delta \rho_b \delta r_b / b$ or $\delta r_b \approx b \delta \rho / \Delta \rho_b = \rho^* / \Delta \rho_b$.

Expansion into spherical harmonic coefficients, $\rho^*{}_{Cnm}$, $\rho^*{}_{Snm}$, follows (Eq. 2.10.25). Conversion into the gravity potential coefficients, i.e. integration, is given by (Phillips & Lambeck, 1980):

$$\left. \begin{array}{c} C_{nm} \\ S_{nm} \end{array} \right\} = \left\{ \begin{array}{c} [3/(2n+1)][b/\underline{\rho}\, \underline{r}](r_b/\underline{r})^{(n+2)} \delta \rho_{Cnm} \\ [3/(2n+1)][b/\underline{\rho}\, \underline{r}](r_b/\underline{r})^{(n+2)} \delta \rho_{Snm} \end{array} \right\}$$

where $\underline{\rho}$ is the Earth's mean density $(5520\,\mathrm{kg/m^3})$ and $\underline{r}$ is the mean radius (in contrast to the major radius, a, in (Eqs. 2.10.23 & 2.10.24)); if δr is known at some point, $\Delta \rho_b$ can be derived or, vice versa, with $\Delta \rho_b$ known, the undulation δr_b follows.

Inversion of the above relation is straightforward (Hide & Horai, 1968):

$$\left. \begin{array}{c} \delta \rho_{Cnm} \\ \delta \rho_{Snm} \end{array} \right\} = \left\{ \begin{array}{c} [(2n+1)/3][\underline{\rho}\, \underline{r}/b](\underline{r}/r_b)^{(n+2)} C_{nm} \\ [(2n+1)/3][\underline{\rho}\, \underline{r}/b](\underline{r}/r_b)^{(n+2)} S_{nm} \end{array} \right\}$$

Of course, also voluminous density variations in mantle space from seismic mantle tomography or thermal convection models can be expanded into spherical harmonics, for example, layer-wise or when the coefficients are expressed as functions of radius r.

(3) A third possibility of integration is Gauss-Legendre quadrature as outlined below.

2.10.7.3 Gauss-Legendre Quadrature

Gauss-Legendre quadrature (GLQ), as a general numerical integration method is most suitable to be applied in spherical coordinates and is supplemented here, after introduction of spherical harmonics and Legendre polynomials (Sect. 2.10.6.1), complementing Sect. 2.9 on integration of gravity effects. The effect of an anomalous mass, say, at Q (Fig. 2.6.1) depends on the distance R and the angular distance (or colatitude in a local spherical coordinate system centred on Q or P), not on azimuth (Eqs. 2.6.8 & 2.6.9); (2.6.10) gives the relation of the local parameter α with the global coordinates. Therefore, GLQ is potentially an exact integration method applying Legendre polynomials (von Frese et al., 1981). This is an advantage over the methods described in Sect. 2.9.

Gauss Legendre quadrature (GLQ) applies interpolation by Legendre polynomials $P_n(x)$ to approximate an integrand $f(x)$ given by a set of n points $f_k = f(x_k)$ in the interval -1 to $+1$; $f(x_k)$ may be an empirical set of points, but here $f(x)$ is calculable as a precisely known function (Eqs. 2.6.8 or 2.6.9). The general summation formula, with the weights A_k, is:

$$\int_{-1}^{+1} f(x)\mathrm{d}x = \sum_{k=1}^{n} A_k f(x_k) \qquad (2.10.26)$$

which renders the exact integral, if $f(x)$ is a polynomial of degree $\leq 2n$ depending on the discretization. The orthogonality of the Legendre polynomials ensures each higher term to fit (and integrate) only the residuals from previous fits by the lower terms. The weights A_k are called Gaussian coefficients given to the $n x_k$ values or nodes (zeros) (see Carnahan et al., 1969):

$$A_k = 2n^{-2}(1-x^2)/P_{n-1}^{2}(x_k) \qquad (2.10.27)$$

Contrary to the standard limits, -1 to $+1$, applications generally have arbitrary integration limits, say y_a and y_b, requiring a mapping transformation of y into x. It involves a scaling factor $1/\Delta_y = 2/(y_b - y_a)$ and a shift of $\Sigma_y = (y_b + y_a)/(2\Delta_y)$; $x = (1/\Delta_y)y - \Sigma_y$ which renders:

$$y = \Delta_y x + \sum\nolimits_y = x(y_b - y_a)/2 + (y_b + y_a)/2 \qquad (2.10.28)$$

The application of GLQ to gravity interpretation is straightforward. However, some special conditions are imposed by the nature of the terrestrial density anomalies. Equations (2.6.8 & 2.6.9) describe potential and gravity effects and can be treated likewise, as can be expressions for the higher derivatives (and for magnetic effects, see von Frese et al., 1981). In the following, f stands for any of these functions of the source parameters r^*, ϕ^*, λ^*. The lower and upper limits are denoted by a and b, which are, of course, not identical for the three parameters. In order to apply GLQ, the coordinates must be transformed to $r^\#$, $\phi^\#$, $\lambda^\#$ with the integration boundaries -1, $+1$. According to (Eq. 2.10.28) the transformations are: $r^* = \Delta_{r^*} r^\# + \Sigma_{r^*}$, with $\Delta_{r^*} = (r_b^* - r_a^*)/2$ and $\Sigma_{r^*} = (r_b^* + r_a^*)/(2\Delta_{r^*})$; $\phi^* = \Delta_{\phi^*}\phi^\# + \Sigma_{\phi^*}$, with $\Delta_{\phi^*} = (\phi_b^* - \phi_a^*)/2$ and $\Sigma_{\phi^*} = (\phi*_b + \phi_a^*)/(2\Delta_{\phi^*})$; $\lambda^* = \Delta_{\lambda^*}\lambda^\# + \Sigma_{\lambda^*}$, with $\Delta_{\lambda^*} = (\lambda_b^* - \lambda_a^*)/2$ and $\Sigma_{\lambda^*} = (\lambda_b^* + \lambda_a^*)/(2\Delta_{\lambda^*})$. Hence GLQ for one parameter, say r^*, is expressed as:

$$\int_{r^*a}^{r^*b} f(r^*)\,dr^* = \Delta_{r^*}\int_{-1}^{+1}[f(\Delta_{r^*}r^\# + \textstyle\sum_{r^*})\Delta\rho]\,dr^* = \Delta_{r^*}\,{}_1\!\sum^n A_k f(r_k^*\Delta\rho_k)$$

The procedures are formally identical for the parameters, so the treatment is nested:

$$\int_{\phi^*a}^{\phi^*b}\left[{}_{\lambda^*a}\!\int^{\lambda^*b}\left\{{}_{r^*a}\!\int^{r^*b} f(r^*,\phi^*,\lambda^*)\Delta\rho\,dr^*\right\}d\phi^*\right]d\lambda^* =$$
$$= \Delta_{\lambda^*}{}_i\sum\left[\Delta_{\phi^*}{}_j\sum\{\Delta_{r^*}f(r^*_k,\phi^*_k,\lambda^*_k)\Delta\rho\,A_k)A_j\}A_i\right] \qquad (2.10.29)$$

Spherical prisms are most easily handled by keeping two parameters fixed when the third one takes on the values of the assumed nodes, say, r^*_k. More complicated arbitrary shapes require special handling of the limits, for example, by approximating them by spline functions; for this and other practical aspects see von Frese et al. (1981).

One noteworthy point is the possibility of extending the GLQ from the above procedure with purely geometrical functions of coordinates (multiplied by constant rock properties $\Delta\rho$), to the combination with a functional property variation with the coordinates $\Delta\rho(r^*,\phi^*,\lambda^*)$. In that case, in (Eq. 2.10.29) $f(r^*,\phi^*,\lambda^*)\Delta\rho \rightarrow f^* = f(r^*,\phi^*,\lambda^*)\,\Delta\rho(r^*,\phi^*,\lambda^*)$ and on the rhs.

$$f(r^*_k,\phi^*_k,\lambda^*_k)\Delta\rho_{ijk} \rightarrow f^*(r^*_k,\phi^*_k,\lambda^*_k).$$

2.10.8 Wavelets

In contrast to the global harmonic functions, wavelets are local, in Cartesian or spherical geometry, approaching zero in the outside regions and thus suited to handle signals of local or regional extent and/or containing discontinuities (Strang, 1992). Thus wavelets overcome the problem of finiteness outside regions of interest (Sect. 2.10.5.1), but note the remark on the equivalent problem of assuming zero mass outside the wavelet region. Similar to the harmonic functions, wavelets permit analysis by different frequency components, with each component matched to its scale. Wavelets were only recently introduced into potential theory (e.g. Freeden, 1999), relevant for scales, small relative to the Earth's dimensions; for features of global or continental dimensions, low-order harmonics are perfectly adequate.

Combination of both is therefore advantageous (Schmidt et al., 2005). An important purpose of applying wavelets is the efficient field representation and analysis, but increasingly wavelets are applied to problems of identifying certain features and interpreting them – with the usual caveats.

The procedure is to adopt or design a basic wavelet prototype function ("analyzing wavelet" or "mother wavelet"), usually built on orthogonal basis vectors (x,y) (as the orthogonal sine and cosine functions in the Fourier series). The original signal is represented in terms of a wavelet expansion using coefficients in a linear combination of the wavelet functions. Given a signal $f(x)$ defined in a domain, say from 0 to 1, scale-varying basis functions divide the signal into functions ranging from 0 to $1/2$ and $1/2$ to 1 or functions from 0 to $1/4$, $1/4$ to $1/2$, $1/2$ to $3/4$, and $3/4$ to 1, etc. By truncating the coefficients below a threshold one achieves a sparse data representation (data compression).

Many standard wavelets exist in the literature (see e.g. Schmidt et al., 2005). In geophysics, the basic functions used depend on the purpose. It is possible to choose them directly from the geophysical problem of gravity effects, as effects of point or line masses; they approach zero outside the scale of interest. But generally one must compromise between effective approximation, orthogonality properties and catching the physics of the problem, as wavelets are not generally exact solutions of the Laplace equation. An example is to emphasize structural edges on any scale; this requires elongated wavelets, the authors called "earth worms". http://www.ned.dem.csiro.au/unrestricted/people/HorowitzFrank/vrml/earthworms. html,

At the time of writing, this topic is quickly evolving, and the reader is referred to the topical literature (see reference under heading "Wavelets", including the relevant page numbers), and to watch out for internet releases, as http://www.cosy.sbg.ac.at/~uhl/wav.html and http://www.amara.com/IEEEwave/IEEEwavelet.html

2.10.9 Stochastic Representation of Anomalies

As mentioned in Sect. 2.10.1, description and interpretation of gravity anomalies can be addressed in a non-deterministic, stochastic or probabilistic manner. The principal question is: Given is a set of observations at points (x_i, y_i) or (x_i, y_i, z_i) in a compact area, limited by a simple concave boundary, for example, the polygon defined by the outer points such that all other points lie inside the polygon. If the area degenerates to a straight or slightly curved profile line, the compact region is defined by the end points: what is known about the anomaly value at any other point between the observations within the limited area? It is assumed that δg exists as a continuum of points $(x, y, \delta g)$ and that the spatial distribution of the observed values is representative of the total or continuous distribution. Such an assumption is not necessarily true or self-evident, and sampling must take the possible spatial variations into account guided by experience and tests. It is, for example, well known that gravity and magnetic anomalies often have very different spectral patterns such that sampling must be different, for instance, much denser in magnetics than in gravity. What do the discrete observations tell about the continuum of points at any (x, y)? The task can be defined as that of estimating or predicting the most probable anomaly value $\delta g(x, y)$ from discrete observations δg_i (not to be confused with stochastic gravity in general relativity; see, e.g., Mowat, 1997).

The properties of the anomaly distribution can be described by the autocorrelation of the data which may be reasonably approximated by the Gauss distribution $\exp(r^2/c^2)$ with $r^2 = x^2 + y^2$ and c called "correlation length". It is thought to depict the probable spatial correlation between anomaly values in a statistical sense. If the anomaly character is wavy, c will indicate the dominant wavelength. A more general approach has been called "geostatistics" and the space behaviour of discrete data sets is analyzed by calculating variograms whose properties are then applied to interpolating intermediate point values and estimating their uncertainties. The method is called "Kriging"; it is described in more detail in Sect. 5.1.5.

A directional structural grain of the data can be approximated by an elliptic autocorrelation detectable by a directional analysis in which the azimuth, λ, about a given station is divided into azimuth intervals, $\Delta \lambda$, and the autocorrelation is calculated in $\Delta \lambda$ brackets. More elaborate azimuthal patterns may be simplified to elliptical patterns $(X/a)^2 + (Y/b)^2 = 1$ or $(bX)^2 + (aY)^2 = a^2 b^2$ where X, Y are orthogonal coordinates rotated in the directions of maximum and minimum length, respectively, and a and b are the corresponding ellipse axes. The axes a and b and the angle λ between X and x (or Y and y) may be optimized by least-squares elliptical fitting. The rotational transformation between the coordinates is described by Eqs. (2.4.4 & 2.4.5).

The stochastic approach is of importance for the geodetic concern about errors and confidence ranges. The geophysical aim to find source distributions emphasizes Newton's deterministic law. The two approaches, stochastic and deterministic functional, do usually converge.

In interpretation and inversion of gravity anomalies, the stochastic approach includes methods applying Monte Carlo techniques: around a starting model or set of model parameters a large number of models with randomly varied parameters are calculated. The random generator usually samples from a Gauss-distribution with an assumed standard deviation. The results are compared to the observations and selected pre-set criteria of fitting. The successful models are collected, assembled and visualized in the form of frequency distributions of the parameters. Efficient computing can thus give a quick overview over the acceptable model range. This approach will be treated again (7.3.3.2).

2.11 Aspects of Magnetostatics

A brief description is included in this text because many methods of gravity interpretation are equally applicable to magnetics. Magnetostatic fields, i.e. magnetic fields with insignificant time variation, are potential fields as the gravity field is. In contrast to gravity whose source is the scalar mass, the magnetic field is a dipole field, its source consists always of dipoles. An idealized dipole is a polarized mathematical point with no volume. It is an intimate, but not annihilating, union of a positive and a negative scalar point source making a vector which, by definition, points from the positive to the negative side. The dipole potential and its space derivatives (field intensity gradients) can be considered as composed of two non-annihilating scalar components. The superposition of these fields does not principally alter the laws governing potential fields. Most of the laws described for gravity therefore hold also for magnetics, except those directly related to the dipole nature.

The physical realization of a dipole is a magnetic needle, e.g. in a compass. It consists of magnetic material and can be described as a rod with magnetic poles at its ends, one of them being attracted to the Earth's magnetic north pole which lies not too far from the rotational north pole; this magnetic needle pole is therefore called "north seeking pole" or "magnetic north pole", its counterpart "south pole". Since north (or south poles) of two needles repel each other, the Earth's pole in the north must be a "magnetic south pole" and vice versa. Unequal poles attract each other.

Originally these mechanical forces were used to define the strength $\boldsymbol{F}$ of the magnetic field, but it was learnt that magnetic effects, as induction, depend also on the magnetic property of the surrounding space, called permeability μ. This is expressed by defining the field through the alternative quantity $\boldsymbol{B} = \mu\boldsymbol{F}$. The permeability μ_o of empty space or vacuum is a fundamental property. Matter does generally change the permeability by magnetic induction, and it has become customary to measure the change by a relative factor called here the "relative permeability" $\mu_r = \mu/\mu_o$, such that, generally, $\boldsymbol{B} = \mu_r\mu_o\boldsymbol{F}$. The affinity of a material to magnetic induction is called susceptibility $\kappa = (\mu - \mu_o)/\mu_o = (\mu_r\mu_o - \mu_o)/\mu_o = \mu_r - 1$ (dimensionless). Many magnetic anomalies of geophysical interest stem from this effect, called "magnetization". It depends on the Earth's present local field strength $B_\oplus$ and on rock susceptibility κ originating in dispersed susceptible minerals as magnetite. Rocks can also be permanently magnetic by remnant magnetization obtained in their genesis and history; this type of magnetization is not affected by the Earth's present magnetic field strength. The ratio of remnant to induced magnetization is called Königsberger ratio Q.

The magnetic flux, $\phi =_S \int \boldsymbol{B} \cdot \boldsymbol{ds}$, with S surface, $\boldsymbol{ds}$ vectorial surface element, as in (Eq. 2.1.1), emanating from a dipole is zero (in this sense, the pole effects annihilate each other) and integration over closed surfaces always gives zero total flux.

The dipole potential follows from two point potentials of equal size and opposite sign superimposed. The pole strength p is a fictitious quantity which cannot be individually realized in nature. The vector $\boldsymbol{\ell}$ from p_- (S) to p_+ (N) approaches zero length while the dipole moment $\boldsymbol{m} = \boldsymbol{LP}$ remains finite (or fixed):

$$\boldsymbol{m} = \lim_{(\ell \to 0)} p \bullet \boldsymbol{\ell} \tag{2.11.1}$$

$$A = \lim_{(\ell \to 0)} (p_+/r_+ - p_-/r_-) \tag{2.11.2}$$

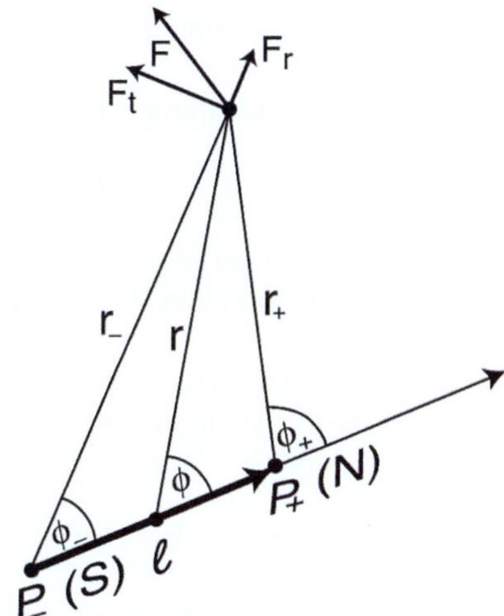

Fig. 2.11.1 The dipole is defined by letting its physical length appraoch zero; definitions of the geometrical quantities

where r_+ and r_- are the distances of both poles to the observation point P (Fig. 2.11.1).

The relations between ℓ, the radii r, r_+ and r_- as well as the angle ϕ, ϕ_+, ϕ_- are expressed by the cosine law:

$$r^2 = r_+^2 + (\ell/2)^2 - 2r_+(\ell/2)\cos\phi = r_-^2 + (\ell/2)^2 - r_-\ell\cos(\phi - \pi)$$
$$= r_-^2 + (\ell/2)^2 + r_-\ell\cos\phi.$$

As ℓ approaches zero, $(\ell/2)^2$ is neglected, r_+, $r_- \to r$, ϕ_+, $\phi_- \to \phi$, and the Taylor expansion of $1/r \to (1 \pm \ell\cos\phi/2r)/r$ for r_+ and r_-. From (Eqs. 2.11.1) and (2.11.2) follows

$$A = -m\cos\phi/r^2, \tag{2.11.3}$$

which can also be written as

$$A = \boldsymbol{m} \bullet \boldsymbol{\nabla}(1/r) \tag{2.11.4}$$

with, in arbitrarily oriented Cartesian coordinates, $\boldsymbol{r} = (x, y, z)$, $\boldsymbol{m} = (m_x, m_y, m_z)$ or $\boldsymbol{m} = m_i (i = 1, 2, 3)$ in index notation. The gradient $\nabla(1/r)$ implies differentiation at the source (two infinitesimally displaced poles).

Physically magnetic fields are connected with electrical currents (including electron spins in mineral magnetization); therefore it is more correct to define the magnetic moment without referring to poles. The dipole moment of a ring current (or a current in a small electrical coil) is

$$\boldsymbol{m} = \mu_o nI\boldsymbol{a} \tag{2.11.5}$$

where nI is the total ring current (current I in n windings) and $\boldsymbol{a}$ is the cross sectional area, i.e., its normal vector if the current direction is dextral (negative electrons would move in sinistral direction). An infinitesimally small coil exactly corresponds to a dipole.

The dipole field strength F is the gradient of the magnetic potential.

$$F = \nabla A = \nabla(m \bullet \nabla)(1/r) \tag{2.11.6}$$

where $m \bullet \nabla. = m_x \partial/\partial x + m_y \partial/\partial y + m_z \partial/\partial z$.

Assume the direction of m to be $\alpha = (\alpha_x, \alpha_y, \alpha_z)$ or $\alpha_i = \cos(\alpha, x_i)$; a is a unit vector.

$m \bullet \nabla. = m\partial/\partial\alpha = m(\cos\alpha_x \partial/\partial x + \cos\alpha_y \partial/\partial y + \cos\alpha_z \partial/\partial z) = m\cos\alpha_i \partial/\partial x_i$ (with the sum convention, i.e. summation over index i appearing twice, $i = 1, 2, 3$), and

$$\nabla(m \bullet \nabla)(1/r) = -\nabla((1/r^2)(xm_x + ym_y + zm_z)) = -\nabla((1/r^2)(r \bullet m)) \tag{2.11.7}$$

Differentiation of $\nabla(xm_x/r^2 + ym_y/r^2 + zm_z/r^2)$ leads to the vector

$$\begin{aligned}
F &= (\partial/\partial x[xm_x/r^2 + ym_y/r^2 + zm_z/r^2], (\partial/\partial y[xm_x/r^2 + ym_y/r^2 + zm_z/r^2], \\
&\quad (\partial/\partial z[xm_x/r^2 + ym_y/r^2 + z\,m_z/r^2]) \\
&= (m_x(x^2 - r^2/3) + m_y xy + m_z xz, m_x xy + m_y(y^2 - r^2/3) + m_z yz, m_x xz \\
&\quad + m_y yz + m_z(z^2 - r^2/3))
\end{aligned}$$

The 9 elements can be ordered in matrix form; the rows represent the field vector components. Thus F can be written as the product of the matrix $\underline{\Delta} = \Delta_{ij} = \{\partial(x_i/r^3)/\partial x_j\}$ with the magnetic moment m_j:

$$\begin{aligned}
F &= \underline{\Delta} \bullet m^{(T)} = \Delta_{ij} m_j = \{\partial(x_i/r^3)/\partial x_j\} \bullet m_j \\
&= \left\{ \begin{array}{ccc}
m_x(x^2 - r^2/3) & m_y xy & m_z xz \\
m_x xy & m_y(y^2 - r^2/3) & m_z yz \\
m_x xz & m_y yz & m_z(z^2 - r^2/3)
\end{array} \right\} \bullet \left\{ \begin{array}{c}
m_x \\
m_y \\
m_z
\end{array} \right\}
\end{aligned}$$

$$\tag{2.11.8}$$

In spherical coordinates: origin at the dipole source; axis parallel to the dipole axis (from $+$ to $-$ or from S to N), radius r, colatitude ϕ and longitude λ:

$$\begin{aligned}
F &= (F_r, F_\phi, 0) \\
F_r &= \partial A/\partial r = 2m\cos\phi/r^3 \tag{2.11.9} \\
F_\theta &= \partial A/r\partial\theta = m\sin\phi/r^3
\end{aligned}$$

The dipole field has axial symmetry and has no λ component. The relation between the spherical and Cartesian coordinates is: $x = r\sin\phi\cos\lambda$, $y = r\sin\phi\sin\lambda$, $z = r\cos\phi, r^2 = x^2 + y + z^2$.

A point mass with a magnetic moment generates a gravity field and a magnetic field which are related to each other. Compare the gravity field $F_g = \nabla U =$

$\text{G}M\mathbf{\nabla}(1/r)$, where M is mass, with the magnetic field (2.11.6) $\mathbf{F} = \mathbf{\nabla}A = \mathbf{\nabla}(\mathbf{m} \bullet \mathbf{\nabla})$ $(1/r) = m\mathbf{\nabla}\partial/\partial\alpha(1/r)$ with m and α constant; hence

$$\mathbf{F} = (m/\text{G}M)\partial(\mathbf{F}_g)/\partial\alpha \qquad (2.11.10)$$

This is *Poisson's relationship* between gravity and magnetics which may be useful in cases where geological bodies are homogeneous in density and magnetization. In such cases the relation holds also for the integral effects $\boldsymbol{\delta F}$ and $\boldsymbol{\delta g} = (\delta g_x, \delta g_y, \delta g_z)$ of extended bodies. Note that the vector gravity effect is required if the magnetic effect is to be derived from gravity. To integrate magnetic effects of extended magnetized bodies for P at O $= (0, 0, 0)$ begin with the potential $A = -_V\int\mathbf{m}^*\mathbf{\nabla}(1/r)\mathrm{d}V$ where $\mathbf{m}^* = \mathbf{dm}/\mathrm{d}V$ is the volume density of magnetization and $\mathbf{F} = -\mathbf{\nabla}_V\int(\mathbf{m}^*\mathbf{\nabla})(1/r)\mathrm{d}V$. If $\mathbf{m}^*$ varies in space no further simplification is possible, but for constant direction $\boldsymbol{\alpha}$, $\mathbf{F} = -\mathbf{\nabla}\partial/\partial\alpha_V\int(\mathbf{m}^*/r)\mathrm{d}V$ which, again, can be compared with gravity.

A generalization of Poisson's relationship has been investigated thoroughly by Holstein and co-workers (Holstein et al., 1999; Holstein, 2002a,b) for the case of uniformly dense and uniformly magnetized polyhedra. Their effects can be calculated by sets of equations which have common features with their gravity effects, such that a common basis for all "gravimagnetic effects" and their numerical computation can be designed.

In practical cases, demagnetization can cause problems. Each magnetic volume element of a body has a magnetizing effect on all other volume elements of the body. Because dipole field lines turn around into the opposite direction outside the source volume the total magnetizing effect is generally a de-magnetization. The effect is the greater, the greater the magnetic susceptibility of the material, and it may be negligible when the susceptibility κ is small. It can be described as a magnetic polarization where the opposite poles (N or S) concentrate at the surfaces where the field lines enter (south poles) or exit (north poles). A special case is an infinite slit in a permeable material homogeneously magnetized; a field-parallel slit is not polarized and no magnetization effect would be measured ($\mathbf{F}$ or $\mu_o\mathbf{F} = \mathbf{B}_o$); a field-normal slit is fully polarized and its observed effect is $\mu_r\mu_o\mathbf{F} = \mu_r\mathbf{B}_o = \mathbf{B}$. $\mathbf{F}$ and $\mathbf{B}$ are different sides of the same physical phenomenon, linked by the factor of magnetic permeability $\mu = \mu_r\mu_o$ where μ describes the effect of matter filling the space.

Since F is the superposition of two polar fields, the Laplace equation $\nabla^2 A = 0$ is automatically satisfied in empty space. Because of the dipole nature the Poisson equation is also $\nabla^2 A = 0$ in homogeneously magnetized matter. Magnetic effects become visible only if magnetization is space-variable such that $\nabla^2 A = \mathbf{\nabla} \bullet \mathbf{m}$. For the same reason, magnetized layers with plane horizontal boundaries generate no external magnetic effects (anomalies) except near their lateral edges.

Integration of magnetic effects for extended magnetized bodies is usually easiest if executed on the scalar potential (often similar to the case of the gravity effect), and then differentiated, for example, with respect to the vertical z.

References

Abramowitz, M., Stegun, I.A.: Handbook of Mathematical Functions with Formulas, Graphs, and Mathematical Tables, 9th printing. *New York, Dover*, 1972

Ackermmann, H.A., Dix, C.H.: The first vertical derivative of gravity. *Geophysics*, 20, 148–154, 1955

Arfken, G.: Mathematical Methods for Physicists, 5th ed. *Academic Press, New York*, 1128pp., 2001

Byerly, W.E.: An Elementary Treatise on Fourier's Series, and Spherical, Cylindrical, and Ellipsoidal Harmonics, with Applications to Problems in Mathematical Physics. *New York, Dover*, 287pp., 1959, 1993, 2003

Carnahan, B., Luther, H.A., Wilkes, J.O.: Applied Numerical methods. *New York, Wiley*, 1969

Çavşak, H.: Dichtemodel für den mitteleuropäischen Abschnitt der EGT aufgrund der gemeinsamen Inversion von Geoid, Schwere und refraktionsseismisch ermittelter Krustenstruktur. *Ph.D. thesis, Mainz*, 1992

Freeden, W.: Multiscale modeling of space borne Geodata. *Teubner, Stuttgart*, 351pp., 1999

Goguel, J.: Calcul de l'attraction d'un polygone horizontal de densité uniforme. *Geophys. Prospect.*, 9., 116–127, 1961

Grant, F.S., West, G.F.: Interpretation Theory in Applied Geophysics. *McGraw-Hill, New York*, 583pp., 1965 [GW65]

Gröbner, W., Hofreiter, N.: Integraltafel, Erster Teil, Unbestimmte Integrale, Springer-Verlag, *Wien, Innsbruck*, 166 pp., 1949

Gröbner, W., Hofreiter, N.: Integraltafel, Zweiter Teil, Bestimmte Integrale, Springer-Verlag, *Wien, Innsbruck*, 204 pp., 1950

Gröbner, W., Hofreiter, N.: Integraltafeln, 2 Volumes, Springer-Verlag, *Wien, Innsbruck*, 1949, 1950

Hide, R., Horai, K.: On the topography of the core-mantle interface. Phys. Earth Planet. Interiors, 1, 305–308, 1968

Holstein, H.: Gravimagnetic similarity in anomaly formulas for uniform polyhedra. *Geophysics*, 67, 1126–1133, 2002a

Holstein, H.: Invariance in gravimagnetic similarity in anomaly formulas for uniform polyhedra. *Geophysics*, 67, 1134–1137, 2002b

Holstein, H., Schürholz, P., Starr, A.J., Chakraborti, M.: Comparison of gravimetric formulas for uniform polyhedra. *Geophysics*, 64, 1438–1446, 1999

Jacoby, W.: Zur Berechnung der Schwerewirkung beliebig geformter Massen mit digitalen Rechenmaschinen. *Z. Geophysik*, 33, 163–166, 1967

Jung, K.: Schwerkraftverfahren in der Angewandten Geophysik. *Akad. Verlagsges. Geest & Portig, Leipzig*, 348pp., 1961 [KJ61]

Klingen, B.: Fouriertransformation für Ingenieur- und Naturwissenschaften (Springer-Lehrbuch). *370 pp. Springer Verlag, Heidelberg*, 2001

Moon, P., Spencer, D.E.: Field Theory Handbook, Including Coordinate Systems, Differential Equations, and Their Solutions, 2nd. ed. *Springer Verlag, New York*, 1988

Morse, P.M., Feshbach, H.: Methods of Theoretical Physics, Part I. *McGraw-Hill*, 1953

Mowat, J.W.: Stochastic gravity. *Phys. Rev. D*, 56/10, 6264–6277, 1997

Nagy, D.: The evaluation of Heuman's lambda function and its application to calculate the gravitational effect of a right circular cylinder. *Pure Appl. Geophys.*, 62, 5–12, 1965 (note: Eq. (1) place denominator before K under square root; Eq. (2): factor must be: 0.096 663 442 59)

Phillips, R.J., Lambeck, K.: Gravity fields of the terrestrial planets: Long wavelength anomalies and tectonics. *Rev. Geophys. Space Phys.*, 18, 27–76, 1980

Pohánka, V.: Optimum expression for computation of the gravity field of a homogeneous polyhedral body. *Geophys. Prospecting*, 36, 733–751, 1988

Sansone, G.: Orthogonal Functions, rev. Engl. ed. *New York, Dover*, XII + 411pp. 1991

Schmidt, M., Fabert, O., Shum, C.K.: On the estimation of a multi-resolution representation of the gravity field based on spherical harmonics and wavelets. *J. Geodynamics*, 39, 512–526, 2005

Strang, G.: Wavelets. *American Scientist*, 82, 250–255, 1992

Talwani, M., Ewing, M.: Rapid computation of gravitational attraction of three-dimensional bodies of arbitrary shape. *Geophysics*, 25, 203–225, 1960

Talwani, M., Worzel, J.L., Landisman, M.: Rapid gravity computations for two-dimensional bodies with application to the Mendocino submarine fracture zone. *J. Geophys. Res.*, 64, 49–59, 1959

Tsuboi, Ch., Fuchida, T.: Relation between gravity values and subterranean mass distribution. *Bull. Earthquake Res. Inst., Tokyo*, 15, 636–649, 1937

von Frese, R.R.B., Hinze, W.J., Braile, L.W., Luca, A.J.: Spherical Earth gravity and magnetic anomaly modeling by Gauss-Legendre quadrature integration. *J. Geophys.*, 49, 234–242, 1981

Weisstein, E.W.: Legendre Polynomial. *CRC Press LLC*, 1999–2004 (*Wolfram Research, Inc.*). http://mathworld.wolfram.com/LegendrePolynomial.html.

Weisstein, E.W.: Helmholtz Differential Equation – Spherical Coordinates, 1999. http://mathworld.wolfram.com/HelmholtzDifferentialEquationSphericalCoordinates.html

References on Wavelets, Not Quoted in Text

Cody, M.A.: The Wavelet Packet Transform. *Dr. Dobb's Journal*, 19, 44–46, 50–54, 1994

Crandall, R.: Projects in Scientific Computation. *Springer-Verlag, New York*, 197–198, 211–212, 1994

Daubechies, I.: Orthonormal Bases of Compactly Supported Wavelets. *Comm. Pure Appl. Math.*, 41, 906–966, 1988

Donoho, D.: Nonlinear Wavelet Methods for Recovery of Signals, Densities, and Spectra from Indirect and Noisy Data, Different Perspectives on Wavelets. *Proceeding of Symposia in Applied Mathematics, 47, I. Daubechies ed. Amer. Math. Soc., Providence, R.I.*, 173–205, 1993

Kaiser, G.: A Friendly Guide to Wavelets, *Birkhauser, Boston*, 300pp., 1994

Meyer, Y.: Wavelets: Algorithms and Applications. *Soc. Ind. Appl. Math., Philadelphia*, 13–31, 101–105, 1993

Press, W.H., Flannery, B.P., Teukolsky, S.A., Veterling, W.T.: Numerical Recipes in FORTRAN, 2nd ed. *Cambridge Univ. Press, New York*, 992pp., 1992

Vetterli, M & Herley, C.: Wavelets and Filter Banks: Theory and Design. *IEEE Trans. Signal Process*, 40, 2207–2232, 1992

Wickerhauser, V.: Adapted Wavelet Analysis from Theory to Software. *AK Peters, Boston*, 504pp., 1994

Chapter 3
Observations and Field Activities

3.1 Introduction

The observation of gravity has become a diverse field of endeavour with many facets and purposes. The emphasis is here on the classical gravity measurements with gravimeters for the purpose of deriving gravity anomalies and interpreting them in geological terms. Anomalies are characteristic gravity variations which are to be defined or constructed from ensembles of observations at point sets, usually at the Earth's surface, and they are related to geological bodies or structures (see, Sect. 1.4). This chapter is a brief introduction into the measurements in order to help assessing their problems and limitations. New developments, for example, in absolute gravimetry, gravity tensor measurements and ship born, airborne or space borne methods will be only briefly presented.

Data collection methods are rapidly evolving, mostly by application of microprocessors. The field procedures, only a decade ago still manual, are automated in modern instruments. But "old fashioned" gravimeters are still much in use, and will be for some time, and their handling is, in many respects, instructive. Details are left to the specialized literature and manufacturer's manuals. Timeless general aspects are survey planning and accuracy of the measurements as well as error search; they represent a through-going topic of the whole data stream (see also Chap. 4).

In most cases gravity surveys have a geological or exploration aim and are regionally limited. Terrestrial measurements reach high precision and spatial resolution, but they remain frequently isolated from regional networks. This causes problems for national and international archives and their users. The situation is improving, for example, by current dedicated satellite missions and airborne gravimetry. Geologically, the aim of gravity measurements is to determine the gravity field in space, i.e. time-invariant gravity values at sets of stations. Slow temporal variations of geodynamic nature are also of growing interest.

In measuring gravity, one must usually find some compromise between the accuracy requirements and economy, depending on the purpose of a survey and on the controls of all the errors influencing the final results. Efforts should be balanced by the gains: a reading accuracy better by more than an order of magnitude than that of common uncontrollable errors is not reasonable; for example, in geologically motivated surveys, 0.1 mGal uncertainties easily result from unknown nearby small

W. Jacoby, P.L. Smilde, *Gravity Interpretation*,
© Springer-Verlag Berlin Heidelberg 2009

density variations in the ground and relief irregularities. On the other hand, if temporal gravity variations are the target of the measurements, precision requirements are much higher for repeated readings at identical stations.

This chapter covers the field work and leads from survey planning and point selection up to the wanted gravity values in space or space and time at given points (coordinates, elevations, possibly epoch) in the form of lists (usually digital) and perhaps graphical representation on maps. Included are the additional necessary field operations as surveying and levelling as well as determination of rock densities and of the neighbouring relief plus error analysis. Reductions and post-reduction treatments as contouring, smoothing, regional – residual separation and derivation of related quantities (KJ61, 93–121) are treated in Chap. 4. The present chapter is divided into an instrumental part, a part covering general characteristics of surveys and survey planning and an overview of auxiliary field measurements.

3.2 Principles of Gravity Measurement and Instrument Types

3.2.1 General Considerations

Commonly instruments are distinguished as relative or absolute. It is not a sharp distinction. "Relative" means that only small differences of gravity can be measured accurately relative to some reference station. Accurate absolute measurements of total gravity acceleration over the whole range from zero to terrestrial g are now realized.

Several physical principles have been applied to measure gravity: the free pendulum period, the calibrated spring balance, spring vibrations, free fall, suspension of a conductor in a temporarily constant, spatially inhomogeneous magnetic field and ultra-precise beam balances for special purposes. In the following, the different types are briefly sketched. In all methods g must be extracted from the primary observations of time and length. More and more of this is done automatically by inbuilt processors that may also calculate the statistics of repeated observations.

Principally all types of instruments show some kind of drift. It means that the observations on any physical scale vary when the measured physical quantity remains constant. Such a drift may have any time behaviour which itself depends on internal instrument states which, in turn, may be affected by external factors as environmental temperature and air pressure, imposed instrument motion, etc. and their time history. Precision instruments are constructed such as to minimize and or control the external influences, but especially for field instruments this can not be fully achieved. Control is possible either by protecting or shielding measures or by calculating the corresponding corrections, if functional relations with observable external parameters are known. However, some uncontrollable rest always remains for which neither theory exists nor experience provides any countermeasure. Therefore nothing can really be said about the time behaviour of the residual drift;

it may contain time-continuous and discontinuous components, and the continuous component may be more or less linear with time. Such components are usually determined by repeated measurements of known (or estimated) physical quantities (see, Sect. 3.3). Since the time intervals between the control measurements cannot be arbitrarily short, little will be known about the short-term or high-frequency drift behaviour.

3.2.2 Pendulums

Pendulums were used earliest to measure gravity, but their accuracy is limited to some 0.5 mGal. A mathematical or ideal pendulum of a given length L swings freely with a period:

$$T = 2\pi(L/g)^{1/2} \tag{3.2.1}$$

i.e. the determination of g requires measuring lengths and times, mostly for many periods which increases the precision. In practice the length measurement is a problem. A mathematical pendulum consists of a point mass suspended by a massless string at a fixed point in constant gravity. Physical pendulums do not distinguish between mass and string, their motion is governed by both effective translation and rotation, and the effective pendulum length must be determined by special procedures. The reversion pendulum exploits the fact that the locations of the effective (mathematical) suspension point and of the effective point mass are exchangeable. However, also the suspension structures and the underground which are not absolutely rigid affect the period. Absolute pendulum apparatuses are particularly sensitive to such uncertainties. The Potsdam absolute result around 1900 was the basis of the world gravity system until its replacement after a systematic error of about 15 mGal had been definitively established. The error was much bigger than the statistical standard error and followed from the above problems and incorrect reductions for them. In relative pendulums the problems are reduced by the fact that the differences of measured periods are smaller and that the uncertain pendulum length drops out of the formulae, by comparing the observed periods T_i at point i with T_o at a reference station where g_o is assumed to be known. From (3.2.1) follows:

$$g_i = g_o(T_o/T_i)^2 \tag{3.2.2}$$

The result is immediately in m/s^2 or converted to mGal.

3.2.3 Spring Gravimeters

The currently dominant instruments are based on weighing a constant mass with an especially calibrated spring balance. Successful vertical gravity meters were invented in the late 1920ies to 30ies for detecting low-density salt domes. Many types of gravity meters have been constructed since (brand names are L&R, Worden, Syntrex, Sodin, etc.). As highly sensitive field instruments gravimeters are small, light, transportable and robust. To protect the delicate sensors from damage, swinging parts are either clamped during transport or limited in their motion. The mostly mechanical devices with electric controls and reading devices are now more and more replaced by control with microprocessors, so that nearly all field operation is automated, from instrument levelling to running programs of repeated measurements and error statistics as well as storage in data loggers. Modern gravity meters have an internal precision of $10\,\mu$Gal or $10^{-8}\,g$ with reasonable effort, some types may achieve $10^{-9}\,g$.

A mathematical spring balance (Fig. 3.2.1) with the spring length l and constant C and the point mass m is elongated by δl by the gravity difference δg as

$$\delta l\, C = \delta g\, m \quad \text{with } \delta g = g_i - g_o \text{ (as above)} \qquad (3.2.3)$$

However, gravity is not constant in the volume occupied, and some kind of average is measured. Neither C nor g are constant within the range of observations, C changes with spring elongation and depends on many environmental effects. The extended mass, the mass of the spring and indeed the whole set-up is buoyant in the surrounding gaseous atmosphere. Furthermore, a spring-mass system performs natural oscillations, so gravimeters must be damped. Marine and airborne gravimeters on moving platforms need especially strong damping. The problems are tackled by, (1) shielding the whole sensor system from environmental influences, for example, by insulating, evacuating and thermostatting it, and (2) by some special appropriate design.

Most instruments are set by nulling: instead of measuring the deflection δl from the null position, it is restored, for example, by shifting the suspension point of a special spring, such as to leave the main spring configuration at identical length. The position is controlled by an analogue or digital device, or it is automatically set by a feed-back system. The observation is then the position taken by the nulling or feed-back system, for example, by a counter of spindle revolutions or a required electric current. In most gravimeters the sensitivity is enhanced by special design of the springs and the geometry of the suspension; such *astastic instruments* are in

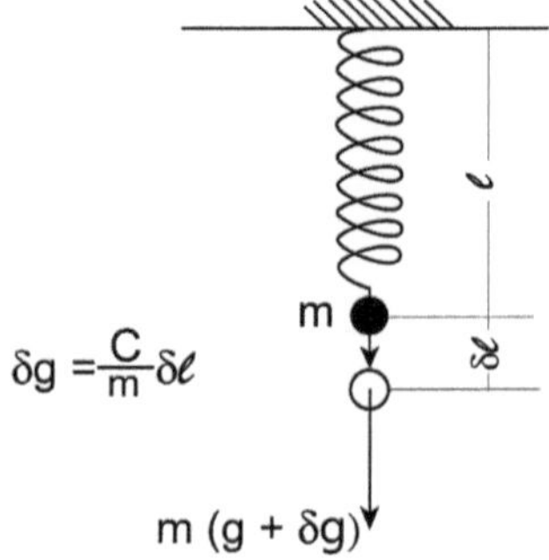

Fig. 3.2.1 Ideal mathematical spring balance to measure $\delta g = (C/m)\delta l$ via measuring δl; C elastic spring constant; m constant mass

equilibrium close to the neutral point where the forces or momenta of weight and of the spring cancel, and in many instruments the sensitivity is adjustable.

Measuring $\delta\ell$ of the auxiliary spring accurately and controlling the sensor orientation are of critical importance; levelling devices, counters, gears, markers, light beams etc. have received much attention and irregularities have been extensively investigated.

The instrument constants C, ℓ and m cannot usually be measured precisely enough. Hence empirical calibration of the whole system is necessary when fully assembled in order to convert the raw observations to the physical unit of gravity. The instrument is taken to a set of points with known gravity differences. The measuring unit is the arbitrary division *div* of an instrument scale; *div* (not to be confused with the divergence operator) is introduced here as the general instrument unit before conversion into the physical unit mGal.

Optical readings of a counter, gauge or meter with a pointer moving along a scale are in arbitrary units; tenths of the finest division can be estimated, sometimes with the aid of a *vernier*. The numbers are noted on a form or in a field book. New instruments have built-in analogue-digital converters which avoid the somewhat subjective fine estimations and gross copying mistakes. And the state of the art is to directly store the readings in a data logger together with any additional information provided by the observer.

A characteristic of spring gravity meters is their drift (Sect. 3.3.2): even with constant gravity the readings change with time. Part of this is predictable or controllable, other effects are not. Especially quartz glass, which is used in several instrument brands (chosen for its thermo-elastic properties), is a visco-elastic or Maxwell fluid, metal is to a lesser extent. This is described as a Newtonian fluid component (dashpot) and a perfectly elastic spring in series (Fig. 3.2.2). Such a material always flows if loaded (by gravity) even though the flow may be very slow. The resulting drift behaviour is a slow spring lengthening, i.e. an apparent gravity increase. If this component is linear and small, it can be assessed and removed. Another kind of material behaviour is elastic afterworking characterized by an elastic spring and a dashpot in parallel: the flow response of the viscous term to short-term loading is gradually transferred to the elastic term so

Fig. 3.2.2 (**a**) Visco-elastic Maxwell body, illustrating the mechanical behaviour of physical springs. (**b**) When the force f acts, the constant elastic lengthening is complemented by a viscous lengthening growing linearly with time. (**c**) If dl is prescribed, the initial elastic force relaxes with time as the lengthening is transferred to the viscous term

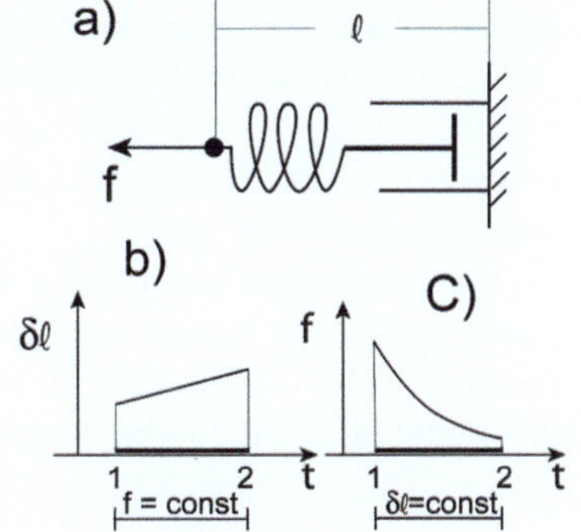

that the system approaches the elastic static limit slowly. Partly crystallized materials may go through sudden recystallizations, cracking or phase changes which make the readings suddenly jump or tear. Additional disturbances may result from malfunctions and voltage change of the power supply. Beside all these internal effects, external environmental influences cannot be shielded off perfectly: temperature, T, and air pressure p change, geomagnetism, vibrations, motions, wind and the history of these states and processes during a survey. Protection is partly achieved by placing the sensor into a thermostatted evacuated Dewar receptacle and by damping its suspension. Bi-metal thermostats can be calibrated such that thermal effects are near zero in the chosen temperature range which is determined in the production process. The sensor device must also be insulated mechanically from deformations resulting from thermal and pressure-induced stresses in the outside housing. Many of these constructional specialities are manufacturer's secrets. Finally, effects of instrument transportation through vibrations and irregular accelerations are impossible to fully control. The drift rate may be rather steady during a period of consistent measurements, and then, for example, during a rest or when transport conditions change, unusual drift may result. It may be caused by internal flow of gas (even though evacuated) or slight friction-generated temperature changes integrated over some time. Such behaviour is very irregular and unpredictable.

Without a satisfactory theory no modelling is possible (see, Sect. 3.2.1). Long series of observations at identical stations or station pairs are needed, and experienced observers know, for example, of the instrumental individuality. Repeated readings at some stations are therefore necessary in order to control or define the drift and possible tears (Sect. 3.3) and to assess errors (after removal of the gravity earth tides). As a rule for exploration surveys, 1 to 2 hour intervals between base station readings are customary, but longer intervals are often enforced by survey size, transport and economy; shorter intervals may be realized in high-precision micro-gravity surveys: Accordingly, the drift behaviour at very short periods remains unknown (shorter than the instrument transport permits; if the instrument is not transported, the test conditions are not relevant). Instrument drift is therefore one of the most serious limitation to the measurement accuracy.

Today much of the manual field work of levelling and adjusting the instruments, reading the scales, noting down the readings has been transferred to automated procedures. Digital storage can be combined with processors which check the performance and calculate the statistics of many repeated measurements such that more reliable and more accurate observations are achieved in affordable time and the statistical errors are obtained. Human errors are reduced. These advantages do, however, also have their price.

The gravimeter has met the needs of mineral and petroleum exploration for 50 years, but it was incapable of airborne operation as the gravitational signals are a factor of $< 10^{-7}$ of the aircraft accelerations. New developments include integrated in-motion GPS and gradient measurements (see, Sect. 3.2.9).

Spring gravimeters can be modified for recording temporal gravity variations, especially the earth tides. During the Apollo 17 mission a spring balance was adapted to the purpose by measuring capacitor plate position (Giganti et al., 1973). The purpose of the lunar surface gravimeter experiment was to obtain the lunar surface gravitational acceleration (to 10^{-11} of lunar gravity and about 10^{-5} relative to Earth gravity) and its temporal variations to determine the magnitude of lunar surface deformation due to tidal forces, measurement of vertical components of lunar natural seismicity, and monitoring of free oscillations of the Moon that may be induced by gravitational radiation from cosmic sources. The equipment consisted of spring mass suspension capacitor plates, electronics and a sunshield. The crew deployed this experiment by levelling and alignment within $\pm 3°$, using the sunshield shadow, and matching the cable to the central station. It was planned for two years operation. Temperature control had to be highly stable. Unfortunately this experiment was only partly successful, but it demonstrated the feasibility of interplanetary gravity measurements.

3.2.4 Vibrating String Gravity Meters

Instruments that exploit the influence of gravity on the vibration frequency of a spring or springs in tension due to a mass (m) weight $(m \cdot g)$ were developed as early as in the 1920ies; such a set-up was designed by Lord Cadman in 1925 and later used by him to detect oil deposits in the Iranian desert. The advantage of vibrating string gravimeters is the wide gravity range (akin to absolute), and they are also less sensitive to platform movements and were deployed in submarines. The principle has been used on Apollo 17 (Traverse Gravimeter Experiment, TGE) with a gimbal mounted, double-stringed Bosch Arma D4E vibrating string accelerometer to establish an Earth-Moon gravity tie. A mass is suspended freely between two springs. The vibration frequency of the mass is measured. For more information see: http://www.geophysics.rice.edu/department/research/manik1/apollo17.html

3.2.5 Beam Balances

Ultra-precise beam balances have been employed to measure near field effects of changing or moving masses, achieved by suspending two masses at different distances from the moving mass, for example, a changing water table at a dam (e.g. Hubler et al., 1995); it is not a method applicable to usual geophysical gravity surveys but aims at constraining the gravitational constant, G, in order to test for non-Newtonian effects, i.e., a suspected distance (r) dependence of G; so far no such effects have been discovered. The use of beam balances is somewhat similar to the torsion balance (Sects. 2.8, 3.2.9), but it is only sensitive to differences in the vertical component of gravitation.

3.2.6 Absolute Gravity Meters

The development of absolute free-fall instruments was delayed until time measurements became sufficiently accurate and methods were invented by which adverse environmental effects could be neutralized. The instruments catapult prisms into a vertical drop about its apex in a high vacuum. Laser pulse and reflection times are measured and converted to length, i.e. location h of the prism for repeated light pulses during flight. Absolute gravity is the acceleration g of the free fall of the prism with the initial height h_o and the initial velocity v_o:

$$h = h_o + v_o t - g/2\, t^2 \tag{3.2.4}$$

High-precision measurement of a series of reflections is carried out with the aid of nano-second pulse lasers, accurate oscillation standards and clocks, assuming constant speed of light c. Many repeated drops can be carried out automatically and the constants of Eq. (3.2.4), h_o, v_o, g, are determined in a least-squares parabola fit to the observations giving also the statistical error. Presently small field instruments exist (Type A10 of Microgsolutions Inc., Erie, CO, USA; for more information see: www.microgsolutions.com) which reach an absolute accuracy of better than $10\,\mu$Gal. Improvements in accuracy, necessary measuring period (< 1 hr), power supply, robustness and costs can be expected in the near future.

Considerable development was invested in controlling environmental disturbances, as gas friction from insufficient vacuum and instability of the platform and structure through microseismic ground noise. Successful constructions place the evacuated free-fall tower on the inert mass of a critically damped long-period seismometer. In the 1980ies the problems were solved in the laboratory, and the first step toward field instruments were bulky and heavy apparatuses which had to be assembled inside rooms and required many hours of usually overnight automatic runs of thousands of drops. They mainly served the establishment of absolute gravity control networks. In the future, combined surveys with absolute and relative gravimeters may become the optimum for controlling the instrument drift and scale and economically obtaining high-precision gravity data (see, Sect. 3.3).

3.2.7 Superconducting Gravity Meters

Stationary instruments of highest precision are the superconducting gravity meters which are not suited for field work. They record very small temporal gravity variations, especially earth tides to $10^{-10} g$ or perhaps $10^{-11} g$. A permanent magnetic field is generated by an electric current in a cooled super-conducting helium ring; a conducting sphere is suspended in the inhomogeneous region of the field and held at a stationary position. Small variations of gravity shift the equilibrium position up or down which is counteracted by a sensor-controlled additional ring current in a coil, recorded and converted to the gravity signal. Beside tides and the effects of small mass changes, changes in Earth rotation are observed.

3.2.8 Artificial Satellites

Satellites are used to measure the gravity field in several ways, for example, by their orbit perturbations and as carriers of radar altimeters and gradiometers. Measuring the gravity field with satellites began with Sputnik in 1957. *Orbit perturbations* are astronomically measured and used to derive models of the geopotential usually expanded into spherical harmonics; the gravity field can be expressed as the gravity disturbance (see, Sect. 4.3). As more satellites became available and observation and determination of orbits became more accurate, more and more detailed solutions were published, also in combination with terrestrial and marine data.

The next step were the *radar satellites*, the first being SEASAT 1978 which had a radar altimeter onboard that measured the sea surface topography. Since, to first order, sea level follows the geoid, its topography is an indirect measure of the gravity potential and gravity field. The altimeter footprint is generally wide enough to more or less average out the waves, but longer-wavelength deviations from the geoid are generated, for example, by ocean currents (Coriolis effect), wind drag and water density variations of thermal and salinity origin. Ever improved accuracy and resolution of the surface topography and mutual control is achieved with the altimeter satellite TOPEX/Poseidon (1992), reinforced since 2001 by Jason-1 flying in tandem.

Current developments (Rummel et al., 2002) are dedicated satellite gravity missions as CHAMP (launched in 2000), GRACE (launched in 2002) and GOCE (scheduled for 2008). CHAMP, on a low, nearly polar orbit, has a pair of hypersensitive accelerometers onboard which measure the gravity gradient or second radial derivative of the Earth's potential W_{rr}. Nearly the whole Earth is repeatedly covered by the satellite tracks. Integration over r and downward continuation to the Earth's surface render the gravity field at geoid or reference ellipsoid level. GRACE is a tandem mission with two satellites chasing each other with the distance between them measured by reflection of laser pulses. The distance variations are analyzed in terms of the gravity variations. The sensitivity is improved over that of CHAMP. A particular feature is the time resolution, as the GRACE repeats the same tracks in shorter intervals, so that temporal gravity variations become measurable. GOCE (ESA, 1999) carries three accelerometer pairs, each mounted 0.5 m apart on bars, one radial, one along flight and one across flight, so that the full gravity tensor (see, Sect. 2.8) will be measured. The orbit is particularly low requiring special feedback and thrusters to achieve drag-free flight and high-resolution gradient measurements. The low orbit limits the flight period of GOCE and the ability to identify temporal field changes, but GOCE is expected to render high spatial resolution. At about 100 km pixel dimension (half wavelength) a precision of about 1 mGal and 2 cm in gravity and geoid height, respectively (GOCE, 2005) promises to distinguish the effects of gravity and of currents on the sea surface topography.

Satellites orbiting the Moon and planets have rendered, mainly by tracking their orbit perturbations, the lunar and some planetary gravity fields related to surface features. For very small celestial bodies as planetoids and comets, estimates of their near-surface gravitation, and hence their mass and density, can be obtained from video-tracking particles that have been thrown up in an artificial impact. This has been tested with some success in the NASA Deep Impact Mission to comet Tempel 1 (Richardson et al., 2005).

3.2.9 Torsion Balance and Gradiometer

The space derivatives of gravity, or better, the second derivatives of the potential are measured with the torsion balance and, more recently, with gradiometers. The measurement of gradients of gravity with the torsion balance was pioneered by Roland von Eötvös (1896); he measured the difference between the gravitational attraction at two points about 1 m apart and achieved high sensitivities. The torsion balance measures some components of the gravity tensor (Sect. 2.8), i.e. second derivatives of the gravitational potential. It has been used in exploration geophysics and geodesy for 5 decades, but the procedures are slow (one measurement per day), and very sensitive to external influences (such as temperature). For such practical reasons the torsion balance was replaced by the gravity meter.

In the 1970s, missile launching began to require exact knowledge of the gravity gradient at the point of launch. This led to a new generation of gravity gradiometers. Moreover, recently ultra-high precision electrostatic accelerometers have been constructed and assembled to devices measuring gravity gradients: gradiometers, capable of rendering the full gravity tensor. Modern devices measure gradients across short distances of, say, 0.5 m. Such gradiometers consisting of displaced accelerometers are now employed in borehole tools, on airplanes and in satellites as CHAMP, GRACE and GOCE. Since 1999, airborne gravity gradiometer systems have been in operation in Canada, Australia, South Africa, etc. Gravity is derived by integration and downward continuation and is comparable to ground gravity collected on a 200m grid; flight lines are typically about 100m above ground level.

Other measurement types are in development, as a superconducting vibrating string gradiometer, a device where the length of the string under tension of a gravitational field is measured by two SQUIDs at the ends of the string (a SQUID is a **S**uperconducting **QU**antum **I**nterference **D**evice for the precise measurement of extremely small magnetic field variations; consult Internet for latest developments).

3.2.10 Special Task Gravity Meters

Beside different methods to measure gravity, special task instruments are of interest. Standard gravimeters are modified for particular purposes, for example, for deployment in boreholes, for underwater measurements and for ship-borne and airborne surveys.

3.2.10.1 Borehole Gravimetry

The deployment in boreholes requires the instruments to be adapted to borehole geometry and environment, remote control etc. As a tool used in possibly hot drilling mud under pressure and in rough transport conditions, the instruments must be mechanically and thermally well shielded. The vertical orientation must be precisely controlled.

3.2.10.2 Underwater Gravimeters

For shallow water surveys, especially on continental shelves and in rivers, land gravimeters have been converted by several measures. The instrument may be fitted into a water-tight pressure casing, which is lowered to the bottom and held there with weights. An automatic levelling device must be added. Remote controls are used for clamping and automatic reading or data logging. Field procedures are time consuming.

3.2.10.3 Sea Gravimeters

Gravimeters carried on moving ships require strong damping, resulting in extended averaging times, and are mounted on gyro-stabilized platforms near the neutral point of the ship which has minimum motion, and elevation is generally very close to sea level. Navigation with GPS has added to ever improving performance and accuracy. Especially the E-W component of ship velocity v_E requires application of the *Eötvös reduction* which takes account of the change in the platform angular velocity and hence its centrifugal acceleration z_o (see, Sect. 2.3, eq. (3.3.3)). Earth angular velocity $\omega \approx 7.25 \ 10^{-5} \, \text{s}^{-1}$ is modified for the moving gravimeter, and in eq. (3.5.7) the ω changes to $\omega + \delta\omega$; $\delta z_0 = 2r \ \omega \ \delta\omega \ \cos^2 \varphi$, and with $\delta\omega = v_E/(r\cos\varphi)$ follows:

$$\delta z_0 = 2 \ \omega \ v_E \ \cos\varphi \tag{3.2.5}$$

If a ship moves 1 m/s (about 10 knots) eastward at the equator, the Eötvös effect would be to reduce z_0 and hence increase the apparent g by about 15 mGal.

Usually surveys are planned such that the crossings of profiles allow some drift control. This way accuracies of 0.5 to 1 mGal have been achieved

3.2.10.4 Aerogravity

Airplanes are similar mobile platforms as ships, however much faster, not at a nearly fixed elevation and more strongly subject to erratic motions. For real-time control of position, velocities and accelerations, GPS is of paramount importance. High frequency data acquisition allows calculation of all the disturbing effects of fast and irregular flight with reasonable precision. The Eötvös reduction is much more critical than for ship borne gravity, it will easily reach 200 to 800 mGal. Obviously the velocity vector must be accurately measured which today is achieved by continuous GPS observations. Aerogravity is an integrated system of gravimetry measurements and real-time navigation. Under certain circumstances, as in mountainous regions, aerogravity successfully competes with land-based gravimetry; the latter suffers from the uncertainties of the near field terrain effects. Airborne gravity gradiometers, on the other hand, are less sensitive to platform movement and are now achieving high accuracies after integration of gravity gradients and downward continuation to ground level. Comparisons have been made that suggest the milligal precision is, or will soon be, reached.

3.3 Scale and Drift of Gravimeters

Traditionally most surveys are still carried out with relative gravimeters. Such instruments must be calibrated because their scale values, i.e., the correct gravity difference per instrument division (mGal/*div*) cannot be precisely determined from the constructional components directly. With field-going absolute gravity meters (Sect. 3.2.6), combined surveys with absolute and relative gravimeters can integrate scale and drift determination with the principal purpose of measuring gravity into one single procedure and a compromise between accuracy and economy.

3.3.1 Instrument Scale

Readings must be generally converted to gravity values, usually in milliGal (mGal), via scale factors in mGal/*div* (or a sliding scale factor across the whole range). Conversion from the scale units *div*, noted in the field, to milliGal is usually done in the office, but automatic digital output can provide the wanted unit immediately. Instrument manufacturers provide their best scale factors, which may, however, change with time. Calibration is then necessary and mostly done by measuring in divisions (Δdiv) known gravity differences Δg between points along a calibration line and may be carried out also in gravity networks (see, Sects. 3.3.2, 3.6.1).

3.3.2 Instrumental Drift

Raw field observations are always perturbed by unwanted instrumental effects as drift and sudden discontinuous changes (tears or jumps) and by geophysical effects as tidal variations. To help discover mistakes, as outliers (see, Sects. 3.7.1, 4.7.2), and correct them by new measurements, some cleaning is best done immediately

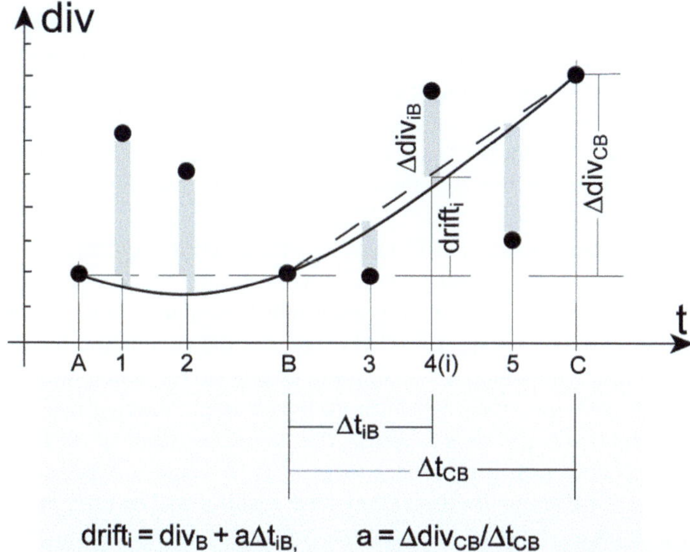

$$\text{drift}_i = div_B + a\Delta t_{iB}, \qquad a = \Delta div_{CB}/\Delta t_{CB}$$

Fig. 3.2.3 Gravimeter drift measured in scale units or divisions *div* and defined by the base station (A, B, …) readings at t_A, t_B,… The field station (1, 2, …) readings at t_1, t_2, … are referenced to the assumed drift curve; note that the smoothest drift behaviour is but an assumption

during a survey. This can be done by drawing a drift curve manually from the field records (Fig. 3.2.3) which may, at this preliminary stage, include both instrumental drift and tidal variations. If several base stations (Sects. 3.4.2, 3.4.3) are used, their mutual gravity differences should be known as accurately as possible. Applying programs as GRAVI (Smilde, unpublished, 1995; see below) is possible also before tidal effects are subtracted, but the tidal reduction (Sect. 4.2) may be included anyway.

Although a practical borderline cannot be drawn, internal instrument effects and external tidal effects are of principally different nature, and the correct way is to first remove the tidal effects, usually the theoretical solid-earth tides (at this stage converted to scale units or divisions, *div*, see, Sect. 4.2).

Removal of instrumental effects (in *div*) is, strictly speaking, not a reduction and is described in this chapter. Fig. 3.2.3 shows the principle of the drift and drift corrections. Two possibilities are shown: linear drift between base readings and a smooth curve which in the case of three base readings may be a parabola. Although there are no reasons for assuming either possibility or any other simple time behaviour, the most practical solutions are usually chosen. Wanted is the difference between the reading div_i at station i and the hypothetical reading $drift_i$ at the same time t_i : $\Delta div_{iB} = div_i - drift_i$. If $drift_i(BC)$ is assumed linear between the readings at B and C, $\Delta div_{iB} = div_i - div_B - a\Delta t_{iB}$ with $a = \Delta div_{CB}/\Delta t_{CB}$ (where $\Delta div_{CB} = div_C - div_B$ and $\Delta t_{CB} = t_C - t_B$). If the drift is referred to station B, $drift_B \equiv 0$ or $\Delta div_{iB} = div_i - a\Delta t_{iB}$.

Similarly, one may assume polynomials or other functionals of time for the drift, but such assumptions are equally arbitrary. One may interpolate or approximate the assumed drift curves between base readings; interpolation implies error-free readings while approximation, for example, by a least-squares fit, implies that reading errors are taken into account, also for the base readings. The fit may be calculated as a Bayesian inversion, which means that the standard errors of the readings are taken as a priori information.

The routine GRAVI (unpubl., Smilde, 1995) assumes no functional of time but is rather probabilistic as it considers correlation lengths (in time) during which point readings are likely to be somewhat correlated to each other; the correlation can be described as a Gauss function $W = \exp(\Delta t^2/\tau^2)$ where τ is the correlation length and W constitutes the weight with which a base reading affects the reference value at the reading time t_i at station i. Any station with more than a single reading is taken into account. The reference value is thus taken as the weighted mean of all base readings at their appropriate times. Since for $|t - t_B| > \tau$, W rapidly approaches zero, a station i outside a correlation window of any base reading will be referenced only to the ordinary arithmetic mean of all reference values. In GRAVI all mutual differences between base station values $drift_k = div_k$ are calculated by solving all the respective observation equations simultaneously: for the reading $div_{j(i)}$ at $t = t_j$ at a repeated station i: $div_{j(i)} - div_i - (drift_i + Ddrift_j) = 0 + e_j$, where e_j is the error of the measurement j and $drift_j$ is a discontinuity during the respective measuring epoch, or in short: $\underline{U}y = 0 + e$. The vector y contains all observations (also non-repeated ones), which naturally include drift and discontinuities, e is the error vector and $\underline{U}$ is the coefficient matrix containing only $0, +1$ and -1 values signifying measurements at i and j. The errors are statistically given by the covariance matrix $\underline{C}(e)$ which is here a diagonal matrix because the errors are assumed uncorrelated. The temporal correlations are given by the matrix $\underline{C}(y)$ containing the above $W = \exp(\Delta t^2/\tau^2)$. The least-squares solution of the system of equations is discussed in detail in Sect. 7.2.1.2.1.

It may be mentioned that the concept of GRAVI is closely related to the geostatistical approach to spatial interpolation between measured points which is realized in the method of "Kriging". It is based on the assumption of a stationary stochastic process, just as in the case of GRAVI. The statistical space behaviour of the (gravity) values in a discrete point set is estimated from variograms which may represent Gaussian or other error distributions. The method is described in more detail in Sect. 5.1.5.

A program as GRAVI should also identify outliers (Sects. 3.8.1, 4.7.2) at repeated stations after the adjustment and include a tidal reduction. If more than one gravimeter is used in a survey, the statistically uncertain scale factors should be adjustable (see above, Sect. 3.3.1) with one instrument chosen as the master gravimeter. Some gravimeters have non-linear scale factors (e.g. the L&R-G spring gravimeters with their piecewise linear scale); this introduces non-linearity into the mutual adjustment procedure which can be solved by iteration. Scale factors can, of course, also be adjusted to fit absolute gravity observations at some points (Sect. 3.2.6).

3.4 Planning a Survey

3.4.1 General

The starting point of any planning is the aim of an investigation. In geophysics nearly always, geological or geodynamic problems are to be solved, and mostly the targets are spatial density structures, sometimes an ongoing temporal change, and the associated gravity variations or anomalies, where a single point would be irrelevant. In geodesy the situation is somewhat different, as a few points with reliable gravity values may serve as base stations for relative networks; such points may be locally isolated or widely spaced. In order to delineate the anticipated anomalies, spatially distributed gravity values are required and observed, for example, along profiles or on point grids. It must be kept in mind that the point distribution should permit a reliable construction of the relevant anomalies; the problem is related to the aim of the later interpretation, as stressed in Sect. 1.4 and will be described in Sect. 5.1.5. The underlying geological mass distributions (that may change with time) prescribe the characteristic length scales and orientations of the anomaly structure, but during surveys unexpected features may be found which will affect or change further planning.

A priori ideas about survey aims determine the planning. There is mostly some preliminary knowledge of regional geology, as rock types, estimated density contrasts, scale and structural strike, or the processes which induce change. The more is known, the more dedicated surveys will be planned.

Besides, logistic and related aspects are important which include the topography of the survey area, the instrument characteristics and error sources related to them; for example, instruments have individual properties that have to be taken into account. The higher the required accuracy, the more stringent these latter conditions are. While regular grids may appear ideal, transport on roads, paths, tracks or the like will usually determine the grid layout and also affect the instrument performance. Repeat times at base stations must be considered in planning in order to recover the instrument drift. Preliminary field inspection and assessment of problems and possibilities are strongly recommendable.

3.4.2 Base Stations

Base stations are chosen for logistic and technical reasons as, for example, accessibility, ground stability and future availability. Hence such stations will have to be well marked, listed and archived (with point sketches). Reliable permanence is most desirable, however, experience has shown that marked stations are quickly destroyed, for example, in road construction and other activities.

Markers must allow a precision of 1 cm in height or better, corresponding to 2 to 3 μGal gravity precision; for measuring temporal change of gravity millimetre precision is necessary. Horizontal requirements are in the decimetre range, but high horizontal gravity gradients (100 E $= 10\,\mu$Gal/m; for the unit E see, Sect. 2.2) should be avoided, this means to stay away from steep slopes and massive structures. For high-precision surveys, stability requires time variations to be minimal in nearby masses as rivers, the sea, the groundwater table, or any man-made structures as piles of material, buildings, tanks, for example, of fuelling stations, mine workings etc. Qualitative or semi-quantitative estimates of such effects (Sect. 5.6) usually suffice.

3.4.3 Base Station Networks

Base station networks are set up by measuring the gravity differences between the stations in several direct ties (where the behaviour of instrument drift should be considered). A recommended scheme is A-B-A-B-A; this can be extended along a sequence A, B, C,... as A-B-A-B-C-B-C-D-C-.... Customarily, the stations are connected in a triangular pattern which somewhat facilitates the network adjustment (see below) and error controls. Several instruments may be used side by side for mutual drift control. In order to reduce any temporal interference with spatial gravity variation, some random sequence is recommendable; if, for example, a profile of points 1, 2, 3,..., n is to be measured up a mountain, an irregular sequence is preferred over a consecutive one.

Adjustment of networks will be necessary because errors in the gravity differences between individual base stations will lead to closure errors. Given, for example, three stations A, B, C in a triangle, the condition is that $\Delta g_{AB} + \Delta g_{BC} + \Delta g_{CA} = 0$. If nothing more is known about the error probabilities, the inconsistency εg is removed by dividing it into equal parts, i.e. by subtracting $\varepsilon g/3$ from each gravity difference. If A, B, C (etc.) are linked together in a larger base station network, there are also more conditions, usually more than one for Δg_{ik}, and a network adjustment with all the given conditions should be carried out. A least-squares procedure with zero closure conditions to be satisfied is described in KJ61, 41–52.

Surveys that aim at measuring temporal change between stations may be carried out in the same way as base station links are measured, in order to describe the instrument drift as accurately as possible. Short repeat times at the stations are essential. Often several instruments are used side by side. As absolute field gravity meters with satisfactory accuracy become available, surveys may combine absolute and relative instruments, or employ absolute gravimeters exclusively. The strategy will, in any case, be determined by the distances between those stations and by the available time.

3.4.4 Field Stations

Field stations are selected in view of the geological aim, but avoiding error influences is also a criterion, as instrument instability, nearby high relief etc. Accurate point position and height are realized by using permanent or temporary survey markers, especially surveyed pegs or simultaneous differential GPS. In unmapped regions, aerial photographs, astronomical surveying and levelling or even barometry were used in the past, and undisturbed single receiver GPS may be sufficient for a reconnaissance survey. In well mapped regions, reliable maps provide locations, but heights must be measured.

Normal gravity varies in N-S direction, i.e. with latitude φ, as $\sim 800|\sin 2\varphi|\mu$Gal/km, and 10 m point uncertainty (0.2 mm on a good 1:50 000 topographic map) implies $< 10\mu$Gal uncertainty. Modern theodolites (total stations with distance measurement to retro-mirrors) and differential GPS will permit cm accuracy in height giving about 3μGal errors in the Free Air reduction and 2μGal in Free Air plus Bouguer reduction.

Nearby high relief, including buildings may have big terrain effects difficult to determine accurately. The station distance to a steep slope should be kept as large as possible; the effect depends on details; for example, a steep 2 m high upward slope 1m beside a gravimeter that is 1m above ground has nearly no effect as it is about level with the instrument; if such a slope is downward it may cause -20μGal terrain effect; even a 25 cm step down, 20 cm sideways, causes a near-field terrain effect of about 10μGal. Hence ditches should be avoided, while walls are less critical. It is obvious that 10μGal accuracy is difficult to achieve. Rough estimates are discussed in Sect. 5.6. Surveyors establishing gravity stations should consider these points.

3.5 Field Procedures

3.5.1 Setting Up Stations

This section is concerned mostly with ordinary land gravimeters. Vertical instrument orientation by levelling is obviously indispensable (in some modern instruments automatic). Gravity meters must be set stably on the ground, usually on a base plate or a tripod. This includes the ground which must be as solid and undisturbed as possible. If a tripod is used, it must stand firmly and the legs must be pushed in deep enough into soft soil; asphalt (tarmac) in sunshine is unsuitable as it slowly creeps. The instrument position must be defined (measured) relative to some permanent or temporary survey marker (or GPS receiver). A plate with no legs requires level ground (sometimes forgotten by surveyors establishing field stations). Special instructions should be taken into account carefully.

The above considerations of station stability and accessibility are especially important for base stations (Sect. 3.4.2) or any stations that are to be measured repeatedly. Underground conditions must be solid and unlikely to change with time.

3.5.2 Surveying Requirements

Field procedures must be considered when planning, but unforeseen circumstances will always force field observers to make spontaneous decisions. These may affect repeated readings at one or several base stations as well as normal field points. They are usually measured along loops between base stations and are carried out as the situation allows. Exact locations may be chosen right then or by the surveyors beforehand. Gravity observers must consider surveying requirements as the lines of sight, visibility of GPS satellites, etc.

3.6 Additional Field Operations

Additional field operations that are important are: instrument calibration, surveying (just mentioned), determining the near field topographic relief and rock densities. This information is needed for reliable measurements and their reductions and evaluation.

3.6.1 Instrument Calibration

Gravimeter calibration by measuring differences (with standard errors s) between points of known absolute gravity values can be done in combination with absolute gravimeters that theoretically need no calibration. If absolute gravity is not known, i.e. is only estimated in the network, calibration may be carried out within given relative networks. In this case calibration means only adjustment to an earlier accepted scale.

Although two end points with a sufficiently large gravity difference often suffice principally for calibration of a linear scale, an assessment of the uncertainties requires more points, for example, along a calibration line or in a point network (KJ61, 50). Such measurements permit to take into account scale variations (e.g. periodic spindle variations in L&R gravimeters) and to better control the instrument drift. Aliasing between the temporal instrument drift and the temporal sequence of measuring the spatial gravity differences along calibration lines is to be avoided; particular observation schedules have been designed which involve certain random elements or sequences; often more than one gravimeter are used simultaneously (see specialized literature, e.g. Kanngieser et al., 1983).

As a simple example, take $g_{abs} \pm s_g$ to be known at two points, where s is the standard error, the difference is $\Delta g_{abs} \pm (s_{g1}^2 + s_{g2}^2)^{1/2}$ or $\pm s_g \sqrt{2}$ (if $s_{g1} = s_{g2} = s_g$). Similarly, $\Delta div \pm (s_{d1}^2 + s_{d2}^2)^{1/2}$ or $\pm s_d \sqrt{2}$. Repeated measurements improve the error by a factor of $n^{-1/2}$. Least-squares adjustment of the gravimeter scale factor f with n absolute stations begins with the error equations for each station i:

$g_{abs,i} = c + f \Delta div_i - v_i$; $i = 1, 2, \ldots n$, $v_i =$ correction for any of the errors at station i. The constant c is necessary because the gravity readings are relative to a station where g_{abs} is not accurately known. The task is that of linear regression with two unknowns, c and f. Writing Σ for sums over $i = 1, n$ with the condition

$\Sigma v^2 = \min$ (i.e. the derivatives with respect to the unknowns set zero), leads to the linear normal equations

$$cn + f\Sigma\Delta div = \Sigma g_{abs} \qquad (3.6.1)$$
$$c\Sigma\Delta div + \Sigma(\Delta div)^2 = \Sigma(\Delta div\, g_{abs})$$

Following Cramer's rule, the solution is, with the determinant D

$$D = n\Sigma(\Delta div)^2 - (\Sigma\Delta div)^2$$
$$c = [\Sigma(g_{abs})\Sigma(\Delta div)^2 - \Sigma\Delta div\Sigma(\Delta div\, g_{abs})]/D \text{ (mGal)} \qquad (3.6.2)$$
$$f = [n\Sigma(\Delta div\, g_{abs}) - \Sigma\Delta div\Sigma g_{abs}]/D \text{ (mGal}/div)$$

With the standard deviation of the fit, $s = (\Sigma v^2/(n-2))^{1/2}$ the standard errors of c and f are given, with the matrix $\{N\}$ of the normal equations, as:

$$s_c = s(N_{11}^{-1})^{1/2}; s_f = s(N_{22}^{-1})^{1/2} \qquad (3.6.3)$$

where $\{N^{-1}\}$ is the inverse matrix $\{N\}$:

$$N_{11}^{-1} = \Sigma(\Delta div)^2/D; \, N_{22}^{-1} = n/D. \qquad (3.6.4)$$

This simple example of inversion has only two unknowns. If the individual differences in g and div have different errors they should be weighted accordingly, and this requires some reasonable assessment of the individual errors. See also Sect. 3.3.2.

3.6.2 Surveying or Levelling and Recording of Earth Tides

Coordinates and elevations are indispensable for reducing and interpreting the gravity observations. With good maps, careful identification of points on the maps and reading the coordinates from them may replace surveying. Levelling, however, is imperative unless accuracy is sacrificed; reading elevations from maps is generally insufficient. Till today, theodolites or total stations are used and this is more time consuming than measuring gravity. In the past, barometers were used in reconnaissance surveys, but GPS has improved the situation. Determination of the near-field neighbouring relief is still time-consuming, even if facilitated with special diagrams (as had often used in industry for the traditional terrain reductions on the basis of topographic maps).

During special precise surveys local earth tides in the region should be known and observations are recommended, because the theoretical tides are often only rough approximations to the local tidal effects. This is especially so in anomalous regions of unusual geology and upper mantle properties and close to the sea shore. Marine tides will be better defined by observation than by gross marine tidal models.

3.6.3 *Rock Densities*

Density ρ is a fundamental quantity in mass reduction, modelling and interpretation. Errors in assumed densities are critical and may interact with other errors, e.g. of elevation. The most direct density information is determination from samples. Another observation is the gamma ray back scatter mostly used in borehole geophysics (gamma-gamma log). Relationships between other geological and geophysical quantities can be used as proxy, as and measured seismic velocities or rock type, identified in the field or taken from sufficiently detailed geological maps. Density-velocity systematics or relationships are frequently used to derive initial densities from seismic models. Special gravity surveys may give representative density values. Note that rock density is a variable property in nature even within apparently uniform rock units or formations and any given numerical value is preliminary and may need adjustment in the process of gravity interpretation.

3.6.3.1 Hand Specimens

Hand specimens (also drill cores) should be routinely collected and measured in the field or lab by weighing samples of volume V in air, $w_a = \rho V g$, and water, $w_w = (\rho - \rho_w)Vg$; the density ratio is $\rho/\rho_w = w_a/(w_a - w_w)$. In view of the natural variations the individual weighing need not be very precise, rather the number of samples or total mass should be large for the average and the natural scatter or standard deviation to become more representative. A spring balance and a pale of water will do, but attention must be paid to possible systematic errors (scales, influence of weathering, dryness or water saturation, weight of sample holders, density of salt water, etc.) which cannot generally be corrected for. Calibration by an exact test mass is recommended. The volume measurement can be done also in a measuring receptacle with a scale for reading the difference when the specimen is inserted.

In many cases, rather small drill cores are taken as samples, even drill chips. While in these cases weathering is less critical, the results tend to vary erratically and the scatter is large. And chips may be highly selective and introduce systematic errors. If the mineral content of the rocks is known, the weighted mean density can be calculated from (precisely known) mineral densities ρ_k and their proportions p_k : $\rho =_k \Sigma \rho_k p_k/_k \Sigma p_k$, however, the void volume must not be neglected.

3.6.3.2 Gamma Ray Back Scatter

Back scatter of gamma rays (Compton scattering) is the basis of the gamma-gamma borehole log and also of some field instruments which are placed on planar rock surfaces. They exploit the fact that electron density is roughly proportional to mass density, but systematic deviations from the relevant bulk density are common. Weak ^{60}Co gamma ray sources are commonly used combined with a caesium iodide detector (CsI); if a spectral analysis is performed (CGG – Spectral Gamma-Gamma), even some information on the elemental composition can be derived. The volume affected by the measurement is as small (order 1–2 liter) as that of hand specimens and thus many individual values are necessary to obtain representative averages.

3.6.3.3 Rock Types

Rock type identification in the field or from geological maps, combined with density-rock type tables often suffices for initial estimates. Such estimated values are sometimes more representative than scattered and biased individual measurements, since rock varies in composition and condition from one specimen to another. Good geological background knowledge of the geophysical interpreter helps avoid gross misjudgements. Table 3.1 (at the end of the chapter), however, demonstrates a high variability of values given for the same rock type.

3.6.3.4 Seismic Velocities

Velocity-density relations or systematics are much used in gravity interpretation, especially where initial models are taken from seismic information. The seismic wave speeds Vp and Vs in elastic solids depend on the relevant moduli $K + 4/3\mu$ and μ, respectively, and density ρ:
$Vp = ((K + 4/3\mu)/\rho)^{1/2}$, $Vs = (\mu/\rho)^{1/2}$, where $K = $ bulk modulus, $\mu = $ shear modulus. While this seems to suggest an inverse relationship between velocity and density, the moduli generally increase with density and the correlation is mostly positive with only few exceptions (velocity-density systematics). An approximate linear relation has been derived from laboratory measurements on mostly crystalline rocks under pressure (Birch, 1960, 1061):

$$\rho \approx a + b\,Vp \text{ or } b \approx \Delta\rho/\Delta Vp \qquad (3.6.5)$$

b is of the order of 300 $(kg/m^3)/(km/s)$ or the inverse b′ is given in $(km/s)/(kg/m^3)$. Similar, though less customary relations exist for Vs. The exact value of b is immaterial because of its large variation and uncertainty, it will only render a rough first guess of density values. The relationship is influenced by chemical, petrological and other parameters as physical state, weathering etc. resulting in considerable scatter. Especially the compactness of a rock as freshness or weathering, jointing, cracking, looseness, water saturation, etc. have a strong influence, usually more on Vp and Vs than on ρ, for example, crystalline rocks, to considerable depth, say 500 m, are mostly cracked with a large influence on Vp and less on ρ. With sufficient data and a wide range of rock types, non-linearity of the ρ-V relationship becomes more evident. The so-called Nafe & Drake (1957) relation, mostly determined for a suite of rocks from unconsolidated marine sediments to basic igneous rocks, is highly non-linear, as evident in the following value pairs (Vp in km/s $-\rho$ in kg/m^3): 1.5–1300; 1.6–1700; 2.3–2100; 7–2800; with the b values: 4000; 570; 150 $(kg/m^3)/(km/s)$, respectively.

Additional parameters, as for example, mean atomic weight $\underline{m}$ can be taken into account (Birch, 1960, 1961, 1969); a rough relationship for crystalline crustal rocks from granite to gabbro is

$$V\mathrm{p} \approx \mathrm{const}\ \rho^{3/2}/\underline{m}^2, \ \text{or inverse}: \ \rho \approx V\mathrm{p}^{2/3}\underline{m}^{4/3} = (V\mathrm{p}\ \underline{m}^2)^{2/3}. \tag{3.6.6}$$

Birch's data also show that ρ of these rocks is correlated with $\underline{m}$, approximately $\Delta\rho \sim 300\ \Delta\underline{m}$ (kg/m^3).

On the other hand, evaluation of density and velocity data from petroleum industry drill holes, i.e. from sedimentary rocks renders (Gardner et al., 1974):

$$\rho \approx 1740\ V\mathrm{p}^{1/4} \text{ in kg/m}^3, \ V\mathrm{p} \text{ in km/s} \tag{3.6.7}$$

Similarly, but in more detail, Darbyshire et al. (2000) gives different expressions for three velocity intervals spanning the whole range of crustal rocks, for $V\mathrm{p} \leq 4.5\,\mathrm{km/s}$ from Zelt (1992):

$$\rho \approx 1000(-0.6997 + 2.2302\ V\mathrm{p} - 0.598\ V\mathrm{p}^2 + 0.0703\ V\mathrm{p}^3 - 0.0028311\ V\mathrm{p}^4)$$
$$\tag{3.6.8}$$

and from Carlson & Herrick (1990)

$$\text{for } 4.5 < V\mathrm{p} \leq 6.6\,\mathrm{km/s}: \rho \approx 1000(3.81 - 6.0/V\mathrm{p}) \tag{3.6.9}$$
$$\text{and for } V\mathrm{p} > 6.6\,\mathrm{km/s}: \rho \approx 1000(5.32 - 15.38/V\mathrm{p})$$

The expressions encompass the effects of composition and compaction. The expressions differ in the power of $V\mathrm{p}$ because they pertain to very different rocks and because, in eq. (3.6.6) $\underline{m}$ appears explicitly, while in the other expressions effects of $\underline{m}$ are implicit.

3.6.3.5 Vertical Gravity Profiles in Boreholes and 3D Surveys in Mine Workings and Tunnels

Vertical gravity profiles directly exploit the gravity effect of the in situ rock masses. Mines, mine shafts, drill holes (with borehole gravimeters, see above) and tunnels can be used for gravity surveys in three dimensions (3D). The largest effects in the observed gravity difference between two depth values are the height effect and twice that of the intervening Bouguer plate $(4\pi G\rho\Delta h)$ which is dominated by the nearby rock density.

Lateral density variations, as for example, the cavity itself (mine shaft and other workings, etc.), topographic relief and geological density variations must be considered in special reductions. Both latter effects vary with position and depth and must be estimated from a set of gravity stations in the 3D vicinity and from a special terrain reduction. For tunnel surveys, complemented by surface stations, the 3D geometry must be fully modelled. The important point is that the gravitational effects of the rock around the observation stations have the largest magnitude varying from station-to-station; hence, the best local density information can be obtained from such 3D surveys by 3D modelling.

Vertical density profiles are obtained also by gamma-gamma density logging (see above) in boreholes, and comparison of both sets of derived density values may reveal specific problems of either method.

3.6.3.6 Nettleton Profiles

The Nettleton (1939) method offers a similar possibility to determine a relevant bulk density from gravity measurements along profiles or on point grids with significant topographic relief as mountains or valleys (KJ61, 90–92). This is another way of introducing 3D aspects into the observations. If the topography (h or TOP) and Bouguer anomaly (BA) calculated with an assumed Bouguer density ρ_B are correlated it may be so because ρ_B differs from the local density value. The method is to calculate the BA with a suite of ρ_B values (reductions for height and terrain, of course, included) and plotted together with h; the curve showing the least correlation is chosen, as the human visual system is well suited to recognize patterns. The Nettleton principle has been modified. The idea can be expressed mathematically in terms of correlation coefficients, the desired density defined by the condition of zero correlation (Jung, 1943):

$$\rho = \rho_o + \Sigma_{(i)}[(BA_{oi} - \underline{BA}_o)(h_i - \underline{h}_o)]/(2\pi G\Sigma_{(i)}[(h_i - \underline{h}_o)^2]) \qquad (3.6.10)$$

with ρ_0 the originally assumed Bouguer density and underlining meaning the arithmetic mean. Regional gravity fields can adversely affect the results and removal of linear fields may be insufficient. Each case would have to be treated individually. In this, the problems of spurious correlations (below) cannot be solved mathematically.

Interference with other gravity components, especially the effects of lateral density variations, may lead to an incorrect Bouguer density identification. The condition of zero correlation will never be perfectly satisfied in nature; indeed, in some cases correlation between BA and TOP is real and caused, for example, through dynamic processes or through erosion.

A related approach is to express the observed gravity as the sum of the elevation related effects with unknown Bouguer density and the unrelated accidental deeper geological effects, hence to treat them as errors in a least-squares sense (Parasnis, 1952). If the distribution of stations (in any map configuration, usually not in profiles) indeed satisfies the condition, the result may be reasonable, but there is little control of this.

Jacoby (1966) attempted to reduce the effects of accidental or geological relations between the full topographic effect (of Bouguer plate and terrain) and elevation (i.e. topography) and to thus estimate both, the Bouguer density and the local vertical gradient (see, Sect. 7.3.2.1.1.2). The idea is to remove linearly related components of BA and TOP by an initial least-squares determination of linear relations; after their removal density and vertical gradients are then computed

by fitting the residual gravity, height and mass effect values. A polynomial fit to
the unknown Bouguer anomaly from deeper sources is included. Steep slopes of
topography are advantageous for stable solutions; indeed, such a geometry approaches the 3D situations discussed in Sect. 3.6.3.5. In the particular case studied,
the island of Helgoland, with near-vertical rock walls, the results agreed remarkably well with carefully measured samples ($> 100\,$kg); the vertical gravity gradient
above a salt-dome is related to the negative gravity anomaly and was found to be
$-0.303 \pm 0.012\,$mGal/m. Of course, more complex relationships are not eliminated
by a linear analysis.

3.6.3.7 Discussion of Densities

3.6.3.7.1 General Considerations

The bulk density including voids (pores, joints, vesicles) is relevant to the gravity
studies, not mineral (or grain) density. Table 3.1 (at the end of the chapter) lists rock
densities as taken from various sources (listed there). The grouping and classification of materials is chosen without attempting a thorough classification. The groups
are: (1) superficial and artificial materials, (2) sedimentary rocks, (3) volcanic rocks,
(4) plutonic and metamorphic rocks, and (5) some minerals and ores. Classification
of some materials is ambiguous, for example, in the case of monomineralic rocks.
Some order has been attempted in the listing.

Individual reliabilities are approximate, mainly because the nature of the errors
or individual scatter of values is complex. The interpreter is warned not to rely too
much on published values. A particular target geological body may easily deviate
from published mean values. On the other hand, the crustal density variability is
moderate, very rarely (and only in small volumes) do densities lie outside the range
from 1000 to $3000\,$kg/m^3 and upper mantle rocks hardly exceed $3500\,$kg/m^3. As
preliminary a priori information for inversion and optimization (Chap. 7) the uncertainties permit estimates of the error bounds.

Knowledge of trends and of the possible exceptions is important for the interpreter.
Different classes of rocks have different characteristics: (1) soils and alluvium are
usually loose superficial layers of low density of minor thickness, from $< 1\,$m to
order $10\,$m; corresponding gravity effects rarely exceed a few tenths of a milligal. (2)
Consolidated sediments of sandstone or conglomerate, clay and shale, and limestone
or dolerite composition are highly variable. (3) Volcanic rocks (which may or may
not be vesicular, jointed and fractured) and (4) plutonic and metamorphic crystalline
rocks as granite, gneiss, gabbro etc. are equivalents with correlated densities.

3.6.3.7.2 Sedimentary Rocks

Sedimentary rocks have densities at the low end of the whole density spectrum. The
effects of moisture and compaction can be large and critical if significant variations

occur in different rock bodies. Their lateral density differences usually vary with depth. In loose rocks wet density may exceed dry density by one third, an effect which decreases with depth. These factors are affected by depth of earlier burial and age which both result in closing and/or filling of pores with minerals. Shales are especially compactable by settling, dehydration and recrystallization. The notion "shale" does not predict density significantly, and the term is indeed ambiguous, including sediments and metamorphic rocks. Limestone and sandstone vary only little in density.

3.6.3.7.3 Volcanic Rocks

Also volcanic rocks strongly vary in density with composition and, near the surface, with texture. The more basic, the denser they are; the corresponding range is about 20%. Loose vesicular and highly broken superficial rocks, as the tops of basalt flows, are quite light, but removal of the loose rock by erosion before burial and filling of the voids with minerals rapidly increase the density. These effects are minor in lava that initially cooled deeper in a flow (relative range $< 10\%$ of mean). Slight variation occurs with crystallinity, amorphous glass being lightest. Mineralization by heavy ores seldom has a large influence on bulk density, and if so, only in small volumes.

3.6.3.7.4 Crystalline Rocks

Density of plutonic and metamorphic crystalline rocks varies with composition and increases with the grade of metamorphism. Since this process affects all kinds of original rock, metamorphic rocks are a highly heterogeneous class, and some exceptions to the relation with composition occur. Metamorphism may involve migration of solutions which can form locally heterogeneous masses and aureoles, also of density variation.

3.6.3.7.5 Assessment of Listed Rock Densities

The published rock density values listed in Table 3.1 (at the end of the chapter) show, by comparison, a large scatter. Some serious errors and discrepancies between different sources occur (some mean values lie outside the ranges given elsewhere). For some materials unbelievably narrow limits are quoted (e.g. loose soil), for others the differences in upper and lower limits can be very large, and one gets the impression that some values are mere guesses. Evidently the scatter need not be symmetric about the mean; large asymmetry of quoted bounds indicates a significant skew in the distribution.

These uncertainties have probably several causes: (1) real density scatter in rock bodies, (2) uncertainties, ambivalences and unreliability in definition of rock types, including authors' insufficient petrological knowledge, (3) neglect of sample characterization, for example, of freshness or weathering, (4) deficient measuring procedures, (5) copying errors from any publications and (6) some arbitrariness of the system of numbers and units used. Indeed, usually no more than two digits may be significant. The conclusion from this state of knowledge is that published values are unreliable information, valid only for initial model assumptions and for approximate guidance after checking.

3.7 Preparing the Data for Reductions and Analysis

A number of operations are best carried out immediately after a day's work. The field notes must be brought into a tractable form with the additional aim to detect large errors. Serious reading errors or mistakes can be corrected by repeating the measurements. Wanted are the gravity values or gravity differences observed from a chosen reference point, and they must be prepared for further post-survey treatment, as for example, refined drift and tidal analysis and reductions. Each individual point must be handled, but also all the measurements of a day or of a whole survey. The following is largely a summary of the aspects discussed above.

Even in the age of portable computers, in preparing or accompanying the digital operations, it is still advisable to draw by hand drift curves that include the earth tides. It can be done in units of scale division instead of milliGal to compare this with computer outputs. Conversion from scale units (div) into mGal, µGal or nm/s^2 is necessary unless provided automatically. Different instrument heights above survey markers (e.g. at identical stations) must be taken into account before constructing the instrument drift where $\Delta div = 3.086\,\mu Gal/cm \times \Delta h(cm)/f$, with $f =$ scale factor in $div/\mu Gal$. Also the earth tides may be reduced in scale units.

The final result will be a set of points with coordinates and elevations as well as the relative gravity values, i.e. differences from the chosen base station value. For the purposes of exploration geophysics the point values are constant in time, but in special cases geodynamic time variations are derived from repeated measurements.

3.8 Error Assessment and Accuracy

Errors of gravity values are essential for assessing the potential of gravity interpretation. The term "error" is related to "inaccurate", "imprecise", "uncertain", "unreliable", and the opposite nouns are: "accuracy", "precision", "certainty", "reliability". Each of these terms carries a somewhat different connotation, especially in contrast to the vague common usage of these terms. Accuracy has to do with the closeness to reality or truth or to the likelihood to be close to the true value, or probably how close. Precision, on the other hand, has to do with the width (or better: narrowness) of the scatter or the standard deviation (see below) or, in other words, the number of significant digits of a quantity, where "significant" refers, for example, to a measuring device; despite its high precision, a value may yet be far off its true value. Reliability and certainty mean nearly the same and imply the degree of trust in the values, which is a fairly subjective assessment. Reproducibility and repeatability, which are self-explanatory, may be added as necessary, but not sufficient conditions for reliable observations.

In view of the subject of gravity interpretation, accuracy and precision gain a special significance. While accuracy is an important aim, precision need not be so. This has to do with the principal ambiguity of potential field interpretation or inversion; a model found to explain the observations precisely can be far from reality. Especially in gravity inversion (Chap. 7), but also in other methods of modelling, standard errors may be calculated which generally tell how precisely the observations are "fitted". In order to underline the restricted significance of such a posteriori standard errors, it is proposed to call them "***apparent standard errors***" (or short: "apparent errors"). This will be demonstrated again in the later chapters.

There are errors or error components of different nature, in any single value of a measured quantity all components occur together and the total error is an integral aggregate. Error components are outliers or gross errors, systematic and random, each being briefly discussed below.

3.8.1 Outliers

Outliers or gross errors stem from misreadings, wrong copying, malfunctions of instruments and the like and are often easily recognized by eye, for example, if plotted in a suitable fashion. In large data sets this may no longer be possible and statistical data snooping may identify gross deviations from the general data distribution which are errors much larger than the established statistical measurement inaccuracy.

However, a single outlier may be a correct value, if the measured quantity has the erratic behaviour of containing in a smooth field a few small-scale local anomalies, as frequently is the case in magnetics. Such local anomalies may or may not be considered significant in a given context; in the first case, more observations may be added in the vicinity, and in the latter case the outlier may be treated as an error, either dismissed or treated as a random error, which has to be decided on the basis of a general judgement of the survey aims.

If the scatter and distribution of the individual values of a quantity is known from repeated measurements, outliers may be defined as those values falling far out of the standard deviation (see below). However, as the Gauss distribution with its standard deviation incorporates few quite large deviations, though with a correspondingly small probability, it cannot be decided unambiguously whether the value in question is part of the set or an outlier. As always in statistics, there is no absolute characterization of a gross error.

3.8.2 Systematic Errors

Systematic errors can be controlled only by comparisons and calibrations, as outlined with the field procedures, for example, in connection with scale and calibration. Any technical malfunction, as for example, a deteriorated sensor vacuum, thermostat problems etc. will affect the measurements, making repeated instrument checks advisable. Levels may be de-adjusted, resulting in apparent scale lengthening by $1/\cos\varepsilon$, where ε is the deviation from the vertical. Some instrument types may have scale irregularities and periodicities revealed by extensive tests (Kanngieser et al., 1983).

3.8.3 Random Errors

Random implies chance, unpredictability. It does not exclude a limitation of the range of scatter. Instrument reading and instrument drift have random components for which no theory exists. Sources are the setting of the instrument, in most cases, the nulling or determining of the internal, i.e. not directly visible null configuration, furthermore the reading of a meter, especially the estimation of the last digit, and instrument levelling as well as auxiliary measurements as that of instrument height etc. which all affect the reductions (Sect. 4.5.2). Incorrect levelling, for example, has a systematic influence on the gravity observation and is thus not of an exclusively random nature, but random errors of levelling produce random effects in gravity reductions.

Individual random errors are thought to be independent from each other, or stochastic, and are customarily assumed to have a normal or Gaussian distribution, in the sense of standard deviations. Given n observations x_i of the true quantity x_0; the individual random errors are $\Delta x_i = x_i - x_0$; the Gaussian frequency or probability distribution for a large number of observations $(n \to \infty)$ is then given by:

$$g(\mathrm{x}) = s^{-1}(2\pi)^{-1/2} \exp(-(\Delta x/2s)^2) \qquad (3.8.1)$$

with the standard deviation s defined as the square root of the variance

$$s^2 = {}_i\Sigma\Delta x_i{}^2/n \qquad (3.8.2)$$

As x_0 is not known, a practical estimate is made by replacing x_0 with the arithmetic mean $\underline{x}$: $\underline{\Delta x_i} = x_i - \underline{x}$ where, henceforth, the underlining is dropped. For the estimate of s^2, the denominator will be $(n-1)$, the number of redundant observations. The normal distribution expresses that in the interval $\pm s$ about the mean, $\underline{x}$, 68% of the individual measurements should lie within the interval $\pm 2s$, 95% ($\pm 2s$ is the so-called "95% confidence interval"). Outside $3s$, only 0.3% of the measurements are expected and outside $5s$ only less than 10^{-6} of the total. Thus the probability of large deviations approaches zero rapidly, but never disappears completely.

In most cases, too few observations exist to test the fundamental assumptions and the limited number of observations may fit a Gaussian distribution more or less; obvious deviations in shape may include asymmetry or skewness, double peaks, a too sharp peak relative to the lateral width; such features suggest the error sources to be investigated more thoroughly and the statistics to be viewed critically. Nevertheless, usually, for want of better alternatives, most of the error treatment, as calculation of standard errors, error propagation, etc. is based on the Gaussian and is believed not to lead to gross misinterpretations.

Many gravimeter types have two levels: the cross level controls the verticality of the instrument, but the longitudinal or sensitivity level can be deliberately set as it determines the point of balance between the moment of weight, lever $\times mg$, and of the spring moment; this point determines the angle of intersection of the two moment curves versus setting, hence the sensitivity. Therefore, the sensitivity level, once set, must be extremely carefully adjusted at each station; it is part of the measurement; and setting of the sensitivity level must not change during a survey. Otherwise it is a source of severe errors. Small errors in cross level setting are less critical as their effects are related to $\cos \varepsilon$ where ε is the error or deviation from verticality. Every instrument type must be carefully assessed for such types of errors.

Errors come in also with the reductions, and each input parameter (height and co-ordinates, densities, topographic relief) is affected by errors. The parameters of the normal reference field, once adopted, are considered true for practical reductions. This definition is principally not erroneous because the local gravity field differs, for example, in vertical or horizontal gradient or in Bouguer density: The corre-sponding effects appear in the calculated Bouguer anomalies and should be treated in the data analysis and interpretation. Densities as the Bouguer density are of a special character; none of the density determinations (Sect. 3.6.3) will render exact densities which are preliminary and will be adjusted in the interpretation process.

The final assessment has to take into account all the above error sources, each estimated carefully and put together to a general number of uncertainty which must be completed by the uncertainties introduced by the various reductions and ulti-mately also by the uncertainties related to modelling. If all the error components are statistically independent, or in other words, orthogonal, the errors add statistically:

$$s_{\text{tot}}^2 = {}_i\sum s_i^2 \tag{3.8.3}$$

This corresponds to the Gaussian error propagation law with the same sensitivity of the resulting s_{tot} to each component s_i, as in a sum. For inversion s_{tot} is of funda-mental importance (Chap. 7).

3.9 Conclusion

In conclusion, the aim is restated: gravity measurements render the g or δg values with their uncertainties for sets of geographically defined points after removal of in-strumental and tidal effects. The gravity values g_i at stations i are assumed to repre-sent the time-averaged or time-independent gravity field in space; the averaging will not necessarily remove certain long-period – geodynamically interesting – compo-nents of the field. The values do depend on the point coordinates, especially height. Predictable dependencies can be calculated, at least for a standard gravity field used for reference and comparison, others are unknown beforehand and are the prime target of interpretation. The standard is described as the normal gravity field and referring the observations to it is called "reductions", the topic of the next chapter.

References

Birch, F.: The velocity of compressional waves in rocks to 10 kilobars, 1. *J Geophys. Res., 65, 1083–1102, 1960*

Birch, F.: The velocity of compressional waves in rocks to 10 kilobars, 2. *J Geophys. Res., 66, 2199–2224, 1961*

Birch, F.: Density and composition of the upper mantle: first approximation as an olivine layer. In: The Earth's Crust and Upper Mantle. Ed. P.J. Hart, *Geophysical Monograph, 13, 18–36, AGU, Washington, D.C. 1969*

Cara, M.: Geophysik. *Berlin, Springer Verlag, XI, 210 pp., 1994*

Carlson, R.L., Herrick, C.N.: Densities and porosities in the oceanic crust and their variations with depth and age. *J. Geophys. Res., 95, 9153–9170, 1990*

Darbyshire, F.A., White, R.S. and Priestley, K.F.: Structure of the crust and uppermost mantle of Iceland from a combined seismic and gravity study. *Earth Planet. Sci. Lett., 181, 409–428, 2000*

Eötvös, R.v.: Untersuchungen über Gravitation und Erdmagnetismus. *Ann. D. Phys. u. Chem., Neue Folge, Vol. LIX, 354–400, 1896* reprint. in **R.v.Eötvös**, Gesamm. Werke, ed. v.Selényi, P., Hungar. *Acad. Sci, Budapest, 25–68, 1953*

ESA (European Space Agency, Johannessen, J.A. ed.): The four candidate earth explorer missions – Gravity field and steady-state ocean circulation mission. *Report SP-1233(1), 217 pp. 1999*

Gardner, G.H.F., Gardner, L.W., Gregory, A.R.: Formation velocity and density – the diagnostic basis for stratigraphic traps. *Geophysics., 39, 770–780, 1974*

Giganti, J., Larson, J.V., Richard, J.-P. Weber, J.: The lunar surface gravimeter experiment for Apollo 17. *Apollo 17 Preliminary Science Report, NASA SP-330, 1973*

GOCE: Mass transport and mass distribution in the earth system. *GOCE-Projektbüro Deutschland, Tech. Univ.München, Geoforschungszentrum Potsdam, 154 pp., 2005*

Gutenberg, B.: Lehrbuch der Geophysik, *Borntraeger, Berlin, 1017 pp., 1929*

Hubler, B., Cornaz, A., Kündig, W.: Determination of the gravitational constant with a lake experiment: New constraints for non-Newtonian gravity. *Phys. Rev. D, 51, 4005–4016, 1995*

Jacoby, W.R.: Die Schwereverteilung auf Helgoland und ihre Auswertung mit Ausgleichsverfahren. *Ph.D. thesis, Kiel, 1966*

Jung, K.: Über die Bestimmung der Bodendichte aus den Schweremessungen. *Beitr. Angew. Geophysik, 10, 154–164, 1943* Further references in KJ61

Kanngieser, E., Kummer, K., Torge, W., Wenzel, H.G.: Das Gravimeter-Eichsystem Hannover. *Wissenschaftliche Arbeiten der Universität Hannover, Nr. 120, 95 pp., Hannover 1983*

KJ61: Jung, K.: Schwerkraftverfahren in der Angewandten Geophysik. *Akad. Verlagsges. Geest & Portig, Leipzig, 348 pp., 1961*

Nafe, J.E., Drake, C.L.: Variation with depth in shallow and deep water sediments of porosity, and the velocities of compressional and shear waves. *Geophysics, 22, 523–552, 1957*

Nettleton, L.L.: Determination of density for reduction of gravity observations. *Geophysics, 4, 167–183, 1939*

Parasnis, D.S.: A study of rock density in the English Midlands. *Monthl. Not. Roy. Astron. Soc., 6, 252–271, 1952*

Richardson, J.E., Melosh, H.J., Artemetva, B.A., Pierazzo, E.: Impact cratering theory and modeling for the Deep Impact Mission: From mission planning to data analysis. *Space Sci. Rev., 117, 241–267, 2005*

Rummel, R., Balmino, G., Johannessen, J., Visser, P., Woodworth, P.: Dedicated gravity field missions – principles and aims. *J .Geodynamics, 33, 3–20, 2002*

Sharma, P.V.: Geophysical Methods in Geology, 2nd ed. *Elsevier, New York, Amsterdam, London, XI, 428 pp., 1986*

Smilde, P.: GRAVI, unpublished computer program for calculation of relative gravity values from field measurements. *Available on request from author, 1995*

Telford, W.M., Geldart, L.P., Sheriff, R.E., Keys, D.A.: Applied Geophysics. *Cambridge Univ. Press, Cambridge, UK, 860 pp., 1976*

TOPEX/Poseidon (multi-authored): A United States/France mission. Oceanography from Space: The Oceans and Climate.

Zelt, C.A.: Documentation of "tramp" and related programs. at http://zephyr.rice.edu/department/faculty/zelt/rayinv.html, *1992*

Bibliography borehole gravimeter

http://wwwbrr.cr.usgs.gov/projects/GW_borehole/bibliography/gravimeter.html

Table 3.1 Rock densities in kg/m^3

Superficial and artificial materials			
Snow, loose	125; 130$-/+$50, 500		
Firm ice, 0°C	917$-/+$37, 3		
Fresh water	1000		
Sea water	1030; 1010–1050		
Brick	1500		
Petrolium	600–900		
Asphalt	1100–1200		
Lignite	1190 $\pm$ 80		
Till, moraine	1800; 1500–2000		
Building rubble, waste, landfill	1300–2000		
Alluvium, wet	1980 $\pm$ 20	dry 1540 $\pm$ 60	
Weathering horizon	1700–2400		
Soils, wet	1920$-/+$72, 48	dry 1460$-/+$46, 54	
Soil	1500–1700		
Soil, loose	1200 $\pm$ 80, 1100–1300		
Soil, compact	2000–2200	dry 1800$-/+$200, 120	"fresh": 2100 $\pm$ 100
Humus layer	1400 $\pm$ 250		
Peat	1050		
Loam, dry	1500$-/+$50, 100		
Loam, wet	1700$-/+$50, 150	1670–1850	
Loam, sandy	1900 $\pm$ 250		
Clay	1700 $\pm$ 200	2200 $\pm$ 400	1800–2600
Clay, wet	2210$-/+$58, 390	dry 1700$-/+$40, 70	
Kaolinite	2530$-/+$330, 100		
Sandy clay, wet	2100 $\pm$ 400		
Silt, wet	1930$-/+$130, 270	dry 1430$-/+$23, 37	

Table 3.1 (continued)

Sand, dry	1500−/+100,150	1500±100	1300–1500		
Sand, moist	1800±100	1600–2000			
Sand, wet	2000±50	2000±100			
Gravel, wet	2000−/+300,400	2000±100			
		dry 1950−/+55,25	1900±100	1500–1800	
Sandy gravel	2100±100				
Loess, wet	1640−/+24,29	dry 1200±45			
Marl	2400±100				
Sedimentary rocks (1200–2700)					
Rock salt, halite	2220−/+120,590	2100−/+100,200	2300−/+200,100	2150±100	2280–2410
Oilshale, bituminous pelite	1100–1300				
Coal, soft	1320−/+120,180				
Shaly coal	1350±150				
Brown coal	1300−/+300,150	1000–1400			
Hard coal	1350±150	1250–1500			
Anthracite	1500−/+160,300	1350±150	1500–1700		
Chalk	2010−/+480,590	2200±400	1800–2600		
Limestone	2490±120	2400–2700	porous: 1650	oolithic: 2000	conglomerate: 2480
Limestone, dense	2580	compact: 2500–2750	"crystalline": 2660		
Limestone, wet	2550−/+620,350	dry 1540±60			
Dolomite	2750−/+310,150	2500–2900	2850–2900		
Dolomite, wet	2700−/+42,20	dry 2300±250			
Marl stone	2210−/+600,400	2300–2500			
Clay marl	2060−/+350,300				
Clay slate	2020−/+250,300				
Slate	2670±30				
Shales, wet	2240−/+63,80	dry 2100−/+54,110			
Shale	2800−/+40,80				

Table 3.1 (continued)

Shale, younger	$2420-/+200,150$	$2400-/+300,200$	2000–2700	
Shale, older	2700 ± 50	2650–2880		
Greywacke	2600			
Sandstone, wet	$2350-/+74,41$	dry $2240-/+64,44$		
Sandstone	$1880-/+250,300$	2000–2600		
Sandst. Mesozoic	2150–2400	Paleoz., older	2350–2650	1890–1920
Arcose	2100–2500	2300 ± 500		
Chert	2700 ± 40			
Diatomite	1800–1900			
Gypsum	$2350-/+150,250$	2250 ± 70	2300	
Anhydrite	$2930-/+30,70$	2960 ± 40	2900	
Bauxite	$2450-/+150,100$	2500 ± 500		
Volcanic rocks				
Trachyte tuff	1400–2200			
Basalt tuff	2000–2300			
Tuffstone	$1300-/+0,1100$			
Calcerous tuff	1640 ± 70			
Pumice	640 ± 280	950		
Rhyolite glass	2240 ± 40			
Obsidian	2300 ± 100			
Rhyolite	2520 ± 180			
Dacite	2580 ± 230			
Phonolite	2590 ± 130			
Hyaloklastite, palagonite, wet	2150 ± 150	2000–2200	dry 1800 ± 200	1700–1900
Quartz porphyry	2400–2600			
Trachyte	2600 ± 190	2470 2600		
Trachyte lava	2400 ± 300			
Andesite	2610 ± 200	porphyric: 2750–2950		

Table 3.1 (continued)

Basalt	2990±30 2700–3100	2950±350 2700–3300	2980±300 ["published" 2915]	2704–	2900±200	2900±100
Basalt molten	2300−/+250,160					
Dolerite	2850−/+400,240					
Diabase (basalt)	2900−/+400,200	2910−/+400,30			2960±150	2800–3000
Melaphyre Basalt	2640−/+40,160	2500–2800				
Porphyry	2700−/+360,200	2600±150				
Plutonic and metamorphic rocks		2600–3400				
Acid igneous	2610−/+300,500					
Nepheline-Syenite	2610±90					
Granite	2640±160	2600±300	2500–2700		2500–2800	2600–2700
Granodiorite	2730±60	2720±60				
Syenite	2770−/+310,140	2770±180	2760±130		2700–2900	2790
Basic igneous	2790−/+700,400					
Quartzdiorite	2790±170	2810±140	2980±140			
Diorite	2850−/+400,250	2850±140				
Anorthosite	2780±150					
Gabbro	3030−/+330,470		2900±300		2950±50	
Hornblende-Gabbro	3080±100					
Essexite	2910±230					
Norite	2920−/+220,320					
Greenstone	2950±50					
Peridotite	3150−/+37,22	3230±60				
Serpentinite	2780±350	2600±100				
Pyroxenite	3170−/+240,170					
Dunite	3280−/+80,30					
Eclogite	3370±170	3390±50				

Table 3.1 (continued)

Quartzite	2650 ± 50	2600 ± 100	2410		
Greywacke	2650 ± 50				
Marble	2750 ± 150	2700 ± 200	2700 ± 160	Carrara: 2720	
Qz-slate	2770 ± 140				
Slate	2790 ± 100				
Chloritic slate	2870 ± 120				
Schist	2640 ± 260	2760–2880			
Schist chloritic	2700–2800				
Micaschist	$2650 - / + 250,350$		2600–2800		
Phyllite	2740 ± 60	2700			
Amphibolite	2960 ± 80				
Gneiss	$2700 - / + 100,200$		$2600 - / + 200,300$	2800 ± 210	2400–2700
Gneiss, schist	$2650 - / + 250,350$				
Granulite	$2650 - / + 130,80$				
Some minerals and ores					
Graphite	$2150 - / + 250,150$				
Diamond	3520				
Corundum	4000 ± 100				
Sulfur	2000 ± 80	1900–2100			
Calcite	$2720 - / + 80,280$	2600–2700			
Quartz	$2650 - / + 150,50$	2650 ± 100			
Flintstone	2700 ± 110				
Feldspar	2550 ± 30				
Orthoclase	2500–2600				
Mica	$2800 - / + 150,350$				
Biotite	2920 ± 250				
Talc	$2710 - / + 10,90$				
Magnesite	$3030 - / + 130,90$				
Fluorite	3140 ± 120				
Hornblende	3000 ± 400				

Table 3.1 (continued)

Epidote	3250–3500			
Olivine	3400 ± 150			
Serpentine	$2600 - / + 200, 100$			
Chalcopyrite	4200 ± 100			
Sulfides	2600–8800			
Pyrite	$5100 - / + 200, 100$	5000 ± 100	4900–5200	
Galena	7500 ± 100	7400–7600		
Carnallite	1600–1700			
Baryte	$4470 - / + 170, 230$	$4500 - / + 500, 200$		
Zircon	$4570 - / + 570, 330$			
Chromite	4500–4800			
Siderite	$3830 - / + 130, 70$			
Haematite	$5180 - / + 280, 120$	$5120 - / + 22, 80$	4700 ± 200	5100
Magnetite	$5100 - / + 200, 100$			
Loadstone	$5100 - / + 140, 300$			
Ilmenite	4670 ± 350			
Iron	7870			
Iron melt	7010			
Copper	8700			
Lead	11340			
Gold	15600–19400			

Densities are from the following sources: KJ61, Telford et al. (1976), Gutenberg (1929), Sharma (1986), Cara (1994) and others. The values are generally adopted at face value from the sources and ordered according to an estimated ranking after reliability, irrespective the sources. The format as generally adopted from the sources can be: single value; range: lower – upper limit; mean $\pm$ symmetric scatter; mean $- / + a, b$: asymmetric scatter ($-a$ to $+b$)
The rock types and materials are ordered according to a rough geological classification in sections, and subsections.

Chapter 4
Gravity Anomalies and Disturbances: Reductions and Analyses

4.1 Introduction

Sets of relative gravity values show spatial patterns, beside a decrease with station elevation, and a poleward increase, also geological effects. The latter are usually obscured by the other influences which must be removed in order to reveal the geological information. This cleaning operation is called "*reduction*" and the results are called anomalies of various kinds which are the target of interpretation. Proper reductions require knowledge of the station coordinates and of the normal gravity field with which the observations are to be compared. In geodesy, the term "disturbance" is more common, this stems historically from different references used, as described below. In both cases the purpose of the reductions is to relate the observations to a reference, and this is, in the proper sense of the term, not a "correction" of errors or mistakes. The customary term "correction" is therefore avoided in this book. This is also true for the tidal effects, removed by a tidal reduction, while, on the other hand, the removal of the instrumental drift can be considered truly a correction. The ultimate accuracy of the anomaly values is affected by both the observations and the reductions.

As the input, and starting point into this chapter, we have lists, computer files and graphical representations (profiles, maps) of observed gravity values with coordinates, possibly also slow geodynamic time variation. Here the procedures and manipulations are described. The manipulations usually include a refined drift and tidal analysis even if tidal effects had already been approximately removed. The reductions are for latitude, elevation and for topographic mass. The normal field is the one considered best suited for the reductions. Purposely for reference a very simple earth model is chosen with standard parameters, even if one knows that local parameters are different. In addition (see below) local values may be used, as determined by observations and measurements (Chap. 3). Further analysis of the gravity anomalies (or disturbances) may involve the so-called regional-residual separation, which, however, is not a "reduction" proper.

This chapter is divided into the following parts: reductions for earth tides, a discussion of terms as gravity anomaly and disturbance, reference earth models; components of gravity, standard reductions for latitude, elevation and topographic mass, preliminary analysis of the anomalies, post-reduction aspects of data analysis,

W. Jacoby, P.L. Smilde, *Gravity Interpretation*,
© Springer-Verlag Berlin Heidelberg 2009

as regional – residual separation, smoothing, contour line construction, use of maps
and derivation of wanted quantities (KJ61, 93-121).

4.2 Earth Tide Reduction

It is advisable to refine preliminary drift corrections and tidal reductions. The
theoretical earth tides must be calculated and removed before treating the instrument drift. Tidal gravity variations reach ± 0.15 mGal. In the field, data analysis is usually in scale units, for example, for 0.1 mGal/div tides reach ± 1.5 div.
Theoretical tidal calculations are based on normal earth models. Since considerable local deviations may exist local recording of the gravity tides is preferred
(Sect. 3.6.2).

The following brief description of earth tides theory is based on KJ61 and begins with the gravitational attraction (on the whole Earth's body) from the moon
and sun (and planets as Jupiter) and the secondary effects of Earth deformation in
the tidal potential field. The lunar effects are about twice the solar effects. Since the
tidal effects are small, the celestial bodies can be treated independently; their effects
are superimposed linearly. In order not to mix up tidal effects with Earth rotation
and the related centrifugal forces, one describes the tidal motions of two bodies, as
Earth and Moon, as translatory revolution about the common centre of gravity, i.e.
without change of individual orientation or rotation. In this motion, gravitational
attraction and centrifugal forces cancel only at the Earth's (or Moon's) centre of
gravity; at all other points the vector sum deviates from zero, varying linearly with
radius r. The tidal force field generates a prolate ellipsoid (Earth approximated as a
sphere, the normal oblate ellipsoidal shape being neglected) and a gravity minimum
at the prolate poles and a maximum along the oblate equator. Only now, the Earth's
self-rotation is considered which means that points rotate through the tidal fields of
deformation and gravity potential, generally twice a day through a maximum and
a minimum. Since, however, the moon revolves about the Earth in a month, i.e. 12
times faster than Earth revolution about the sun, the larger lunar effect shifts phase
in such a way that the tidal period is prolonged from 12 to nearly 13 hours. During
full and new moon, the lunar and solar effects are added (spring tides), while at half
moon they are subtracted (neap tides). Programs are available to do the proper calculations of the tidal effects (see GRAVI: 3.3.2; Wentzel, 1995: http://www.gik.uni-
karlsruhe.de/~wenzel/hw95/hw95.txt;   http://www.astronomynotes.com/gravappl/
s10.htm).

Ocean tides are also non-negligible especially at points near (and high above) the
sea shore and near tidal rivers. There are, however, variable phase shifts between
solid earth tides and ocean tides, especially along island coasts, in bays and estu-
aries; these are regularly published as tidal tables in almanacs of the hydrographic
offices. Ocean tidal models are available, but near coasts special measures are advisable.

4.3 The Time-Invariant Gravity Anomalies and Their Fundamental Properties

Gravity anomalies after tidal reduction are the time-invariant deviations from reference values and the object of geophysical interpretation. "Anomaly" is a typically geophysical term. The reference or norm is a model, i.e. a mental image of reality, which must be kept in mind. Therefore such models must be simple, for example, in the form of the reference ellipsoid and the Bouguer slab for the topographic mass (see below).

One *anomaly* value, i.e. an observation minus the theoretical norm refers to an individual point in space (and time) and, alone, cannot be interpreted. Geophysicists usually call "anomalies" sets of points in space. These are features in map view or along a profile imagined to be continuous fields, related to some mass distribution. The relation between the ideal continuous fields and the discrete sets of observations is not trivial and requires special attention (see Sect. 5.1.5).

In *geophysics* gravity anomaly must be specified, for example, as Bouguer anomaly (BA), Free Air or Faye anomaly (FA) or Isostatic anomaly (IA), and some specialized kinds. The differences lie in the purpose and scope of the measurements and the underlying reference models to which observations are compared. Definitions must be unique for proper treatment of the observations. However, in practice some insignificantly small or apparently irrelevant quantities are customarily neglected for economic reasons. MicroGal precision is unnecessary if some of the errors are of the order of a tenth of a milligal; such errors stem, for example, from unknown nearby mass inhomogeneity. In other cases, for example, if temporal gravity change is aimed at, some usually neglected quantities are significant.

The "anomalies", as commonly used in geophysics, are inconsistently referenced to normal gravity on the ellipsoid and height relative to the geoid. Anomalies in this sense are abbreviated in this book by the letter "A", i.e., FA, BA or IA.

In *geodesy*, requirements of accuracy, both in principle and in practice, are generally strict. Observations pertain to global reference systems that are subject to specific conditions such as to preserve mass of the Earth; thus geodetic usage and definition of the term "anomaly" differs from geophysical practice. Indeed, geophysics and geodesy use two related, but differently defined quantities, partly for historical reasons: *gravity anomaly* (geophysics) and *gravity disturbance* (geodesy).

The "gravity disturbance", as used in geodesy and described in Sect. 4.3, has received little attention in geophysics, and thus no customary notation seems to exist. It appears in order to suggest a notation analogous to the above: FD, BD or ID, where the Free Air gravity Disturbance FD is synonymous with simply the gravity disturbance as such, while the BD is the (hardly used) geoid-ellipsoid reduced BA, and the ID is the (even less used) geoid-ellipsoid reduced IA.

In *geophysics and geology* gravity measurements aim at the study the Earth's interior. Geophysics is the physics branch of geology in broad sense, and gravity reflects density variations in the interior and is used to find them as emphasized here. *Geodesy* aims at accurately measuring the shape of the Earth, defining its surface and establishing geodetic point networks or reference systems into which future measurements can be integrated. Gravity is intimately linked to the Earth's figure. The different aspects have consequences on how gravity observations are analyzed, and have led to the differently defined anomaly and disturbance (see Hackney and Featherstone, 2003). For both communities it is useful to understand the mutually used terms, to bridge the gap between the two approaches and to make use of the somewhat differing concepts to their advantage.

In geophysics *gravity anomaly* is the difference of observed and normal gravity, both related to two different locations. (1) The observation is *reduced to the geoid* (mean seal level). Historically, this choice stems from the easy availability of the orthometric height of a point above sea level determined by levelling relative to the local horizontal (or vertical) with orthometric reductions (height error $\delta h \approx \delta W / g_0$), and is listed for benchmarks and contoured on topographic maps. (2) Normal gravity is defined on the *normal rotating ellipsoid* which is fitted to the geoid (Heiskanen & Moritz, 1967) and is the basis of the geodetic reference system GRS80 (Moritz, 1980). The ellipsoidal height, i.e., height relative to the reference ellipsoid was poorly known, if at all, and for exploration surveys this difference introduces in limited areas no more than a nearly constant shift of the reference and as such is not relevant to the purpose.

The *gravity disturbance* is defined as the difference of observed and normal gravity at the same point P. Today, ellipsoidal heights are known to considerable precision and can be directly calculated for GPS positioned points. In many cases, it would then be easy to replace the gravity anomaly by the gravity disturbance for which the observed and normal values are taken at the observation point P reduced to the corresponding point on the ellipsoid. For large-scale regional and global gravity studies this is mandatory (see below). In such scales one must also worry about the horizontal datum (latitude) and generally the geocentric latitude ϕ describing the ellipsoid is chosen (not to be mixed up with the colatitude, also called ϕ, to the geographical latitude φ).

Gravity anomalies and gravity disturbances are principally vector quantities, but especially in geophysics, gravity data are based on scalar measurements. Both, gravity anomalies and gravity disturbances can be defined also as scalar magnitude quantities.

The gravity vector is normal to an equipotential surface such as the *geoid*, which in the oceans, approximately coincides with the mean sea level. It is extrapolated into continents as though physically realized in channels. The geoid is an irregular surface; in empty space it is described by spherical harmonics as the solution of the Laplace equation (see Chap. 2). The density inside the Earth, on the other hand, has an irregular form, not mathematically-analytically describable, and this is true also for gravity at the surface. Nevertheless, spherical harmonics are used to approximate gravity at the surface, and in spite of some theoretical limitations, convergence does not normally present a problem. In the early 21st century, EGM96 (Lemoine et al. 1998) represents the best geoid, and satellite missions as GOCE promise to achieve cm and mGal accuracy on 100 km space resolution, many orders of magnitude better than only few decades ago. Therefore gravity anomalies and disturbances can be much more readily converted into each other. For the geophysical definition of gravity anomalies the irregular geoid need not always be the reference surface, but it is relevant to the geodetic definition of the gravity disturbance. Gravity anomalies (at geoid level) are easily reduced to gravity disturbances at ellipsoid level by the same reductions used for the observation level to the geoid level. For large-scale problems the rotational ellipsoid, best-fitting the geoid, is the preferred reference surface.

Note that *reductions do not physically move the observations* to another level – as incorrectly expressed frequently. Reductions compare the observation with normal values by estimating the latter at observation locations or by downward or upward continuation of observed values *in the normal gravity field*, not in the anomalous field. Anomalies are determined at the observation points, not at the reference level. If the anomalous effects, for example, the anomalous vertical gravity gradient can be estimated from the observed gravity distribution by applying Laplace's equation, and if they are used in the reductions, the anomalies may be transferred to the reduction level, however with incalculable problems, especially in downward continuation through topographic mass. It is simpler to consider and interpret the anomalies or disturbances at the points of observation and to take that into account in subsequent modelling.

Normal gravity refers to an idealized **normal earth** as reference for the real Earth with interior irregularities. The norm must be simple, easily calculable and imaginable. Otherwise it would be confusing and useless. The equilibrium figure of a homogeneous or layered self-gravitating rotating fluid body with no lateral (horizontal) density variations has been chosen; it is an oblate ellipsoid of rotation (Clairot's theorem) constrained by the astronomically observed moment of inertia. The parameters of the body are adjusted to fit the Earth; they depend on the basic dimensions, the mean density-depth distribution and the angular velocity. The mean radial density function of a sphere is modified, in the case of the rotational ellipsoid, to surfaces of constant density conforming to internal equipotential surfaces which are all ellipsoids with flattening decreasing with depth. The definition is not quite unique gravitationally, but it is further constrained by seismological data, especially by eigenperiods.

The theme of this chapter is the **reductions** and their calculation. It begins with listing the various calculable influences affecting the gravity measurements. They are calculable for the chosen normal reference earth and removed before further data analysis also because they generally mask the effects of interest. The removal is called reduction, usually resulting in more uniform and tractable data for imagination and interpretation. Calculation of the reductions requires knowledge of the coordinates, including height, and observation time. The term "reduction" means both a definition and a computational rule. The expression *"gravity effect"*, on the other hand, should be clearly distinguished from the term *"anomaly"*. An effect is usually computed for a given mass distribution in order to be compared with an anomaly based on observations. Calculation is forward and straightforward. Interpretation of anomalies is the inverse problem with the aid of forward solutions.

4.4 Components of Observed Gravity

Gravity at a given point is the sum of several components, both time-invariant (or averaged) and time-varying, the former mostly much larger than the latter. The latter components are partly removed in the process of data collection and analysis, but may also be the target of research. The permanent or lasting components are effects of latitude, elevation, topographic mass and interior mass anomalies. Observed gravity is always an integral effect of all masses.

The permanent components are:

- g_n, normal gravity on the normalized earth ellipsoid, depending on latitude;
- g_h the change predicted for the point elevation h above the reference surface (ellipsoid or geoid, see above);
- g_{top}, the change predicted for the topographic mass, i.e. the mass located between the Earth's physical surface and the reference surface, depending on elevation and the density distribution within the topographic mass; in the oceans it is the change corresponding to the change from sea water to rock replacing the sea water;
- g_{geol}, effects of any interior deviations from the normal earth below the physical surface, i.e. not only below the reference surface; it is usually the target of research; g_{geol} may be divided into several components, of which some are believed to be known well enough to be calculated separately.

$$g_{obs} = g_n + g_h + g_{top} + g_{geol} \qquad (4.4.1)$$

The small time-varying components are:
- Periodic variations caused by the relative astronomical positions of Earth, Moon, Sun and other cosmic bodies,
- periodic variations caused by the solid earth tidal deformation
- more or less periodic ocean tide effects,
- irregular variations, due to changes in water level (groundwater, rivers, lakes, ocean), ice volume (e.g. by melting or growing of glaciers) and resulting solid-earth deformation,
- irregular variations, caused by mass movements in the atmosphere and resulting earth deformation (marine, solid earth),
- irregular variations, caused by mass movements of endogenic origin (volcanism, earthquakes, tectonics) and of exogenic origin (erosion, sedimentation),
- and finally any acceleration in an inertial system; such effects have, so far, turned out to be unknown or negligible in geophysical surveys).

The above effects have a wide spectrum of frequencies or quasi-frequencies. The periodic effects, often summarized under the term "tidal" may be taken into account when the time-invariant gravity values are determined, for example, when constructing the gravimeter drift, and/or calculated from solid earth tidal models; they reach several tens of a microGal and have been treated above

Aperiodic variations related to groundwater, atmosphere and solid earth movements as considered here are changes of gravitational attraction and related earth deformation; short period variations as in earthquakes with associated point accelerations are not the topic of this book. The slow gravitational variations must be empirically estimated and are important if repeated observations are taken at the same stations, or are the very object of investigation. They can be treated in a similar way as static gravity anomalies in terms of inversion and interpretation. What is termed "irregular" depends on the particular research.

4.4.1 Normal Gravity

Normal gravity would be observed on the oblate reference ellipsoid of rotation that approximates the equilibrium figure best fitted to Earth. The ellipsoidal parameters depend on the dimensions, the mean density-depth distribution and the angular velocity. The latitudinal gravity variation depends only weakly on the density-depth distribution (Clairot), but the Earth's moment of inertia θ is strongly affected by it ($\theta \approx 0.33 MR^2$ instead of $0.4 MR^2$ for a homogeneous sphere, where $R = 6371\,\mathrm{km}$ is the mean radius; see Task 4.1).

The currently (2008) used reference ellipsoid is part of the Geodetic Reference System GRS80 (Moritz, 1980) incorporated in the World Geodetic System WGS84 and has the following parameters:

equatorial radius $a = 6\ 378\ 137\,\mathrm{m}$,
flattening $f = (a - c)/a = 1/298.257\ 222$, where $c = $ polar radius (calculable from a and f);
gravitational constant $\times$ mass $Gm = 3986005\ 10^8\,\mathrm{m}^3\,\mathrm{s}^{-2}$,
angular frequency $\omega = 7292\ 115\ 10^{-11}\,\mathrm{s}^{-1}$.

(By the way, the volume of the reference ellipsoid is $1.083\ 10^{12}\,\mathrm{km}^3$, its mass is, with $G = 6.6742\ 10^{-11}\,\mathrm{m}^3\,\mathrm{kg}^{-1}\mathrm{s}^{-2}$, $M = 5.972\ 10^{24}\,\mathrm{kg}$, and its mean density is, accordingly $\rho \approx 5513.5\,\mathrm{kg/m}^3$.)

The corresponding normal gravity g_n at the ellipsoidal surface, the International Gravity Formula, is usually given, with geocentric latitude ϕ, as

$$g_n = g_{eq}(1 + c_2 \sin^2 \phi - c_4 \sin^2 2\phi) \qquad (4.4.2)$$

with the equatorial normal gravity $g_{eq} = 9.780237\,\mathrm{m/s}^2$

$$c_2 = 0.005\ 3024$$

$$c_4 = 0.000\ 0058$$

A more accurate formula (with a relative error of 10^{-10} or absolute $10^{-4}\,\mathrm{mGal}$) contains terms of several even powers of $\sin\phi$ with the coefficients:

power 0: 1,
2: +0.005 279 0414,
4: +0.000 023 2718,
6: +0.000 000 1262,
8: +0.000 000 0007.

http://www.gfy.ku.dk/~iag/HB2000/part4/grs80_corr.htm

The coefficients are determined by fitting Eq. 4.4.2 to terrestrial and satellite observations. Eq. 4.4.2 directly renders the normal horizontal gradient or derivative with respect to ϕ which, for many purposes, is transformed into $\partial g_n/\partial x$, where x is distance (km) pointing north. Its value is $\sim 0.8|\sin 2\phi|\,\mathrm{mGal/km}$ and is 0 at the equator and at the poles.

In geodesy the standard for computing the magnitude of normal gravity on the surface of the geocentric reference ellipsoid is the closed *Somigliana-Pizzetti* formula:

$$g_n = g_{eq}(1 + k\sin^2\phi)/(1 - e^2\sin^2\phi)^2 \tag{4.4.3}$$

where $k = (bg_p/ag_{eq}) - 1 = 0.001\ 931\ 851\ 353$, $e^2 = (a^2 - b^2)a^2 = 0.006\ 694\ 380\ 022\ 90$, and a = ellipsoidal semi-major axis, b = ellipsoidal semi-minor axis, g_p = polar normal gravity and g_{eq} = equatorial normal gravity. Equation (4.4.3) is generally accurate to $1\,\mu\mathrm{Gal}$.

The corresponding normal vertical gradient of gravity is a function of ϕ and elevation h above the ellipsoid. For practical purposes, $\partial g_n/\partial h \approx -0.3086\,\mathrm{mGal/m}$ is a good approximation at the Earth's surface. However, inclusion of the latitude and elevation dependence leads to a quadratic expression for the *FA* reduction in $\sin\phi$ and in h (based on Eq. 4.4.3, Featherstone and Dentith, 1997):

$$\delta g_F = 2g_{eq}(1 + f + m - 2f\sin^2\phi)h/a - 3g_{eq}h^2/a^2 \tag{4.4.4}$$

where the geodetic parameter $m = z_e/g_e = \omega^2 a^2 b/(GM\sin^2\phi)^2 = 0.003\ 449\ 786\ 003\ 08$; z_e, g_e, respectively, are the centrifugal and gravitational acceleration at the ellipsoidal equator, ω is the Earth's angular velocity, and M is the Earth's mass. The normal vertical gradient varies with latitude from -0.3088 at the equator to $-0.3083\,\mathrm{mGal/m}$ at the poles, and it varies with h as $\partial g_0/\partial h \approx \partial g_0/\partial h|_{h=0} + 0.000036h$ [km], due to the quadratic height correction term, which corresponds to an absolute decrease of the gradient with h; the effect integrates to about $7\,\mathrm{mGal}$ for $10\,\mathrm{km}$ height.

Because, in the past, the elevation of a point was best known above sea level, not above the ellipsoid, the elevation reduction and associated anomalies in geophysics traditionally refer to the geoid, while in geodesy the gravity disturbance refers to the ellipsoid. The geoid is described by a spherical harmonic series truncated at some finite degree and order. The normal earth ellipsoid is expressed by the zonal harmonics and is as such a constituent of the geoid, and the best fit to it.

The irregular surface of the real Earth is treated as a simple addition to the internally regularized normal reference earth, i.e., the mass between the physical surface and the geoid or ellipsoid is simply placed on top above sea level in continents or excavated below sea level in oceans and replaced by water: the topographic mass is added or subtracted and this is an integral part of the reference earth model. Normal gravity is thus supplemented by the effects of the topographic masses. Their treatment is discussed with the Bouguer reduction (Sect. 4.5).

For some local problems (small relative to the Earth's radius) a Cartesian homogeneous or horizontally layered half space may be taken with the local values corresponding to the latitude on the ellipsoid.

4.4.2 Deviations From the Normal Earth

The above list of the gravity components includes that of the interior mass anomalies, the usual target of gravity research. Sufficiently well known geological density structures may be considered belonging to the reference earth or to the deviations from it, and the corresponding geological reductions may be computed as a forward problem, similar to the treatment of terrain or mass model effects. The temporal gravity variations, likewise, may belong to both the normal earth (e.g. the theoretical solid earth tides) and to deviations from it (variations in earth elasticity, ground water, atmosphere variations, geodynamic processes, etc.).

In summing up, it is obvious that the terms "normal earth", "normal gravity" and the various anomalies are used in different ways. Mostly, only the time-invariant regularized ellipsoidal earth model is considered the norm, defining g_n and g_h, while the actual morphological-topographic shape and mass is considered a deviation. Here it is preferred to include the latter into the reference earth (Bouguer earth) with their assumed parameters, and on that basis the topographic mass reductions g_{top} are carried out. Whether isostatic and some of the geological effects should be included in, or excluded from, the normal earth, is a matter of the given project aims and of personal choice.

4.5   The Reductions

From the list of gravity components Eq. (4.4.1) immediately follow the definitions of the reductions and the instructions how to calculate them:

$$g_{geol} = g_{obs} - g_n - g_h - g_{top} \tag{4.5.1}$$

For values relative to some base station:

$$\delta g_{geol} = \delta g_{obs} - \delta g_n - \delta g_h - \delta g_{top} \tag{4.5.1a}$$

where all values are referenced to the same station. Any constant shift of all reduced values is irrelevant for most geophysical tasks. Remember that the vertical gradient in δg_h is negative, so that the reduction is positive for $h > 0$. If a geological effect is estimated in advance, as for example an effect of an isostatic model, it may be written also on the rhs of Eqs. (4.5.1, 4.5.1a) and g_{geol} and δg_{geol} will then represent the effects of the rest of the geological mass distribution (see Sects. 4.5.3.3, 4.5.3.4). An exercise of gravity reductions is offered as Task 4.2.

4.5.1 The Normal Reduction

In practice this reduction will depend on the size of the survey region; globally and regionally g_n is calculated from Eqs. (4.4.2) or (4.4.3) and subtracted from the values in an absolute gravity system, or the theoretical latitudinal variation must be taken relative to the same reference point which introduces a constant shift of the reduced values. For local surveys of small spatial dimensions, often a constant latitudinal or south-north horizontal gradient is assumed. In mid latitudes, the gradient is close to +0.8 mGal/km northward. Again, reduction and observations can use the same reference point.

Since earlier versions of the international normal gravity formula (1930 and 1967) slightly differed in the numerical parameter values, comparing gravity data or maps from different periods requires knowledge of the corresponding normal gravity differences. These are

$$\delta g_{\mathbf{n}_{(1980-1930)}} = -16.3 + 13.7 \sin^2 \phi \, (\mathrm{mGal})$$

$$\delta g_{\mathbf{n}_{(1980-1967)}} = 0.8316 + 0.0782 \sin^2 \phi \, (\mathrm{mGal})$$

4.5.2 The Height Reduction

Usually elevation above or below the reference surface is reduced by assuming the normal vertical gradient, often simply the mean value at the ellipsoidal surface of about $-0.3086\,\mathrm{mGal/m}$. Systematic errors from their magnitude can be estimated for the range of elevation differences in a survey region and are avoided if necessary, see Eq. (4.4.4). The exact value to be taken depends on the accuracy requirements and elevation differences in view of the largest error influences, for example, those of density uncertainties (below). In applied geophysics $-0.3086\,\mathrm{mGal/m}$ usually suffices in non-mountainous terrain with less than 1 km height variations. The reference surface, i.e. the ellipsoid, the geoid or an approximation has been discussed above. The remaining anomaly after normal and height reduction is usually called the Free Air anomaly (*FA*), see Sect. 4.6.1.

4.5.3 Topographic Mass Reduction

The topographic mass between the Earth's surface and the reference surface can have a substantial effect on the observed gravity value. The effect is estimated and subtracted. As mentioned, the topographic mass is considered added to, or subtracted from, the reference ellipsoid; hence the corresponding gravity effect at the point of observation, to be mass-reduced, is subtracted (land above sea level) or added (ocean). The customarily assumed Bouguer density is $2670\,\mathrm{kg/m^3}$ estimated in the past as the weighted average for surface rocks. Application of the Bouguer reduction results in the Bouguer anomaly (*BA*). The topographic mass is usually divided into two parts: (1) the Bouguer slab, mostly as a horizontal, laterally infinite slab or plate of rock between the observation level and the reference level (sometimes the Bouguer plate reduction is exclusively called Bouguer reduction) and (2) the deviation of the topographic surface from the plane surface, i.e. the relief; its effect is removed in the terrain reduction and then one calls the result "complete Bouguer anomaly".

4.5.3.1 The Bouguer Reduction

The infinite horizontal Bouguer plate of thickness h and density $\rho_B = 2670\,\mathrm{kg/m^3}$ has the effect $\delta g_{BP} = 2\pi G \rho_B h$, about $0.1117\,\mathrm{mGal/m}$ (for $\rho = 2390\,\mathrm{kg/m^3}$ the effect is $0.1\,\mathrm{mGal/m}$; remember: 10 m rock has 1 mGal effect). That the Bouguer plate is horizontally infinite does not present a problem because, as the solid angle approach (Chap. 2) demonstrates, the effect of distant parts of the plate vanishes quickly, for example, for $r > 10h$ it is $< 0.05\,\delta g_{BP}$. Except in high alpine mountains, Bouguer plates often represent a sufficiently precise representation of the topographic mass. Mountain slopes are smoother than it looks and rarely exceed $10°$ in low to medium mountain ranges.

The difference between Free Air anomaly and gravity disturbance must also be taken into account in the Bouguer plate reduction. The height h is the geoidal height in the first case, the ellipsoidal height in the second case.

It can also be shown that an equivalent *spherical Bouguer plate* of thickness h, i.e., with the same effect $2\pi G\rho_B h$, extends laterally to where the tangential plane through P" on a reference sphere below P intersects the tangential sphere through P with angular width θ_B. According to KJ61 (p. 73–77), the effect of a spherical (radius R) Bouguer plate of thickness h and angular width is $\delta g_{BPsph} \approx 2\pi G\rho h$ $(1 - [(R+h)\cos\theta - R]/((R+h)^2 + R^2 - 2(R+h)R\cos\theta)^{1/2})$.

The Bouguer reduction density $\rho_B = 2670\,\mathrm{kg/m^3}$ is part of the assumptions describing the reference earth. For cases where, locally, the real rock densities significantly differ from this value, it is recommended to, nevertheless, calculate the standard Bouguer anomaly. This will make it compatible with published standard Bouguer anomalies. A more correct density can be taken for a second calculation of an adapted Bouguer anomaly. The deviations of densities from the standard values can be treated as a geological reduction or deferred to the interpretation, dealt with as any other buried density anomalies. Incorrect Bouguer density values affect the error estimates of the Bouguer anomalies, also in connection e.g. with elevation errors.

The main effect of the Bouguer reduction is to remove large gravity differences between nearby points at different elevations. In spite of rather smooth topography in medium relief terrain, elevation differences cause significant Bouguer effects, and the short-wavelength behaviour contrasts with average long-wavelength behaviour of, say, $> 100\,\mathrm{km}$ dimension in large mountain ranges as the Alps. The regional mean Bouguer anomalies are highly correlated with mean elevation. This clearly hints to systematic deviations of the Bouguer model of topography on top of, or excavated from, the normal ellipsoid. In the real Earth isostasy is a dominant feature.

4.5.3.2 The Terrain Reduction

The relief, often called "terrain" or "topography" deviates from the plane (or spherical) surface of the Bouguer plate, and the gravity effect of the mass between the irregular surface and the plane must be estimated. The same density of $2670\,\mathrm{kg/m^3}$ is assumed because the terrain effect is a correction to the Bouguer slab effect calculated with that density. Since density tends to increase with depth, systematic errors may be introduced by these assumptions, but mostly the real density variation below the surface is unknown. The calculation methods for the terrain effect are a forward problem of the same nature as encountered in any gravity modelling (Chaps. 2, 5, 6). A few traditional methods are sketched below.

In the infinite plane case, the terrain effect is always negative, hence the reduction is positive. Any part of the topographic mass above the level of the observation acts to reduce the gravity value; any depression relative to the observation level means that the plate effect that had been subtracted is too large. In the case of the spherical Bouguer plate the situation is more complicated: the sign of the reduction is positive only if the relief rises above the tangential plane through P or if it is deeper than the curved level through P; mass lying below the tangential plane and above the Earth's curved level through P has a positive effect and has also not yet been taken into account by the spherical plate; hence the reduction is negative.

Usually, the terrain is divided into the near field where insufficiently detailed map information exists, and the far field taken from maps or digital terrain models (DTM). Mostly the terrain effect quickly decreases with distance from the observation point. Often the only non-negligible effects come from near-field topography which should be avoided in point selection. Sketches, for example, of two-dimensional sections, may permit estimates of the near-field effects (see e.g. Sects. 2.9.7.4, 5.6.4). Special schemes (to $100\,\mathrm{m}$, or so) have been designed by companies into which the near topography is to be sketched such that the effect can be estimated by counting points. It may require local surveying, possibly using simple instruments as inclinometers.

For calculating the far-field terrain effect approximate methods were applied in the past which were based on cylinder ring segments and precalculated diagrams and or nomograms (Sect. 2.9.2.2). It required for each observation point at height h_P the estimation of height h_top at grid points or for surface segments, one of the most tedious tasks in traditional gravity reduction. Templates or overlay charts, i.e. transparent reticules (after Helmert) with rings and radii were used in combination with nomograms or tabulated values of the contributions $\Delta\delta g_\mathrm{top}$ (to the total topographic terrain effect $\delta g_\mathrm{top} = \Sigma \Delta\delta g_\mathrm{top}$) as function $f(\Delta h, r)$ of distance r between P and grid point and of $\Delta h = h_\mathrm{top} - h_\mathrm{P}$. Sometimes, 2D approximations are appropriate for the terrain, especially in the near field, and a special case is briefly sketched in Sect. 5.6.4 .

An analytical method due to Schweydar (KJ61, 179–183) fits circular mean Δh^2 values (around P) by second-order polynomials in horizontal distance R through any consecutive triples of mean Δh^2 values to render coefficients which, with the aid of tabulated values, are converted to the terrain effects δg_top; for more details on terrain reductions see KJ61, 172–188 or the original literature quoted there.

Today, terrain effects are usually calculated from digital terrain models on rectangular grids. These need not be re-read for each station, such the computational work is left to digital programs. Effects of spherical curvature can be easily built into the programs, but the plate and terrain reductions must be applied consistently, either plane or spherical. Terrain effects can be calculated also on the basis of digitized contours, but care must be taken where a contour passes very close to a gravity station (Jacoby, 1967).

4.5.3.3 Geological Reductions

Sometimes additional geological reductions are applied, as mentioned. Precondition is sufficiently reliable a priori knowledge of the geological structures in question. Part of the geological anomaly g_{geol} (or δg_{geol}) is moved to the right hand side of Eqs. (4.5.1) and (4.5.1a) by computing, with any suitable method, and subtracting the effects of the known geological bodies.

4.5.3.4 Isostatic Reductions

A special geological aspect is isostasy, i.e. the tendency of vertical mass balance in the upper few hundred kilometres inside the Earth's crust and mantle. The effect is of interest especially in large regions or globally. The way how the balance is achieved was, and is, a matter of debate and is generally assumed in some idealized fashion. Isostatic reductions have been applied to Bouguer anomalies since many decades and generally this was done with the aid of simplified isostatic model assumptions as those of Airy or Pratt, and the Earth around a station (or a grid point) P was divided into compartments by circles and radii through P (Airy-Heiskanen, Hayford). Details are not the subject of this treatise, except saying that the topographic mean compartment height is directly translated into either a crustal thickness (Airy) or an upper layer density (Pratt: usually 100 km thick) and then into the corresponding gravity effect at P. The effects were tabulated and were added up. The resulting anomalies are called "isostatic anomalies", i.e. the deviations of observed gravity from normal gravity (normal, height and mass reductions) modified by the effects of topographic/bathymetric mass and its isostatic compensation. One may also imagine the process as redistributing the topographic mass to the compensating volume (KJ61, 80). A related method is that of condensating the topographic mass onto the geoid or the ellipsoid, the gravitational effect on P is usually small, so that condensation anomalies hardly differ from free air anomalies.

The *indirect effect of geodesy* refers to geoid calculations on the basis of reduced gravity anomalies; the Earth's mass, hence the potential and the geoid are effectively changed by the reductions, and to compensate this change, the reduced mass is condensed onto the ellipsoid, and the effect of this compensation is called the "indirect effect". This term is hardly used in geophysics, but occasionally it describes the reduction from the geoid to the ellipsoid to arrive at the gravity disturbance.

The atmosphere is usually neglected in any of the reductions. Estimated by the flat Bouguer effect, the atmosphere exerts an upward pull of the order of $400\,\mu$Gal, and air pressure variations do have effects; for example, a $50\,$hPa (hectopascal) high (low) would cause about $20\,\mu$Gal gravity decrease (increase) which may be removed in the drift correction. Globally the observation is inside an approximately homogeneous spherical shell which has no effect at the Earth's surface, but this is not true for high elevations on mountains, in airplanes or at satellite level.

4.6 The Result of the Reductions: Gravity Anomalies and Gravity Disturbances

The above reductions lead to a set of gravity anomalies at points with coordinates and elevations. These are sometimes called "absolute" if referred to the geoid or to the ellipsoid itself, and "relative" if referred to an arbitrary base station value. The point values are considered time-invariant exception for special geodynamic effects. Primarily the anomalies are digital lists of numbers, for example, in computer files or in printed form. Such lists are the basis of further analysis, numerical modelling and interpretation, but as such they are uninspiring. The reduced data sets become useful for interpretation (Chaps. 5, 6) only after some preparatory manipulations, as visualisation. Traditionally visualisation has been realized by contouring with lines of equal anomaly values or by drawing profiles, now mostly executed by computer programs including colouring. Visualisation is undoubtedly important for guiding interpreters' ideas (see Sect. 5.2). It must, however, be kept in mind (see Sects. 1.4, 5.1.5) that the results are principally distinct from the observed discrete values and that the methods of deriving the continuous fields, functions and pictures, as interpolation, functional fitting, kriging, etc., influences the essence of the anomalies and hence their interpretation. This is especially relevant if in the interpretation the observed values are replaced by calculated points on grids or profiles. In interpretation it is essential to keep in mind the definitions of the different kinds of anomalies, i.e. the reference models on which they are based and the reductions applied to the observations (above: Sect. 4.5). We now have the Free Air anomaly (FA), the Bouguer anomaly (BA) and various isostatic anomalies (IA). If so-called geological reductions (to which isostatic reductions belong) are applied, various modified anomalies result. Reductions can be considered as part of the interpretation.

4.6.1 FA: Free Air Anomaly, Faye Anomaly

The Free Air anomaly (FA) is defined by the observation after normal and height reduction referred to the geoid or to an arbitrary relative datum. According to Eqs. (4.5.1 & 4.5.1a):

$$FA = g_{\text{obs}} - g_{\text{n}} - g_{\text{h}} \quad \text{or} \quad \delta FA = \delta g_{\text{obs}} - \delta g_{\text{n}} - \delta g_{\text{h}}$$

If the geoid undulation relative to the reference ellipsoid is taken into account in the height reduction, i.e., if δg_h is taken relative to the ellipsoid, the result is the gravity disturbance or Free Air gravity disturbance FD. Terrain effects in mountainous regions strongly affect the FA values and make them highly variable and irregular. The terrain reduction removes such effects of local mass irregularities, but not the dominant effect of the topographic mass in the form of the Bouguer slab. If the terrain reduction is applied, the result is called *Faye anomaly*.

4.6.2 BA: *Bouguer Anomaly*

The Bouguer anomaly results from complete application of Eqs. (4.5.1 or 4.5.1a)

$$BA = g_{obs} - g_n - g_h - g_{top} \quad \text{or} \quad \delta BA = \delta g_{obs} - \delta g_n - \delta g_h - \delta g_{top}.$$

It should thus render the geological effects usually addressed in interpretation, if all assumptions correspond to reality. The effect g_{top} or δg_{top} of the topographic mass is meant to include all masses above the reference level, i.e. in the usual reduction procedure, the Bouguer plate and the terrain. If only the Bouguer plate reduction is applied (Bouguer reduction), the "*simple Bouguer anomaly*" results; if the effects of deviation of the topographic terrain or relief from the plane surface are removed by the terrain reduction the result is called "*complete Bouguer anomaly*". Any density deviations from the Bouguer earth assumptions below *and* above the reduction level (i.e. between geoid or arbitrary level and Earth surface) are still contained in the BA values. Nevertheless, the BA has the advantage to be not much dependent on the local scales of topography and to bring out geological effects more clearly than the FA (including the Faye anomaly) does.

As mentioned, the height and topographic mass reductions can be referenced to the normal ellipsoid, although this is rarely done. In that case the result is the Bouguer gravity disturbance BD.

4.6.3 IA: *Isostatic Anomalies*

The isostatic anomaly IA is calculated from the BA by adding an isostatic reduction, $-\delta g_{im}$, based on schematic isostatic models (*im*). The most obvious feature balancing large-scale topographic mass is crustal thickness variation, but other kinds of density distributions have also been assumed:

$$IA = BA - g_{im} = g_{obs} - g_n - g_h - g_{top} - g_{im} \quad \text{or}$$
$$\delta IA = \delta g_{obs} - \delta g_n - \delta g_h - \delta g_{top} - \delta g_{im}.$$

The isostatic model must be specified. The tendency of mass balance in the upper few hundred kilometres inside the Earths crust and mantle is a geological aspect of large regions. The Airy and Pratt models as well as combinations have been taken to describe this tendency (Sect. 4.5.3.4). Isostatic anomalies (*IA*) represent deviations from theoretical gravity fields composed of normal gravity plus height and mass effects of topography / bathymetry and of its isostatic compensation as assumed to approximate reality. Reality, however, varies regionally on Earth. Again, the basis of the isostatic reductions may be the Bouguer gravity disturbance, and in that case the result should be called the isostatic gravity disturbance *ID*.

4.7 Preliminary Data Analysis

4.7.1 General Aspects

Preliminary data analysis aims at rendering reliable and interpretable anomalies. It is necessary (1) to identify outliers, i.e. gross errors or mistakes generated somewhere from observation to data transfer and processing and, beyond the reductions for geophysical comparability, (2) to make the anomalies amenable to the human mind to form model ideas or pictures. The classical way of visualizing spatial data is drawing profiles and contouring, which has a strong relationship to forming interpretative ideas (see Sects. 5.2 and 5.3). In preparation, discrete anomaly values are converted to continuous representations, often via gridding, i.e. the derivation of regular point grids, in one and two dimensions, from irregularly spaced observations, for example, by mathematical and numerical procedures. Functional fitting of the data is a customary intermediate step, and upward or downward continuation essentially requires it. Some functional expansions, as standard polynomials, have only formal value, others have a physical meaning. Preparatory steps include attempts to separate different field components and may extend to the calculation of field quantities (Sect. 2.10.2) not directly observed and the investigation of relationships with other geological and geophysical quantities. These tasks are achieved by manual, statistical and analytical methods. Data analysis nearly always has a filtering character. As such, filtering is a formal or neutral operation, but it may affect interpretation (see Sect. 1.1.1). Statistical methods are used to emphasize certain aspects (e.g. gradients), directions and relations with other data. Special wavelets are suited to isolate certain features (e.g. edge effects). Field quantities have strong filtering aspects and emphasize different field components. Modellers and interpreters should test the consequences and investigate them in view of the reliability and consistency of the whole data body, gravity and non-gravity (see Chap. 7). This is particularly important in regional-local separation. Data analysis therefore cannot be principally disconnected from interpretation. Various possibilities and applications are discussed also in this Chapter, below, but the choice of methods will always depend on the target.

4.7.2 Data Snooping or Identifying Outliers

Outliers are considered to be mistakes made at any stage. Erratic effects that can be estimated, as those of buildings, are better removed before data analysis, but effects from unknown sources may appear to occur at 'random' and may be treated like observational errors. Very localized anthropogenic magnetic effects embedded in a smooth geological anomaly (widely differing length scales) may be considered outliers. There is no absolute definition, it depends the aim of a survey. Outliers are better removed and/or corrected before smoothing; otherwise the errors affect the results; for example, high-pass filters only partly remove data scatter and errors. An approach would be to remove the outliers by fitting a smooth function and repeat the process in an iterative way.

Outliers may be obvious by inspection and intuitively interpreted as mistakes. If less obvious to the eye, they can be searched by statistical criteria for the likelihood of belonging to a stochastic or Gaussian distribution or a distribution with some autocorrelation characteristics. Assume a space (or time) function to be defined by inaccurate data points; if the function is known or anticipated, it can be determined by parameter adjustment with the residuals approximately Gauss distributed and rendering the standard deviation. An analysis of the distribution may reveal the probability of certain residuals not to belong to the error set. Using the median in comparison with the arithmetic average may identify outliers, too.

A function of unknown kind is more difficult to determine; this is the case in the usual Fourier analysis or when step functions at unknown positions are searched. Then also the residuals are less certain. A statistical analysis of residuals $r(x) = f(x) - f^*(x)$ should be independent from the function $f(x)$ assumed, where $f^*(x)$ is the smoothed function. The analysis is affected by the autocorrelation function $A(b)$ between $r(x)$ and $r(x - b)$, shifted to the right; for a stochastic error $A(b)$ should be a delta function at $b = 0$. An outlier may be defined as a residual r where $|r|$ is greater than the value of an assumed limiting probability. KJ61, 268, 284 treats the case $f(x) = $ const and $r(x)$, or better r_i (discrete and equidistant), with 1st order autocorrelation (only direct neighbours are correlated). However, the identification of an outlier cannot be absolute but may have only some probability. The analysis may be extended to directional analysis and statistics (see below: Sects. 4.7.3 and 2.10.9).

4.7.3 Smoothing, Averaging, Filtering

Filtering begins, before any formal data treatment, in the field with point spacing and distribution, affecting the drawing of contour lines or gridding. Data series are affected by random or stochastic errors with a certain distribution, more or less Gaussian. It is assumed that outliers and jumps which do not belong to that distribution have been correctly removed by data snooping; however, it is never guaranteed.

Explicit smoothing can be carried out in many ways, for example, manually by inspection and drawing or by averaging, functional fitting and statistical methods. The choice of a smoothing method depends on what is considered significant and what irrelevant. Some methods emphasize short wavelengths, some long ones, and each method has its own character and associated errors. The simplest filtering method is to average point values of spatial (or temporal) data series in usually overlapping areas with weighting functions (see Sect. 2.10.2). The averaging region around a given point P may be a circular area of radius r around P or a ring between r and $r + \Delta r$ with $r =$ distance from P; it is a convolution with a weighting or filter function $w(r)$ which may be smooth in the continuous case or a box function in the case of a constant weight. In the spectral domain, it corresponds to multiplication with the Fourier transform $W(k)$ of $w(k)$. In practice it is given by a finite number of discrete coefficients.

Several schemes for equidistant data points were designed (see e.g. KJ61) to reduce tedious manual calculations. The simple box has a spectrum with unfavourable side lobes. The 'triangular' filter is somewhat more appropriate: $w(r) = 1 - ar$ $(a = 1/r_{max})$; its spectrum has very subdued side lobes (see Sect. 2.10.2). Fourth differences correspond to a parabola fit through P and two neighbours at each side. In binomial filters the coefficients are based on Pascal's triangle whose every second row is used; the bell shape of the function is expressed as

$$\delta g^*_i = 2^{-2n}[\delta g_i + \sum\nolimits_{1,n}(2n!)(k!(2n-k)!)^{-1}(\delta g_{i-k} + \delta g_{i+k})] \tag{4.7.1}$$

It approaches a probability distribution, and for large n, it becomes Gaussian (3.8.1):

$$\delta g^*_i = (n\pi)^{-1/2}[\delta g_i + \sum\nolimits_{1,n}\exp(-k^{-2}/n)(\delta g_{i-k} + \delta g_{i+k}) \tag{4.7.2}$$

For continuous functions:

$$\delta g^*_i = 1/(L\sqrt{\pi})[\int_{-\infty}^{\infty} \delta g(x)\exp(-[(x-r)/2L]^2)dr \tag{4.7.3}$$

with the spectrum (see Sect. 2.10.2)

$$\delta g^*(k,L) = \exp[-k^2L^2]; k^2 = k_x{}^2 + k_y{}^2 \tag{4.7.4}$$

$2L$ characterizes the width of the Gaussian.

Before filtering, irregularly distributed data points are mostly interpolated onto a regular grid or profile. Alternatively one may fit a continuous function to the irregular discrete data set (see below), but this also can introduce errors difficult to assess. Furthermore, the functional fitting introduces smoothing of its own character, combining in some way with the explicit filtering.

4.7.4 Functional Fitting

4.7.4.1 Linear Regression and Polynomial Fitting

Linear regression is fitting a linear function, as a straight line or a plane, to data points. Average gradients of the gravity data are estimated, permitting the integration of torsion balance and gradiometer observations into gravimeter data. Before frequency analysis, subtraction of the linear trend is mandatory. An extension of linear regression is *polynomial fitting* (in one or two or more dimensions; see Sect. 2.10.3). It usually employs ordinary non-orthogonal polynomials and is a purely descriptive method of questionable value, for example, for regional-local separation. The coefficients are formal and depend on the degree chosen for the polynomial; for example, in a quadratic fit the linear term (or terms in x and y) does no longer represent the mean gradient (or its x and y components).

The residuals from least-squares fitting will add up to zero (at the point set fitted) but not for local anomalies. A problem is also that within data holes with no observational constraints, the polynomial may grow excessively and affect the marginal data region, as the best fit concerns only the data points. *Orthogonalized polynomials* (Sect. 2.10.3) are principally better to handle, but the extra work involved is not normally justified.

4.7.4.2 Fourier Expansion and Spectral or Wave Number Filtering

Fourier expansion (theory: see Sect. 2.10.5) is widely applied in spectral or wave number filtering. The desired spectral windows can be chosen arbitrarily. Fourier expansion is a decomposition into a series of spectral components or harmonic functions $\sin kx$ and $\cos kx$ ($k = 2\pi/\lambda$), in two dimensions $\sin k_x x \sin k_y y$ and $\cos kx \cos k_y y$ ($k_x = 2\pi/\lambda_x, k_y = 2\pi/\lambda_y$). Again, it is a formal operation, but has often interpretational motivation and is applied to upward or downward continuation, differentiation, regional-local separation etc. The decision about filter properties, as cut-off frequencies, is up to choice. Calculation of derivatives of the anomalies from the irregular data sets and upward or downward continuation are conceptually and practically best carried out in the spectral domain.

Discrete Fourier Transforms (DFT) estimate the Fourier transform of a function from a finite number of sampled points, as is always the case if applied to empirical anomaly values:

$$F = {}_l \sum\nolimits_1^m \delta g(l\Delta x)\mathrm{e}^{-il\Delta x} \tag{4.7.5}$$

Frequencies $> k_N = 1/(2\Delta x)$ (N for Nyquist) cannot be resolved. A related problem is that of aliasing, i.e. unwanted interference of the sampling intervals with any real waves (λ_r) of the observed fields (e.g. a λ_r close to Δx can lead to kind of a beat (see Arfken, 2001; Byerly, 1959, 1993, 2003).

Usually the FFT (Fast Fourier Transform) is used. It requires 2^n de-trended samples at constant intervals, i.e. the data set must be chopped or extended to a number of 2^n where n is a positive integer (e.g., $n = 10$, $2^n = 1024$). The 2^n data at Δx intervals define the fundamental wavelength $\lambda_o = (2^n - 1)\Delta x$. A linear regression is mostly carried out for de-trending on the full data set. Extension of data sets to 2^n samples creates problematic modification. The section of anomaly of length L may have ends that do not meet. They are forced to meet with sharp contrasts. Such contrasts cause problems of convergence of the Fourier series. Principally,

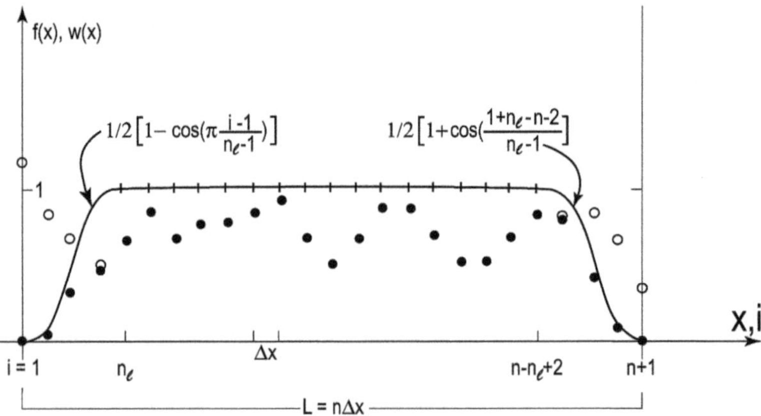

Fig. 4.4.1 Illustration of the application of a cosine window to taper a time or space series at its ends. Symbols (Δx, L,i, n, n_1) used in text are explained

convergence of finite series is guaranteed, but corners require inclusion of high frequencies to avoid artificial large oscillations with no meaning.

Windowed Fourier Transforms (WFT) attempt to solve this problem by windowing sharp transitions so that the input data converge at the endpoints via weighting functions, for example, the cosine window. As reality is often not known, caution is generally needed. As shown in Fig. 4.4.1, a data series ($i = 1$, $n+1$; such that $x = 0$, $n\Delta x = L$; $\varphi = 0, 2\pi$) is tapered at its ends, by multiplying it for $i \leq n_1$ and $i \geq n - n_1 + 2$ with cosine functions $w(x)$ (in between $w(x) = 1$) according to:

$$w(x) = \begin{cases} 1/2[1 - \cos(\pi/(i-1)/(n_1-1))], & i \leq n_1 \\ 1, & n_l < i < n - n_1 + 2 \\ 1/2[1 + \cos(\pi(i+n_1-n-1)/(n_1-1))], & i \geq n - n_1 + 2 \end{cases} \quad (4.7.6)$$

The series $w(x)f(x)$ is, of course, not the same as the original $f(x)$, but the tapering is hoped to essentially leave unchanged the important part of its spectrum and to only cut off the high frequency end which stems from the arbitrary cut-off of the data section. For this the width of the taper must not be too broad and tests are always recommended to assess the effects of the tapering.

4.7.4.3 Other Functions

Examples of functions that can be used for smoothing and filtering are the *Hermite polynomials*, the *sinx/x* function and geophysical effects expressed as functions of the coordinates of *point and line masses* (see Sect. 2.10.5). The aims of applying such functions goes beyond filtering and usually involves aspects of inversion and interpretation (see KJ61). The application of these functions to describing or fitting certain types of gravity anomalies usually involves their Fourier expansions and Fourier transformation. *Wavelets* (Sect. 2.10.8), likewise, allow spectral decomposition as series of special wavelet functions, for example, those describing and emphasizing edge effects.

4.7.5 Statistical Approach, Correlations, Regression

Correlation and regression analysis are two common and closely related statistical methods (see Sect. 2.10.9). Regression analysis has the quality of (linear) functional fitting (Sect. 4.7.4.1), which itself is either carried out manually, but usually involves least-squares fitting, hence statistics as well. Statistical methods are treated also in Sects. 5.1.5 and 5.5.3. They may guide the interpretation, though correlation does not automatically imply a causal relationship. Such methods can, however, reveal relations with features or patterns and with other relevant data, for example, geology, elevation, Moho depth, etc. Cross correlation between $\delta g(x)$ and a quantity $q(x)$, or between $\delta g(x,y)$ and $q(x, y)$, implies a linear relationship to exist when the corresponding values, each at identical coordinates x_i (and y_i), are plotted versus each other, irrespective of any value of i. Correlations can bring out features of special interest and these can then be investigated further. The term "autocorrelation", often denotes correlation between signals offset in the same data set, $\delta g(x)$ and $\delta g(x-u)$. Usually, relations are obscured by data scatter and/or non-linear relations, and a common measure of the degree of correlation is the correlation coefficient r_{gq} between δg and q.

A brief introduction to correlation and regression analysis follows. Assume two variables, x and y, say, Bouguer anomaly and Moho depth, that are suspected to be related to each other, but with considerable scatter. The x_i and y_i values may, for example, pertain to the same geographical points i or areas $i(\varphi_i, \lambda_i)$. Correlation usually implies a linear relationship; more involved relations may be concealed by the data scatter. The correlation coefficient between n data pairs x_i, y_i (n = length of data vector) is

$$r_{xy} = r_{yx} = s_{xy}/(s_x s_y) \tag{4.7.7}$$

with the covariance

$$s_{xy} = \sum_i (x_i - \underline{x})(y_i - \underline{y})/(n-1), \tag{4.7.8}$$

the mean values

$$\underline{x} = \sum_i x_i/n, \underline{y} = \sum_i y_i/n \tag{4.7.9}$$

and the standard deviations

$$s_x = \left(\sum_i (x_i - \underline{x})^2/(n-1) \right)^{1/2} \quad \text{and} \quad s_y = \left(\sum_i (y_i - \underline{y})^2/(n-1) \right)^{1/2}. \tag{4.7.10}$$

Generally $-1 \leq r_{xy} \leq +1$, but even extremely high correlation, i.e. r_{xy} very close to 1 or -1, may be accidentally caused by a third parameter, z. This can be tested with the aid of a partial correlation coefficient

$$r_{xy(z)} = (r_{xy} - r_{xz}r_{yz})/\left((1-r^2_{xz})(1-r^2_{yz})\right)^{1/2} \qquad (4.7.11)$$

where r_{xz} and r_{yz} are defined equivalently to r_{xy}. If $|r_{xy(z)}| < |r_{xy}|$ the latter is probably overestimated, due to parameter z, and r_{xz} may be more significant or relevant; if $r_{xy(z)} \approx 0$, while $r_{xy} > 0$, the latter may be only apparent, attributed to z. If $|r_{xy(z)}| > |r_{xy}|$, z weakens the correlation between x and y, and if $r_{xy(z)} \approx 0$, r_{xy} is probably meaningful. Insignificant influence of z on the x, y correlation is indicated by $|r_{xy(z)}| \approx |r_{xy}|$.

Multiple correlation analysis can be extended to more parameters, say z and w, and so on, by calculating analogously:

$$r_{xy(z)} = (r_{xy(w)} - r_{xz(w)}r_{yz(w)})/\left((1-r^2_{xz(w)})(1-r^2_{yz(w)})\right)^{1/2}. \qquad (4.7.12)$$

The results are generally not invariant to the sequence of analysis.

A correlation need by no means be one of cause and effect. Even an important statistical significance of a relation found need not be great. The reader is referred Knight (1999), Rice (1995), Sachs (1984), Electronic Textbook (2005). It can be tested by the *T-test* (Rubin, 1994) for discriminating of data sets x_{ik} with differing mean values $\underline{x}_i$ and standard deviations $\pm s_k$ (i counts sets, total number n_i, k counts individual samples). Is the difference $\Delta\underline{x} = \underline{x}_i - \underline{x}_j (i \neq j)$ statistically significant? It depends obviously on the standard deviations or scatter of the sets, $s = (s_i^2/n_i + s_j^2/n_j)^{1/2}$. If $T = \Delta\underline{x}/s > T^*(\alpha, f)$, the difference is significant on the α level, say at $\alpha = 95\%$, where f is the number of degrees of freedom; T^* has been calculated and tabulated for stochastic (not auto-correlated) distributions for different pairs of α, f values (see texts; tables can be easily found also in the internet). For correlation coefficients the T-test is:

$$T = r(n-2)^{1/2}/(1-r^2)^{1/2} > T^*(n, \alpha) \qquad (4.7.13)$$

where r is any of the above correlation coefficients and n = number of data pairs. A few values of T^* for $\alpha = 95\%$ and n are, as $\boldsymbol{n}$, T^*: $\boldsymbol{3}$, 6.61; $\boldsymbol{4}$, 2.92; $\boldsymbol{5}$, 2.35; $\boldsymbol{7}$, 2.05; $\boldsymbol{10}$, 1.86; $\boldsymbol{30}$, 1.70; $\boldsymbol{100}$, 1.66; ∞, 1.65.

The F-test of significance of multiple correlation coefficients r_{mult} is similar to the T-test. F is calculated with the assumed level of significance, say 95%, and the number of degrees of freedom f (equal to the difference between the dimension and the number of independent variables, for linear regression $f = 1$) and the excess number of observations $q = n - f$.

$$F = r_{mult}^2 q/(1 - r_{mult}^2)f. \qquad (4.7.14)$$

$F^* = F(f, q, r_{mult})$ is tabulated, and F as computed must exceed F^* to indicate significance.

In *regression analysis*, usually one variable, say x, is considered independent and precise, the other, y, is considered a function of x, i.e. dependent and error-affected. The order may, of course, be reversed, or both may be considered imprecise. These assumptions influence the statistics and the results and can be tested to some extent. Regression analysis has the quality of functional fitting (Sect. 2.10.3), but here, instead of functions of the space coordinates, it is functions of, for example, some geological parameters. Only linear regression is considered. Assume the observation y_i to be described by:

$$y_i = a_o + a_1 x_i + v_i = l(x_i) + v_i \qquad (4.7.15)$$

y is linear in x: a_o, a_1 are constants, and v_i is a deviation or residual from the assumed straight line. This is generalized to a multi-parameter relation or a plane or hyperplane:

$$y_i = a_o + a_1 x_{i1} + a_2 x_{i2} + \ldots + a_m x_{mi} + v_i. \qquad (4.7.16)$$

The coefficients a_k are calculated by least squares, i.e. by minimizing the sum of the squared residuals, $_i \sum v_i^2 = \min$, leading to the symmetric set of the linear normal equations:

$$N_{kj} a_k = r_j \qquad (4.7.17)$$

with $N_{oo} = n, N_{kj} =_i \sum x_{ki} x_{ji}, r_j =_i \sum x_{ji} y_j$.

The variance (3.7.2), here: $S_{xx} = {}_i\Sigma_i{}^2/(n-m-1)$, and the standard deviation $s_x = S_{xx}{}^{1/2}$ are measures of the statistical closeness of the fit $y(x)$. Note that the solution is possible (or reliable) only if the parameters x_{ki} are mutually linearly independent, otherwise N_{kj} becomes singular (or nearly so). A less critical condition is also that the residual distribution is Gaussian.

For small scatter, i.e., s_x small, the linear relation can be analogously expressed as $x(y)$, if $|a_1| \approx 1$ the corresponding coefficients would be $a^*_o \approx -a_o/a_1, a^*_1 \approx +1/a_1$, and $v^*_1 \approx -v_i/a_1$. The larger the scatter and the more $|a_1|$ differs from 1, the greater the deviations from these values, because now $v^*_i = x_1 - l^*(y_i)$, instead of $v_i = y_1 - l(x_i)$ is minimized, and the fit approaches $x_i \approx -a_o/a_1 - y_i/a_1 + v^*_i$, where v_i can take on quite different values. A useful visual indication of the reliability of regression analysis is to plot both relations together (at least easy in the one-dimensional x, y case); if both straight lines intersect at an acute angle the regression is strong, if they intersect at nearly right angles it is meaningless.

A measure of the success of introducing a new variable in multiple regression is the variance reduction:

$$b =_i \sum (l(y_i) - \underline{y})^2 /_i \sum (y_i - \underline{y})^2 \qquad (4.7.18)$$

The multiple correlation coefficient is, in this case, $r_{\text{mult}} = b^{1/2} (0 \leq r_{\text{mult}} \leq 1)$; r_{mult} grows always with variables added if they are independent and the residuals are Gaussian, but only if $r_{\text{mult}} > 0.6$, the correlation is strong and significant, and if $r_{\text{mult}} < 0.4$ it is insignificant. A test of significance of the multiple correlation coefficient is also the F-test (see above).

In summary, the possibilities of statistical analysis are to reveal the existence or non-existence of relationships. Practical applications and a more thorough study of the theory is left to the readers. Statistics can guide, but not replace, further geophysical and geological investigations.

4.7.6 Derived Field Quantities

Preliminary data analysis may involve the derivation of field quantities as the potential, gravity gradients and higher derivatives (Sect. 2.10.4), usually calculated from gridded anomaly values with the aid of function fitting (Sects. 4.7.4.1–4.7.4.3). Derivatives can be estimated from the spatial variation of δg_z at and above the surface of the massive Earth. This operation enhances short wavelengths and subdues longer-wavelength components. It is sometimes taken as removal of a regional field (below). The reverse, i.e. integration from the vertical derivative to the potential enhances the longer wavelengths. Satellite observations of the sea surface give an indirect estimate of the potential variation. All these operations involve continuous fields derived from discrete observations and are subject to errors, difficult to estimate (see Sect. 5.5.1).

4.7.7 Regional-Residual Separation

4.7.7.1 General

It is very often the case that the gravity anomaly observed and derived by the reductions, contains parts of no immediate interest, either the smooth long-wavelength regional background field, or the very short-wavelength components of only very local or no significance. To enhance small-scale features, downward continuation or higher derivatives can be calculated. The motivation of regional-local separation may also be that the anomalies of interest are too small to be recognizable, perhaps only as wiggles of contour lines. After the smooth "regional" is removed the residuals are taken to be the "local" anomalies.

There is no unique definition of smoothness; it is not simply a matter of spectral filtering. The act of regional-residual separation is principally interpretational suffering from the ambiguity problem. Succinctly speaking, separation requires knowledge of the mass distribution and cannot be done before the interpretation. Indeed, in cases where large-scale geological masses (e.g. a seismically determined Moho depth variation) with long-wavelength gravity effects are known, they should be better included in the interpretation proper or may be computed beforehand as a geological reduction and removed rather than a purely mathematical definition. Many methods have been used for separation. All methods include arbitrary choices or assumptions.

The residual or local values δg_{res} are defined as

$$\delta g_{\mathrm{res}} = \delta g_{\mathrm{obs}} - \delta g_{\mathrm{reg}}. \tag{4.7.19}$$

Defined by $(\delta g_{\mathrm{reg}} - \delta g_{\mathrm{obs}})$, δg_{res} has the opposite sign of associated causative mass anomalies.

4.7.7.2 Manual Smoothing

Manual smoothing of anomaly profiles is perhaps the simplest and most obvious method. Similar procedures may be applied to a contoured map, directly by smoothing contours or by constructing parallel profiles, smoothing them and re-contouring. The degree of smoothing will depend on the case at hand. Individual smooth curves drawn by hand through the same observed profiles differ, especially when drawn by different interpreters. The uncertainties must be kept in mind and possibly several attempts should be compared to each other. Manual operations are subjective in the sense that they can never be repeated exactly.

4.7.7.3 Averages and Functional Fitting

The various ways of averaging (Sect. 4.7.3) or linear fitting (see Sects. 2.10.2, 2.10.3) may be applied and the results may be declared regional δg_{reg}. These mathematical methods are formally more objective, as they avoid the uncontrollable variations of manual drawing, but their very choice is subjective mostly without objective criteria. Admittedly, the results of the numerical operations are exactly repeatable if applied in the same way to the same data set. A case in point is the popular polynomial fitting, where the choice of the degree is arbitrary, including the linear regression in one or two dimensions.

In all methods of functional fitting, least squares (lsq) is the principal criterion, implying that the residuals behave as random errors. This is not realistic. Harmonic functions (Sect. 2.10.5) permit any spectral filtering, as low pass, band pass and high pass filtering and are thus very flexible. Their orthogonality is an important advantage, but the caveats are the same, and the choice of wavelengths for regional and local is arbitrary; shorter wavelengths may be taken for local. Similar remarks pertain to wavelets; their advantage for the regional definition is that they are spatially limited. Akin to functionals are finite elements (Mallick and Sharma, 1999) where the shape function for the regional are constructed from few regional points of the field. In all cases the error sum (in large data sets) should be zero which is usually ensured by the formal smoothing schemes. However, there is always a conflict with the possibility of suppressing significant details or the possibility of misinterpreting significant smaller-scale anomalies with non-zero integral effects.

4.7.7.4 Upward and Downward Continuation, Derived Field Quantities

A physically defined smoothing method based on the Laplace equation is upward continuation (see Sect. 2.10.5.3); to what height is arbitrary. Upward and downward continuation apply harmonic functions because they naturally define their space behaviour. Analytical upward continuation is an effective means to smooth a given anomaly distribution because small scale anomalies decay more rapidly with height than longer scales. The opposite approach of downward continuation through hopefully homogeneous layers enhances the small-scale local features. In the spectral domain the amplitudes vary exponentially, as e^{kz}, with the vertical z and the wave number k (where $z < 0$ upward). This is not a true separation of wavelengths, only one of emphasis.

The different field quantities (Sect. 2.7.2), from the potential to the higher derivatives emphasize different parts of the wave number spectrum or wavelengths and can also be calculated with the aim of recognizing different field components termed regional and residual. As with field continuation, the numerical procedures of differentiating and integrating are best carried out in the spectral domain (Sect. 2.10.5.2).

4.7.7.5 Interpolation

Another similar method is to cut out the presumably anomalous region and to interpolate the field through selected marginal points around the anomaly region. To be sure, one must know the undisturbed regional field and the anomalous area. The boundaries must be drawn arbitrarily and marginal anomalous effects cannot be totally excluded. The regional field is then interpolated, manually (see above) or by some numerical interpolation routine.

4.7.7.6 Iterative Approach from Simple to Complex Modelling

Regional-residual separation can be included into the inversion-interpretation procedure by proceeding from simple to detailed modelling of gravity anomalies. Such an iterative approach to anomaly separation is described by Boschetti et al. (2004) and Strykowski et al. (2005); the first step is an approximate removal of a local feature, fitting only the regional with simple models and subtracting the effect from the original anomaly; in the second step the isolated local feature is model-fitted and removed. The aim is to reduce feature interference and the process is repeated until satisfactory convergence. A similar approach consists of the following steps. Assume a complex gravity anomaly to be composed of different components and try to isolate them step by step by fitting them with simple equivalent source geometries (which need not resemble a priori geological and geophysical knowledge) optimizing, for example, their locations and then subtracting their effects. The procedure is repeated until no further adjustment occurs. Only then the isolated anomalies are subjected to a geologically or geophysically guided modelling and interpretation. Formally, the method will be successful, but there is no guarantee for being geologically correct.

4.7.8 Directional Analysis

Finally, classical contour lines are amenable to directional analysis if showing clear directional preference; if less visible but important for the analysis, it is useful to emphasize the directional variation by directional statistics (e.g. Buchheim and

Lauterbach, 1954), generally presented as a rose diagram. A full circle (360°) is divided by n $\Delta\alpha$ segments ($n = 360°/\Delta\alpha$), and in each angular window ($\alpha, \alpha + \Delta\alpha$) the length sum of all line elements within that window is plotted as a proportional length. The connection of the end points is the rose diagram or a polar diagram. In the past usually carried out manually, it is now mostly done by numerical routines, for example, as part of the contouring procedure by directional statistics (Sect. 2.10.9). Certain geometrical constraints are discussed in Sect. 5.3.

4.8 Evaluation of Reduction Errors

Reductions and related manipulations introduce additional errors to those carried over from the observations (see Sect. 3.7). Errors accumulate along the whole path and ultimately influence the interpretation. Various error types arise, for example, from inaccurate auxiliary observations, as surveying and rock densities. Again, gross errors, systematic errors and random errors can occur and are briefly discussed below.

Gross errors (Sect. 3.7.), if detected should be removed. Data analysis (Sect. 4.7.2) includes inspection, statistical analysis and may lead to re-measurement as the best remedy.

Systematic errors (Sect. 3.7.2) from wrong reduction parameters are incorrigible if not recognized; they must be avoided by all means. The standard reductions do not represent systematic errors (Sect. 4.1). Local deviations from the normal reference assumptions, as vertical gradients or rock densities lead to anomalies, not errors. If this is taken into account properly by calculating the model effects for the original point locations, it represents no error. R*eduction errors* result if the points are *thought to move* in space *and* model effects are calculated or modelled at the shifted locations (heights). Similarly, a topographic mass density different from the standard $2670\,\mathrm{kg/m^3}$ must be modelled. Isostatic reductions are especially susceptible to such systematic errors due to unrealistic model assumptions. Systematic errors may be introduced also when continuous anomalies are derived from discrete values of the *BA*, *FA* or *IA*, when the data points have a space distribution unfavourable for recovering the features of interest (see Sect. 5.1.5). Systematic errors cannot be detected by a posteriori standard errors. This emphasizes the fact that the latter are only "apparent" (Sect. 3.8) and may be deceiving.

Random errors (Sect. 3.7.3) are independent from each other, but as has been discussed, this is not always the case, for example, density errors interfere with elevation errors in the Bouguer reduction. No errors but anomalies are introduced if reference, reduction and modelling are understood properly. Random errors in the tidal reductions (Sect. 4.7.2) are relatively small. The normal reduction is usually also uncritical because the horizontal derivative of normal gravity is small. Larger errors may come in through the height and mass reductions (Sect. 4.7.5) resulting from random errors of point elevations and from density fluctuations (even if $2670\,\mathrm{kg/m^3}$ is a correct global average) and from inaccurately mapped or digitized terrain.

Beyond the reductions as such, data analysis, functional fitting and regional-local anomaly separation will also introduce errors, however these remain essentially undefined because the basic assumptions are arbitrary and regional-local separation as such cannot be made in an unambiguous manner (see Sect. 4.7.5). One can only make the mistake to take the separation literally and thus misinterpret the regional or local anomalies. There is no guarantee against errors and mistakes in such procedures.

4.9 Conclusion

At the end of this Chapter, station listings and images of gravity anomalies with coordinates, elevations and perhaps additional characteristics exist to be handed over to be interpreted, which is the original purpose of the observations. In carrying out that task it is important to take into account the fundamental assumptions of the reductions and the related manipulations and data analyses.

References

Arfken, G.: Mathematical Methods for Physicists, 5th ed. *Orlando, FL, Academic Press, 1128 pp., 2001*

Boschetti, F., Therond, V., Hornby, P.: Feature removal and isolation in potential field data. *Geophys. J. Int., 159, 833–841, 2004*

Buchheim, W., Lauterbach, R.: Isanomalen-Richtungsstatistik als Hilfsmittel tektonischer Analyse. *Gerl. Beitr. Geophys., 63, 88–98, 1954*

Byerly, W.E.: An Elementary Treatise on Fourier's Series, and Spherical, Cylindrical, and Ellipsoidal Harmonics, with Applications to Problems in Mathematical Physics. *New York, Dover, 287 pp., 1959, 1993, 2003*

Electronic Textbook: http://www.statsoft.com/textbook/stathome.html, 2005

Featherstone, W.E., Dentith, M.C.: A geodetic approach to gravity reduction for geophysics. *Comp. Geosci., 23, 1063–1070, 1997*

Hackney, R.I., Featherstone, W.E.: Geodetic versus geophysical perspectives of the 'gravity anomaly'. *Geophys. J. Int., 154 (1), 35–43, 2003*

Heiskanen, W.H., Moritz, H.: Physical Geodesy. *Freeman, San Francisco, 403 pp., 1967*

Jacoby, W.R.: Zur Berechnung der Schwerewirkung beliebig geformter dreidimensionaler Massen mit digitalen Rechenmaschinen. *Z. Geophys., 33, 163–166, 1967*

Knight, K.: Mathematical Statistics. *Chapman & Hall, 504 pp., 1999*

Lemoine, F.G., Kenyon, S.C., Factor, J. K., Trimmer, R.G., Pavlis, N. K., Chinn, D. S., Cox, C. M., Klosko, S. M. Luthcke, S. B., Torrence, M. H., Wang, Y. M., Williamson, R. G., Pavlis, E. C., Rapp, R. H., Olson, T. R.: The development of the joint NASA GFSC and the National Imagery and Mapping Agency (NIMA) geopotential model EGM96. *TP-1998-206861, NASA Goddard Space Flight Center, Greenbelt, Maryland, 1998.*

Mallick, K., Sharma, K.K.: A finite element method for computation of the regional gravity anomaly. *Geophysics, 64, 461–469, 1999*

Moritz, H.: Geodetic reference system 1980. *Bull. Géodés., 54, 395–405, 1980*

Rice, J.A.: Mathematical Statistics and Data Analysis, 2nd ed. *Duxbury Press, Pacific Grove, CA, 672 pp., 1995*

Rubin, J.: Handbook of Usability Testing. How to Plan, Design, and Conduct Effective Tests. *Wiley, New York, NY, 356 pp., 1994*

Sachs, L.: Applied Statistics. A Handbook of Techniques, 2nd ed. *Springer Verlag, Berlin etc., 707 pp., 1984*

Strykowski, G., Boschetti, F., Papp, G.: Estimation of the mass density contrasts and the 3D geometrical shape of the source bodies in the Yilgarn area, Eastern Goldfields, Western Australia. *J. Geodynamics, 39, 444–460, 2005*

Wentzel, H.G.: http://www.gik.uni-karlsruhe.de/~wenzel/hw95/hw95.txt; http://www.astronomy-notes.com/gravappl/s10.htm, 1995

Chapter 5
Qualitative Interpretation

5.1 Fundamental Ideas, Principles

5.1.1 Qualitative and Quantitative Interpretation

Qualitative interpretation is at the root of interpretation. "Qualitative" is contrasted with "quantitative" and means something like "principally right" or also "of the right order", while "quantitative" means "numerically correct". "Principally right" implies that an interpretation envisages the right type of model; "of the right order" denotes a different aspect, namely "semi-quantitative" or "approximate" with low requirements on accuracy, efficient. "Quantitative" is not absolutely definable, because numerical correctness can never be absolute or error-free, but the requirements on accuracy are high, or better: appropriate to a given problem.

Good interpretative ideas or models are provoked by given anomalies, their patterns and relationships with geology or by analogy with related features. Forming such ideas obviously largely feeds on experience which teaches this better than words can do. The qualitative models have then to be quantified and tested and either verified or falsified. The testing requires calculations, but especially in potential fields with their principal ambiguity of inversion, basic ideas must be first. This requires creative imagination and a good geological background. In large scales, experience with dynamic models is important; on small scales, also technological, engineering and economic insight is necessary. At the idea stage, simple approximate, semi-quantitative estimates of gravity effects can teach what can be learnt and what not. The interpreter must be used to effortless estimate and exclude a large number of possibilities immediately. In all cases the question is (1) what is the principal cause of the observed anomaly and (2) what is the approximate dimension, depth, geometry and strength? "Interpretation" has a broader meaning. Beside finding a suitable density model, it includes also methodological aspects and an assessment of geological implications, for example, prospects of economic resources, and relations in the widest sense and in the whole field of study.

The two sides of qualitative interpretation are: (1) forming principal ideas and (2) aiding the idea forming process. It comprises visual, geometrical, physical,

W. Jacoby, P.L. Smilde, *Gravity Interpretation*,
© Springer-Verlag Berlin Heidelberg 2009

geological and geophysical aspects, and it includes the aspect of semi-quantitative estimates of model effects in the form of simple rules to easily convince oneself that an idea makes sense. (1) Sects. 5.1–5.5 discuss fundamentals and qualitative constraints of interpretation and (2) Sect. 5.6 presents simple approaches to gravity effect estimation; a final Sect. 5.7 illustrates this with a few examples.

5.1.2 The Ambiguity Problem and a priori Information

The starting point is the notorious principal ambiguity of the inverse problem. A given gravity anomaly can be explained by infinite numbers of density models teaching nothing by themselves. Preconceived ideas can be refuted or confirmed by gravity data, and quite different possibilities may be overlooked. An example is the bell-shaped effect generated by a point mass (or a homogeneous sphere) that may also be generated by a shallow flat lens. Often the existence of a compact sphere-like mass at some appropriate depth is then taken for granted (Fig. 5.1.1). A case in point is the interpretation of large-scale gravity anomalies related to mantle convection; when the first long-wavelength satellite solutions of the global gravity field had been derived, gravity lows were interpreted with hot low-density rising flow and highs by down going flow. However, convection disturbs both the temperature field associated with voluminous density anomalies *and* the static equilibrium position of density contrast surfaces, for example, the Earth's surface, representing thin lens-shaped mass anomalies, and both must be considered (see Sect. 5.7.8; McKenzie, 1977).

The starting point of interpretation can be a geological problem or a given gravity anomaly. In the latter case, for example, a new data set from poorly known areas or in satellite-derived gravity fields of other planets, little geological a priori information may exist. In such a case one may be guided by comparing the anomalous features with similar known instances or by applying very general ideas of planet formation and dynamics. In the more usual cases, where the starting point is a geological problem, the feature of interest which one wants to know, may be the depth structure, and a dedicated gravity survey has thus been conducted. In such cases, usually a large amount of a priori knowledge exists which can be directly fed into the interpretation procedure.

Because of the principal ambiguity, blind search is futile. If there is no automatically natural solution of a problem, it is reasonable to set the stage in a plausible and well founded and constrained framework. Qualitative interpretation is the important step of making conscious choices of the range of possible models, better than being unconsciously guided by preconceived ideas. It is essential for determining the direction of the search for probable mass distributions that cause the observed gravity anomalies. One may say: a real density distribution exists which is not known but is looked for; the best possible success of interpretation is the density distribution which most probably equals the real one or most closely approaches it, and the probability distribution around it. However, what is that: the density

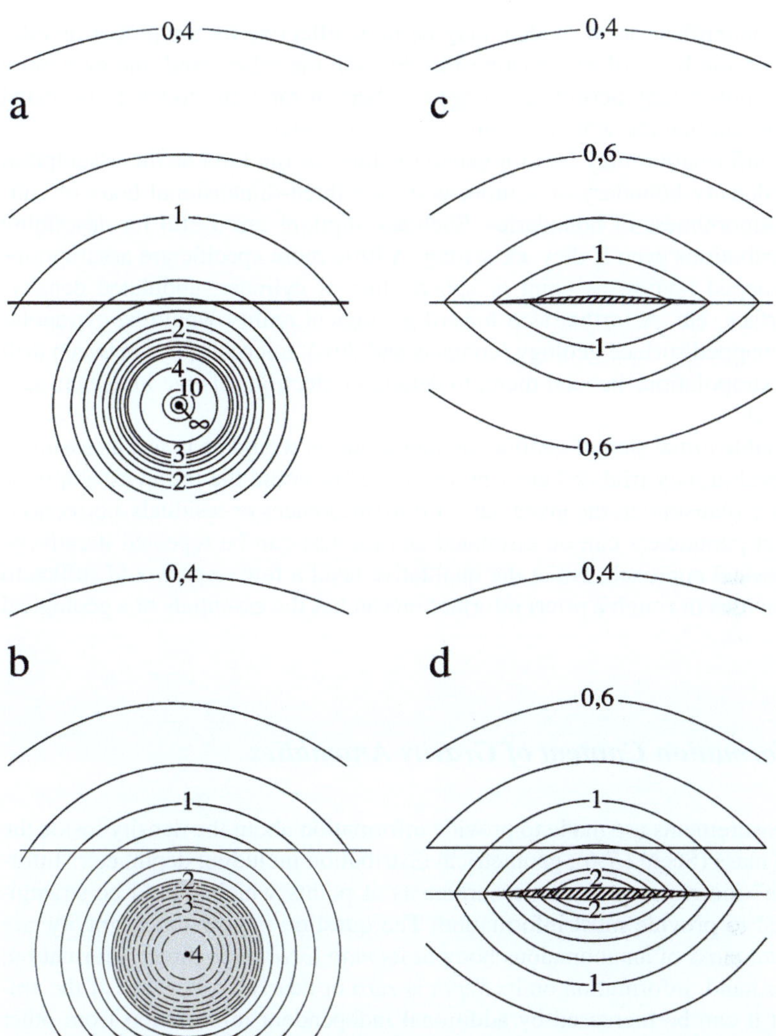

Fig. 5.1.1 An infinity of mass distributions produces the same gravitational field in the source-free outer space, depicted by equipotential surfaces intersecting the paper plane; numerical values in arbitrary units: (**a**) point mass, (**b**) homogeneous sphere (inside, the equipotential surfaces are shown by *dashed lines*) (**c**) the equivalent stratum at the surface where the thickness symbolizes the surface mass density, (**d**) an equivalent stratum at some depth or a thin lens-shaped mass (after KJ61, 208)

distribution which most probably equals the real one or most closely approaches what is principally unknown? It can only be replaced by models on the basis of a priori information and must be judged by some simple rules relating the models to their effects. This has also an aspect of work economy. A well informed search for appropriate and reliable density models is more economic than blind search.

Arriving at quantification of models may be most effective by the proper qualitative design on the basis of pre-existing insights. On the other hand, the ever more efficient computers may permit very large numbers of random models to be tested without contradicting the work economy of the interpreter.

A priori information may be of a general nature, of the kind as the assumption of a single density boundary or compactness of a three-dimensional body or simplicity and smoothness of boundaries. Such assumptions are useful for describing density distributions even if they are wrong. A little more specific are assumptions of simple-shaped bodies, as point or sphere, line or cylinder, undulated density-contrast surface, etc. A further step toward geological reality would be extrapolations from mapped surface geology (contacts and dips), interpolations between drill holes and extrapolations beyond them, to detailed reflection seismic images in sections or in 3D.

A reasonable initial step of qualitative interpretation would be to test an assumed body by a preliminary trial and error procedure: How could the initial assumptions be changed to plausibly fit the given data? From differences or residuals, corrections of the model parameters can be estimated and the test can be repeated iteratively. It is the classical approach, but at the qualitative level a few steps would suffice to find out if an idea or rough a priori information catches the essentials of a geological problem.

5.1.3 Information Content of Gravity Anomalies

Gravity measurements are made to provide information about the density inside the Earth, total mass (Sect. 2.7.6), geometrical distribution including depth, etc. "Information gain" motivates gravity measurements at points which should be strategically located to provide most information. The question is how much gravity *can* tell. The *total mass* of an anomalous body or its *map location* is direct information; on the other hand, information on its *depth* is zero or nearly so, because of the ambiguity, but it can be increased by additional independent information from other sources.

The logical steps are: (1) questions or aims, (2) measurements at sets of discrete points, (3) derivation of anomalies as functions of the space coordinates and (4) the interpretation of the anomalies, i.e. a description or model of the density distribution, with the aid of additional a priori information. At a single observation point, g carries little information, even less than the vector $\mathbf{g}$. Obviously the recognition of anomalies from discrete point sets is inseparable from the act of interpretation or extraction of information about the inaccessible Earth's interior (see Sect. 5.1.5). Basically information is provided by the measurements and concentrated in the anomalies. The interpretation is the ultimate information gained, but it is limited by the inherent uncertainties. Therefore the information content of gravity data can not be easily defined or quantified, but it is definitely enhanced by additional a priori information. It also depends on the resolution and accuracy of the data. Any complete representation of the gravity field carries exactly the same amount of information which is not changed by mathematical operations. Each field quantity, however, emphasizes different spectral aspects, and hence, different components of the source distribution. For example, the spatial frequency content tells something about the probable depth of the density variations and may permit some separation of different density components.

The limited information content of scalar (i.e. incomplete) gravity data can be completed with the aid of the Laplace equation ((2.7.1), see Sects. 2.7.4, 2.10) since the vector quality can be retrieved from values at the boundary. If known at the surface, $g(x,y)$ completely determines the g field anywhere above, because the vector field of g is determined in the source-free, zero-density space. The limitation in practice is the limited number of observations and their errors. New information is gained by new observations. One can obtain vector information by measuring components of the gravity tensor (Sect. 2.8, eq. (2.8.1)), for example, deflections of the vertical with torsion balances or with gradiometers (borehole instruments, GOCE satellite: Sects. 3.2.8, 3.2.9) and by analysis of horizontal accelerations of satellites.

5.1.4 *Data Representation and Interpretation Constraints*

Data are primarily digital: numbers or symbols mostly connected implicitly with physical units. Qualitative ideas generally present themselves as images. Graphical and digital representations are equally important, they complement each other. Bringing the two sides together calls for graphical visualisation of anomalies and models on the basis of the digital data (Sect. 5.2). Other means to help forming ideas on qualitative interpretation are: simple geometrical constraints (Sect. 5.3), supplemented by physical constraints, as plausible limits to density (Sect. 5.4) and, most importantly basic geological knowledge (Sect. 5.5). Correlation, regression and other statistical analyses of geological and/or geophysical data are valuable tools. Helpful are also simple semi quantitative rules or formulae (Sect. 5.6) on the basis of Newton's law. With simple model geometry a feeling can be developed of what is essential and what is negligible.

5.1.5 *Anomaly and Model Effect*

A strict distinction is made between the observed anomalies, δg_o, and the calculated effects, δg_m, by which the observations are to be explained. The two notions should not be mixed up, as unfortunately is common. The present chapter partly deals with the estimation of effects. Some simple methods of a semi-quantitative character are needed to quickly test ideas. Matching model effects with anomalies is the process of interpretation.

An anomaly is a mental picture or pattern and forms intuitively and subconsciously by imagination from few discrete observations, a deviation from the reference earth. It is subjective and vague; it has no definite error bounds or uncertainties and is part of the interpretation. How imagination works obviously depends on the interpreter's experience. Where not observed, some kind of interpolation would render values less certain than observed values. The purist view is that only the observations have a factual value and nothing is known, where not observed and only the observations should be used in interpretation. This contradicts the experience of the existence of patterns or anomalies in space, not arbitrary scatter; their acceptance is a realistic view. Anomalies may be more or less obscured by random scatter as found in any data set in geology, geophysics and elsewhere, and caution is obviously necessary to protect the interpreter from rash and uncritical conclusions.

If the anomaly is to be constructed from the limited data, objective methods are needed to get somewhat reliable estimates for points in data gaps between the data points, i.e. the most probable values and their probability distributions (values with error bars). This is especially important if the task is to derive strategic points (extrema, half width coordinates, inflection points, etc.) or a regular grid of points from irregularly spaced data; grids are the basis of graphic presentations or may be applied in numerical interpretation and inversion methods. The success of such methods depends decisively on the point distribution where the errors are to be minimized; unsuitably located points may lead to gross misinterpretations.

A plausible quantitative estimation method is based on the geostatistical spatial behaviour of the observations, under the assumption that the estimated points belong to the same statistical family. It can be characterized, for example, by variograms of the data or by their autocorrelation. It is assumed that the data can be modelled as a stationary stochastic process, i.e. independent from the location within the observation region. In gravity this will never hold exactly, and different variograms may be calculated for different neighbourhoods. Variograms are a measure of the spatial relation between the values at (pairs of) points and are defined by the variance S. In profiles (coordinate x, $\Delta x = x_i - x_j$):

$$S(\Delta x) = (1/n)_i \sum^{n} (\delta g(x_i) - \delta g(x_j))^2 \qquad (5.1.1)$$

where the pairs x_i, x_j are selected to fall into a class or interval $\Delta x \pm dx$; their number n will generally depend on Δx. For marginal points calculation may include the assumption of periodicity of the pattern in the interval of x considered. Generally, the variogram is a monotonous function of Δx growing from zero or a finite value (observational error of δg) to an asymptote (sill); beyond the range (where the sill is approximately reached) the gravity values are considered independent from each other. Variograms can be calculated for gravity maps (coordinates x, y) analogously where Δx is replaced by $\Delta d = (\Delta x^2 + \Delta y^2)^{1/2}$. Mostly gravity maps show some directional variation, which can be investigated by variograms for azimuthal intervals; they may render varying ranges, roughly describing an ellipse about the point x,y; two-dimensionality can then be defined by a ratio of its long over its short axis being above a certain number (however, see remarks in Sect. 5.3.1 on depth extent in relation to radius of curvature of models).

The experimental or empirical variogram S is, due to errors and irregularities of the data, approximate. Therefore S may be fitted by a theoretical variogram which describes a certain family of stochastic processes, and the fit will also render the characteristic values of sill, range and possibly the observational error; the Gaussian model based on the Gaussian probability distribution and error function is an example.

The interpolation method is called "Kriging" (after D.G. Krige), "Gaussian process regression" or "optimal prediction" which is a form of Bayesian inversion (see Wikipedia: "Kriging"), applied to profile and map data sets. As the unmeasured points are assumed to have the same statistic as the measured ones, the gravity value $\delta g(x)$ at an arbitrary point x and its error or probability distribution are estimated on the basis of the variogram $S(\Delta x)$ and the location relative to the observed points $\delta g_i = \delta g(x_i)$ at locations x_i, as a linear combination or weighted average, the Krige estimator:

$$\delta g(x) = {}_i\sum w_i \delta g_i \tag{5.1.2}$$

where the weights w_i are calculated from the appropriate variogram (the weights in other interpolation schemes are less appropriate to the data set). Naturally the w_i decrease with the distance $|x - x_i|$ from the observations and the uncertainties grow accordingly (Fig. 5.1.2). The w_i are determined, subject to the condition that the mean error of the estimates is zero and its variance is minimal. The solution of this extreme value problem takes the stationarity of the stochastic process into account and leads to a system of linear equations for the w_i relating to the location x (which is different for each x) to the locations of the neighbouring data points. If all neighbours' distances are outside the range (of the variogram) $\delta g(x)$ becomes their unweighted arithmetic mean and the error variance is undetermined. For more details and the theory underlying Kriging see, e.g., Sachs (1984); Armstrong (1998).

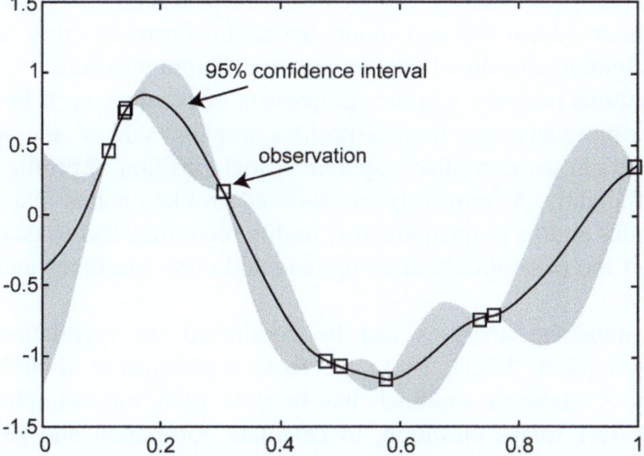

Fig. 5.1.2 Example of Kriging interpolation with confidence intervals (after http://en.wikipedia. org/wiki/Kriging). For interpolation the data (*squares*) are assumed to be error-free, otherwise the curve need not exactly fit the data where the 95% error band would assume the data error width

Kriging seems to be the most appropriate method to define an anomaly from discrete data sets. However it should be kept in mind that any estimate of unmeasured values may have large improbable errors (low probability as calculated by Kriging). There is no insurance against undetected anomalies, for example, of small scale relative to the average inter-point distance or in larger data gaps; such features may be overlooked in calculation the variogram. In this sense the sceptical view is true that nothing is known where not observed, but this "nothing" has probability limits to be carefully assessed, and not any gravity value in between is equally probable.

5.2 Digital and Visual Modes of Representation

Qualitative gravity interpretation can be circumscribed as creative integration of data distributions or gravity anomaly fields with geology and physics. The mental process is highly complex and involves imagination. This remains true in the age of digital modelling and is greatly aided by easy transformation between the digital and visual data representations. Qualitative interpretation happens largely in the analogue world, and quantitative interpretation is then only the translation into digital computation. There is no strict separation, and "qualitative" includes semi-quantitative estimates by the use of rules of thumb.

The importance of visualisation cannot be over-emphasized; digital modelling methods cannot be applied without knowing what one is doing. Details of visualisation techniques are not presented here, and the reader is referred to specialized books (Natl. Acad. Press, 1997; Nakamura & Nakamura, 2001; Schneider & Eberly, 2002) and is encouraged to implement and apply such techniques. Data are much more easily visualized along profiles (corresponding to 2D models) than in map view (corresponding to 3D models); profiles and 2D models are popular and useful even where the source distribution is distinctly three-dimensional. In that case the two-dimensional step is only preliminary, and methods of constructing 3D models from 2D sections facilitate interpretation. Colours are useful for recognition of patterns and relations, facilitating also direct comparison of different data sets.

Conscious study of relations includes graphic comparison of parameters. It begins with looking at two or more analogue figures (profiles or maps) side by side or comparing different data sets in some similar way. Direct juxtaposition of profiles can reveal relationships efficiently. A frequently encountered problem is that maps are usually in different scales and/or projections. It is highly recommended to take some extra trouble, exploit the possibilities of computers and copy machines and achieve scale compatibility.

In digital data sets, statistical analyses can be conducted, as correlation (Sect. 5.5.2) and regression (Sect. 5.5.3). These are more sophisticated, though often not more revealing. A classical approach has been to plot, for example, gravity anomalies (*BA* or *FA*) versus elevation, to calculate correlation and regression coefficients and to deduce properties of the isostatic compensation (e.g. Woollard, 1962). Multiple regression has turned out informative about different factors affecting gravity anomalies (Jacoby et al., 1991). Regionalisation of such analyses can uncover important difference of structures and processes. In a recent study of the Iceland crust, a similar approach has resulted in new insights into geo-dynamic processes of ridge-plume interaction (Fedorova et al., 2005).

5.3 Geometrical Constraints: Patterns

Ideas form when geometry and patterns are viewed; pattern is a visual concept and thus especially requires visual techniques, from simple sketching to sophisticated computer graphics. Locations and sizes of maxima and minima, zones of gradients and their shape and orientation will be noted and highlighted. Spatial spectra may be helpful as well.

5.3.1 Three-Dimensional – Two-Dimensional

A fundamental distinction is whether an anomaly feature is elongated or irregular in map view. It is then classed as two-dimensional (2D) or three-dimensional (3D). These somewhat misleading terms imply, without gross errors, model geometries of horizontally infinite extent or compact point-centred symmetry in three-dimensional space. The former case is simpler to treat as it can be visualized on a 2D sheet of paper. 3D cases may be handled at a preliminary step in 2D sections, keeping the differences in gravity effects (see Chap. 2) in mind. The 2D approach is appropri-ate generally if the ratio of vertical (z) over horizontal (r) scale is small: $z/r \ll 1$ (Sect. 2.9.3.1), i.e. the geological object has a large radius of curvature relative to its depth extent. Figure 5.3.1a shows the gravity effects at the edge of a 1 km deep step (2D) and of circular cylinders (3D) of 10 and 3 km radius; the visual impression near the edge is very similar, but the amplitude is significantly affected, especially for the 3 km cylinder; in an enlarged scale, also the differences of the step effect from the two cylinder effects are shown; note their weak maxima near the edge; here the lateral difference in mass of the 3D cylinders versus the 2D step is gravitation-ally more effective than nearby; the effect of the 3 km cylinder has its maximum at $x = -3$ km beyond which the difference $\delta g_{step} - \delta g_{cyl}$ grows rapidly. Figure 5.3.1b shows a least-squares fit of a 2D step effect to the anomalies generated by the two cylinders. The resulting densities are, correspondingly, smaller (in detail they de-pend on the point set chosen for the fit), and the residuals, shown in extended scale, are not large, except where the opposite edge of the 3 km cylinder is approached.

5.3.2 Spatial Frequency

The dominant spatial frequency content is depth-dependent, but there is no unique relation between wavelengths and source depths. Gravity effects of laterally limited

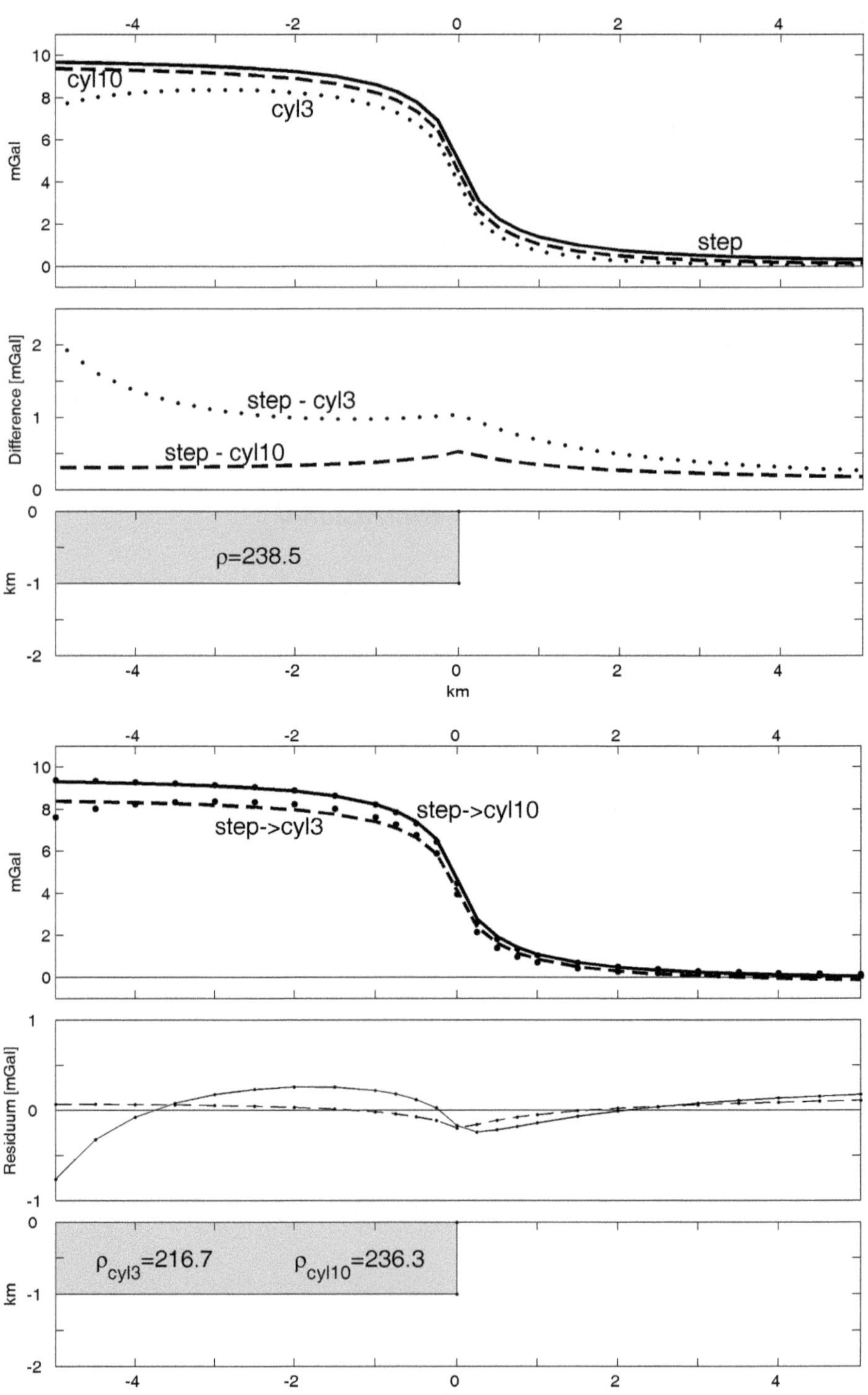
cyl10
cyl3
step
Difference [mGal]
step - cyl3
step - cyl10
ρ=238.5
km
mGal
km
step->cyl3
step->cyl10
Residuum [mGal]
ρ_cyl3=216.7
ρ_cyl10=236.3
km
km

bodies are attenuated and broadened with depth, the inverse is not necessarily true: long-wavelength gravity anomalies may have shallow long-wavelength sources; there are also exceptions to attenuation and broadening of the effect: (1) a constant anomaly of infinite wavelength does not attenuate and may be caused by an infinite horizontal homogeneous slab (Bouguer slab) at any depth, and (2) a harmonic sinusoidal source generates an equally sinusoidal effect (as $\sin kx$) which, while attenuating, does not broaden (see Sect. 2.10.5.3), but wavelength and phase are unchanged and equal to those of an equivalent surface mass at constant depth (or a more complicated source).

Interpretation can gain from considering the frequency domain by expanding spatial distributions in a harmonic series (Fourier series in Cartesian geometry or spherical harmonics) for which powerful numerical routines exist. The spectral components of low frequencies may, or may not, be caused by deep sources, and the high frequencies are probably caused by shallow sources. Maximum depth criteria (Sect. 5.6.7) can be applied, however with caution. Applications are also the separation of regional and local components of a gravity anomaly, but the problem is non-unique (see Sect. 4.7.7).

5.4 Physical Constraints: Realistic Limits, Integral Relations

There are physical limits to acceptable models. Mass densities of voluminous crustal materials will not exceed values of 3000 to $4000\,\text{kg/m}^3$ or be less than $1000\,\text{kg/m}^3$ although small volumes, for example, of ore concentrations and cavities do occur beyond these limits. Usually it is the mean density of some large volume which is of interest. It depends on the target, aims and scale how best to define densities and their limits.

In gravity nearly always only density contrasts relative to a reference count: $\Delta\rho_i = \rho_i - \rho_0$; ρ_0 can be a constant or some spatial function, for example, one of depth only $\rho_0(z)$; $\Delta\rho$ can, of course, be negative causing negative gravity anomalies, perhaps surprising to some physicists.

Surface integrals over the gravity anomaly (Sect. 2.7.4) permit useful estimates of the total causative mass and its lateral centre of gravity prior to quantitative

Fig. 5.3.1 Comparison of 2D and 3D effects: edge, $d = 1\,\text{km}$, density $\rho = 238.5\,\text{kg/m}^3$, step and circular disks, $R = 3$ and $10\,\text{km}$. (**a**) Gravity effects and differences ($\delta g_{\text{step}} - \delta g_{\text{disk}}$); the effects are similar, but the amplitudes vary significantly, especially for the 3 km disk. (**b**) Least-squares fit of the 2D step effect to the anomalies generated by the two disks, by adjusting the density; given for the discs: $\rho_0 = 238.5\,\text{kg/m}^3$; equivalent step density for 10km disk: $236.3\,\text{kg/m}^3$ ($<1\%$ smaller), for 3km disk: $216.7\,\text{kg/m}^3$ ($<10\%$); the residuals $r(x)$ are shown in an extended scale; for the 10 km disk, generally $r(x) \leq 1\%$ of the disc anomaly except near the edge ($\sim 2\%$); for the 3 km disk, mostly $r(x) \leq 2.5\%$, beyond $x < -4\,\text{km}$, $r(x)$ drops rapidly toward the opposite edge of the disk

modelling, where the centre of gravity can often be approximated. Strict separation from other body effects is, however, not possible, nor can the surface integrals be extended to infinity which they theoretically should; both neglects can introduce errors. They can be reduced with assumptions of simple idealized masses and fitting the observations to the mass effects, but this is then quantitative interpretation.

5.5 Geological Constraints: Visual and Statistical Analyses, Structures, Densities

5.5.1 General Remarks

Basically, qualitative interpretation is an art, and geological relations generally enter into modelling ideas, often subconsciously, and thus evade systematic discussion. An experienced interpreter has a feeling for, or a good guess of, what is looked for. It cannot be acquired by conscious acts of learning, but the subject needs conscious deliberation.

5.5.2 Scale

Gravity-geology relations depend somewhat on scale. Most structures subject to geophysical exploration are small or locally embedded in larger features as continental shelves, rift zones or mountain belts. The corresponding scales are typically from <1 km to 10 km or a bit more.

Geological structures up to about 100 km scales may be called "quasi-local" and more or less independent from other structures. This is not literarily true in terms of geological processes, but such small masses do not necessarily influence the whole density structure which enters the interpretation; it may be treated as some form of a regional anomaly possibly removed (see Sect. 4.7.7).

From about 100 km scales upwards, the mass distributions quite probably have generated compensating structures which may mutually balance their effects, at least in part, a mechanism called "isostasy. An isostatic reduction may be applied, but it depends on the specific mode of isostatic compensation, which need not correspond to reality. There are many deviations from the classical Airy assumption of isostasy realized by Moho undulations.

5.5.3 Gravity in Relation with Other Geological Quantities

Geological quantities will be understood in a broad sense to include any relevant Earth features. Particular gravity-geology relations may become obvious by inspection and graphic comparison. Quantitative relations may be investigated statistically by correlation and regression between specific parameters, for example, gravity and topography, Moho depth etc., and more involved multi-parameter analysis; for large continental and global data sets of gravity and topography by regression analysis (Sect. 4.7.5) the crust-mantle density contrast for an isostatic model may be thus derived. Multivariate linear regression of many parameters (gravity, topography, seismic Moho depth, vertical motion, teleseismic station arrival time residuals, heat flow density and geological regionalisation) can isolate the important from unimportant influences on the gravity field and the models and processes that probably formed the structures.

According to the brief introduction into correlation and regression analysis (Sect. 4.7.5), correlation between $\delta g(x)$ and a quantity $q(x)$, or between $\delta g(x,y)$ and $q(x,y)$, implies a linear relationship to exist when the corresponding gravity and q values, at identical coordinates x_i (and y_i), are plotted versus each other. A measure of the degree of correlation is the correlation coefficient r_{gq} between δg and q (4.7.7). Correlations can bring out features of special interest and these must then be investigated further. Examples of such studies are Woollard (1962) for North American Bouguer anomalies and topography and for some global data; a multivariate and regionalized investigation of central Europe (Jacoby et al., 1991), and tectonically regionalized regression of topography, Bouguer anomaly and seismically determined Moho depths of Iceland and the surrounding North Atlantic (Fedorova et al., 2005), used to construct a crustal model by combining sparse seismic data with a regular topography and gravity point grid.

5.6 Some Simple Estimates of Gravity Effects

Ideas are only useful if they carry also some semi-quantitative information. For the purposes of qualitative interpretation or preliminary tests one needs more than just ideas, one must also have a feeling for the approximate size or amplitude of the associated effects. A few simple rules and formulae are useful for easy estimation of the effects of proposed models. The theoretical basis is Newton's law of gravitation and the fundamental relationships that have been described in Chap. 2. As suggested in Sect. 5.1.2, a preliminary trial and error test will quickly show, whether a chosen assumption is a viable prospective starting point for more sophisticated modelling.

5.6.1 Bouguer Plate

For a first, very rough estimate of an expected model gravity effect, the *Bouguer plate* can be taken, with a thickness similar to the vertical dimension of the model (Sect. 2.9.3.3). As a rule of thumb, a rock slab of 10 m thickness (and density, $2390 \, \text{kg/m}^3$) has a Bouguer effect of 1 mGal; usually an approximate density contrast will suffice: the simple expression is thus, in units mGal, m, and kg/m^3 (Task 5.1):

$$\delta g \approx 0.1 \times d \times \Delta\rho/2390. \tag{5.6.1}$$

The somewhat surprising fact that such a coarse mass approximation works for a reasonable gravity estimate, has a simple reason: the effect δg very rapidly decreases and vanishes as $\sim z/r^3$, more rapidly than $\sim 1/r^2$, when a mass element Δm is removed horizontally by the distance r; the distant parts of an infinite horizontal plate (or an infinite mass line or cylinder, in the 2D case, below), have little influence on the approximation of a laterally limited body.

The estimate will be the better, the wider the body is laterally, but simple adjustments to the estimate can be made by roughly assessing the *solid angle Ω* expanded from the central point (maximum anomaly) to the edges of the body (remember: for infinitely distant edges $\Omega = 2\pi$). Such an assessment can be easily done by a quick sketch. It is, again, simpler in the two-dimensional case, in which $\Omega = 2\alpha$, with α the plane angle extended from any point P to the edges to an infinite stripe of surface mass ρ^*. Figure 5.6.1 shows diagrammatically for a 2D strip and a 3D circular disc at depth z, how the effect increases with increasing body *width w^** from zero ($w^*/z = 0$) to that of a Bouguer plate ($w^*/z \to \infty$).

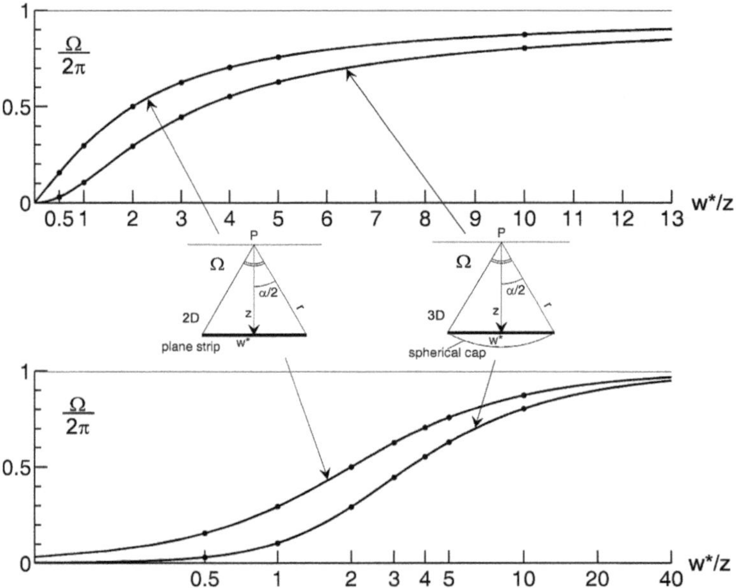

Fig. 5.6.1 Gravity effect increasing with increasing body *width w^** from zero to that of a Bouguer plate ($w^*/z \to \infty$) for a 2D strip and a 3D circular disc ($w^* = 2r$, r = radius) at depth z

5.6.2 Scale Rule

The scale rule states that gravity effects are proportional to the model length scale (2.9.1.1). It helps to form ideas of expected model effect relations. The rule can also be expressed as similarity of shapes and gravity effects, i.e. similar shapes (with the same density) generate similar gravity effects δg in space: define a model in units of meter and calculate the effects in units of μGal – magnify the whole by, for example, a factor of $1000(\text{m} \rightarrow \text{km}, \mu\text{Gal} \rightarrow \text{mGal})$. More important is the consequence that one can define scale-independent, i.e. only shape-dependent parameters on the basis of some features of the observed anomalies as maximum, half width, etc. and derive from them some shape parameters as depth, maximum depth, etc. The use of the solid angle in interpretation belongs to this class of shape characterisation.

5.6.3 Half Width Rules

A gravity anomaly can be characterized by its *extreme value*, δg_e located at x_e, and *half width*, w, defined as the distance $\Delta x = |x_{h1} - x_{h2}|$ between the two points where $\delta g_e/2$ at x_{h1} and x_{h2}. Three points thus characterize the limited gravity anomaly and are used to estimate mass or density anomaly and its depth. The relations depend on the assumed or suggested geometry. Several instances are given below. Two meanings are used in the literature, either as defined here or half of it.

5.6.3.1 3D: Point Mass or Sphere

Simple estimates can be based on *point or spherical masses*, i.e. directly on Newton's law. Of course, a point, even a homogeneous sphere, is ungeological, but sufficiently compact or sphere-like bodies can be approximated this way fairly well. A homogeneous (or radially symmetrically dense) sphere has the same effect as the anomalous mass concentrated at the sphere centre. Therefore (2.5.1): $\delta g_x = G\Delta mz/r^3$, is the basis for estimates and $\delta g_e = G\Delta m/z^2$ is its maximum at $x_e = 0$ and $\Delta m = \Delta \rho V$, the anomalous mass or excess mass of volume V and density contrast $\Delta \rho$, $z = $ depth to the centre point and $r = $ distance from the mass point to the station.

The half-maximum values $\delta g(d_{1/2}) = \delta g_e/2$ lie at the horizontal distance $d_{1/2}$ from the location of the mass point, and from (2.5.1) follows that

$$w = 2(4^{1/3} - 1)^{1/2}z = 2 \times 0.76642 \cdot z \approx 3z/2 \rightarrow z \approx 2w/3 \qquad (5.6.2a)$$

From δg_e and z follows the total anomalous mass $\Delta m = \Delta \rho V$:

$$\Delta m = \delta g_e z^2/G \approx (4/9)\delta g_e w^2/G \qquad (5.6.2b)$$

For a given anomalous density $\Delta \rho$ in a spherical volume $V = (4/3)\pi R^3$ (radius R), with all quantities in SI units, except δg in mGal $= 10^{-5}\,\text{m/s}^2$ and $G = 6.67\,10^{-11}\,\text{m}^3\text{kg}^{-1}\text{s}^{-2}$, V and R can be calculated, with minor numerical approximations, in $[\text{m}^3]$:

$$V = \Delta m/\Delta \rho \approx (4/9)\delta g_e w^2/(G\Delta \rho) \approx (2/3)10^5 \delta g_e w^2/\Delta \rho,$$

and

$$R \approx 25(\delta g_e w^2/\Delta \rho)^{1/3} \qquad (5.6.2c)$$

or the density, if V or R is given:

$$\Delta\rho_R \approx (4/9)\delta g_e w^2/(GV) \approx (2/3)10^5\delta g_e w^2/V \qquad (5.6.2d)$$

Letting, on the other hand, the sphere extend to the surface, i.e., $R = z$, in [m^3]:

$$\Delta\rho_z = \Delta m/(4\pi z^3/3) \approx 9/(8\pi G)\delta g_e/w \approx 5.5\,10^4\delta g_e/w \approx 3.6\,10^4\delta g_e/z \quad (5.6.2e)$$

The geological plausibility of $\Delta\rho_z$ must, of course, be assessed. For comparison, the thickness of a Bouguer plate of the same density $\Delta\rho_z$ with the effect δg_e (5.6.1) would be $d \approx 2z/3$ or $2R/3$, i.e. the deepest point of the anomalous spherical mass would extend to $3d$, three times the Bouguer plate thickness.

For relations of similar nature with other field quantities, see KJ61, 218–219.

5.6.3.2 2D: Mass Line or Cylinder, Vertical Rod, Vertical Fault step

The equivalent 2D case is the (infinite) *horizontal mass line or uniform cylinder* (Sect. 2.9.3.1), which may be a reasonable approximation for an elongated body of roughly circular cross section and whose z/l ratio is small (z = depth, l = length). The gravity effect (2.9.3.1) is simply: $\delta g = 2G\rho^+ z/r^2$, where ρ^+ is the mass per unit length of the line in kg/m and r is the distance of P from the line. The effective density contrast is $\Delta\rho \approx \rho^+/(\pi R^2)$, R = cylinder radius. The *half width* rule is, in this case, depth $z = w/2$ or 50 % of the half width between the two points of half-maximum gravity value. For simple estimates, the maximum gravity effect may be sufficient, i.e.: $\delta g_e = 2G\rho^+/z \approx 4G\rho^+/w$, and with the above relations, $\Delta\rho \approx w\delta g_e/(4\pi GR^2)$ or $R \approx (w\delta g_e/(4\pi G\Delta\rho))^{1/2}$. With δg_e in mGal, lengths in km and $\Delta\rho$ in kg/m^3, the above expressions take the form: $\Delta\rho \approx 12\delta g_e w/R^2$ or $R \approx 3.5(\delta g_e w/\Delta\rho)^{1/2}$.

In Sects. 2.9.3.1 and 2.9.7.1.1, the effect of, and the correction for, finite length of the horizontal mass line, instead of ∞, has been shown to be $a = \lambda/(x^2 + \lambda^2 + z^2)^{1/2}$ where $0 < a < 1$. The so-called 21/2D approach (Sect. 2.9.8) can, of course, be applied here as well.

Note that an elongated map shape is not a necessary condition for approximate two-dimensionality; the condition is that the ratio z/l of depth over horizontal extent is small. In the case of a curved density boundary or edge with radius of curvature r_c, the necessary condition for satisfactory two-dimensionality is $z/r_c \ll 1$.

For a narrow mass extending steeply from near the surface to great depth z, such as a volcanic chimney, a simple formula is that of the *vertical rod* (Sect. 2.9.3; eqs. (2.9.10) and (2.9.11)): $\delta g^{(l)} = G\rho^+[1/r_1 - 1/r_2] = G\rho a[1/r_1 - 1/r_2]$, where a is the area of the horizontal cross section and r_1, r_2 are the distances from P to the upper and lower end of the rod, respectively. This remains correct if the rod extends above the point elevation (the effects of the parts of the rod, symmetric about the point elevation, cancel). The effect of a vertical rod is not easily distinguishable from that of a point or sphere, although the half width w is greater, for infinitely deep bottom depth of the rod, $w_{rod}/d_{top} \approx 2\sqrt{3} \approx 3.5$, versus $w_{point}/d_{point} \approx 2.7$, where d = depth; a depth to the point mass $d_{\text{point}} \approx 4/3d_{\text{top}}$ would make the half widths about equal, but as the rod effect decays laterally very gradually, it is difficult to estimate the zero level. If the latter is unknown, the two effects (or curves with horizontal distance R) $\delta g_{\text{point}}(R/d_{\text{top}})$ and $\delta g_{\text{rod}}(R/d_{\text{top}})$ for $d_{\text{point}} = 2d_{\text{top}}$ are highly similar for $0 < R/d_{\text{top}} < 6$.

A **vertical wall** or **dyke model of infinite strike**, called in GW65, 294 "vertical ribbon", extending in depth from z_1 to z_2, and of thickness t and density contrast $\Delta\rho$ is characterized by the maximum anomaly δg_{e}, the half width w and the anomaly integral I to estimate z_1, z_2 and $\Delta\rho t$. With $A = 2G\Delta\rho t$ and the shape factor $\beta = z_2/z_1$: $\delta g(x) = A\ln(r_2{}^2/r_1{}^2)$, $r_i{}^2 = x^2 + z_i{}^2$, $i = 1, 2$

$$\delta g_{\text{e}} = A\ln\beta \tag{5.6.3}$$

$$w = 2z_1\sqrt{\beta} \tag{5.6.4}$$

$$I = \int_{-\infty}^{\infty} \delta g(x)\,\mathrm{d}x = A\pi(z_2 - z_1) = 2\pi G\Delta m^*, \tag{5.6.5}$$

where the 2D mass $\Delta m^* = \Delta\rho t(z_2 - z_1)$ [kg/m]. The scale-independent shape function $f(\beta)$ can be calculated: $f(\beta) = \delta g_{\text{e}}w/I = 2\beta\ln\beta/\pi(\beta - 1)$; $f(\beta)$ can be estimated from the observations, and the transcendental equation for β is solved graphically (see shape function $f(\beta)$ for vertical dyke in Fig. 5.6.2) with the aid of the graph of $f(\beta)$. Having found β, it is easy to find z_1 from (5.6.4), z_2 from β, and $\Delta\rho t = A/2G$ with A from (5.6.3) or (5.6.5).

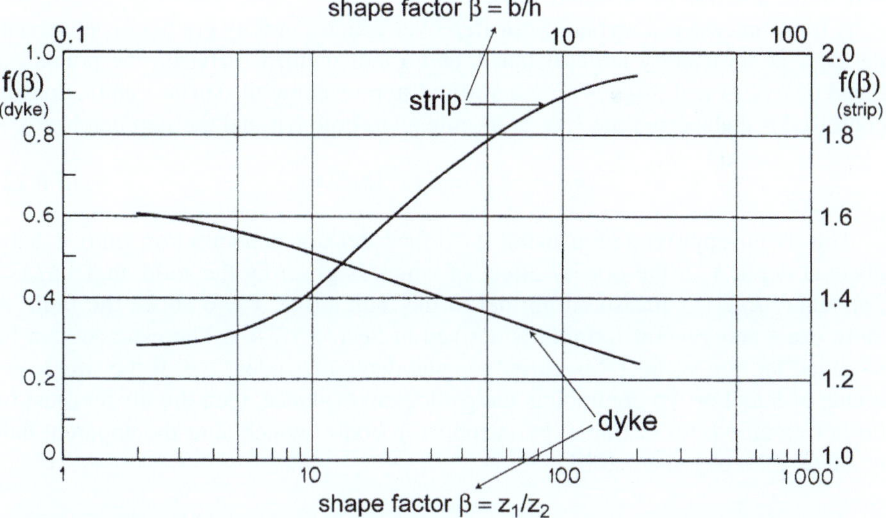

Fig. 5.6.2 Shape functions $f(\beta)$ of 2D vertical dykes and horizontal strips (after GW65, 294) for estimating the geometrical and physical parameters (dyke: depths z_1 and z_2 to *top and bottom*, respectively; strip: width b and depth h) from $\beta = z_1/z_2$ (dyke) or $\beta = b/h$. The scale-independent shape function can be calculated from the data: $f(\beta) = w\,\delta g_{\text{e}}/I$, where w = anomaly half width, δg_{e} = extreme value, I = integral over the anomaly (5.6.5), (5.6.8)

Two similar cases of GW65, 295 are the "strip anticline", i.e. the thin 2D **horizontal strip** of width 2b and the circular disk of radius a, both of mass density $\Delta\rho t$ and at depth h. With A as above, and $\beta = b/h$: $\delta g(x) = A \arctan[(x+b)/h] - \arctan[(x-b)/h]$

$$\delta g_e = 2A \arctan \beta \tag{5.6.6}$$

and taking into account the periphery angle

$$w = 2h(1+\beta^2)^{1/2} \tag{5.6.7}$$

$$I = \int_{-\infty}^{\infty} \delta g(x)\mathrm{d}x = A\pi b = 2\pi G\Delta m^*, \tag{5.6.8}$$

As above, $f(\beta) = \delta g_e w/I = 4(1+\beta^2)^{1/2}\arctan\beta/\beta$, and the solution for h, b and $\Delta\rho t$ proceeds analogously (see $f(\beta)$ for horizontal strip in Fig. 5.6.2).

The **circular disk** leads to elliptical integrals (the solid angle is an elliptical cone) and cannot be treated in elementary terms, but with $\beta = a/h$ and A as above

$$\delta g_e = A\pi(1 - (1+\beta^2)^{1/2}) \tag{5.6.9}$$

$$w \approx h(1.24 + 1.07\beta), \quad \text{valid for } 0.5 < \beta < 2.5 \tag{5.6.10}$$

$$I = \int_{-\infty}^{\infty} \delta g(x)\mathrm{d}x = 2A\pi h((1+\beta^2)^{1/2} - 1) = 2\pi G\Delta m^* \tag{5.6.11}$$

Analogously, $f(\beta) = \delta g_e w/I = (0.62 + 0.53\beta)/(1+\beta^2)^{1/2}$, proceeding from here to the solution for h, a and $\Delta\rho t$ likewise.

A frequent case is a **vertical fault step** over which a gravity gradient is observed, i.e. Δg_∞ at sufficiently large distance, and a half width w between the points (or lines) of $\delta g_\infty/4$ and $3\delta g_\infty/4$. For a vertical step reaching the surface and a density contrast $\Delta\rho$ the appropriate *half width* rule gives both $\Delta\rho$ and the step height:

$$h = w, \Delta\rho = \delta g_\infty/(2\pi G w) \tag{5.6.12}$$

This is an approximation using a surface density condensation onto a half-plane at depth $h/2$, the gravity effect of which is given by the solid angle $\Delta\Omega = 2\arctan(x/[h/2])$, x measured normal to the fault strike, $x = 0$ above the fault. A more exact and general formula is derived in Sect. 2.9.7.4.1. The estimate can be modified for non-vertical dip, buried top and depth dependent $\Delta\rho$. If the simple estimate of h and/or $\Delta\rho$ contradicts the geological evidence, then the model must be further modified, for example, by introducing bodies which alter the apparent half width (e.g. Jacoby et al., 2001).

5.6.4 *Use of the Solid Angle Ω: Vertical Templates*

The relation with Ω can be exploited and carried out by sketching for which one may need a ruler and a protractor (possibly a compass). The basic expression for the gravity effect Δg to be evaluated is $\boldsymbol{\Delta g \approx G\Delta\Omega\Delta\rho\Delta h}$, where $\Delta\rho$ is the relevant density contrast and Δh is the appropriate layer thickness; it may not always be clearly defined and must be assessed according to the geometrical situation. $\Delta\Omega$ must be estimated from a sketch, for example. The estimate is rather easy in 2D models where the solid angle is given by the plane angle $\Delta\alpha$ as $\Delta\Omega = 2\Delta\alpha$. For the theory see Sect. 2.9.1.1, rule 1, Sect. 2.9.7.2, (2.9.7), (2.9.49) and Fig 2.9.1.

A special case is the effect of a *sloping terrain surface*. Consider a flat valley bottom and a flat top plane, connected by a constant slope (angle α, height h, density ρ) and find that all points on the slope experience the constant terrain effect of $2\alpha G\rho h$; beyond the slope, on flat ground, the terrain effect (Sect. 4.5.3.2) decreases towards zero.

For a quick semi-quantitative assessment of a two-dimensional model (cross section) one can sketch a *transparent overlay* (KJ61, 152–154) with equidistant (Δz) horizontal lines, and radii through P by dividing the angle π into equal parts $\Delta\alpha$. Each compartment thus defined has a gravity effect on P of $\Delta g \approx 2\Delta\alpha G\Delta\rho\Delta h$ (Sect. 2.9.7.2). To get an estimate of the assumed 2D model section, one only needs to count the compartments, perhaps including fractions, covering the geological body. A graphical comparison of the theoretical effect with the observed data will render a density contrasts estimate and suggest shape improvements. – The method can be refined by finer divisions $dzd\alpha$ produced for office use (see Sect. 6.2.1.1 (3)), but computer programs have generally replaced the use of templates.

5.6.5 Undulated Boundaries

It is also useful to remember that if a buried mass, for example, an approximately horizontal density contrast surface is undulated (wavelength λ), the gravity effect decays with vertical separation z as e^{-kz} (Sect. 2.10.5, Task 5.2). Obviously, in such an undulating case, no half width rule exists.

5.6.6 Diagrams

Many specialized *diagrams* for simple estimates have been published. Simple diagrams are useful for a quick look, but some diagrams are complicated and are appropriate to quantitative interpretation. A simple example may be one for thickness or depth-to-bottom estimation of isolated exposed rock bodies (Jacoby, 1970), based on the ratio of the maximum effects over that of the Bouguer slab of equal thickness; beside the shape of the exposure, the inclination of the contacts should be known. Double-logarithmic diagrams make use of the scale or similarity rule, permitting the relations to be presented by a single-parameter set of curves for any scale and density contrast (parameter may be the contact slope α, for example). The diagrams (GW65) treated in Sect. 5.6.3.2, are further applications of exploiting the scale rule (self-similarity), the solid angle and the half width w for actual scale.

5.6.7 Maximum Depth Rules

The aim of maximum depth rules is semi-quantitative. There is interest, for example, in exploration, at what depth a geological body may be encountered at most, however, maximum values are relatively uncertain since the results of such estimates always depend on assumptions and they are not extremes in a mathematical sense. Undulations of buried density contrast surfaces introduce an additional complication, because they represent mass anomalies with alternating sign which affects the half width of the gravity anomaly; a compact target mass at some depth has its characteristic gravity half width which is, however, narrowed by side masses of opposite sign. Hence, what appears to be the maximum depth of a single point mass can be exceeded. Indeed, some geological processes do generate just such mass distribution (e.g. thinning of a salt layer beside a salt dome or convecting systems).

Downward continuation is a related method; the maximum depth would be reached where the lateral extent of an undulation contracts to a delta function, i.e. a point mass. The method is, however, fraught with problems due to data noise and general heterogeneity of geological masses.

More sophisticated methods have been proposed which take into account the maximum possible density contrasts (Jung, 1927; Dürrbaum, 1974). The method involves application of variational calculus and diagrams and often exceeds the wanted simplicity.

5.6.8 Edge Effects

Edge effects are observed frequently. They result from abrupt lateral density contrasts and occur even over edges with the same vertical mass sum on both sides over a limited depth interval, for example, juxtaposed crustal structures in perfect isostatic balance. Geological examples are continental margins and oceanic transform faults or inactive fracture zones where lithospheres of different ages abut against each other (Figs. 5.6.3, 5.7.5, 6.5.5, 7.4.2). An inexperienced interpreter may think of uncompensated mass anomalies.

The example of Fig. 5.6.3 consists of juxtaposed layers of different vertical density distribution, but of equal vertical mass sum, i.e. isostatic at crustal depth (30 km), somewhat reminiscent of a continental margin. The vertical contact at $x = 0$ is abrupt; the right-hand block is homogeneous, the left one is divided into two $d = 15$ km sublayers of opposite density: top: $\Delta\rho = +180 \, \text{kg/m}^3$, bottom: $-180 \, \text{kg/m}^3$. The individual Bouguer effect of each sublayer would be δg_B of $+$ or -113 mGal. The edge effect of this boundary is $\delta g_m \approx +/-20$ mGal, at both sides of the edge at $x = -/+10$ km; above the boundary the effect is zero. It is a typical mass dipole effect, the integral of which is $_\infty\int^\infty \delta g(x)\mathrm{d}x = 0$. The maximum gradient is $grad \approx 6$ mGal/km. The outer half width" w_o, i.e. the distance between the

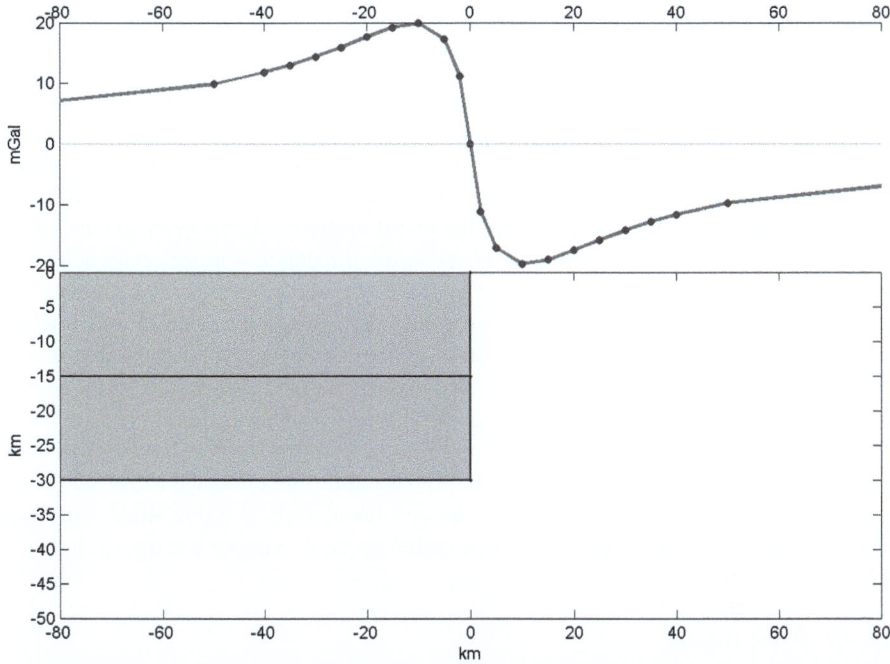

Fig. 5.6.3 gravity effect over the edge of two 30 km thick blocks of equal vertical mass sum; 321 homogeneous, left-hand block consisting of two $d = 15$ km sublayers, *top* $\Delta\rho = 180$, *bottom* -180 kg/m^3. The effect reaches $\sim$15 % of the Bouguer effect δg_B (113 mGal); the characteristic features are discussed in the text

two positions (*x*-values) where the effect falls back to half the extreme, $|\delta g_\mathrm{m}/2|$, is about 2×50 km $= 100$ km; this is remarkably large for a depth extent of only 30 km.

The effects scale linearly with the geometrical dimensions, i.e. the slab depth d (15km), and with the density contrast $\Delta\rho$ (180 kg/m^3). The edge effect reaches $\sim 15\%$ of the Bouguer effect δg_B (113 mGal): $\boldsymbol{\delta g_\mathrm{m}} \approx \mathbf{0.15}\ \boldsymbol{\delta g_\mathrm{B}} \approx \boldsymbol{\delta g_\mathrm{B}}/\mathbf{6}$ and the distance between the extreme values is $\boldsymbol{b} = 2\boldsymbol{x_\mathrm{m}} \approx \mathbf{4/3}\ \boldsymbol{d}$ (20 km). The extreme gravity gradient, $\boldsymbol{grad} \approx \boldsymbol{(9/2)}\ \boldsymbol{\delta g_\mathrm{m}}/\boldsymbol{d}$ (6 mGal/km), occurs above the boundary. The outer half width (see above), $|\delta g_\mathrm{m}/2|$, is about $\boldsymbol{w_o} \approx \boldsymbol{(20/3)}\ \boldsymbol{d} \approx \mathbf{7}\ \boldsymbol{d}$ (100 km).

Of course, in reality the lateral change of vertical structure is usually somewhat transitional which moderates these effects in amplitude and stretches them laterally; moreover, the vertical density distribution may be different, affecting the geometrical relations and amplitudes. It is also likely that the anomalous structure is laterally limited and may approach the vertical point or line dipole, treated below (Sect. 5.6.9). The above numerical values are only rough guides. But in principle, the gravity edge effect depends on the difference between the half widths of two equal but opposite density contrasts at different depths. In practice, the scaling laws (Sect. 5.6.2) are always applicable.

5.6.9 Vertical Dipoles

Vertical dipoles consist of vertically displaced, approximately equal masses of opposite sign, but in reality, equality of mass may be only approximate. Idealized cases are point masses (or spheres) and horizontal lines (or cylinders). Dipolar situations are common in nature, very often consisting of one voluminous part and one being flat or surface-like. Vertical dipoles are the result of the Earth tendency toward equilibrium and the mechanical situation where vertical balance is more easily achieved than lateral balance. Examples are isostasy, mantle plumes (an aspect of convection) and generally mantle convection in a layer with deformable upper and lower boundaries, where uplift by a low-density mass generates an excess at the surface and other density transitions, and where heavy mass generates depression and a deficit at the surface. The situation is further complicated by an equivalent deformation at the bottom of a convecting layer, such that the structure is a double dipole. However, the lower part is often located so deep that it has negligible effects at the observation level. For simple estimates, point and line dipoles as well as plume-like structures of a thin positive surface mass and an extended vertical cylinder are treated here.

5.6.9.1 Point Dipoles

The point dipole (Fig. 5.6.4) consists of the mass m_+ at $R = 0$ and depth $Z_+' = 1$ and the mass m_- at $Z_-' = a > 1 (|m_-| = m_+ = m)$ in non-dimensional vertical cylinder coordinates; dimensional scaling by the depth Z_+ includes the gravity effect (at the surface $Z = 0$), and primes are dropped from here on. For $R = 0$:

$$\delta g_0 = Gm(1 - 1/a^2). \qquad (5.6.13)$$

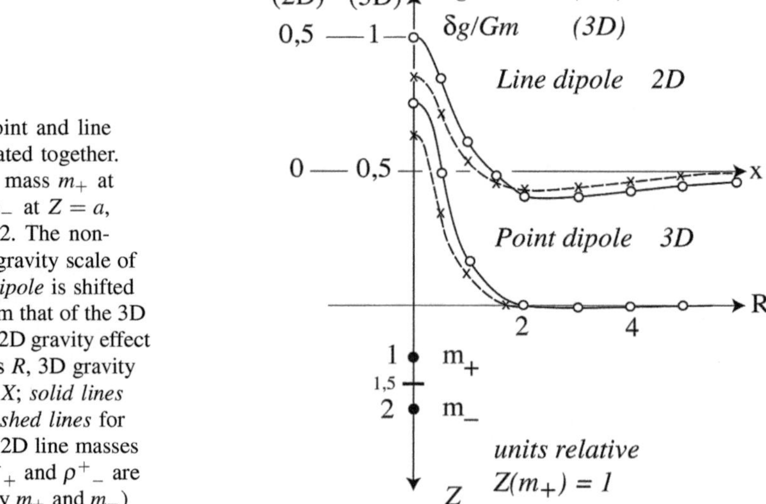

Fig. 5.6.4 Point and line dipole illustrated together. *Point dipole*: mass m_+ at $Z = 1$ and m_- at $Z = a$, $a = 1.5$ and 2. The non-dimensional gravity scale of the *2D line dipole* is shifted up by 0.5 from that of the 3D point dipole; 2D gravity effect plotted versus R, 3D gravity effect versus X; *solid lines* for $a = 2$, *dashed lines* for $a = 1.5$. The 2D line masses of density $\rho^+{}_+$ and $\rho^+{}_-$ are represented by m_+ and m_-)

The expression describes the central maximum: $f = \delta g_0/Gm$, increasing with a; for $a = 1$, $f = 0$ (m_- and m_+ cancel); $a = 1.2$, $f = 0.31$; $a = 1.5$, $f = 0.61$; $a = 2$, $f = 0.75$; $a = 4$, $f = 0.94$; $a = 10$, $f = 0.99$; $a \to \infty$, $f = 1$.

At $R \neq 0$, after some rewriting:

$$\delta g = Gm[1 - a(R^2 + 1)^{3/2}/(R^2 + a^2)^{3/2}]/(R^2 + 1)^{3/2} \tag{5.6.14}$$

From the square bracket of (5.6.14) follows for the radius, R_0, of $\delta g = 0$ (change of sign):

$$R_0 = \left[(1 - a^{4/3})/(a^{-2/3} - 1)\right]^{1/2} \tag{5.6.15}$$

R_0, after some irregular behaviour between $a = 1$ and 1.1, grows with a, but more and more slowly as a increases (e.g. $a = 2$, $R_0 \approx 2$; $a = 10$, $R_0 \approx 5$; $a = 1000$, $R_0 \approx 50$, etc.); in other words, the region of $\delta g > 0$ expectedly expands when the mass m_- drops to greater depth, except when only 10% and less deeper than m_+.

From the maximum at $R = 0$, the effect δg decreases to 0 at R_0. For $R > R_0\,\delta g$ is dominated by the lower mass m_-. The gravity effect outside is generally very small and hardly reaches 5% of the maximum δg_0 in the case of $a = 2$; for $a < 2$ the positive and negative sources are so close to each other that partial cancellation is considerable everywhere; for $a > 2$, the inner area is more and more strongly dominated by the upper mass, while outside the effect of the lower mass, though dominant, is spread out farther and farther. The total area integral of δg (see Sect. 2.7.6; eq. (2.7.12)) is zero because the total anomalous mass is zero. These properties of the dipole effect are illustrated in Fig. 5.6.4 for $a = 1.5$ and 2. It means that deep compensation is easily overlooked in gravity interpretation and lack of compensation or isostatic equilibrium may be only apparent, especially in 3D situations.

5.6.9.2 Line Dipoles (2D)

The analogous 2D case is the line dipole. It applies to many elongated terrestrial features as ridges, trenches, continental margins, etc. In 2D, the expressions are simpler than for the 3D point dipole. The line and point dipoles shown in Fig. 5.6.4 have identical depths, but the effects are different. The line masses have line densities $\rho^+_{\ +}$ and $\rho^+_{\ -}$ (in kg/m) at $x = 0$, $z_+' = 1$ and $z_-' = a$ ($a > 1$, $|\rho^+_{\ -}| = \rho^+_{\ +} = \rho^+$) in non-dimensional Cartesian coordinates (in Fig. 5.6.4, the point and line dipoles are both marked identically m_+ and m_-). Dimensional scaling by the depth z^+ includes the gravity effect (at the surface $z = 0$), and primes are dropped from here on. For $x = 0$:

$$\delta g_e = 2G\rho^+(1 - 1/a). \tag{5.6.13a}$$

At $x \neq 0$, after some rewriting:

$$\delta g = 2G\rho^+(x^2 - a)(1 - a)/\left[(x^2 + 1)(x^2 + a^2)\right] \tag{5.6.14a}$$

From (5.6.13a) set zero, follows for the coordinate x_0 of $\delta g = 0$ (change of sign) is:

$$x_0 = a^{1/2} \qquad\qquad\qquad (5.6.15a)$$

x_0 grows with a as $a^{1/2}$. Here, too, the region of $\delta g > 0$ expands when the mass ρ^+_- drops to greater depth. The 2D expressions also describe a central maximum $f = \delta g_e/2G\rho^+$, increasing with a ($a = 1.5$, $f = 0.33$; $a = 2$, $f = 0.5$; $a = 4$, $f = 0.75$; $a = 10$, $f = 0.9$; $a \to \infty$, $f = 1$). The effect decreases with distance x to 0 at $x = x_0$ where δg changes sign. Outside $(x > x_0)$ δg is dominated by the lower mass ρ^+_-. The weak 2D gravity effect outside is relatively stronger than in the 3D case; it may exceed 10% of the maximum δg_0. Also in 2D the outside negative effect is spread out farther and farther, and the total area integral of δg (see Sect. 2.7.6; eq. (2.7.12)) is zero. Integration in 2D is linear with x, while in 3D it is proportional to R^2. Deep" compensation is therefore more easily recognized than in 3D geometry.

5.6.9.3 Crustal Root Dipoles, Mantle Plume Dipoles

More specialized, though realistic, vertical mass dipoles are the combinations of topography with crustal roots and mantle plumes of low-density material. As above, the dipole effects are the more distinct the narrower the structures; they depend on the ratio d/w of depth extent over width as discussed in connection with the anomaly half width (Sect. 5.6.1); if d/w grows large, the effects become edge effects (Sect. 5.6.8). Principally the *FA* is a dipole effect and the *BA* is only one part of it, but as dipole" effects they are generally underrated, although familiar in magnetics with its principal dipole sources. We estimate the effects and their semi-quantitative magnitude. The knowledge of the effects is necessary for qualitative interpretation of observed gravity anomalies.

The present vertical dipoles" are still strongly idealized, though less than point and line dipoles (Sect. 5.6.9.2) are. The low density bodies uplift the surface to achieve an isostatic" balance which is here assumed complete. Two models are presented, crustal root and mantle plume, each with some variants. The mass dipoles consist of vertically displaced positive and "negative" masses equal in magnitude, but of different volume and effective density contrasts. The effects depend on geometry, and the present results are examples only. Cartesian geometry will suffice for the estimates. Details of complex density distributions within crust and mantle are neglected and deferred to quantitative interpretation. Fig. 5.6.5a–c shows idealized schematic crustal models (a) in a 2D cross section of a mountain belt with a locally compensating root in three versions, and a 3D cylinder-shaped plume and uplift (b and c), both exactly compensated, called "Airy plume", in two versions.

a) *Estimation of BA and FA effects.* What is meant with *BA* and *FA* effects are the calculated effects which correspond to the definitions of the Bouguer and Free Air anomalies. A very rough estimate can be realized by taking the Bouguer slab (Sect. 5.6.1; eq. (5.6.1)) for the mean topography and the solid angle approach (Sect. 5.6.4) for the root effect.

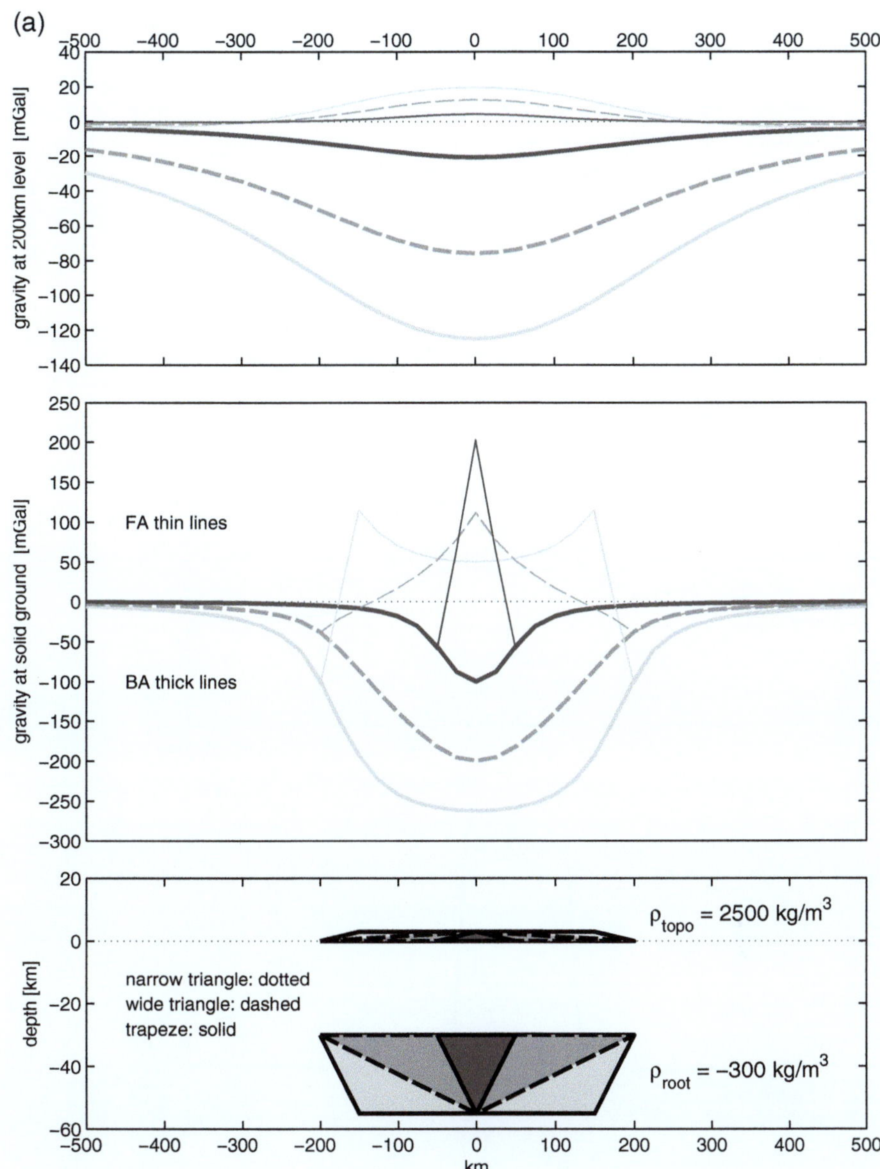

Fig. 5.6.5 (continued)

(b)

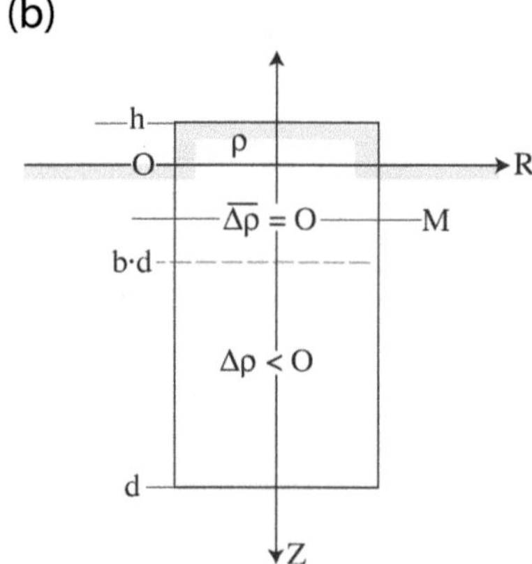

(c)

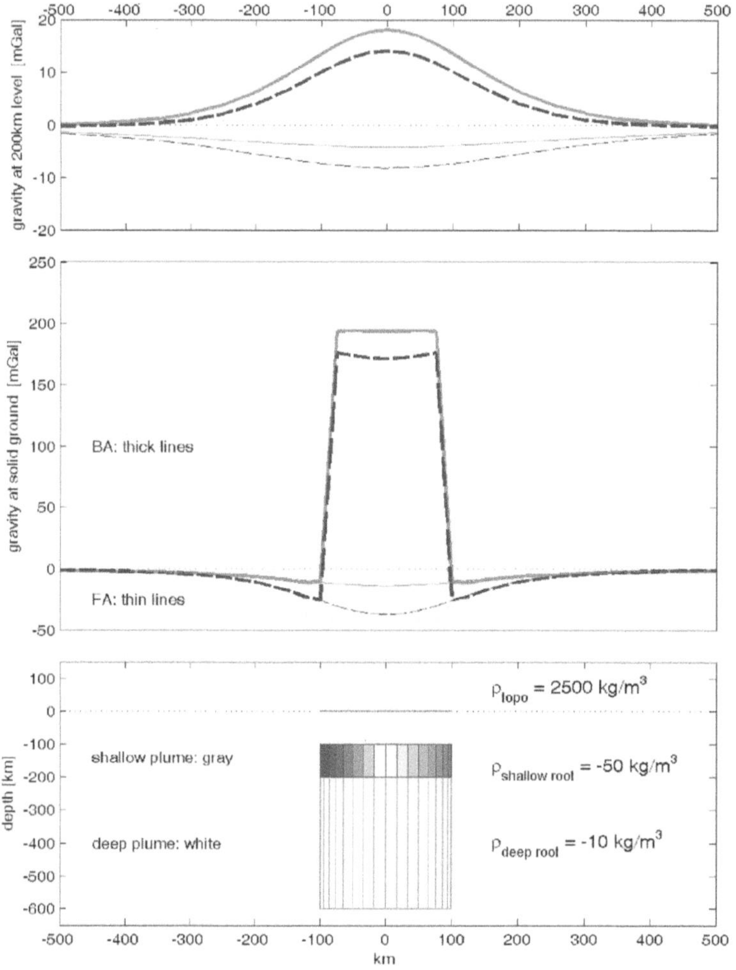

Fig. 5.6.5 (continued)

For the theoretical calculation of idealized effects, assume a triangular 2D section of width w and vertex height h, a background crustal thickness b and a triangular root of depth extent d (bottom vertex at $z = b + d$). The isostatic densities are ρ for topography and $\Delta\rho = -\rho h/d$ for the root relative to the mantle. Widths, $w = 100\,$km and $400\,$km, demonstrate the width-dependence of the effect either by laterally stretching the triangular mountain range or by shifting apart the unchanged flanks leaving a flat top in between (trapezoid shape). Fig. 5.6.5a shows the FA and BA effects calculated at points on the solid surface for the three cases with $h = 3\,$km, $b = 30\,$km, $\rho = 2500\,$kg/m^3, and consequently $\Delta\rho - 300\,$kg/m^3.

For the triangular case with $100\,$km base, the BA effect, i.e. the effect of the root only, reaches about $-100\,$mGal or 32% of the Bouguer slab effect ($314\,$mGal), and its half width w' ($\sim110\,$km) exceeds the structural width w by 10 %. The FA effect (sum of topography and root effects) has a maximum at the apex of $\sim200\,$mGal (65% of Bouguer slab effect) and negative flanks (minima about -20% slab effect at the edges); the average FA over the triangular topography is of the order of $+70\,$mGal (22 % slab effect).

For the $400\,$km triangular structure the BA effect of the root (only) doubles ($-200\,$mGal, 65%) and the FA effect of topography and root decreases to $111\,$mGal (35%, average of a central $200\,$km wide strip: 16%) and $-39\,$mGal (12%), respectively, for the maximum and the flank minima, while the flank effects decay more slowly with horizontal distance.

The trapezoid structure reaches a BA effect (root only) of about -84% of the slab effect. The FA effects has a two-peaked shape (edge effects) with maxima of 37% of Bouguer slab (averaged: 25%) over the upper plateau edge and a central plateau of 17%. The positive FA effect is flanked by deep minima at the foot of the slope of -33% of the Bouguer slab effect; this suggests that the flanking minima, as over the Alpine Molasse and the Ganges plains, are not caused only by thick sediments.

The amplitudes decrease, of course, with increasing observation level (e.g. of satellites) where the decrease is stronger for the topographic excess mass and its

---

Fig. 5.6.5 Crustal and plume dipoles and calculated gravity effects: BA as the effect of the crustal roots or the mantle roots alone, FA as the dipole effect of topography and root together; observed at two levels, at the surface of topography and at $200\,$km elevation (satellite)
(a) Schematic 2D crustal cross sections of mountain belts and locally compensating roots (Airy-type isostasy) in three different versions: two triangular sections of width $w = 100$ and $400\,$km and one trapezoid section (w 0 $400\,$km); vertex height $h = 3\,$km, depth extent d of the root (bottom vertex at $z = b + d$), background crustal thickness $b = 30\,$km and $\rho = 2500\,$kg/m^3, crustal root density contrast $\Delta\rho - 300\,$kg/m^3. (b) Airy plume: vertical circular cylinder (radius R_c, density contrast $\Delta\rho_p < 0$ versus mantle, vertical extent bd, between $Z = (1 - b)d$ and $Z = d$; $0 < b < 1$, surface ($\rho = 2500\,$kg/m^3) uplifted by h, Airy mass balance: $\Delta\rho_p \approx -\rho h/(bd)$; neutral layer ($\Delta\rho = 0$) between $Z = 0$ and $(1 - b)d$; b is calculated from mass balance between Moho uplift by h, density $\Delta\rho_M$, and top part of plume (depth extent bd, density contrast $\Delta\rho_p$). (c) BA and FA effects in two versions, calculated (at surface elevation and at $200\,$km height) along $1000\,$km profiles across a given cylindrical plume ($R_c = 100\,$km, vertical extent of $\Delta\rho$: 100 to 600 or 100 to 200 km, $\Delta\rho = -10$ or $-50\,$kg/m^3, respectively, $2\,$km uplift with $\rho = 2500\,$kg/m^3)

gravity effect is much smoother than at the surface. The (negative) *BA* effect (root only) does, naturally, increase with the size of the root (approximate amplitude for the narrow triangle: -20 mGal, broad triangle: -75 mGal, trapezoid: -125 mGal), and so does the composite *FA* effect (central value for the narrow triangle: $+4$ mGal, broad triangle: $+13$ mGal, trapezoid: $+20$ mGal) which is remarkably low for the narrow structure.

Observed *FA* (see Sect. 4.6.1) over mountain ranges are generally positive, although large local variations in mountainous terrain make averaging difficult and estimates of the terrain effects have large uncertainties. Aerogravimetry and, even more so, satellite gravimetry have less spatial resolution, and downward continuation to geoid level will give a more representative smooth picture; recent results from the GRACE mission render the following approximate amplitudes: central Andes: $+70$, Himalaya: $+50$, Caucasus-Iran: $+50$ and Alps: $+30$, Southern Alps, New Zealand: $+20$ mGal, all of these with uncertainties of about ± 10 mGal. These values are quite similar to the above simple estimates. The negative *BA* values are long known (Andes and Himalaya: -400 to -500 mGal, Alps, Caucasus-Iran: -200 to -300 mGal, Southern Alps: -100 mGal) again with strong variations and uncertainties. Also these values are similar to those predicted above, if the average elevation of the Himalayas and high Andes is taken into account (at least twice that assumed for the present models).

b) What kind of gravity effects are generated by a simple *compensated Airy plume structure*? Mantle plumes foster melting and thus thicken and alter the crust generating vertical dipoles with two depth components, a crustal root and the deep plume itself (example Iceland plume: Sect. 5.7.7, 6.5.7). For the simplest case, assume a vertical circular cylinder (radius R_c, density contrast $\Delta\rho_p < 0$ versus mantle, vertical extent bd, between $Z = (1-b)d$ and $Z = d$; $0 < b < 1$ (Fig. 5.6.5b). It uplifts the surface by the amount h; Airy mass balance requires $\Delta\rho_p \approx -\rho h/(bd)$, where ρ is the density of the uplifted mass. A neutral layer is assumed to exist between $Z = 0$ and $(1-b)d$ with effective $\Delta\rho = 0$ and no gravity effect. It may simply represent some arbitrary depth to the plume top, or it may consist of material of zero average lateral density contrast by internal compensation of density anomalies, for example, of an uplifted Moho ($\Delta\rho_M > 0$) and other density increases above equilibrium, compensated by the subjacent top part of the plume ($\Delta\rho_p < 0$). This vague, though necessary, assumption is made here to estimate the value of b. If the Moho is uplifted by the amount of h (same as the surface) and the mass compensation is achieved by: $h\Delta\rho_M \approx (1-b)d\Delta\rho_p = -\rho h(1-b)/b$ (insert: $\Delta\rho_p \approx -\rho h/(bd)$); this leads to $b \approx \rho/(\rho + \Delta\rho_M) \approx \rho/\rho_m$ where ρ_m is the upper mantle density; hence $b \approx 0.8$ to 0.9, i.e. the upper 10 to 20% of the vertical plume extent. These assumptions must be checked or specified in concrete cases, but there is the additional difficulty that the deeper parts of the plume have a small gravity effect, hardly recognizable at the surface.

For this simple estimate of the *axial* gravity effect δg_e, the dipole is divided into three parts: (1) above the surface treated as a thin disc by taking the Bouguer plate effect (5.6.1): $\delta g_{(1)} \approx 2\pi G\rho h \approx 0.1 \times h \times \rho/2390$; (2) $\delta g_{(2)} = 0$ for the cylinder between $Z = 0$ and $(1-b)$d. (3) The lower part of the plume cylinder between

$Z = (1 - b)d$ and d may be approximated by a vertical rod (see Sect. 2.9.3.2: Eqs. (2.9.10) and (2.9.10)). The rod or line density ρ^+ is given by $\pi R^2 \Delta \rho_p$ (kg/m) and the axial gravity effect at the disk surface is given by

$\delta g_{(3)} \approx \pi G R^2 \Delta \rho_p [1/(h + (1-b)d) - 1/(h+d)]$, which, in view of $h << d$, after some arithmetics, is $\delta g_{(3)} \approx \pi G R^2 \Delta \rho_p f/d$, where $f = b/(1-b) \approx \rho/(\rho_m - \rho)$ (from $b \approx \rho/\rho_m$, see above). Summing gives the total axial effect:

$$\delta g_e = \delta g_{(1)} + \delta g_{(2)} + \delta g_{(3)} \approx 2\pi G \rho h \left[1 - (R/d)^2 \rho_m/(2(\rho_m - \rho))\right] \qquad (5.6.16)$$

With the Bouguer plate effect of the disk, numerically simplified:

$$\delta g_e \approx 0.1 \times h \times \rho/2390 \left[1 - (R/d)^2 \rho_m/(2(\rho_m - \rho))\right]. \qquad (5.6.17)$$

For reasonable values of R (50 to 100km), d (500km), ρ (2600 to 2800 kg/m^3) and ρ_m (3300kg/m^3) this expression renders (positive) values of $(0.9 \pm 0.05)\delta g_{plate}$, which means that the plume itself has only a very small influence, which will be even smaller for plumes with a deeper top and smaller diameter.

The above estimate is supported by calculations of *BA* and *FA* profiles (Fig. 5.6.5c) across a given cylindrical plume ($R_c = 100$km, 100 to 600 km depth, $\Delta \rho = -10$kg/m^3, 2 km uplift with $\rho = 2500$kg/m^3). At the plume centre, the Bouguer plate effect δg_{plate} of the 2 km uplifted disk (205 mGal) is reduced by only 7%, and beyond the disk edge the plume effect is -5% of δg_{plate} at most, decaying in amplitude to $<1\%$ at 500 km distance. It is also interesting to calculate the plume dipole effect at satellite level of 200 km, shown as well. Here the amplitudes of the disk and plume effects decay to $+11\%$ and -2% of δg_{plate} leaving $+9\%$ for the maximum total effect over the plume centre and a radius of the positive effect of about 500 km, beyond which the negative effect nowhere exceeds -1% (here -1 mGal).

The influence of the depth extent of the Airy plume (same uplift) and a correspondingly changed density contrast is also shown in Fig. 5.6.5c for the extreme case of a plume extending from 100 to only 200 km depth and $\Delta \rho = -50$kg/m^3. Here the plume effect at the surface is more than twice that of the deep plume; the central maximum and the rim minimum are $+83\%$ and -12% of δg_{plate}, respectively. At 200 km level of observation, the effects of the uplifted disk and of the plume more strongly counteract than in the case of the deep plume, such that the amplitude drops from 9% to 7% of δg_{plate} and the radius of the positive effect is about 390 km.

As mentioned, the Airy plume can give only a rough idea of effects which can be expected. In a more complete analysis, variants should be considered to include features as plume heads (lateral spreading of low density material, tapered or somewhat conical uplifts, combinations of plumes with magmatically thickened crust, flexure of the lithosphere, and calculations have to be done in spherical geometry. Ultimately it is also necessary to extend the analysis to the dynamic cases of plume convection (see e.g. Schubert et al., 2001, pp. 537–543). None of these or other variants are treated here, but must be taken into account in a thorough analysis.

In practice, qualitative interpretation of plume anomalies will involve trial and error, starting with the observables as h and δg_o and with guesses of the parameters ρ, d, R and b, to predict δg_e. Expression (5.6.17) can guide the search for better fitting parameters: increasing h, ρ, d and decreasing R and the density factor $\rho_m/(2(\rho_m - \rho))$ will increase δg_e and vice versa (at least if the initial estimate of δg_e was realistic). The sensitivity of the result to h and ρ is direct in the range of realistic values of d, R and densities (see Sects. 5.7.7 and 6.5.7 for examples).

5.7 Examples

Several examples demonstrate how simple models, mostly 2D, can catch various characteristic geological situations. Semi-quantitative interpretation goes beyond the first qualitative guesses and may even employ some optimization showing how the simple initial assumption can lead better results. The modelling does not aim at fitting particular very detailed features. Some models are taken up in the following chapters (6, 7) for more detailed interpretation. In all the cases, the geological problem, data with errors, initial assumption including error estimates for the model parameters and results are briefly sketched.

Most cases belong into the category of large-scale tectonic features, as a graben, a shelf edge, an oceanic ridge, an oceanic trench, a diapir of granite or salt, mantle convection, a mantle plume. Most cases were investigated by the authors and their students. The associated gravity anomalies were taken from different sources, as published maps and/or profiles or digital data sets and are considered more or less typical for the structures. The examples are illustrated by Figs. 5.7.1, 5.7.2, 5.7.3, 5.7.4, 5.7.5, 5.7.6, 5.7.7, 5.7.8, 5.7.9.

5.7.1 Messel Maar Crater and Fault Zone (MFZ) and Meerfeld Maar

Maars are volcanic craters excavated by phreatomagmatic explosions aided by the collapse of wall rock. They are often filled with light sediments and deeper brecciated diatreme rocks, thus a promising target for gravity studies, for example, of the crater depth and deeper structure. Young maars have a tuff wall around them with characteristic morphology and magnetic anomalies which aid their identification. Older maars are deeply eroded and less conspicuous but can be detected by geophysical investigations, especially by reconnaissance gravity surveys.

5.7.1.1 The Messel Maar

The Messel pit of oil shale, NE of Darmstadt, Germany, was formerly mined for energy and chemical exploitation and is now a UNESCO world heritage site for its

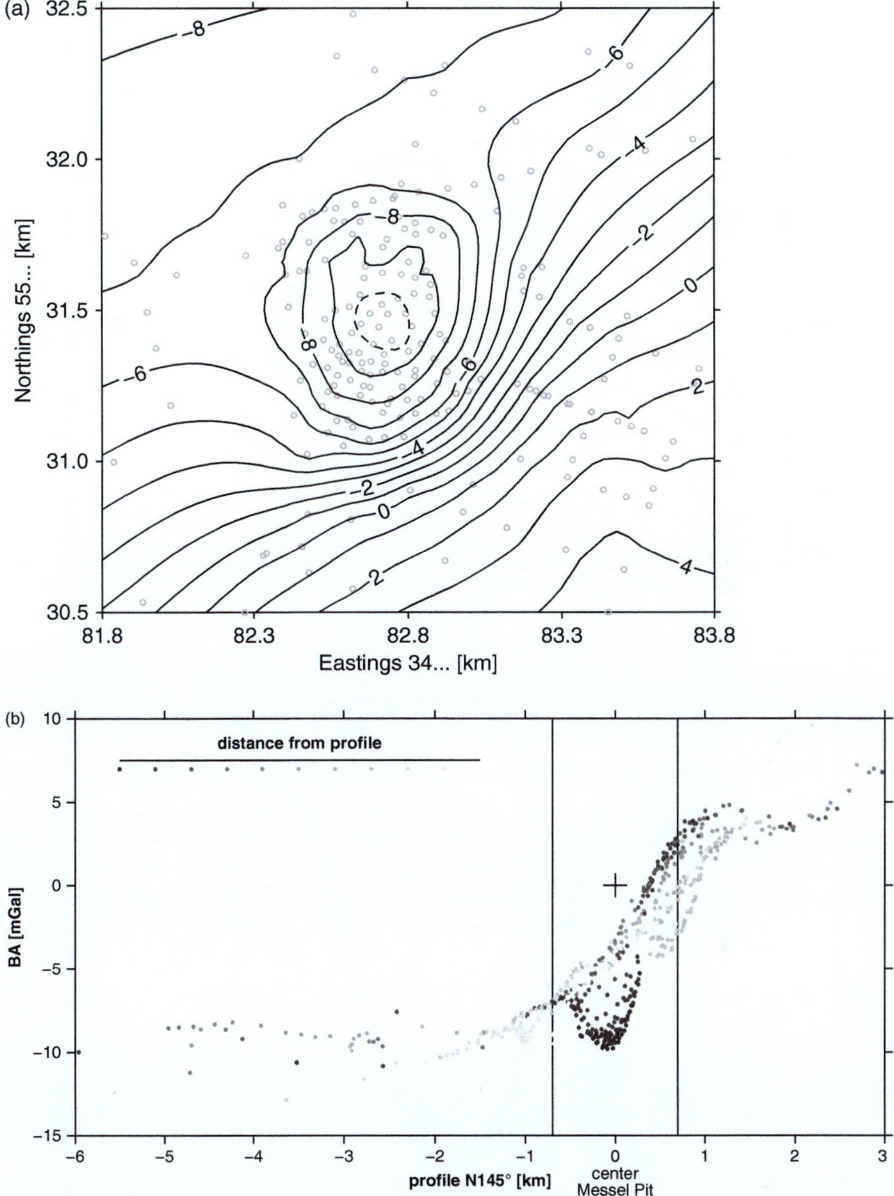

Fig. 5.7.1 (**a**) *BA* observed in and around the Messel pit; the maar-related anomaly is superimposed on the regional anomaly related to the Messel Fault Zone (MFZ); (**b**) *BA* values projected onto a plane normal to the linear MFZ of gravity gradients; the *grey* tones symbolize the distance from the vertical gravity plane through the centre of the Messel pit (*the lighter grey*, the farther away); the stations in and around the Messel Maar are included in the projection and clearly outline the corresponding gravity anomaly; (**c**) Meerfelder Maar, *BA* shown as a contour map (*top*) and as a section along the line AA′ (see map)

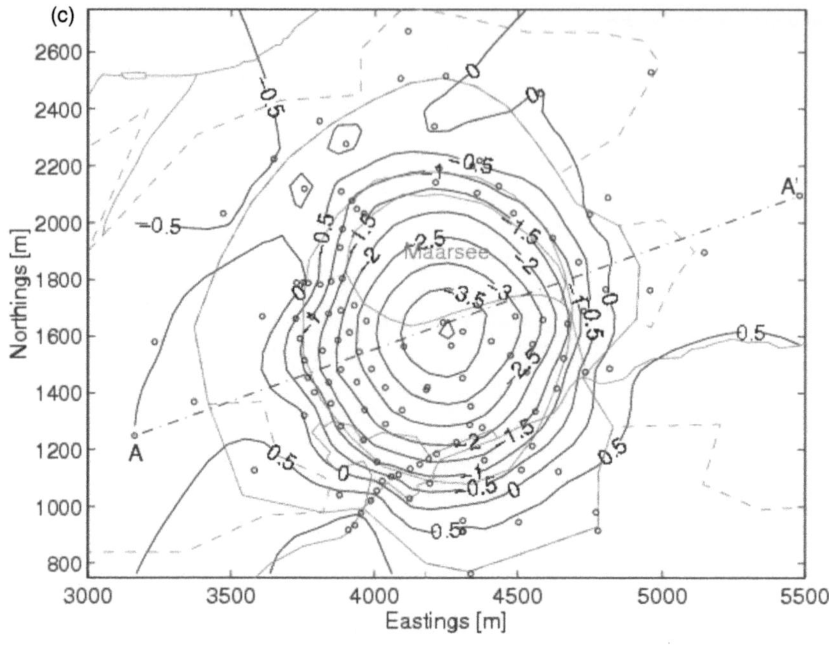

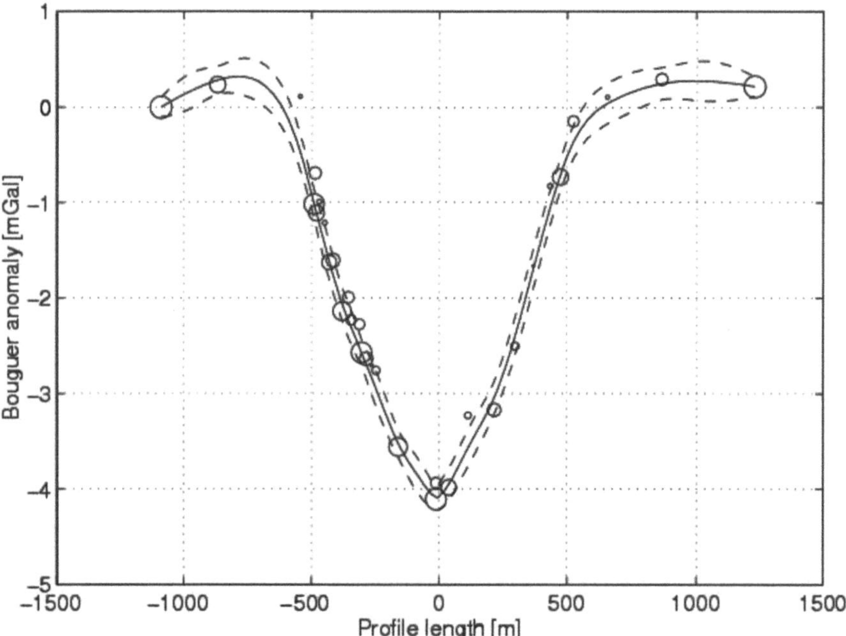

Fig. 5.7.1 (continued)

excellent Eocene fossils. Figure 5.7.1a shows the *BA* as embedded into the regional field which reflects the Messel Fault Zone (see below). Bituminous laminated sediments fill an approximately sigmoid hole in Variscan crystalline rocks of $0.7 \times 1\,\mathrm{km}^2$ dimension with apparently steep walls. At the time the geophysical investigations started, the origin of the oil shale basin was debated, even its nature as a crater. The power of gravity as a qualitative exploration tool was demonstrated as the anomaly, complemented by magnetic and other data, could hardly be interpreted other than by a maar-like crater (Jacoby et al., 2001, 2003), which was subsequently proven by drilling.

The relative *BA* minimum is approximately $-7\,\mathrm{mGal}$; this can be well estimated from the map even without separating "the" regional anomaly; it is also evident in the data profile across the Messel Fault Zone (see below, Fig. 5.7.1b). The half width $w < 1\,\mathrm{km}$, i.e. smaller than the geological basin, suggests $z < 600\,\mathrm{m}$ (see Eq. (5.6.2a)). However, the crater fill has an extremely low density, evident in direct measurements on samples, especially drill cores. With an oil shale density $\rho \approx 1200\,\mathrm{kg/m}^3$ and $> 2500\,\mathrm{kg/m}^3$ for country rock, $\Delta\rho \approx -1300\,\mathrm{kg/m}^3$, and the simple Bouguer effect gives $d \approx 10 \times \delta g \times 2390/\Delta\rho \approx 130\,\mathrm{m}$. If the solid angle at the centre is taken into account, d may be 10 to 20% larger; on the other hand, the somewhat low density of the brecciated diatreme rocks below the oils hales, contributing to the total gravity anomaly, would roughly compensate the solid angle effect. These estimates are very well confirmed by more quantitative modelling and ultimately by drilling (see Sect. 6.5.1).

5.7.1.2 Messel Fault Zone (MFZ)

The Messel Maar is one of at least 5 similar, though smaller, structures and a few volcanic extrusions aligned along a NE trending strip of a distinct southeastward gravity gradient, called the Messel Fault Zone (MFZ); Fig. 5.7.1b shows the *BA* projected onto a plane normal to the MFZ. Gravity drops across it to the NW by 15 mGal and the maximum gradient is nearly 10 mGal/km. If for a first estimate, the gradient is explained by a vertical fault step (Sect. 5.6.3.2), the fault properties, throw h and density contrast $\Delta\rho$, follow directly (5.6.12) from the gravity difference on both sides ($\sim 15\,\mathrm{mGal}$ across 3 km distance) and its half width w of $\sim 1\,\mathrm{km}$. The results are $h \approx 1\,\mathrm{km}$ und $\Delta\rho \approx 360\,\mathrm{kg/m}^3$. These do, however, contradict the geological evidence for a much smaller throw h, as the Permian sandstones to the NW are seldom thicker than 100 m (Jacoby et al., 2001), and a more detailed quantitative interpretation is discussed in Sect. 6.5.1.

5.7.1.3 Meerfeld Maar

The Quaternary Meerfeld Maar is located in the Eifel, the type locality of maar structures. It is a nearly circular morphological depression with steep walls and a flat bottom of sediments and a shallow lake filling half of the present crater bottom. The main question is the depth extent of the sediments and subjacent diatreme rocks. Figure 5.7.1c shows the *BA* as contours and as a section along the line AA'. The *BA* amplitude is $\delta g = 4 \pm 0.3$ mGal and the half width $w \approx 1.6$ km. From eq. (5.6.2a) follows for the equivalent mass point the depth $z \approx 1$ km. The corresponding sphere volume and radius, according to eq. (5.6.2c), would be for $\Delta\rho \approx -200$ or -600 kg/m^3: $V \approx 3$ or 1 km^3 and $R \approx 0.9$ or 0.6 km, respectively. A fitting sphere of 1 km radius (grazing the surface) would have a density of -150 kg/m^3. Quantitative interpretation (Smilde, 1998) takes into account the conical shape of maar structures and also the densities of the lower diatreme rocks. This is demonstrated in Sect. 7.4.1.

5.7.2 Salt Diapir: Helgoland

Salt diapirs of low density and soft rheology are structurally and dynamically buoyant masses in the crust and of a characteristic scale of a few kilometres. They occur in sedimentary basins of economic interest. The qualitative aspect of gravity primarily concerns the discovery of salt structures mainly as oil traps, and interpretation first asks for size, depth extent and density contrast which depends on the country rocks while mono-mineralic rock salt has a well defined density. The example chosen is the North Sea island of Helgoland, built of Bunter sandstone uplifted and tilted by a salt dome.

A simplified gravity map of the Helgoland area and a profile across strike are shown in Fig. 5.7.2. The island is situated on the NE flank of an elliptical gravity low, peaking at -26 mGal. The NW strike of the long axis corresponds to the strike of the island and the tilted sandstones. Below the map, a profile is shown roughly across strike; it is asymmetric with a sharp peak less than 1 km west of the Helgoland rock. The regional field is irregular and to the SW of the gravity minimum only few data points exist; hence, a reference for the effect of the target is difficult to define; very roughly, the amplitude δg is -7 mGal. The half width is even more difficult to define, the apparent value of 5 to 6 km is not representative for the sharp peak anomaly. Where measured in detail on land, steeper gradients are apparent. Furthermore, it is a case intermediate between 2D and 3D. According to Sect. 5.6.3.2, z should be between $w/2$ and $2w/3$, hence between 2.5 and 4 km and for a 2D cylinder of radius 2 km, $\Delta\rho \approx -120$ kg/m^3, however, if the density contrast of rock salt versus the sediments, $\Delta\rho_{\text{salt}} \approx -200$ kg/m^3, the cylinder radius would be only a little smaller, ~ 1.8 km, extending the body to 4–5 km depth. On the other hand, the available information of the shape of the anomaly and from a drill hole which encountered the rock salt at 0.7 km depth, suggests that part of the anomaly should be directly attributed to the shallow part of the salt dome west of Helgoland, such that the deeper parts would have a somewhat smaller effect and the above estimates must be modified. The shape of the anomaly, indeed, suggests a triangular cross section as depicted in Fig. 5.7.2, together with the simple model. The estimates are corroborated by detailed 3D modelling (see Sect. 6.5.2; Jacoby, 1966).

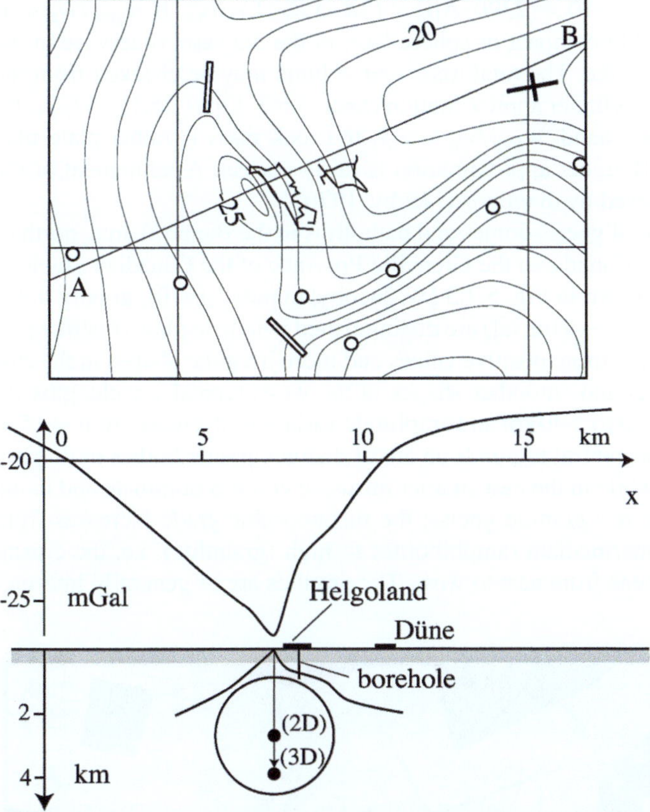

Fig. 5.7.2 Bouguer anomaly in the area around Helgoland (*top*); profile AB across the anomaly due to the Helgoland salt dome (*middle*); section across the model (*bottom*) intermediate between a 2D cylinder and a 3D sphere and geological section based on drilling results

5.7.3 Granite Batholiths: Bancroft Area, Canada

Igneous rock bodies of intrusive or metasomatic origin are wide spread and frequently exposed in deeply eroded crust. Their genesis, petrology and structure vary widely, from basic (e.g. gabbro) to acid (granite), from sheet complexes, laccoliths and plutons to "rootless" batholiths. Questions are typically the volume and depth extent, the depth variation of the roof rocks, the density contrast $\Delta\rho$, etc. The latter parameter is best estimated from petrological information as well as from field and laboratory measurements (Sect. 3.6.3). The principal questions can often be clarified in advance, and volume V and depth extent d remain the basic questions which gravity surveys can answer. Semi-quantitative estimates require from gravity mainly the anomaly patterns and amplitudes δg and the half widths w_g relative to the widths w_{geol} of the rock mass exposures. The depth extent roughly follows from the Bouguer effect (5.6.1), easily modifiable by an approximate solid angle

$\Delta\Omega_{est}$ instead of 2π: $(d \approx 10 \times (2390/\Delta\rho) \times (2\pi/\Delta\Omega_{est}) \times \delta g)$. If $w_{geol}/w_g < 1$, the rock mass is probably compact or sphere-like, in the 3D case, otherwise in the 2D case, more cylinder-like. The total mass and volume may be checked from the point – sphere or line – cylinder approximation (Sect. 5.6.3: Eqs. (2.5.1), (5.6.2a, b), (2.9.7)). If, on the other hand, $w_{geol}/w_g \gg 1$, the rock mass is rather plate-like, $\Delta\Omega_{est} \to 2\pi$, and the Bouguer approximation is fairly correct. A refinement of the estimates may be achieved by diagrams (Jacoby, 1970).

For the granitic arch of gneiss domes and batholiths in the Bancroft area, north of Peterborough, Ontario, Canada, in the Grenville Province of the Canadian Shield, a simplified *BA* map is shown in Fig. 5.7.3 (W. Jacoby, unpubl., 1970); gravity highs (>-40 mGal) and lows (<-40 mGal) are distinguished which display a distinct pattern: a general trend from more positive values and more irregular shapes in the east to more negative values and smoother shapes in the west; typical wavelengths increase from ~ 15 to ~ 30 to >40 km and amplitude variations decrease from ~ 15 to ~ 10 to <10 mGal. The central region is an arc of distinct granite bodies of diapiric to intrusive character while in the east smaller intrusive granites dominate and in the west the whole terrain is a granitic gneiss; the metamorphic grade increases from low (greenschist) to intermediate (amphibolite) to high (granulite), i.e. the crustal level of exposure deepens from east to west. The granites are of generally interme-

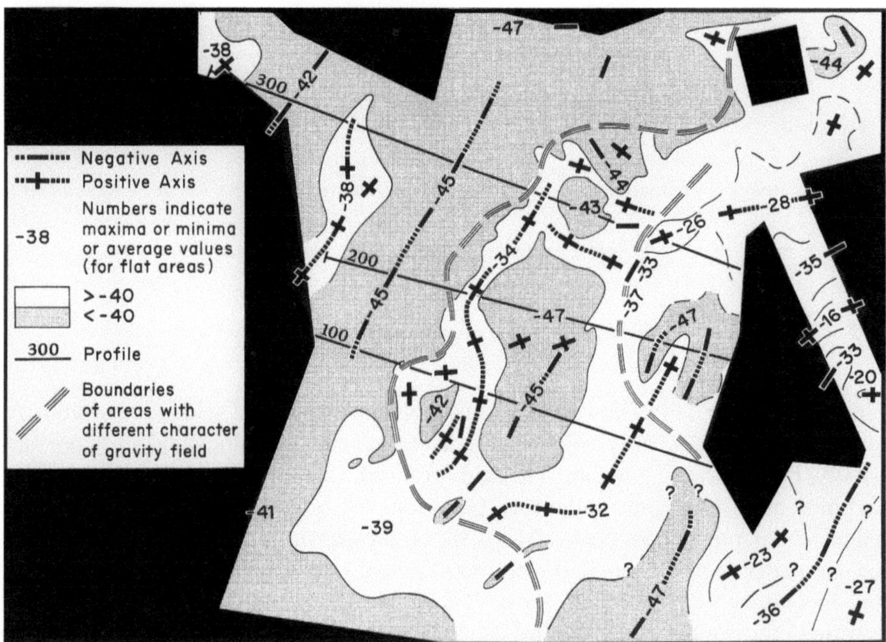

Fig. 5.7.3 Simplified gravity anomalies over granite batholiths in the Bancroft area, Ontario, Canada (Jacoby, unpubl.); explanations in insert. Of the three profiles δg the central one is treated in more detail in Sect. 6.5.3

diate composition, and the bodies contain accumulations of mafic minerals which outline the internal structure.

Concentrating on the central arc, diameters or widths are about 15 km (w_{geol}) with associated relative gravity (*BA*) minima $\delta g \approx -10\,\text{mGal}$. Generally the contacts have been mapped to dip away from the centre. The density contrast $\Delta \rho$ of the granitic bodies (s.l.) against the metasedimentary, largely amphibolitic country rock is estimated to be about $-150\,\text{kg/m}^3$. It is likely that the depth extent $d < w_{\text{geol}}$ such that $2\pi/\Delta\Omega_{\text{est}} \leq 1.4$. The above expression for d renders, with these figures $d \approx 2.2\,\text{km}$. This is a plausible estimate. Since the *BA* values over the larger bodies are all similar ($-45 \pm 1.5\,\text{mGal}$), while the *BA* strongly varies over the intervening rocks (about $-33 \pm 6\,\text{mGal}$), the question is really for the depth variation of the pendants of metasedimentary roof rocks above a largely granitic layer. Hence, the pendants probably reach mostly to depths $\leq 2.2\,\text{km}$. As generally density varies more gradually from granitic to country rock, variations will extend deeper than abrupt contrasts suggest; this must be taken into account in a more quantitative and detailed interpretation (Sect. 6.5.3). It appears that the structure can be interpreted as an undulating density boundary with two dominant wavelengths: about 30 km of the major granite updomings and 6–7 km. The 30 km wavelength is probably the fastest growing undulation of a Rayleigh-Taylor instability during deep burial, while the shorter wavelength may reflect remnant folds.

5.7.4 Rhine Graben

Grabens are rifted crustal depressions with down-faulted blocks originating from stretching, irrespective of the dimension, from meter to 100 km. Grabens can be symmetric or asymmetric with only one master fault and a tilted block. A lasting debate has centred on symmetry versus asymmetry, especially in deeper crustal levels. Depth is generally related to width.

The Rhine graben is a classical example, $\sim$300 km long and $\sim$30 km wide and affecting the whole crust. The depression is filled with Tertiary sediments. The main questions are the depth of the sediment fill and crustal thickness, the density contrasts, the dip of the master faults and their continuation at depth. The *BA* profile (Fig. 5.7.4) across the southern graben near Strasbourg (Prodehl et al., 1995) shows an at least 250 km wide asymmetric arch up by 30 mGal from Lorraine in the west to a value of about $-10\,\text{mGal}$ and down to the east by at least 70 mGal to the Bavarian Molasse and the Alps; a few 10 to 20 mGal undulations of 20–30 km wavelength are superimposed, including the Rhine graben minimum of $\delta g \approx -20\,\text{mGal}$ near the crest of the arch; the uncertainty is estimated to be less than $\pm 5\,\text{mGal}$ on the basis of a smooth regional, drawn manually and subtracted. For a semi-quantitative estimate this appears acceptable (see Sect. 4.7.7).

A very simple rectangular or trapezoidal 2D section is a natural approximation; for the sediment fill, with a thickness most likely much smaller than the morphological graben width, the Bouguer effect will suffice (density contrast as-

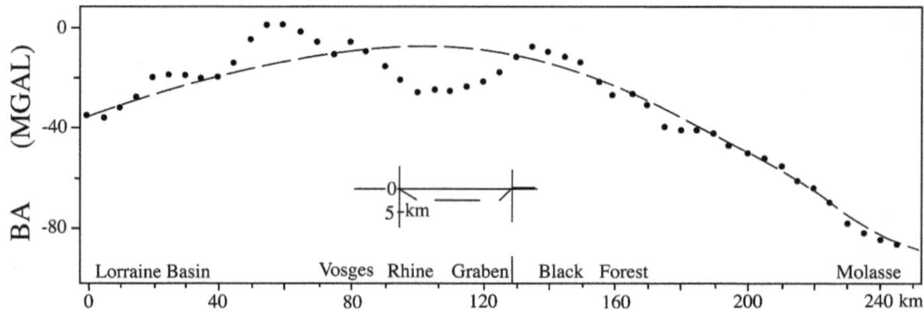

Fig. 5.7.4 Bouguer anomaly (*BA*) profile across the Rhine graben at about the latitude of Strasbourg; insert shows a simple scale model of the sediment fill (assumed $-200\,\mathrm{kg/m^3}$) estimated from the Bouguer plate effect to be 2.4 km thick. W to E Moho depth variation (not shown) is estimated in the same manner to rise 2.5 km to below the graben and drop ~ 6 km to the Bavarian Molasse

sumed: $\Delta\rho = -200\,\mathrm{kg/m^3}$): $\delta g \approx 0.1 \times d \times \Delta\rho/2390$ [units: mGal, m, $\mathrm{kg/m^3}$] $\rightarrow d \approx 10 \times 20 \times (2400/200) \approx 2400\,\mathrm{m}$, which is of the right order. One may calculate the depths corresponding to different values of the density contrast. The results are compared to 2D block models of limited width and inclined master faults (Fig. 5.7.4). If the gravity arch is caused by crustal thinning, i.e. Moho up arching, a first-order estimate of its amplitude may be based on the Bouguer effect: applying the same expression as above and assuming $\Delta\rho \approx 300\,\mathrm{kg/m^3}$, we estimate 2.5 km Moho rise from the west and a $\sim$6 km drop toward the Alps; however, here the increasing sediment thickness will also affect gravity. This example will be continued in Sect. 6.5.4.

5.7.5 SE Iceland Shelf Edge

An example for a transition from thin oceanic crust to thicker crust is the steep SE shelf edge of the Iceland Plateau. It is a first-order morphological and structural element. A profile is shown in Fig. 5.7.5. The *BA* rises from Iceland to the ocean basin by 160 mGal (error about 10%). The *BA* appears as a superposition of a monotonous increase from Iceland toward the ocean basin, reflecting crustal thinning, and a dipolar feature, relatively positive above the upper shelf break and negative above the foot of the slope, suggestive of an edge effect as treated in Sect. 5.6.8. It is much more conspicuous in the *FA* and in the residual *rBA* after subtraction of the Moho effect as calculated in a preliminary way (see Sect. 6.5.5). The double amplitude of *FA* is about 70 mGal, of rBA about 50 mGal, and the maximum *FA* approximately coincides with the zero of *rBA*.

The residual *BA* is used here for a semi-quantitative estimate. The edge seems to be very close to the shelf edge. The Iceland Plateau is not continental, but the shelf edge is similar to many continental margins and may even incorporate a continental splinter, originally from the Greenland shelf (Fedorova et al., 2005). Shelf edges

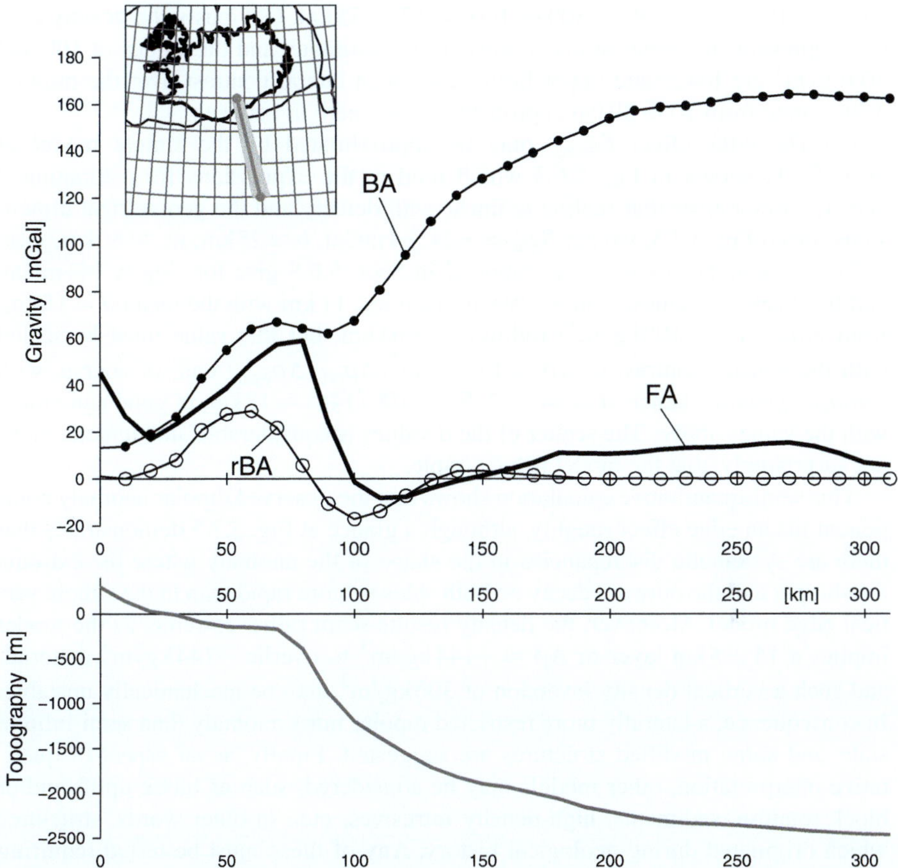

Fig. 5.7.5 Gravity profiles across the SE Iceland shelf edge: (1) *BA* profile from near the Iceland SE coast to the Atlantic basin toward SSE (Bouguer reduction density of rock mass $2600\,\mathrm{kg/m^3}$), (2) residual *BA*: rBA (after subtraction of a Moho effect), (3) *FA*, and (4) bathymetry, topography. Data from Fedorova et al. (2005) and references quoted there. For several models see Sect. 6.5.5, Fig. 6.5.5

usually result from continental rifting and splitting; this may have occurred during the Oligocene when the North Atlantic spreading axis jumped from the Norwegian basin westward to the Greenland margin.

The principal questions are: (1) crustal thickness contrast Δd and densities $\Delta\rho_{\mathrm{Moho}}$ from the increase of the *BA* between Iceland and the Atlantic basin, which, however, cannot be separated without independent information, and (2) from the dipolar component of the *rBA*, the properties of the edge, as depth extent and lateral density contrast $\Delta\rho_{\mathrm{edge}}$; for this, the very simple edge of Sect. 5.6.8 is taken.

(1) From $\Delta g = \Delta BA \approx 160\,\mathrm{mGal}$ between land and sea, the simple Bouguer effect is taken with a customary $\Delta\rho \approx 300\,\mathrm{kg/m^3}$; however, for Iceland a value of $\sim100\,\mathrm{kg/m^3}$ was derived (e.g. Fedorova et al., 2005). The Bouguer effect then ren-

ders $d \approx 10 \times 150 \times 2390/(300 \text{ or } 100) \approx 12 \text{ or } 35$ km. Obviously the density contrast represents the greatest uncertainty of this estimate, and the values of 100 and 300kg/m^3 are lower and upper limits; the mean lateral contrast over the implied depth range from 10 to 50 km is probably somewhere in between.

(2) The edge effect δg_{edge} may be approximated by the simple model of Sect. 5.6.8, shown in Fig. 5.6.3 which renders the expressions for estimating d and $\Delta \rho$. Remember that scaling is linear with density and the geometrical dimensions. From Fig. 5.7.5, we get $\delta g_{\text{m}} \approx \pm 24 \pm 4 \text{mGal}$, $b \approx 25$ km, $w_{\text{o}} \approx 80$ km, $grad$ $\approx 2 \text{mGal/km}$. The expressions supplied in Sect. 5.6.8 give for $\delta g_{\text{B}} \approx 144 \text{mGal}$, and for d several values: from b: 19 km, from w_{o}: 11 km with the mean $d \approx 15$ km; from $grad$ ($\Delta \rho = 100 \text{kg/m}^3$) follows $d \approx 64$ km; but this value must be scaled with the density contrast to $\Delta \rho = 144 \text{kg/m}^3 (\Delta \rho_{\text{ref}}/\Delta \rho_{\delta\text{gm}})$ and, of course, with $(grad_{\text{obs}}/grad_{\text{ref}})$; hence, $d \approx 64 \times (2/5) \times (100/144) \approx 18$ km, in good agreement with the values above. The scatter of the d values is considerable and demonstrates, not surprisingly, that the model is too simple.

This semi-quantitative calculation shows that the observed dipolar anomaly component fits an edge effect roughly, although a glance at Fig. 5.7.5 demonstrates that there are systematic discrepancies in the shape of the anomaly where the extrema are sharper and the outward decay on both sides is more rapid than in the simple vertical edge model. Moreover, the density results seem rather extreme, as the model implies a 15 ± 5 km layer of $\Delta \rho \approx +144 \text{kg/m}^3$ to overlie -144kg/m^3 material, and such a vertical density inversion of 300kg/m^3 may be mechanically unstable. In consequence, a laterally more restricted dipolar mass anomaly than semi-infinite slabs and some modified structures are suggested. Finally, as an aspect of qualitative interpretation, other models may be considered, such as block uplift and/or block rotation, sediments, high-density intrusives, etc., in other words, structures which originated during geological history. Any of these must be tested requiring more sophisticated quantitative modelling, see Sects. 6.5.5 and 7.4.2.

5.7.6 Spreading Ridges, Reykjanes Ridge

Ridges form an Earth encircling system of divergent plate margins where new oceanic lithosphere is formed from hot, partially melting rising mantle material. Here a first, very simple estimate of densities and mass anomalies is given. For density, very simple estimates employ the Bouguer effect and Cartesian 2D approximation for the elongated ridges and trenches. Typically, but with considerable scatter, the axial FA is about $+10 \pm 10 \text{mGal}$, for the height relative to deep sea basins, take $d = 2500 \pm 500$ m, for the density contrast $\Delta \rho = 1800 \pm 200 \text{kg/m}^3$. The Bouguer plate effect (Sect. 5.6.1) of such an under-water ridge is $\delta g \approx 0.1 \times 2500 \times (1800/2390) \approx 185 \pm 40 \text{mGal}$, nearly 20 times as large as the observed FA. Hence, the average mantle density anomaly should produce about -175mGal, about the average relative BA; with, say, $d = 100$ km follows $\Delta \rho \approx 10 \times BA \times 2390/d \approx -40 \pm 10 \text{kg/m}^3$, and with the thermal expansivity $\alpha \approx 3.10^{-5} \text{K}^{-1}$ and $\Delta \rho / \rho = -\alpha \Delta T$, the mean thermal anomaly $\Delta T \approx 400 \pm 100$ K. The same estimate, of course, follows from a simple isostatic argument: $\Delta \rho_{\text{ridge}} d_{\text{ridge}} \approx -\Delta \rho_T d_T$, with the above values $\rightarrow \Delta \rho_T \approx -40$ kg/m^3. These numerical estimates are not altogether unreasonable.

The Reykjanes Ridge SE of Iceland may serve as a special example for the divergent plate margins. It has unusual features: it is about $30°$ oblique to the usual direction normal to spreading and is for 1000 km nearly straight, not interrupted and offset by significant transform faults; instead of a central rift it has a central horst obliquely traversed by narrow volcanic ridges oriented approximately normal to spreading (further to the S, an axial rift develops, more typical of slow spreading ridges). The positive morphology is accompanied by both a positive *FA* rising to nearly $+60$ mGal and a *BA* which is -60 to -80 mGal relative to the adjacent basins ($\delta g_e \approx -70 \pm 10$ mGal). The half width w of the gravity anomaly is of order 150 to 200 km (Fig. 5.7.6). The anomaly may be wider, if the anomalously shallow sea and the regional gravity high are taken into account which both are probably related to the Iceland plume (see Sect. 5.7.9). A negative density anomaly underneath originates from the hot rising mantle material which cools as the plates diverge sideways. The mantle anomaly may be considered an upward pointing hot low-density asthenospheric wedge under the cooling and thickening lithosphere. Alternatively, the temperature anomaly ΔT of the hot region may be taken; it is the high ambient local temperature $T(x, z)$ from which the equilibrium geotherm $T_\infty(z)$ is subtracted that is reached as the plates form from hot mantle cooling with time and distance from the axis: $\Delta T = T(x,z) - T_\infty(z)$ describing onion-shaped ΔT isotherms. Near Iceland, hot plume inflow must also be considered.

Here a simple estimate of the hot, low density mantle is based on an approximation by a 2D mass line or cylinder (Sect. 5.6.3.2 and the expressions therein). The half width $w \approx 150$–200 km suggests the depth of a mass line or a cylinder that may extend to a depth of twice its radius, i.e. $R \approx w$, about 150 km. A shal-

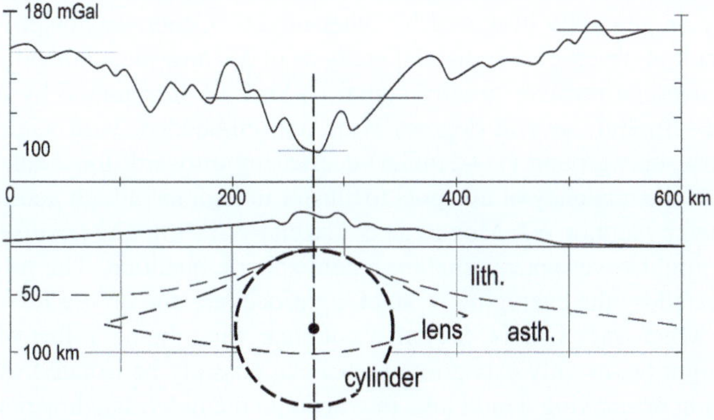

Fig. 5.7.6 Section across Reykjanes Ridge at about $62°$N. Gravity (*BA*, reduction density of ridge topography $(2600$–$1000)$ kg/m^3) and morphology from Jacoby et al. (2007) and references quoted there; gravity amplitude and half width indicated. Section shows sketch of lithosphere-asthenosphere structure, 2D line mass, cylinder of 75 km radius and lens-shaped body

lower flatter density anomaly is likely, concentrated toward the ridge axis and gradually thinning with distance (Fig. 5.7.6). The mean density may be roughly estimated from $\Delta \rho \approx w \delta g_o / (4 \pi G R^2)$ (all quantities in equivalent SI units) rendering $\Delta \rho \approx -22 \, \text{kg/m}^3$, which would correspond to a mean excess temperature of $+230 \, \text{K}$ and an isostatic submarine ridge topography of $+2 \, \text{km}$ (while 100% shallow compensation may not exist). These values are of the right order, but they may refer to only part of the ridge elevation because even the more distant reference basins may be affected by the Iceland plume. More quantitative modelling and a more thorough discussion of this are deferred to Sect. 6.5.6.

5.7.7 *Plumes, the Iceland Plume*

Plumes underlying the hotspots of non-plate-margin volcanism (Morgan, 1971) are a feature of high Rayleigh-number convection, as demonstrated by laboratory and numerical modelling. They are envisioned as narrow stem-like upwellings from hot bottom boundary layers heated from below. Many experiments suggest that the buoyant material should accumulate above the CMB into diapirs that evolve into big rising plume heads (super-plumes) leaving a tail or stem behind which may be further fed from below regularly or episodically. Plumes are a dynamic process affecting the density distribution and deflecting density surfaces. Plumes were, and are still, debated, and other mantle heterogeneities, for example, relicts from former subduction should also be considered, but seismic mantle tomography now (2008) can resolve some plumes rather clearly, and large low-velocity hot regions above the core-mantle boundary are observed, as well.

As shown in Sect. 5.6.9.3, Airy plumes in static vertical equilibrium with the uplifted excess mass above them have gravity effects with a central high and a negative rim. But deep plumes with small density contrasts have very small effects at the surface which may be generally obscured by other effects. Generally hotspot-plume effects are difficult to observe. A statistical analysis of 33 intra-plate hotspots (Fig. 5.7.7) led to an average positive gravity signal ($+5 \, \text{mGal}$) surrounded by a distinct negative rim ($-3 \, \text{mGal}$) several degrees wide and embedded, with some oscillations, in a positive background ($< +2 \, \text{mGal}$) decreasing outward; this background probably reflects the tendency of hotspots to cluster in regions of high geoid and gravity (unpublished by author WJ; Monereau & Cazenave, 1990). The positive feature tapers off and might cover approximately a hemispheric quadrant. The numerical values are affected by the averaging or stacking procedure and do not refer to individual hotspots which vary widely. It is also not clear what the net effect of an "average plume" might be, as only exceptional ones can possibly be isolated. A better way of convolution or stacking would take into account the individual hotspot strengths, possibly leaving the signal shape constant.

Compared to the model of Sect. 5.6.9.3 (Fig. 5.6.5b) the observed signal (Fig. 5.7.7) is broader, but this reflects the coarse $5°$ spatial resolution. The low resolution implies averaging over a corresponding area which, if applied to the mod-

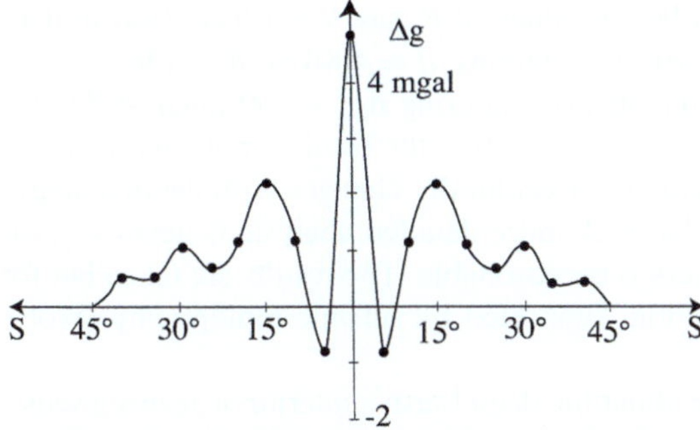

Fig. 5.7.7 Plume gravity effects, radially averaged for 33 intra-plate hotspots and plotted versus angular distance S; symmetry simulates a section across a hotspot. The characteristic central high surrounded by a distinct low is embedded in positive background reflecting the tendency of the hotspots to occur in regions of high geoid and gravity. The fit of the 33 hotspot anomalies achieves a variance reduction of the gravity field of no more than 1.5 % (unpublished W. Jacoby)

elling results, renders a central maximum of about $+10\,$mGal, or so; the averaging less effectively suppresses the amplitude of the negative ring. Finally, with the 3D half width $w = 2R_{1/2}$, a plume should extend to order $z = 2/3\,w$, amounting to upper mantle depths at least, but in view of the minute gravity effect of deep parts of the plume, it can be argued that greater depths can hardly be resolved by gravity.

As a special case, the *Iceland* plume is chosen which occurs on, and interacts with, the Mid-Atlantic Ridge. The *FA* rises to about $+60\,$mGal and topography to about 700-800 m above sea level. If these figures are compared, a 10–20% influence of the plume is implied relative to the topographic Bouguer plate effect, more than the 5–10% estimated in Sect. 5.6.9.3b, which may reflect the special situation of a plume beneath a spreading ridge and extra crustal thickness. A large area of several thousand kilometres dimension is characterized by relatively shallow seafloor culminating in Iceland at an elevation of $>+1000\,$m which exceeds the average height of ridge crest by >3000m; the surrounding ocean basins are >1000m shallower than normal basins of comparable age. The *FA* is positive in a similarly large area and reaches $+60\,$mGal near the supposed plume centre (NW Vatnajökull), while ridge axial values are generally of order $+10\,$mGal, such that the excess due to the Iceland plume is of order $+50\,$mGal. A negative gravity ring (see above) is missing. The *BA* has a minimum of $<-40\,$mGal and is only about $+200\,$mGal over the ocean basins. The geoid is also high, reaching $+60\,$m. From which depth in the mantle the plume originates is debated, but tomography sees a weak anomaly at least to middle mantle depths, possibly to its bottom (Bijward & Spakman, 1999).

The parameters are fairly uncertain. The critical quantity is h; as said above, 700–800 m mean elevation suggest a plume influence of roughly 10–20% of the uplift effect. However, the reference for defining the uplift h is uncertain, hence a trial and error search (Sect. 5.6.9.3) may start with $h \approx 2$km, $\delta g_e \approx +60\,$mGal, uplift densities $\rho \approx 2000\,$kg/m^3 (partially inundated), R (100 km?), d (500 km?). From (5.6.17) δg_e would be $\sim 160\,$mGal which seems too large by a factor 3. This argues for lower values of h, ρ and d and larger values of R, say $h \approx 1.5\,\mathrm{km}$ (for the Iceland plateau), $\rho \approx 1800$ (due to partly inundation), $d \approx 400\,\mathrm{km}$, $R \approx 130\,\mathrm{km}$ (still much smaller than radius of Iceland plateau), rendering $\delta g_\mathrm{e} \approx 100\,\mathrm{mGal}$, still twice the free air anomaly. As pointed out in Sect. 5.6.9.3, the result is not very sensitive to changes in R and d; the plume gravity effect hardly changes with depth changes when the starting value is already large. A more detailed analysis is necessary, although none of the chosen parameters is unreasonable. The results are taken but for a preliminary qualitative interpretation, supported by seismic tomography (Wolfe et al., 1997; Foulger et al., 2000).

At the present state of ignorance about the deep Earth's interior, it seems a sensible approach to probe the Earth's gravity field by systematically shifting correlation analyses with several model plume effects in the search of their possible existence in the field. This can be a step toward the problem of quantitative interpretation and inversion.

5.7.8 *Tonga-Kermadec Trench, Subduction and Back arc Basin*

Deep sea trenches mark convergent plate boundaries and are the expression of an oceanic plate subducting under an upper plate which may be continental (e.g. western South America, Japan) or oceanic (e.g. Mariana island arc, Tonga-Kermadec island arc). The subducted plate or slab is geophysically visible mainly by its seismicity resulting from deformation under the subduction stresses. Cooled oceanic lithosphere, differentiated into crust and uppermost mantle by seafloor spreading, is denser than the surrounding mantle and subducts under the force of gravity (or negative buoyancy). This is a classical case of the dual role of gravity generating density anomalies (active role) that, in turn, produce big gravity anomaly signals (passive role). Slab density is thus of decisive geodynamic importance.

A complex of phenomena accompanies plate convergence, beginning from a seaward forebulge due to bending, a deep-sea trench, often a sedimentary arc, a volcanic arc and either an Andean-type volcanic mountain range or a back-arc basin, also called marginal sea. Back-arc basins form by subducting plate roll-back, i.e. ocean-ward trench migration leaving a gap filled by mantle upwelling. Furthermore, water in the oceanic crust is set free enhancing melting and uprise of the melts to form a buoyant mantle wedge and back-arc seafloor spreading. If on the other hand, the upper plate is moved toward the retreating trench, an Andean mountain range develops. Hence, the integral gravity signal is also very complex and subject to detailed quantitative analysis and interpretation. A 'simple' conceptual model should distinguish at least three components: the lithospheric slab, the buoyant diapir above, and the surface masses, down dragged in the trench and up-bent in the forebulge and pushed up above the low-density diapir: the latter masses are partly taken into account in the Bouguer reduction and will be neglected in the present estimates. Semi-quantitative gravity interpretation, here, will consider only the subducted plate, and it is asked if its thickness, density contrast, and temperature can be roughly estimated.

The profile across the Tonga-Kermadec trench (Fig. 5.7.8: FA, *BA* and morphology; from Jacoby, 1975, and references therein) extends from the Lau Basin to the Tonga trench. The FA rises from near zero, 500 km from the trench like a ramp to about $+180$ mGal from where it abruptly drops to below -100 mGal over the trench; the ramp is characterized by $\sim$50 km long oscillations. The *BA* shows a minimum-maximum-minimum pattern of >250 km wavelength, undulating between 180, 80, 180, and 130 mGal and rising again toward the trench where the 180 mGal plateau over the Lau Basin also shows $\sim$50 km long undulations. Qualitatively, the *FA* pattern suggests a general mass accumulation in the upper few hundred kilometres to which the slab certainly contributes significantly, while the *BA* seems to mainly reflect the top-positive dipole of the diapir with its uplifted roof; the slab effect is hardly identifiable, perhaps only in some weak asymmetry of the *BA* rising toward the trench. From Fig. 5.7.8, the FA half width of $w \approx 250$ km and $\delta g_e \approx 150$ mGal, at least, follow. For the *BA* $w \approx 130$ km which is the width of the Lau Basin, and $\delta g_e \approx 70$ mGal, both values about half those of the *FA*.

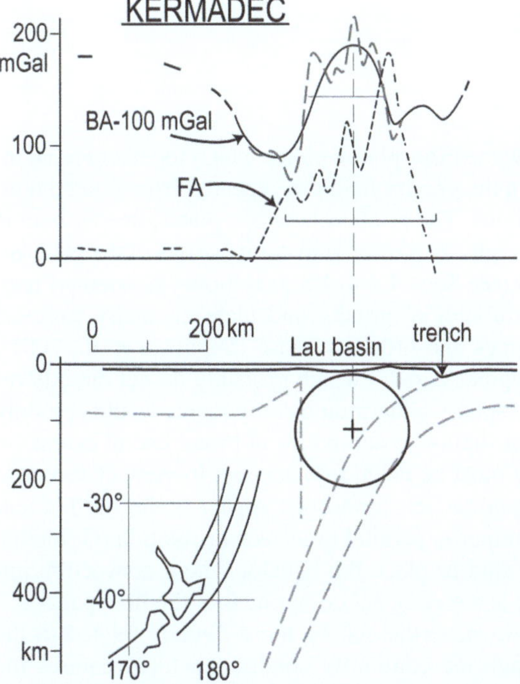

Fig. 5.7.8 Section through Tonga-Kermadec subduction region; location shown by insert at bottom (Jacoby, 1975, see references quoted there). *Top*: *dashed line*: FA; *grey dashed line*: *BA* – 100 mGal; *solid line*: smoothed *BA* – 100 mGal; *horizontal line*: approximate half width. *Bottom*: spherical cylinder section from *BA* half width (see text); *light-grey dashed contours*: approximate lithosphere (including slab) – asthenosphere structure, not known in the early sixties

If, in this situation, the basic semi-quantitative model is restricted to the mass anomaly of the slab, errors follow from the neglected components. Inclusion of the diapir into a dipole model is complicated by the vertical sequence: positive uplifted top, negative diapir, positive bottom of the slab; this cannot easily explain the positive *FA*. Ignoring all these complications and approximating only a slab by a 2D mass line (Sect. 5.6.3.2), taking the half width of the *FA* as representative of the total mass disturbance, but only a third of its amplitude, 50 mGal, because the slab constitutes only part of the mass accumulation, we find from the expressions in (Sect. 5.6.3.2) a depth z of 120 km, and with a cylinder radius $R = 100$ km assumed, $\Delta\rho \approx 150\,\text{kg/m}^3$; the cross sectional area of the slab, sketched in Fig. 5.7.8, is about twice the area of a circle (cylinder) of 100 km radius, suggesting a slab density contrast of $+75\,\text{kg/m}^3$, not a totally unreasonable value. The example is presented to demonstrate, that semi-quantitative estimates in cases of complex structures can give orders of magnitude and that more quantitative interpretation must encompass accurate bathymetry and upper and lower crust, as well as of mantle structure from seismic data (see Sect. 6.5.8).

5.7.9 Mantle Convection

Mantle flow is largely hidden below the lithospheric plates which together are an integral convecting system. Effects in the gravity field are expected from related temperature and, hence, density variations. The relation between plates, deeper mantle flow and gravity is, however, not simple, as is evident in comparisons of the plate geometry and the global gravity field (see Sect. 1.5.4, 1.5.5; National Academy Press, 1997). Expansion in spherical harmonics of gravity and plate geometry suggests some correlation in harmonic degrees and orders 4 and 5 (Schubert et al., 2001), but the largest amplitudes are of degrees 2 and 3 which probably do not directly reflect plate-related flow. Low degree spherical harmonic components would certainly suggest mantle wide anomalies, but shallower anomalies of broad lateral extent are not excluded. Seismic tomography must be taken into account in view of velocity-temperature-density relations. A qualitatively important aspect is the Earth's tendency to orientate its axis of largest inertia parallel to its axis or rotation (Goldreich & Toomre, 1969) and this would tend to place the equator where convection upwellings predominate, thus hinting at a significant component of the flow pattern.

Some component of gravity may, nevertheless, be more directly related to the moving plates, for example, through the continuity condition which requires the plate motion to be linked with flow. Respective data are difficult to isolate from the global gravity field which is an integral of all masses of Earth. Stacking gravity profiles along plate motion trajectories toward the trenches may enhance such a signal while suppressing other components. Examples are shown in Fig. 5.7.9a (after Seidler et al., 1983) where the *FA* is plotted along the trajectories of plate motion relative to the hotspot frame of reference versus distance (coordinate X_B in degree) from the trailing edge ($X_B = 0$) to the leading edge; the figure shows

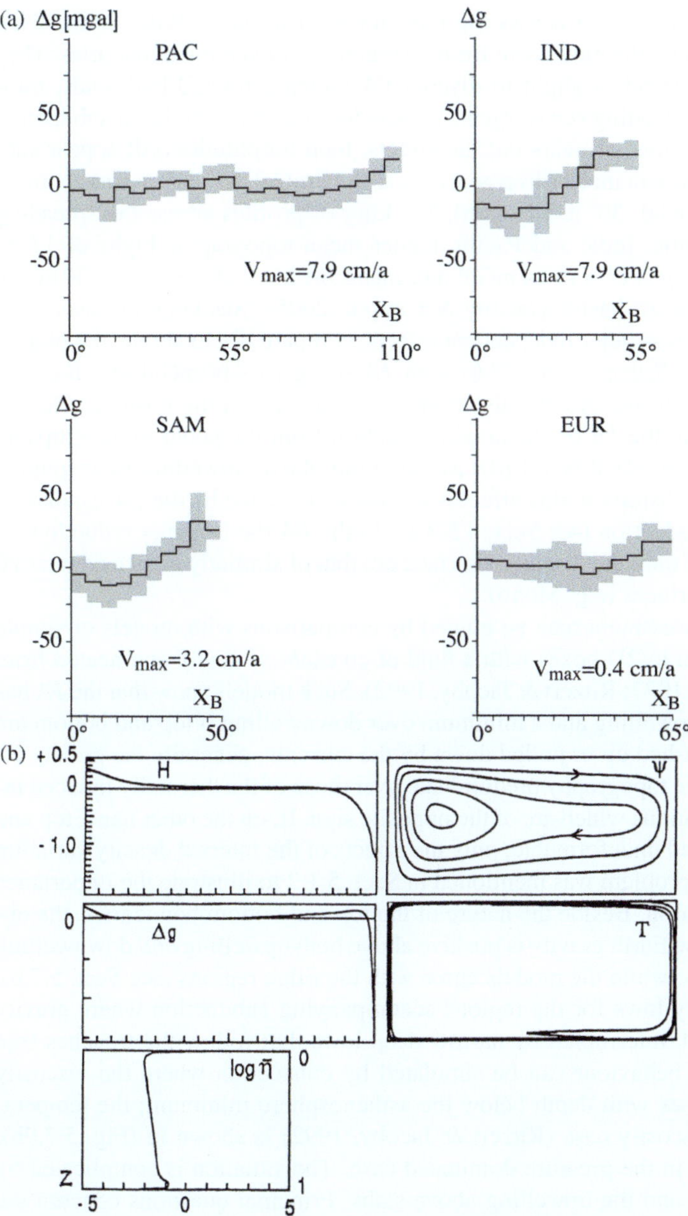

Fig. 5.7.9 Mantle convection. (**a**) FA plotted along trajectories of plate motion in hotspot frame of reference versus distance X_B (in degree) from trailing edge ($X_B = 0$) to the leading edge; *FA* averaged for X_B = const (across the trajectories) shown as a bar, standard deviation shown in *grey*. Plates PAC, SAM, IND, EUR move from divergent to convergent boundaries. (**b**) Numerical model of convection in Cartesian box (Rayleigh number Ra $= 10^6$ with temperature dominated p, T-dependent viscosity (Ritzert & Jacoby, 1992); *boxes* show calculated elevation H (below water), *flow lines* as stream function contours Ψ; the *FA* gravity anomaly Δg; the temperature field T; horizontally averaged viscosity as log η. (scales are dimensionless; note especially the relative variation of H and Δg)

the averages for X_B = const across the trajectories. The plates PAC, SAM, IND, EUR have in common the movement from divergent to convergent boundaries. They show a consistent trend of slight to distinct FA increase toward both ends, trailing (divergent) and leading (convergent, i.e. subducting or overriding a subducting plate). Since the analysis smears out the effects, their amplitudes will appear subdued and the widths enhanced (divergence: $\sim 10 \pm 10\,$mGal, $15° \pm 5°$ width; convergence: $20 \pm {>}10\,$mGal, $30° \pm 5°$ width). Stacking of profiles across the spreading ridges in the Atlantic, Indic and Pacific render mean topographic highs of 1.0 to 1.6 km, mean FA highs of 6 to 14 mGal and mean BA lows of -80 to $-130\,$mGal, relative to the adjacent basins (Jacoby & Çavşak, 2005). Stacking and averaging does not fully suppress other independent effects; compare, for example, the plume-affected Reykjanes Ridge (Sect. 5.7.6) with FA rising to $+60\,$mGal and BA only -60 to $-80\,$mGal. In the gravity disturbance (see Sect. 4.3) the positive effect is enhanced relative to the FA by the height reduction from the geoid to the ellipsoid $(N\partial g_n/\partial h \times (-1) \approx +0.3086N$ [m]), as the geoid above upwelling flow (plume, ridge) is positively disturbed; this effect is somewhat lessened by the corresponding geoidal Bouguer reduction (see Sect. 4.5.3.1). In the BA the Bouguer reduction removes the effect of only the displaced surface not that of similarly displaced internal density contrast surfaces (e.g. Moho).

A qualitative assessment may be guided by comparisons with models of simple cases of convection in 2D boxes with a fluid of constant viscosity and heated from below (McKenzie, 1977; Ritzert & Jacoby, 1992). Such models show that the FA has a maximum over upwelling and a minimum over downwelling if top and bottom are deformable and pushed up or pulled down by the currents; generally the gravity effects of the dynamic topography predominate over those of the thermally induced internal density variations which are of the opposite sign. If, on the other hand, top and bottom are fixed and undeformable, only the effects of the internal density variation are observed. The problem was mentioned in Sect. 5.1.2 to illustrate the importance of a priori information. Beside the nature of the top and bottom boundaries, the observation that on the Earth gravity is positive above both upwelling *and* downwelling presents a problem: while the models agree with the ridge regions (see Sect. 5.7.6), they predict gravity lows for the regions accompanying subduction where gravity highs are observed, interrupted by narrow deep lows that follow the trenches (see Sect. 5.7.8). Such behaviour can be simulated by convection where the viscosity of the fluid increases with depth below the asthenosphere minimum; the temperature dominated viscosity case (Ritzert & Jacoby, 1992) is shown in (Fig. 5.7.9b); the effect is larger in the pressure dominated case. The situation is complicated by stagnation of flow and the upwelling above slabs. Principal questions concern the rheology of the mantle, the size or volume and depth of the convecting system, its lateral extent and the temperature variations. The anomaly half widths cannot exclude shallow convection but certainly permit mantle-wide flow which is supported by the tomographic evidence of subduction tending to penetrate the whole mantle.

Obviously, the whole problem involves careful quantitative estimates of the complex relationships, it goes far beyond any simple estimates and always involves a tightrope walk of dynamic balances between the various processes and mass anomalies involved. This is beyond the scope of the present treatise (see e.g., Thoraval et al., 1995; Bunge et al., 1996; Čadek & Fleitout, L., 1999; Steinberger & Holme, 2002). Qualitative interpretation is largely restricted to such considerations as those above.

5.8 Error Discussion and Conclusions

Errors of qualitative and semi-quantitative interpretation are naturally made also in the three categories, gross, systematic and random (Sects. 3.7.1, 3.7.2, 3.7.3), but they are of a more general nature than those discussed in the previous chapters. They have more to do with notions as reliability and confidence. Interpretation is an act of associating meaning with the observations or understanding which is the primary aim of measuring and analyzing gravity. Semi-quantitative estimates of gravity effects serve the purpose of testing the ideas, but "crazy ideas" cannot always be proven wrong by calculation.

Gross and systematic errors in the initial stages of interpretation include, beside the usual ones as mistaken numerical values, computational mistakes and copying errors, the use of wrong formulae and wrong preconceived ideas. Ideas may be partly right guided by fantasy and imagination and right or wrong refers to consistency or inconsistency with structures and processes existing inside Earth and with the laws of physics (e.g. mechanical stability), chemistry, geological history etc. The best safeguards against gross mistakes are experience, controls, rethinking and review by fellow scientists, in other words: never to contend oneself.

The above is related to the notion of the "apparent a posteriori standard error" introduced in Sect. 3.8. The principal ambiguity of gravity interpretation implies that wrong models which are at variance with "reality" may be, nevertheless, quite compatible with the gravity observations. Estimated numerical uncertainties or standard deviations may be very small, but as long as not excessive, they are rather irrelevant for the purpose of qualitative interpretation. The same is true for any random errors because they cannot even be statistically quantified like observations and the desired estimates are not meant to be precise.

At the end of this chapter, the interpreter of a given gravity anomaly has clear and tested ideas of the model or models in mind. Preliminary estimates have shown the way to thoroughly quantifiable or quantitative modelling and interpretation.

References

Armstrong, M.: Basic Linear Geostatistics. *Springer Verlag, 155 pp., 1998*

Bijward, H., Spakman, W.: Tomographic evidence for as narrow whole mantle plume below Iceland, 1. *Earth Planet. Sci. Lett., 166, 121–1126, 1999*

Bunge, H.P., Richards, H.P., Baumgardner, J.R.: Effect of depth-dependent viscosity on the planform of mantle convection. *Nature, 379, 436–438, 1996*

Čadek, O., Fleitout, L.: A global geoid model with imposed plate velocities and partial layering. *J. Geophys. Res., 104, 29055–29076, 1999*

Dürrbaum, H.-J.: On the estimation of the limiting depth of gravitating bodies. *Bundesanstalt Geowissenschaften Rohstoffe, Hannover, Geol. Jb. E2, 23–32, 1974*

Fedorova, T., Jacoby, W.R., Wallner, H.: Crust-mantle transition and Moho model for Iceland and surroundings from seismic, topography and gravity data. *Tectonophysics, 396, 119–140, 2005*

Foulger, G.R., Pritchard, M.J., Julian, B.R., Evans, J.R., Allen, R.M., Nolet, G., Morgan, W.J., Bergsson, B.H., Erlendsson, P., Jakobsdottir, S., Ragnarsson, S., Stefansson, R. and Vogfjord K.: The seismic anomaly beneath Iceland extends down to the mantle transition zone and no deeper. *Geophys. J. Int. 142: 1–9, 2000*

Goldreich, P., Toomre, A.: Some remarks on polar wandering. *J Geophys. Res, 74, 2555–2567, 1969*

GW65 Grant, F.S., West, G.F.: Interpretation Theory in Applied Geophysics. *McGraw-Hill, New York etc., 583 pp., 1965*

Jacoby, W.R, Çavşak, H.: Inversion of gravity anomalies over spreading oceanic ridges. *J. Geodynamics, 39, 461–474, 2005*

Jacoby, W.R.: Schweremessungen auf Helgoland – Auswertung mit Ausgleichsverfahren. *Z. Geophys., 32, 340–351, 1966*

Jacoby, W.R.: Gravity diagrams for thickness determination of exposed rock bodies. *Geophysics, 35, 471–475, 1970*

Jacoby, W.R.: Velocity-density systematics from seismic and gravity data. *Veröff. Zentr. Inst. Phys. Erde, Nr. 31, 323–333, 1975*

Jacoby, W.R., Czerwek, D., Pakirnus, H.: Lithosphere structure, heat flow, gravity, and other geoparameters in Central Europe. in Terrestrial Heat Flow and the Lithosphere Structure, ed. V. Cermák, L. Rybach. *Springer, Berlin, pp. 112–132, 1991*

Jacoby, W.R., Wallner, H., Smilde, P.: Tektonik und Vulkanismus entlang der Messel Störungszone auf dem Sprendlinger Horst: Geophysikalische Ergebnisse. *Z .dt. Geol. Ges., 151, 493–510. 2001*

Jacoby, W.R., Mertz, D., Wallner, H.: Students' field research extends knowledge of origin of a Unesco world heritage site in Germany. *EOS Transactions Am. Geophys. Union., 84, 39, 393,396–397, 2003*

Jacoby, W., Weigel, W., Fedorova, T.: Crustal structure of Reykjanes Ridge near 62°N, on the basis of seismic refraction and gravity data. *J. Geodynamics, 43, 55–72, 2007*

Jung, K.: Über die größte mögliche Schwankung der Schwereintensität und die Dichte eines engmaschigen Netzes von Pendelstationen. *Z. Geophys., 3, 137–156, 1927*

KJ61: Jung, K.: Schwerkraftverfahren in der Angewandten Geophysik. *Akad. Verlagsges. Geest & Portig, Leipzig, 348 pp., 1961*

McKenzie, D.P.: Surface deformation, gravity anomalies, and convection. *Geophys. J. Roy. Astron. Soc., 48, 211–238, 1977*

Monereau, M., Cazenave, A.: Depth and geoid anomalies over oceanic hotspot swells: A global survey. *J Geophys. Res, 95, 15429–15438, 1990*

Morgan, W.J.: Convection plumes in the lower mantle. *Nature, 230, 42–43, 1971*

Nakamura, Sh. & Nakamura, Sh.: Numerical Analysis and Graphical Visualization with Matlab, 2nd ed. *Prentice Hall, 544 pp., 2001*

National Academy Press: Satellite Gravity and the Geosphere. *Nat. Res. Coun., Rep. 1997, Nat. Acad. Press, Washington D.C., NRC (Comm. Earth Gravity from Space, US Geodynamics Committee, Board Earth Sci. & Resources, Comm. Geosci., Envir. Resources), 1997* http://www.issi.unibe.ch/workshops/SatGrav/Book.htm

Prodehl, C, Mueller, St., Haak, V.: The European Cenozoic rift system. In Continental Rifts: Evolution, Structure, Tectonics, K.H. Olsen, ed. *133–212, Elsevier, 466 pp., 1995*

Ritzert, M., Jacoby W.R.: The geoid signal of 2-D convection with temperature- and pressure-dependent rheology. *J. Geodynamics, 16 , 81–102, 1992*

Sachs, L.: Applied Statistics. A Handbook of Techniques, 2nd ed. *Springer Verlag, Berlin etc., 707 pp., 1984*

Schneider, P., Eberly, D.: Geometric Tools for Computer Graphics. *Publ. Morgan Kaufmann, 1056 pp., 2002*

Schubert, G., Turcotte, D.L., Olsen, P.: Mantle Convection in the Earth and Planets. *Cambridge Univ. Press, 940 pp. 2001*

Seidler, E., Lemmens, M., Jacoby, W.R.: On the global gravity field and plate kinematics. *Tectonophysics, 96, 181–202, 1983*

Smilde, P.L.: Verwendung und Bewertung von a priori Information bei potentiell singulären Inversionsproblemen am Beispiel der gravimetrischen Bestimmung von Dichteverteilungen. *Deutsche Geodät. Komm., C 490, München, 1998*

Steinberger, B., Holme, R.: An explanation for Earth's gravity spectrum based on viscous mantle flow models. *Geophys. Res. Lett., 29, 219 doi 1029/2002GL015476, 2002*

Thoraval, C., Machetel, P., Cazenave, A.: Locally layered convection inferred from dynamical models of the Earth mantle. *Nature, 375, 777–780, 1995*

Wolfe, C.J., Bjarnason, I.Th., VanDecar, J.C., Solomon, S.C.: Seismic structure of the Iceland mantle plume. *Nature, 385, 245–247, 1997*

Woollard, G.P.: The relation of gravity anomalies to surface elevation, crustal structure and geology. *Final Report, Contract AF 23 (601)-3455, Research Report Series No. 62-9, Aeronautical Chart and Information Center United States Air Force, St. Louis, Missouri, 1962*

Relevant Literature, Not Quoted

Davies, G.F.: Role of the lithosphere in mantle convection. *J. Geophys. Res., 93, 10451–10466, 1988*

Insergueix, D.-Filippi, D., Dupeyrat, L., Tric, E., Menvielle, M.: The influence of plate kinematics, convection intensity, and subduction geometry on the Earth's upper mantle dynamics in the vicinity of a subduction zone. *Geophys. J. Int., 138, 275–284, 1999*

Kaula, W.M.: Theory of Satellite geodesy. *Blaisdell, Waltham, 124 pp., 1966*

Lowman, J.P., King, S.D., Gable, C.W.: The influence of tectonic plates on mantle convection patterns, temperature and heat flow. *Geophys. J. Int., 146, 619–636, 2001*

Richards, M.A., Bunge, H.-P., Lithgow-Bertelloni, C., Baumgardner, J.R.: Mantle convection and plate motion history: Toward general circulation models. In "History and Dynamics of Global Plate Motions", M.A. Richards, R.G. Gordon, R.D. van der Hilst, eds. *AGU Geophys. Monograph Ser., 121, 398 pp., 2000*

Chapter 6
Quantitative Interpretation

Quantitative interpretation means quantification of qualitative models. "Qualitative" is not "purely qualitative". All models are approximate estimates. There is no clear distinction between "qualitative" and "quantitative". "Quantitative" is never "absolutely accurate" or error-free, and model parameters must be complemented by their error bounds. What is considered sufficiently accurate depends on the study aims, data situation, numerical realisation etc.

The present chapter concentrates on how the *forward problem* is best solved by calculating gravity effects δg of given or assumed mass anomalies. The different strategies for organizing the calculations are emphasized, based on qualitative models of Chap. 5. The ultimate aim of gravity interpretation requires detailed routines; however, they are usually available or should be written for the given problems. This chapter tries to show up the possibilities on the theoretical basis laid down in Chap. 2. Trial and error, optimization and inversion build on this chapter.

6.1 Introduction: From Qualitative to Quantitative Interpretation

Complex geological bodies must be constructed on the basis of simple body types. Specific parametrizations offer themselves which are most appropriate to the description based on the coordinates of points and the associated densities: quadruplets $\{x_k, y_k, z_k, \rho_k\}$ or quintuplets $\{x_k, y_k, z_k, \rho_k, t_{kl}\}$ where the index k signifies a countable model parameter and l an instant in time. Any kind of coordinates are in use. Frequently only *two* geometrical coordinates are specified, the third is assumed infinite; such models are called "two-dimensional" or 2D (see Sect. 2.9.7). Generally the basic model bodies are uniform in density.

Since the gravity effects of geological bodies are always small relative to standard earth gravity $\boldsymbol{g}_o$ in a study area, the associated vertical deflections are generally negligible and in a fixed (x, y, z) coordinate system with z parallel to $\boldsymbol{g}_o$. The gravity effect is thus always assumed to be the z component of the gravitational vector effect

W. Jacoby, P.L. Smilde, *Gravity Interpretation*,
© Springer-Verlag Berlin Heidelberg 2009

although the disturbing vector will generally have also x, y components and thus deflect g from g_0 and the fixed z direction.

6.1.1 Principal Considerations: Qualitative and Quantitative Interpretation

In quantitative gravity interpretation essentially the locations, depths, shapes and density contrasts of geological bodies are to be defined as "accurately" as possible. It may be desirable to first isolate those parts of the observed gravity anomalies which are caused by the target bodies, and this is part of interpretation. It requires independent additional a priori information, which, of course, may be ambiguous in itself. The whole task, thus, will be to approach "the truth" by combining all available information and adjust it mutually within the error limits of each. The compromise between all pieces of conflicting evidence is largely a matter of judgement and experience. The traditional manual approach by trial and error is necessarily subjective and uncertain. Its quantification is Bayesian inversion described in Chap. 7 which, thus, is the top aim of quantitative interpretation.

Due to the ambiguity problem gravity can never define or prove a "true" model but can definitively prove an assumed model wrong – in line with Karl Popper's view that science advances by falsification of preliminary hypotheses. One can learn how wrong a model is. Gravity interpretation is never final: any new data warrant better quantifying the models. Another valuable aspect of gravity interpretation is suitability to aid interpolation between gaps in other types of information, for example, between boreholes or seismic surveys. Standard methods as seismic reflection may also leave "blind spots" where gravity modelling can be applied profitably to fill such gaps. Thus, gravity is and remains a unique and economic tool for exploration.

Non-ambiguous quantitative information, contained in gravity anomalies, is the total amount of the anomalous mass and its horizontal centre of gravity (see Sect. 2.7.6). The accuracy of the information depends on the accuracy of the data and, to a large extent, on the definition of the zero level of an investigated anomaly. In a somewhat similar fashion, maximum depths may also be estimated. Downward continuation of gravity anomalies through homogeneous layers can also give some direct insights by computing the idealized mass anomaly in the form of the equivalent stratum, but it is only a guide to quantitative interpretation.

A model is as accurate as the given error bounds permit, i.e. errors from the data, from neglected model aspects, from additional information. The best to be achieved is good estimates of most likely model parameters and of their error bounds. They are essentially of statistical nature, but that is too simple because of the ambiguity problem, and the possibility to construct totally wrong models that fit within arbitrarily narrow error bounds.

6.1.2 General Methodological Aspects

It is "*anomalies*" that are interpreted. Anomalies are "reduced" observations. In contrast, it is gravity "*effects*" which are calculated for given assumed or known mass anomalies. The height reduction (Sect. 4.5.2) does not shift the observation points onto a reference level below which all disturbing mass anomalies lie. The anomalous gravity effects at the field locations are left after the reduction. This is critical in mountainous terrain, and it is generally recommendable to interpret anomaly values at the observation locations, i.e. to compute model effects for these locations. With present computing facilities this does not present any difficulty.

Q*uantifying the error bounds* is an essential aspect of quantitative interpretation. The mathematical approach to this problem is optimization and inversion (Chap. 7). It is the last step to arrive at quantitative models and their uncertainties by matching the gravity observations and adjusting the model parameters that generate the matching effects. It is not an aspect of gravity alone, but also of any other a priori information.

Customarily one distinguishes *direct and indirect methods*, taken here as an ordering principle. An interpretation method is called "*direct*" if some parameters describing a model (e.g. density, dimensions, location, depth etc.) are calculated directly from characteristic features of the observed anomaly $\delta g(x)$ (e.g. amplitude, "half width", some ratios, etc.) by the use of formulae or diagrams or "characteristic curves". Such direct methods are based on rather simple models with only few parameters. Otherwise the direct relation between observation and model parameters will not be direct. Some of the direct methods belong into the category of estimates or quick semi-quantitative methods described in Sect. 5.6. More involved methods are treated here and some will be only quoted from the literature.

The *indirect* approach is the traditional *trial and error* procedure. It is iterative and starts with assuming a preliminary or initial model, calculating its effects δg_{m} and continues with comparing them to the observations δg_{obs}. The residual field (r_{i} or $r = \delta g_{\mathrm{obs}} - \delta g_{\mathrm{m}}$) is then examined for systematic space variations which may be, again, interpreted by modifications to the previous model, i.e., taken to estimate the model changes (or the difference model) which lead to reduce the residuals or to better fit the observations. This is repeated until the fit is considered satisfactory. The aim is to match the observed anomalies by the computed effects, however it suffices within the error bounds of the observations. The final residuals are (from the original observed data) are considered small enough and their variation is considered sufficiently random in space. If trial and error is carried out manually it is guided mainly by experience (see remarks in Sect. 6.4). By it one finds an *acceptable model*, i.e. one that is compatible with the observations, but not all variants which are also data-compatible. This must be considered within the geological constraints; otherwise an infinite space of irrelevant models would be included. Systematic approaches are probably better than arbitrary ones, but the ever more efficient computers permit also a random (Monte Carlo) search.

There is no clear-cut distinction between direct and indirect methods of interpretation, they are end members between simple direct calculations and involved mathematical procedures rendering more sophisticated and detailed models on the basis of the whole observational set. Besides, at any interpretatative step simple direct methods and more complex indirect model improvement can be combined. Computers provide additional support through effective graphics (Sect. 5.5). The different skills of humans trained in different fields as mathematics, geophysics and geology, in combination, promise the best results.

The differences between all these compatible models, say, $\Delta_\lambda(x_k, y_k, z_k, \rho_k) = \Delta_\lambda \rho_k(x, y, z)$ have no significant gravity effects (i.e. they are effectively zero) at the stations considered, where λ stands for any possible model combination. It is called the "nullspace" of models. One can formulate the interpretation task as that of finding the nullspace of models beyond finding one model by trial and error. It involves a priori information as formulated as Bayesian inversion (Chap. 7). Monte Carlo routines with random variation of the parameters and genetic or evolutionary algorithms can find solutions. Combination of these procedures with straightforward iterative least-squares routines is a successful strategy. If an anomaly, i.e. a confined field variation is identified, certain parameters can be determined (e.g. the half width etc.) and then some simple formula may render a preliminary solution which can be iteratively refined (demonstrated for magnetic anomalies by Bosum, 1981; Hahn & Bosum, 1986). Euler deconvolution has been stressed as a tool combined with a cluster analysis, called "RODIN".

6.1.3 Philosophy of Modelling: Detailed Description Versus Catching the Fundamental Features and Their Uncertainties

Geological structures are generally of high complexity at all scales and models will always be approximations. It is an art to design approximations which depict the essential aspects of a geological density distribution while simplifying the rest, for example, small structures of insignificant gravity effect. It is essentially the data noise and work economy which determine what is "small". Relevant scales depend on distance from the observation points. Near-surface inhomogeneities introduce such noise. No rock body is perfectly homogeneous in density, but only some of the internal density variations will be considered important in modelling. Examples of approximations are the neglect of the dominant long dimension (2D) and of purely vertical components of density variation $\rho(z)$. However, different vertical density gradients in neighbouring rock units (e.g. from compaction of sediments or from diagenetic rock alteration) may lead to not-negligible lateral density contrasts.

Two principal approaches to modelling exist. One is to refine the geometrical description to great precision and detail, depending on the kind of a priori information available. Detailed reflection seismic and drill hole data permit complex 3D parametrizations, e.g., by polyhedral triangulation, as in the interactive modelling program IGMAS (Götze & Lahmeyer, 1988). The opposite approach is to assume a set of relatively simple adjustable shapes of which models can be composed and which can be changed in an easy adjustment procedure. This approach is behind the program INVERT (Smilde, 1998). The criterion of changes to have measurable effects calls for a hierarchy of scales and a strategy from crude to refined. Sensitivity tests reveal the minimum size of significant density variations (in space and density) in the whole model region. Systematic and summation effects must be taken into account. For optimization and inversion (Chap. 7) the number of model parameters must be kept low (to be determined by tests).

The path to quantitative interpretation requires first decisions as to what is essential and what can be neglected, and then adequate geometrical model descriptions, approximations or parametrization, for which gravity effects can be calculated. Many possible geometrical descriptions exist. Useful practical methods are presented in a systematic fashion, for example, combining elementary bodies (Chap. 2 and Sect. 5.6). The present chapter is mainly on methodological aspects of gravity calculation for the various tasks common in gravity interpretation.

6.1.4 Model Types: Two and Three-Dimensional; Large Model Bodies Versus Small Mass Elements

Geology is *three-dimensional* (3D). If, however, one coordinate describes the dominant extent of a body with little variation in that direction only two geometrical coordinates are needed to describe the body in its *two-dimensional* cross section (2D, see Sect. 2.9.7). The term 2D is not quite correct, but it is descriptive and generally accepted. A planar surface mass (kg/m^2, see Sect. 2.7.3) is also 2D (if laterally unlimited).

The loss of accuracy when reducing the geometry form 3D to 2D must be estimated. Some seem to believe that only 3D modelling is acceptable, but often simple models are better, and sophistication is not a sufficient criterion of intelligent interpretation. Regard for the human imagination is essential, and it is generally more adapted to simpler models. The so-called *two-and-a-half-dimensional models* (21/2D) take the non-infinite extent of model units into account by end corrections (Sect. 6.2.3; KJ61, 144, 163–164; GW65, 292), but to be adequate the density beyond the end must be considered.

The geometry of geological masses can be approximated by few large uniform volumes which catch the essentials or by many very simple small mass elements assembled to larger bodies. Large bodies, e.g., prisms, polyhedra, disks and their 2D equivalents, are usually describable by relatively few parameters. This is an important advantage for inversion. Small bodies, as points or spheres, small prisms, etc., and their 2D equivalents, have several advantages: few parameters for each and simpler mathematical expressions allowing to shift, alter, generate or annihilate such elements. Disadvantages are the extremely small effects creating rounding errors in large sums, and inversion is not directly possible (see Sect. 7.2.2.3.1.2). Combinations of large bodies and small mass elements are feasible, for example, by combining many small elements to bigger ones depending on their distance from the calculation point. Reasonably efficient estimation was formerly possible also for the effects by "individual" volume division into point P-related compartments by the use of overlay templates. This now nearly forgotten procedure has the advantage to work for shapes which are not amenable to analytical treatment and will also be treated here (Sects. 6.3.2.2 & 6.3.3.2

The following sections on 2D and 3D modelling, will follow the same scheme: large model bodies, direct and indirect methods, then the same order for small mass elements. The order is repeated for 3D models. Some unavoidable repetition is, hopefully, acceptable.

6.1.5 Density

6.1.5.1 Reference Density

Although for the gravity calculations it is not essential whether densities of the model parts are taken as the absolute or relative values, it is generally advantageous to use relative density contrasts, because the observed gravity anomaly variation reflects only the lateral variations of density, referred to some reference density ρ_{ref} (see Sect. 2.5, Eq. 2.5.4), for the model unit k:

$$\Delta\rho_k = \rho_k - \rho_{ref} \qquad (6.1.1)$$

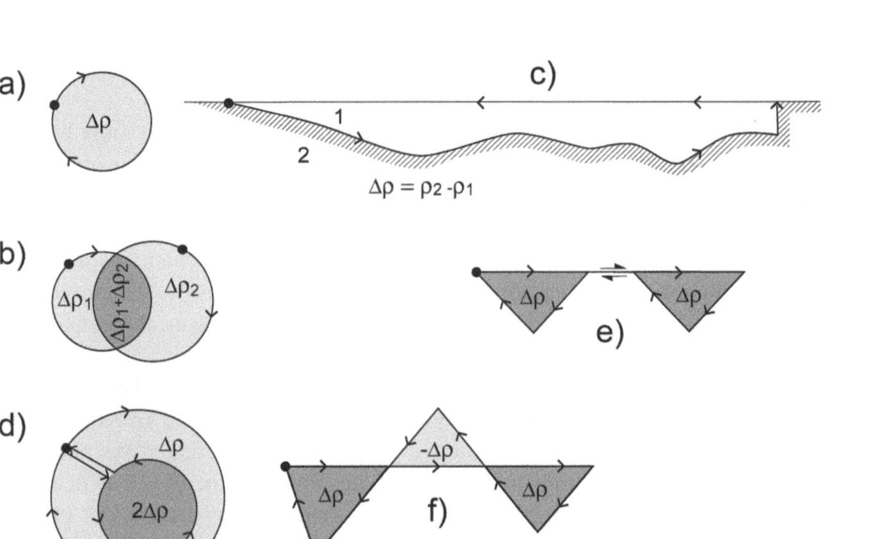

Fig. 6.1.1 On the definition of density contrasts in gravity modelling: schematic illustration of the different situations discussed in the text: (**a**) Simple imbedding, (**b**) overlap, (**c**) undulating layer boundary, (**d**) nesting, (**e**) disjoint bodies and (**f**) bodies of alternating sign of $\Delta\rho$

ρ_{ref} can be defined arbitrarily, for example, mutually between two bodies or as a constant background of all bodies considered (see Sect. 6.2.1.2; Fig. 6.1.1). In the latter case, $\rho_{ref} = \rho_0$ must not have lateral variations; it may be a constant or depth-dependent, $\rho_0(z)$, for example, horizontally averaged over the model extent. Each such an assumption of ρ_0 leads to another constant value g_0 of the gravity effect:

$$g_0 = 2\pi G_0 \int^{z0} \rho_0(z)dz \qquad (6.1.2)$$

equivalent to the effect of a Bouguer slab (Sect. 2.9.3.3); at greater depth than z_0 no density variations are taken into account. If only the corresponding $\Delta\rho$ values enter the modelling, g_0 does not appear in the calculated gravity effects. The observations and the theoretical effects can be directly related (e.g. by their correlation coefficient) if and only if both sets have zero mean. This is usually not the case, and any interpretational algorithms should take care of subtracting the mean automatically. The mean density may be taken as reference density ρ_0 to be first subtracted from the individual values, however, since their final values are not known a priori, the reference density ρ_0 is an important notion which must be carefully considered.

There are several possibilities (Fig. 6.1.1).

(1) The body is *embedded in "country rock"* or a reference mass, which may be another model body (Fig. 6.1.1d) or even several bodies; in the latter case the assumption is not constant density, but only constant density contrast, because the background varies in density; the same is true if the reference density is depth dependent $\rho_0(z)$ which will remain principally undetermined, i.e. independently established at best. Related seismic velocities offer indirect information (see Sect. 3.6.3.4).

(2) If all non-overlapping bodies k of a model are simply assigned their individual density contrasts $\Delta\rho_k$, (Fig. 6.1.1a) these refer to the *same background*, possibly difficult to identify; e.g. if the model volume comprises the whole crust, the background would be the underlying mantle if the crust-mantle boundary is undulated; otherwise ρ_0 is arbitrary and irrelevant for the interpretation. The individual relative density contrasts $\Delta\rho_k$ are meaningful, also mutually.

The reference ρ_{ref} or ρ_0, if not depth-dependent, may be estimated with (Eq. 6.1.1) on the basis of a priori estimates of the individual absolute densities ρ_k^{prior} and a posteriori density contrasts $\Delta\rho_k^{post}$ (e.g. as obtained in an inversion as described in Chap. 7) for one or several geological model bodies k; from Eq. 6.1.1 follows $\rho_{ok} = \rho_k^{prior} - \Delta\rho_k^{post}$. Several such determinations for different model bodies will generally lead to conflicting values of ρ_{ok}, and an average, possibly weighted, may be calculated as $\underline{\rho}_o = {}_k\Sigma(w_k\rho_{ok})/{}_k\Sigma w_k$, where w_k may be chosen, for example, such as to reflect the importance of the model bodies k. A posteriori absolute densities ρ_k^{post} can then be calculated again by (Eq. 6.1.1): $\rho_k^{post} = \underline{\rho}_o - \Delta\rho_k$.

(3) For models consisting of undulated layers, the *inter-layer density contrasts* can b chosen for the model parametrization (Fig. 6.1.1c): Consider a model, for example, of the upper and lower crust plus mantle, or several superimposed sedimentary layers, and only the boundaries are of consequence. There are two options to describe this.

(3.1) Each layer is defined as in case (2) (Fig. 6.1.1a) by all the points describing its top and bottom and the density contrast is defined against a background ρ_0 common to all layers; the disadvantage is that all boundary points must be defined twice in reverse order, because layers in contact have each to be traced around in the same (usually clockwise) direction.

(3.2) Each boundary between two layers is defined only once, and the relevant density contrast is that between layer $k+1$ and layer k: $\Delta\rho_{k+1,k}$. The volume is given now by points defining, for example, the upper boundary of layer $k+1$ (realizable as a polyhedron in 3D or as a polygon in 2D), completed or "closed" by a horizontal surface (3D) or line (2D) which does not contribute to the lateral gravity variation (Fig. 6.1.1c). The body must be extended to lateral distances from the observation points far enough to avoid end effects. The region outside the relevant model space and the far ends, so to speak, represents the lateral continuation of Earth structure. The procedure is repeated from layer to layer, the reference for the first boundary is the density of the top layer 1, and as the density contrasts are defined for each boundary relative to the overlying layer, the layer densities ρ_k are calculated by summing the intervening density contrasts (calling $\Delta\rho_{k,k-1} = \Delta\rho_k$, and $\Delta\rho_1 = \rho_1$):

$$\rho_k = {}_1\Sigma^k \Delta\rho_k \tag{6.1.3}$$

In cases, some layers may be laterally limited lenses, such that in the vertical sequence such layers may not exist everywhere. In a region where, say, layer $k+1$ is missing it is required to extend the layer boundary $k, k+1$ (between layers k and $k+1$) and repeat it as the boundary $k+1, k+2$ (identical to boundary $k, k+2$, such that the layer $k+1$ thickness is zero). By following such a procedure, it is ensured that the calculated density contrasts $\Delta\rho_k$ will be correctly taken into account by vertical summation according to (6.1.3) in the whole model volume, even where layer $k+1$ is missing. Similar situations (e.g. if more layers than one are missing) are treated analogously.

(4) Bodies can be *nested*. If a body is placed into another, the density contrast is relative to that one (Fig. 6.1.1d). If a body intersects the boundaries of another, i.e. overlaps it, its uniformly defined density contrast implies a non-uniform absolute density because the absolute density at any point in the model space will be the sum of all "local" density contrasts at that point and the reference density, say ρ_0. Care must be taken when composing complicated models of partly overlapping bodies.

(5) Separate, *non-compact bodies of identical $\Delta\rho$* can be linked to one by connecting a point of each closed boundary by a line (2D) or an edge or plane (3D) in both directions which cancels any possible formal gravity effect (Fig. 6.1.1e).

(6) Bodies are usually "compact" or "closed" in the sense of their boundaries not intersecting themselves, but this need not be so (Fig. 6.1.1b,f). In partial volumes where parts of the circumscribing *boundaries intersect each other* the sense of direction is reversed (most obvious in the 2D case, (Fig. 6.1.1f) such that the sign of the calculated gravity effect of these parts is reversed, i.e. $\Delta\rho$ reverses its sign.

(7) Of course model construction can include *combinations* of all the above possibilities.

6.1.5.2 Errors of Density

The calculation of gravity effects is subject to errors related to the assumptions of densities. Because, for reasons of computational efficiency, the "calculation radius" must always be limited (depending on machines, model parametrization etc.) it is important to take into account the effects of model edges, and there can be a considerable difference whether absolute or relative densities are chosen; in the first case the edge is one between absolute density and zero, in the latter it is only between "density anomaly" and reference density which can be zero if the density anomaly is confined to the central model volume.

The mean density value depends on the "sampling distribution" within the model space; model bodies of different volumes should be weighted accordingly, but since the effects of "unit volumes" depend on their relative locations with respect to the data points, there is no simple rule for defining the "best". Data points are usually distributed irregularly. If the model effects are computed at the data point locations, this is no serious problem. For interpretative schemes involving least- squares adjustments, the point distribution affects the results, which is implicit weighting.

6.2 Two-Dimensional (2D) Approximations and Modelling

The "two-dimensional" case simpler than 3D. Though 2D is a much stronger "idealisation" than most 3D descriptions, it is "reasonable" for horizontally extended density structures if along y, little relevant changes occur and if the depth extent of the mass anomalies is much smaller than the horizontal scales of variation perpendicular to the section (see Sect. 5.3.1, Fig. 5.3.1). 2D masses (in kg/m^2) correspond to volume density (kg/m^3) in a vertical slab of unit thickness. The fundamental mass element of 2D models is the infinite horizontal mass line or the element $dm^* = \rho^* dx\,dz$ with ρ^* in kg/m^2. It corresponds to the point mass m or dm in 3D space. Integration over y from $-\infty$ to $+\infty$ has been carried out (Eq. 2.9.7). Alternatively, in the solid-angle approach, the infinitesimal horizontal ribbon of surface mass, $dx\rho^+$, can be chosen with ρ^+ in kg/m, and integration is then equivalent to the projection of the ribbon onto the unit sphere around P (Eqs. 2.9.49 & 2.9.50). The theoretical expressions for calculating 2D effects are treated in Sect. 2.9.7. Many numerical codes are freely available with instructions how to organize the data and model input. In the program INVERT (Smilde, 1998), a 2D subroutine is called PROFILE based on the "Talwani method" (2.9.7.4.3).

6.2.1 Few Large 2D Model Units

6.2.1.1 Direct Interpretation Methods with Few Large 2D Bodies

Models are made amenable to direct interpretation by drastic idealisation. The unknown quantities include mass or density, dimensions, dip angles, etc. Examples of methods are: (1) total mass anomaly, centre of gravity and higher moments; (2) the use of characteristic curves for simple models as ribbons and steps; (3) use of templates or overlay charts; (4) determination of one or a few horizontal circular mass lines; (5) determination of rectangular cross-sections as thick and thin 2D dykes; (6) the equivalent stratum (see Sect. 2.7.2) which can be calculated by downward continuation (Sect. 2.10.5.3); (7) calculation of a density contrast boundary parameterized at regular intervals. These methods are between the direct and indirect approach.

(1) *Mass, centre of gravity and higher moments* can be directly found from expansions of the gravity field into Legendre polynomials which correspond to the so-called "reduced multipole moments" of a body, depending on its shape. For a detailed account of the method the reader is referred to the text GW65, pp. 222–234, for 2D especially pp. 232–234. It involves integrations of the gravity anomaly (see Sect. 2.7.6) and interpretation on the basis of simple bodies. A practical problem is that anomalies are hardly ever isolated (see Sects. 2.10.2 and 2.10.3, 4.7.7). A recent application of the technique is presented by Sailhac & Gibert (2003) on the basis of the continuous wavelet transform (see Sect. 2.10.8).

(2) The use of *characteristic curves* for simple models as fault steps has been extensively treated by GW65, 282–287. Though 2D geometry is mostly justified, the task is complex enough. Three "free" parameters are the density contrast $\Delta\rho$, the dip angle α and the ratio of h/d ($h =$ depth of burial, $d =$ vertical offset) the forth unknown is the absolute length scale which follows from the dimension of the gravity anomaly. Relatively simple cases of this kind, taken from GW65, 294–295, are presented in Sect. 5.6.3.2 as a qualitative or semi-quantitative method.

(3) The horizontal mass line or equivalent *circular cylinder* is the fundamental 2D mass element, but nevertheless a "large body", characterized by four parameters: horizontal location x (after the x axis has been oriented normal to the cylinder axis), axial depth, cylinder radius R and density ρ. Without a priori information on R and/or ρ only three independent parameters describe the effect. As shown in Sect. 5.6.3.2, amplitude δg_0 and half width w give the parameters depth z and line density ρ^+;

$$z = w/2; \rho^+ = \delta g_0/(4G); \rho^+ = \pi R^2 \Delta\rho \qquad (6.2.1)$$

A more quantitative interpretation takes all observations into account to optimize w and δg_0 by fitting the mass line effect (Eq. 2.9.7) to the observations with the above three parameters as the unknowns. The reference gravity value can also be resolved (see Sect. 5.6.3.2).

(4) Rectangular cross-sections represent *"thick" and "thin" 2D dykes. "Thick" dykes* have thicknesses resolvable by gravity effects, otherwise dykes are called "thin". Parameters of thick dykes are the corner coordinates (again after defining the x and y axes) x_1, z_1, x_2, z_2 ($\Delta x = x_2 - x_1$, $\Delta z = z_2 - z_1$) and $\Delta\rho$. The extreme anomaly δg_e (perhaps after subtraction of a trend) may define the "location", for example, $x_0 = (x_2 - x_1)/2$. $\Delta\rho$ may be estimated in advance or found in connection with z_2, z_1 and Δx. The basic theory is treated in Sects. 2.9.3.3 & 2.9.3.4, and for the 2D situation in Sects. 2.9.7.1.2 and 2.9.7.4.2. The most important expressions are (Eqs. 2.9.50 & 2.9.51 and 2.9.59 & 2.9.60). If the body orientation is arbitrary,

i.e. not vertical, coordinate rotation must be invoked. A number of related cases (vertical and oblique dykes of finite and infinite depth extent, steps, staircases, horsts, anticlines) are treated with the appropriate formulae by KJ61, 190–199.

The *thin dyke* requires fewer parameters. Thickness, d, and density contrast, $\Delta\rho$, cannot be resolved independently: with $d = \Delta x$, the parameters are z_1, z_2 and $d\Delta\rho$. The location x of the anomaly extreme value is, again, determined first. The basic expressions are given in Sect. 2.9.3.4 (Eqs. 2.9.21 & 2.9.22) and in Sect. 2.9.7.3.1 for "hollow" rectangular pipes of which each side can be taken. Non-vertical dykes can be derived by coordinate rotation (Sect. 2.9.7.3.2).

The mathematical expressions are generally too complicated for simple direct solutions. An approach is that of (2) with characteristic curves or the design of specialized methods, for example, by diagrams which can be useful if special questions arise routinely.

(5) *Downward continuation* (see Sect. 2.10.5.3) is the calculation of the gravity anomalies at some depth below the observation level from the observed $\delta g(x, z = 0)$. The gravity variation at the lower level z, $\delta g^*(x, z)$, can then be converted to an equivalent stratum (Sect. 6.2.7.2) by the expression $\rho^*(x) = \delta g^*(x, z)/2\pi G$. With a density contrast $\Delta\rho$, $\rho^*(x)$ may then be converted to a thickness variation of a "thin" layer or to a depth variation of a density contrast surface by the relation $\delta h(x) = \rho^*/\Delta\rho$. Depth or thickness variations must remain relatively small to avoid non-linear effects, presenting a limiting condition It must be checked in specific examples. In practice, downward continuation is carried out via the Fourier transformation of given discrete gravity profiles. It involves a wave number (k) dependent amplitude amplification by e^{kz}. The profile at depth z, $\delta g^*(x, z)$, is then obtained with the inverse Fourier transformation.

(6) Other ways to compute a *single density contrast boundary* in a 2D section from measured gravity above it, are subject the same limiting conditions as above (5): the boundary must be a unique function of the horizontal coordinate x (no folding over) and the density contrast assumed must be sufficiently large to allow fitting the gravity variations within the limits of vertical freedom of the surface undulations. The maximum "uplift" will be by h, the mean depth, and the gravity effect of local depressions may be too small such that no solution may exists. The method, thus, represents a form of maximum depth rule. The parametrization can be realized by the "Talwani method" (Sect. 2.9.7.4.4; Talwani et al., 1959) where the section is approximated by a polygon of straight line segments with corner points (x, z) with the ends connected by a horizontal line (see below and Fig. 6.1.1c). The expressions for the numerical evaluation are (Eqs. 2.9.61, 2.9.64 & 2.9.65). Another parametrization is with small segments (width Δx) of the depth curve and approximate their anomalous 2D mass as rectangular 2D mass elements of $\Delta\rho\Delta x\Delta h$, where h is measured from some arbitrary constant depth (e.g. the average depth). Depending on the desired accuracy, the effect of the mass elements can be calculated, for example, by mass line approximation $\Delta m^* = \Delta\rho\Delta x\Delta h$ with the line at the element centre $(x, h/2)$: $\Delta\delta g = 2G\Delta m^*\Delta h/r^2$, where $\Delta h = (h_o - h/2)$ and $r^2 = x^2 + \Delta h^2$; or, if Δx is chosen wider, by the solid angle (see Sect. 2.9.7.2) expanded from P

to the segment Δx between x_{k+1} and x_k at depth $h/2$ from the reference depth: $\Delta \delta g = 2G\Delta m^*(\arctan(\Delta h/x_{k+1}) - \arctan \Delta h/x_k)$. Due to the non-linearity with depth and the influence of neighbouring elements, the procedure will have to be iterative (see Sect. 6.2.1.2 (5)).

6.2.1.2 Indirect Interpretation Methods with Few Large 2D Bodies

Indirect interpretation by trial and error model adjustment can be realized with any of the above parametrizations. The geometrical model description should be flexible without requiring a very large number of parameters. Rectangular cross sections may be useful in some cases. The classical method is based on the polygonal section or the "Talwani method". It permits an efficient description, and change, of arbitrary shapes by relatively few parameters (corner coordinates). "Thin" horizontal layers, also vertical or oblique "dykes" and "thick" rectangular cross sections provide practical approximations to 2D bodies. The undulated density contrast surface is well suited for the trial and error approach.

(1) The expressions for numerical evaluation of the effects of 2D *oblique steps and polygons* ("Talwani method") are Eqs. (2.9.61, 2.9.64 and 2.9.65). They contain angles and distances which generally must be calculated from coordinates (x_i, z_i) of observation points $P_i(i = 1$ to $n)$ and (x_k, z_k) of corner points k ($k = 1$ to m; where the last point $k = m$ is identical to the first point $k = 1$). Tests should always be made before "imported" routines are used for "production runs".

A specific polygon is assigned its constant density contrast $\Delta \rho$ (Sect. 6.1.5.1 and Fig. 6.1.1). The corner points are read in sequence, usually clockwise along the polygon; programming then takes care of the calculated effects δg to be positive if $\Delta \rho$ is positive and P essentially lies above the main part of the body. Changing the direction to anticlockwise, changes the sign of the effects. Complex models are built of several bodies which may be apart from each other, in contact or overlapping (see Fig. 6.1.1). Nesting or multiple wrapping (Fig. 6.1.1d) is an easy way to realize small stepwise or nearly continuous density variations, and an example is the calculation of the thermal expansivity, for example, of the cooling lithosphere at spreading ocean ridges (Jacoby & Çavşak, 2005).

(2) *"Thin" layer-like or "dyke-like" bodies* are, in a sense, large and small at the same time, large in area and small in thickness; large 2D volumes are represented by many elements. Such bodies are thus, intermediate and might be treated also as "many small bodies" (Sect. 6.2.2.2). The effects of horizontal layers (Sects. 2.9.7.1.2; 2.9.7.2) is calculated by (Eqs. 2.9.50 and 2.9.52). Irregular 2D geometries of geological bodies can be composed of horizontal layers with individual thicknesses $d_k = (z_k - z_{k-1})$, where $z_0 = 0$, increasing with depth, and defined by horizontal boundaries at, say, x_{k1} and x_{k2} for each layer k. This procedure can be modified to include definitions of x_{k1} and x_{k2} as functions of depth z or z_k, and may be incorporated into iterative algorithms of model self-adjustment akin to optimization.

The expressions needed for vertical and oblique dykes (Eq. 2.9.51) are given in Sect. 2.9.7.1.2 and in Sect. 2.9.7.3.2 (where they are parts of elements of 2D polygonal hollow cross sections and require rotation of coordinates). Especially "thin" vertical dykes (density $\Delta\rho$, thickness Δx, where only $\Delta\rho\Delta x$ is specified) may be used to define general body and layer boundaries where each "dyke" extends from some fixed bottom depth z_n (possibly $z_n \to \infty$) to the depth z_k of the boundary k. The gravity effect of such a "dyke" is, according to (Eq. 2.9.51), $\delta g_{nk} = 2G\rho\Delta x\ln(r_n/r_k)$, where r is the distance from P to the dyke top or bottom, and the effect of a layer or body bounded by z_{k-1} and z_k is $\delta g_k = \delta g_{k-1} - \delta g_k = 2G\rho\,\Delta x\ln(r_k/r_{k-1})$. The computational scheme will involve the proper definition of all body or layer boundaries k as sequences of z_k values, i.e. counted as z_{kj} by the additional counting variable j, summation for the contributions of each body or layer and summing all body effects. For "large" r values several "thin dykes" may be combined.

(3) *"Thick" rectangular cross sections* are large units of any size of which only few are needed to approximate arbitrary geological bodies. The use of rather few 2D blocks is especially suitable for the trial and error approach because the gravity effects change in a rather simple and obvious way with the possible changes of such blocks in size and location. The rectangular shape restricts the flexibility. The needed theoretical expressions (2.9.59 & 2.9.60) are derived in Sect. 2.9.7.4.2, from combinations of vertical steps (Sect. 2.9.7.4.1).

(4) *Vertical 2D templates* are mentioned, mainly for historical reasons, as tools for the estimation of the gravity effects **δg** of irregular cross sections, if not approximated by polygons. Such templates or overlay charts are centred on the observation point P and consist of a pattern of intersecting lines such that area segments or compartments are delineated. The segments are defined to contribute each the same incremental effect, say $\Delta\delta g$, if the segment is "filled" by mass, or in other words, if the template compartment covers the 2D body of density $\Delta\rho$. The contributions are counted and summed to render δg at P; the procedure is repeated at each station. Such templates (similar to the horizontal ones for the estimation of the terrain reduction, see Sect. 4.5.3.2) had been common before efficient computers became available.

A classical template (KJ61, 152) consists of equidistant horizontal lines (Δz) and radii ($\Delta\varphi$), making trapezoids contributing each $\Delta\delta g = 2G\rho\Delta z\Delta\varphi$ where $2\Delta\varphi$ represents the solid angle (see Sect. 2.9.7.2). Coarse versions of such a template can be drawn even by hand and may be used for quick semi-quantitative estimates of gravity effects (see Sect. 5.6.4). The "occupied" compartments can be counted and only partly occupied fractions can be estimated, the sum rendering the wanted effect δg at P. Another type of vertical 2D template consists of equidistant circles and radii of specially calculated angular increments around P, and similar templates exist for the higher derivatives δW_{zx}, δW_{xx}, and δW_{zzz} (KJ61, 157–161).

(5) Improvements to an *undulated density contrast profile* can be iterated with the expressions given in Sect. 6.2.1.1 (6). The convergence depends on the ratio of $h/\Delta x$ (elevation of point above contrast profile over width of 2D mass element). If $h/\Delta x < 1$, the influence of the laterally displaced elements is small and convergence will be fast.

6.2.2 Many Small 2D Model Units

Small 2D mass elements are infinite horizontal mass lines (Sect. 2.9.3.1) or thin rods, cylinders or beams of rectangular cross section. The elements are assumed so thin that only the product of density and cross sectional area will represent a line density ρ^+ (kg/m). They represent a finite approximation to integration by summation suited to the forward calculation of gravity effect of arbitrary 2D bodies but not suited to direct interpretation. Thin layers and "dykes" of Sect. 6.2.1.2 (2) ("large" elements) are in one dimension also "small" elements characterized by their surface density (kg/m^2). The advantage of the many small elements is arbitrarily high spatial resolution and flexibility, but the disadvantage is the summation of small, but very many rounding errors which must be carefully checked in all instances. The expression describing the gravity effects of line elements (Eq. 2.9.7) is the simplest, with or without "end corrections" (Eq. 2.9.8).

6.2.2.1 Direct Interpretation Methods with Many Small 2D Bodies

Large sets of elements, for example, in regular grids, would have to be strongly simplified for direct interpretation. If the individual ρ^+ values in a grid are prescribed by functions of the coordinates x, z, a continuous density variation could be approximated so that certain parameters of $\rho^+(x, z)$ might be calculated directly from given gravity anomaly values. In that sense, inversion of given gravity anomalies for the parameters (Chap. 7), may be considered a "direct" method. Another scheme may be iterative adjustment along the lines of Sect. 6.2.1.1 (6), where boundaries define regions of different line densities ρ^+ which can be shifted, for example, by addition and subtraction (annihilation) or by conversion of mass lines.

6.2.2.2 Indirect Interpretation Methods with Many Small 2D Bodies

The indirect approach of trial and error with many "small" line elements seems more "natural". However, practical routines must be designed to efficiently handle large sets of such elements when their densities are to be changed to better fit observed gravity profiles. There is, indeed, no real difference between the indirect and the direct approaches.

6.2.3 Two-and-a-Half Dimensional ($2\frac{1}{2}$ D) Models

In order to avoid the theoretically unrealistic, strictly infinite structures, end corrections (Eq. 2.9.8) for finite length of the structures can be applied, in which case the models are called "two-and-a-half dimensional" or $2\frac{1}{2}$D (KJ61, 144, 163–164; GW65, 292). They are much in use, e.g., for building starting models for a full 3D investigation. Beside the 2D cross sections of geological structures their lateral extent must be defined. For mass lines or thin rods the individual end correction are uniquely given by the density beyond the ends (see Sects. 6.1.4 and 6.1.5). For complicated, e.g., polygonal cross sections, it is not quite so simple, especially if the real structural ends are variable and not abrupt.

Free software from various sources is available in the public domain which can be found and downloaded through the internet (see, e.g. http://www.rockware.com catalog/pages/grav2dc.html, http://www.wits.ac.za/science/geophysics/software. htm by Gordon Cooper; useful for teaching purposes (P. Keating, pers. comm., 2006). Another program is "Potent": http://www.geoss.com.au/ which, after a few days, will run in a demo mode, restricted to two bodies. The result can be saved as an x, y, z file to be important into software as Geosoft; P. Keating (pers. comm., 2006) considers it one of the best modelling package. Authored by Steven Sheriff, is the 3D software http://www.umt.edu/geosciences/faculty/sheriff/Sheriff_Vita_abstracts/ Sheriff_software.htm. Internet-based information is short-lived, and the quoted webpages may be outdated or disappear any time, while new software will be offered probably in quick succession. No software should be applied blindly.

6.3   Three-Dimensional (3D) Approximation and Modelling

6.3.1   Few Large 3D Model Units

Clearly, fixed body shapes are not very suitable for accurate quantitative interpretation, since they are not flexible enough to approximate complex forms. Direct methods should render location, size, depth and possibly shape and orientation of large bodies from features of the gravity anomaly. In indirect iterative trial and error more parameters can be incorporated.

The possibilities to parameterize 3D model geometry are treated in Chap. 2 and expressions are presented for polyhedra, rectangular prisms, dykes, horizontal layers bounded by depth contours, vertical cylinder ring sectors. 2D Fourier transforms or wavelets and downward or upward continuation may be applied.

The above parametrizations can be adapted to direct or indirect interpretation strategies. Direct methods require fixing of many model parameters to leave only very few free if volume description requires too many parameters. Parametrization by polyhedra by which any geometry can be approximated, requires compromises between resolution and economy and is better suited to indirect methods. The opposite is true for specialized, more idealized forms as, for example, rectangular prisms or cuboids; though less flexible in shape, they permit direct determination of quantities as depth, depth extent and lateral dimensions and are economically described, possibly in combination with smaller elements of the same type. In smallbody parametrization, more and more of these may be combined to larger ones, depending, for example, on distance from the observation point.

6.3.1.1 Direct Interpretation Methods with Few Large 3D Bodies

Any geometrical parametrizations, to be capable of direct parameter determination, must be simplified to leave only one or few parameters free. Several interpretational diagrams by GW65 are expressedly not very suitable. There are two exceptions to this with some practical value.

(1) On the basis of the integral relationships (Sect. 2.7.6), GW65, 222–234, propose the expansion of given anomalies $\delta g(x,y)$ into *Legendre polynomials* (see Sect. 2.10.6.1) which can be related to characteristic quantities of 3D mass distributions as total mass and mass moments. These can be related to certain simple geometrical shapes as rectangular prisms and tri-axial ellipsoids. This theoretically interesting aspect has not gained practical importance, probably because geological density distributions fit such simple models only poorly; it is also due to the country rock inhomogeneneity. Moreover, in exploration too much idealization is not useful. The total mass, say, of an ore body, $M_{\mathrm{ore}} = \rho_{\mathrm{ore}}V = M_{\delta g}(1 + \rho_0/\Delta\rho)$ follows from the gravitationally identified total anomalous mass $\Delta M_{\delta g} = V\Delta\rho$ where ρ_0 is the density of the country rock.

(2) *Diagrams* for evaluating limited dykes are extensively treated by GW65, 273–282. Free parameters are, beside location and dimension (scale), density times thickness $\Delta\rho^* = \Delta\rho d$, the ratio $h/l(h = $ depth to top, $l = $ length of dyke), dip angle α, and horizontal length y^*. The diagrams shown by GW65 suggest that they are applicable only in special situations.

(3) Depth and mass of large spheres and/or of the equivalent *point masses* can be determined directly from anomaly amplitudes and half widths (Sect. 5.6.3.1), and the horizontal coordinates follow from those of the anomaly centre of gravity. The corresponding equations are given in Sect. 5.6.3.1. Separation of several such bodies is, however, ambiguous (see Sect. 4.7.7).

(4) "Thick" and "thin" laterally limited dykes are 3D *rectangular prisms or cuboids* (Webster's New Collegiate Dictionary: "*Math.* A rectangular parallelepiped") equivalent to 2D "thick" rectangular cross sections and "thin" dykes. The "thin" case is treated in (Sect. 6.3.1.1 (2)), the "thick" case is hardly amenable to direct interpretation.

(5) A more attractive possibility of direct 3D interpretation exists for *smooth undulations of density contrast layers*, e.g. in sedimentary basins. Application of downward continuation of gravity anomalies (Sect. 2.10.5.3) in two dimensions and of the equivalent stratum analogy makes it possible to calculate undulations for given density contrasts. The ambiguities, for example, in the presence of several density contrast surfaces, can be reduced if borehole density logs are available (see GW65, 255–263).

(6) A *single density contrast surface* can also be calculated in 3D, as in the 2D case (Sect. 6.2.1.1 (6)), under the same limiting conditions. The direct method is, by definition, to calculate only one improvement to the initial surface, which can be a step in an iterative procedure, as sketched in the section on indirect methods (Sect. 6.3.1.2 (5)). In analogy to the 2D situation (polygonal section, "Talwani method" (Sect. 2.9.7.4.4; Talwani et al., 1959), parametrization of the

density contrast surface can be in the form of polyhedra with plane triangles between the given points (x, y, z) ("triangulation") expanded from a horizontal polygonal rim. However, only one or very few points can be adjusted directly at a single step to better fit the observations. The expressions for the numerical evaluation are Eqs. (2.9.41, 2.9.42, 2.9.43).

Another possibility is to take small segments, for example, squares of dimension a^2 and to approximate their anomalous mass as rectangular discs of $\Delta\rho\Delta h a^2$, Δh is the depth improvement. It is assumed that the gravity anomaly values δg form a regular grid with the points centred above the square mass elements at an elevation h (which may vary from point to point). The "linear" simplification may be that the effect of the surface improvement is limited in the calculation to only that of the directly subjacent square, for which $\delta g = G\Delta\Omega\Delta\rho\Delta h$, where $\Delta\Omega$ depends on h and a, i.e. $\Delta h = \delta g/(G\Delta\Omega(h,a)\Delta\rho)$. For a "reasonable" range of h/a values, $\Delta\Omega(h,a)$ can be approximated on the basis of Eq. (2.9.34) in Sect. 2.9.4.4. If $h << a$, $\Delta\Omega \to 2\pi$, and $\Delta h \approx \delta g/(2\pi G\Delta\rho)$, for $h > a$ or $h >> a$, a better approximation would be

$$\Delta h \approx \delta g(4/3)(h^2/a^2)/(\pi G\Delta\rho) \tag{6.3.1}$$

(derive this approximation in Task 6.1) but with increasing h, the influence of the neighbouring mass elements becomes detrimental to the purpose of direct interpretation. (see Sect. 6.3.1.2 (5)).

(7) For any task of routine application it may be useful to develop *special direct methods*. As an example, diagrams were developed for cone-shaped craters as diatremes and maars (Sebazungu, 2005) described by parameters as depth of burial, vertical extent, inclination of the walls, top and bottom radius (where these parameters are not mutually independent). Input parameters are amplitude of δg, half width or some equivalent quality etc. Although the directly determined values are only preliminary, they may serve as a priori information for a more detailed optimization (Chap. 7).

6.3.1.2 Indirect Interpretation Methods with Few Large 3D Bodies

Indirect interpretation by trial and error cannot be standardized for the determination of depth, shape and density of 3D mass anomalies. The analytical expressions for the forward calculations are presented in Sect. 2.9.6. For some purposes, graphical methods with templates were used before the advent of efficient computers (see Sect. 6.1.4). Methodological possibilities are briefly sketched here. The most flexible parametrizations, suitable for analytical and numerical evaluation and approximation of arbitrary shapes are probably the polyhedra (Sect. 2.9.6.2) and stacks of horizontal polygonal discs (Sect. 2.9.4.2) by which given contour lines can be exploited; for special cases, as "thin dykes" of laterally limited extent, equations for planar elements (Sects. 2.9.3.3 & 2.9.3.4) can be derived by coordinate rotation (Sect. 2.4.3.1). Cuboids and other regular (Sect. 2.9.6.1) bodies are less flexible to fit realistic 3D shapes. Cylinders or cones can be taken for crater-like bodies.

(1) *Massive polyhedra* (Sect. 2.9.6.2) with arbitrary complexity are generally applicable. One way is to first derive a set of vertical polygonal sections of anomalous masses from geology or geophysical models. Then triangulation can connect the sections. The gravity effects are finally calculated with expressions given by several authors (IGMAS: Götze & Lahmeyer, 1988; Çavşak, 1992; Holstein et al., 1999, Holstein, 2002a, b). Such methods permit a highly detailed description of 3D shapes, but they require large numbers of geometrical parameters (coordinates) and the sensitivity of the gravity effects to details and changes in detail may be low. Furthermore, detailed parametrization leads to the numerical evaluation of very many, very small contributions to the total gravity effect of a polyhedron, such that rounding errors may become a problem (see Holstein et al., 1999). Large numbers of parameters restrict the possibilities of formal inversion (Chap. 7).

(2) *"Thin" layers* as horizontal sections through a body or layers are especially suited for elevation or depth contours, for example, for caylculating terrain effects (Sect. 4.5.3.2) or from 3D seismic reflection studies, which, however, are affected by inaccurate velocities. The classical method is the approximation of the boundary by a polygon (Sect. 2.9.4.2), and the calculation is done for straight line segments (Eq. 2.9.30; Talwani & Ewing, 1960; GW65, 302–303). For some smooth curved contour lines that can be described mathematically, analytical integration of Eq. (2.9.33) may be possible. Alternatively, for contours given by sets of x, y points, approximation can also be done by computing the azimuth λ from P to any x_i, y_i and $\Delta\lambda$ consecutively. For each triangle, thus defined, mean $z/r(\lambda)$ values (Eq. 2.9.33) are taken (instead of averaging r or R). Problems arise for points P near a contour where special measures have to be taken Jacoby (1967). Earlier methods (Goguel, 1961; Baranov, 1953, see also KJ61, 168–170) are based on the geometry of the solid angle and corresponding diagrams.

The effects of oblique finite rectangular "thin" bodies or dykes can be calculated from Eqs. (2.9.18 & 2.9.20) for the plane-normal and plane-parallel components, respectively. The complete set of x, y, z components of the gravity effects is given in (Eqs. 2.9.66 & 2.9.67). This applies to vertical dykes, as demonstrated in (Eqs. 2.9.62 & 2.9.63), and requires coordinate rotation (Sect. 2.4.3.1) for oblique dykes. Coordinate rotation is illustrated in the case of the oblique polygon (Sect. 2.9.4.3) and is to be applied to a special rectangular plate (Eqs. 2.9.31 & 2.9.32). An equivalent expression is also provided by GW65, 226 (8–24).

(3) Large *rectangular prisms* (expressions given in Sect. 2.9.6.1) offer another way of composing 3D models, however, less efficient for approximating geological bodies. Trial and error in changing the prisms will concentrate on few big prisms. A special form of bodies with round outline and steep walls is the vertical circular or more general cylinder.

(4) *Templates* have been presented e.g. by KJ61, 155–160 (where the theoretical background and applications to estimating δg and some higher derivatives are derived). The templates can be used to "manually" or "graphically" calculate effects of irregularly shaped bodies. The templates consist of intersecting lines such that compartments are formed which have identical "incremental" gravity effects $\Delta\delta g$. "Graphic" means that the body must be drawn, for example, in sections, and

"manual" means that template compartments are counted, in which the body (i.e. density $\Delta\rho$) is present. Application to 3D problems is akin to, but more complicated than, 2D templates (Sect. 6.2.1.2 (4)).

(5) The direct calculation of improvements to an *undulated density contrast surface* (Sect. 6.3.1.1 (6)) is also the basis of trial and error. The convergence depends on the ratio of h/a (a = side length of square mass element, h = elevation of point above element). If $h/a < 1$, the influence of the laterally displaced elements is small and convergence will be fast, but it slows or deteriorates with h growing to $h/a \gg 1$.

(6) Some interesting formulae for *special geometries* are documented by KJ61, 154–155, 189–190; they include circular disks and rings, cones and the paraboloid. The solutions are based on the solid angle as described in Sect. 2.9.1.1 and some of them are treated in Sect. 2.9.3. They are simple only for axial observation points, otherwise they lead to elliptical integrals.

6.3.2 Many Small 3D Model Units

Small elements have advantages and disadvantages for modelling and may be rectangular prisms, small thin mass plates, thin vertical rods, spheres, mass points, etc. Mathematical expressions for their individual gravity effects may be simple or involved (see Sect. 2.9). The numerical routines can be economized in many ways. Instead of calculating a mathematical expression for each element, it is feasible to first calculate tables of the effects versus some parameters; in the actual computation, they can be interpolated from the table which may be even more accurate than evaluating complex expressions each time.

Generally a compromise has to be found between local resolution and efficiency. Elements close to a station P can be subdivided and elements far removed from P may be combined. Elements may be skipped altogether, or different expressions can be used. It is necessary to analyse the errors of the computations, especially rounding errors, with the correct expressions and with various possible approximations. Approximations must be checked for their systematic errors in connection with the large sums involved. Exact total mass of an element and depth of its centre of gravity are the most critical aspects. Handling many mass elements requires organisation of the input, for example, of density contrast boundaries, such that each element is characterized correctly and/or can be easily changed.

6.3.2.1 Direct Interpretation Methods with Many Small 3D Bodies

"Direct methods" in the classical sense cannot handle large numbers of parameters. The only thing which may be called "direct" is "automatic" algorithms leading to a model without human interference and which might also be considered an automated "indirect" method. *Monte Carlo* search might also be included (Sects. 2.10.9; 7.3.3.2).

6.3.2.2 Indirect Interpretation Methods with Many Small 3D Bodies

Also for manual trial and error, elements must not be chosen too small. To be tractable, handling of many small elements must be programmed. Elements can be changed for density by random number routines applied to regions of density contrasts in a way to "move" the contrast boundaries, and large sets of models may be generated from which only "good ones" are accepted. Two possibilities to parameterize 3D models are briefly discussed: (1) point grids and (2) grids of vertical rods. Other possibilities can be invented with semi-small mass elements, as layers (see 2D case, Sect. 6.3.1.2 (2)) defined by their limiting contours.

(1) *Point grids* can realize any resolution, replacing integration by finite element summation. The elementary prisms $\Delta x \cdot \Delta y \cdot \Delta z$ may be approximeted by point masses m (Sect. 2.9.2.5) where density is given by $\rho_k = m_k/(\Delta x \cdot \Delta y \cdot \Delta z)$ or the full theory of rectangular prisms (Sect. 2.9.6.1) can be applied. With efficient computers, the possibly large numbers of elements need not present a problem. The computation of the point effects δg_i can be speeded up by applying efficient interpolation with functions in the parameter space of distance r and "sight angle" φ. Densities can be assigned by functions of the coordinates, $\Delta\rho(x,y,z)$. The gravity effect at an observation point $P(x_i,y_i,z_i)$ is then obtained by numerical convolution of $\delta g(\Delta x_{ik},\Delta y_{ik},\Delta z_{ik})$ with $\Delta\rho(x_k,y_k,z_k)$ where $\Delta x_{ik}=x_k-x_i$, etc. Spectral convolution methods (Sect. 4.7.4.2) permit fast calculations. The spectral treatment offers upward or downward continuation (Sects. 2.10.5.3 and 6.2.1.1 (5) & (6)), especially for layers at $z=z_l$ of the present 3D grid where density is given by $\Delta\rho(x,y,z_l)$.

(2) Arbitrary bodies can be described also by *vertical rods* along which the density varies, for example, in discrete steps. Assume that in a suitable model "box" all relevant density contrasts $\Delta\rho_k$ versus a common background density ρ_o are described by a regular x, y grid of vertical rods or mass lines: $(x_i=i\Delta x, y_j=j\Delta y; i=1$ to $n, j=1$ to m); each rod intersects the density contrast surfaces at a sequence of z_{ijk} values ($j=1$ to p) where all $\Delta\rho_k=\rho_{k+1}-\rho_k$ values are specified. If at some location a layer k (boundaries k, k − 1) is missing, $z_k=z_{k-1}$. This permits a unique treatment as long as there are no overturned folds. Cases where a given density contrast is met by a rod more than once have to be treated separately.

The fundamental mass element for the calculation is the rod extending from depth z_k to infinity: according to (Eq. 2.9.10) $\delta g^{(1)}=G\rho^+/r_o$ where ρ^+ is the line mass (kg/m) and $r_o=(\Delta x^2+\Delta y^2+\Delta z^2)^{1/2}$ (Δ stands for the difference between rod top and P) is the distance between the top of the rod and P. The expression is simple, and the computational efficiency can be increased further by tabulating r values versus $\Delta x, \Delta y$ and Δz and interpolating them. In the "far field" elements can be combined. In the "near field" a refinement of the standard rods is achieved by interpolating neighbouring z_{ikj} values onto a finer grid, so that the errors of the mass line approximation never exceed a preset threshold. Efficient interpolation methods or fitting z_k surfaces to the z_{ijk} values can be applied. Alternatively, the nearby rods my be replaced by rectangular prisms.

6.4 Summary: Strategies of Model Building; from Trial and Error to Inversion

Quantitative interpretation requires calculation of gravity effects of arbitrarily shaped bodies to any desired accuracy. However, interpretation goes beyond that. The aim is to limit the "model space" to geologically probable models and their error bounds. Here strategies and methods to approach or approximate geological structures are treated. The emphasis is on the possibilities that exist to build the available parametrizations into interpretation schemes. Several programs are available, often in the public domain. Here, the basis for understanding existing programs, but also for designing new methods is provided.

Errors or limits of certainty characterize probability distributions; they are often assumed to be Gaussian. What is known is the goodness of fit composed of the residuals encountered in modelling and interpretation; they contain data and model errors. Direct methods largely ignore the residuals; they usually render as many model parameters (e.g. mass and depth) as data points are used (e.g. amplitude and half width), but they are selected from larger sets of observations. If averages are used also their standard deviation can be estimated. The indirect methods of trial and error largely build on experience and the "error" or "misfit" of the residuals are taken as the basis for subsequent trials. Visualization of the residual patterns and a priori knowledge of geology are important. Trial and error can be idea provoking and leads to optimization and inversion toward the minimum sum of squared residuals or to Monte Carlo methods. It is always an advantage if the interpreter combines the "manual" approach with mathematical methods. Human creativity is in many respects superior to algorithms and machines. Apparently small standard errors of the a posteriori model variables are *not a sufficient* condition for the results to correctly describe the real geology, or in other words, precision is not accuracy.

6.5 Examples

Examples explain concepts better than any theoretical discourse. Qualitative and semi-quantitative interpretation (Sect. 5.7) prepared the way to quantitatively models with more detail to be derived with the outlined methods. The cases of Sect. 5.7 (except Sect. 5.7.9) are taken up again with an equivalent numbering, mostly by trial and error (since optimization and inversion are left to Chap. 7). The examples are chosen mainly from published work, with the aim to give insights into how the models were derived, without using "fully computerized" and "automated" methods, and they are only sketched, and readers are referred to the original work and references cited therein. Classical trial and error does not hide the working mechanisms. To narrate the search can be instructive. Obviously one generally goes from simple models with few components to more and more detail and complexity, following geometrical, physical and geological guiding principles (Sects. 5.3, 5.4, 5.5).

The aspects treated in the present chapter may differ in direction and emphasis from the previous chapter. Two examples will be followed up in the last chapter.

6.5.1 Messel Maar Crater and Fault Zone (MFZ)

The Messel pit of oil shale, NE of Darmstadt, Germany, was formerly mined for energy and chemical exploitation and is now a UNESCO world heritage site for its excellent Eocene fossils. Figure 5.7.1a shows the *BA* as embedded into the regional field which reflects the Messel Fault Zone (see below). Bituminous laminated sediments fill an approximately sigmoid hole in Variscan crystalline rocks of $0.7 \times 1\,\mathrm{km}^2$ dimension with apparently steep walls. At the time the geophysical investigations started, the origin of the oil shale basin was debated, even its nature as a crater. The power of gravity as a qualitative exploration tool was demonstrated as the anomaly, complemented by magnetic and other data, could hardly be interpreted other than by a maar-like crater (Jacoby et al., 2001, 2003), which was subsequently proven by drilling.

The Messel oil shale pit and the question of its origin were introduced in Sect. 5.7.1. The aim of the gravity study was to clarify the nature and origin of the basin: a relic of a once much larger lake? But of steep-walled crater shape? Of volcanic or impact origin? Gravity has the potential to reveal shape and depth extent, and accompanying magnetic measurements can test for volcanic products as magnetized crater fill and relic tuff deposits.

A few more facts;

- 1 km wide, rather deep "basin" filled with lake sediments deposited under anoxic conditions about 50 Ma ago in Variscan country rocks;
- laminated sediments, at depth with clastics, to 230 m depth below the original surface;
- shallow features and Quaternary sediments removed by erosion and mining;
- located on the Messel Fault Zone (MFZ) expressed as a linear gradient zone (Fig. 5.7.1b) accompanied by mapped volcanics;
- observed negative gravity anomaly of about −7 mGal (Fig. 5.7.1a) coinciding with the basin, accompanied by a smooth negative magnetic total field anomaly.

Qualitative estimates of Sect. 5.7.1 pointed to a maar crater on the MFZ (oil hale density contrast $-1300\,\mathrm{kg/m}^3$ and thickness $\sim$130 m (below the pit bottom), and the existence of a fault zone expressed by the linear north-easterly trending gravity gradient of 15 mGal drop to the NW. Quantitative gravity modelling and interpretation were carried out by Moos (1994). 2D and 3D modelling was combined to simultaneously fit the Messel anomaly (1) and the gravity gradient zone (2). (1) To test the crater hypothesis, the location, depth extent and dimension of cylinder-shaped low-density bodies (Sects. 6.3.1.2 (3), 2.9.2.2) were adjusted to fit gravity as well as the surface and drilling information. Several stacked cylinders of radii, decreasing with depth, can approximate the crater shape, and the density could be varied from one to the next to simulate downward densification or decreasing negative density contrast, but such a variation was hardly resolvable in view of the correlation between radius and density. Cylinders were parameterized as polygonal horizontal discs that could approximate circular, elliptical, sigmoid and other shapes. Figure 6.5.1 presents an acceptable example: three nearly co-axial circular cylinders have a density contrast versus country rock of $-1170\,kg/m^3$ (a priori estimate: -1300) and a total thickness of 170 m. The MFZ was fitted by a fault dipping $\sim 50°$ NW and extending to 560 m depth between two blocks with a density drop of $600\,kg/m^3$ to the NW (a priori 1 km vertical fault, $360\,kg/m^3$; Sect. 5.7.1). However, geological evidence is for a much smaller fault drop. The problem was resolved by introducing a steep narrow body of slightly reduced density $(-100\,kg/m^3)$ traversing below the Messel crater and extending with SW-NE-strike to several kilometres depth; the low-density body separates slightly denser rocks $(+100\,kg/m^3)$ to the SE from those to the NW, on which a layer of Permian sandstones rests, which are $250\,kg/m^3$ less dense and, on average, a few tens of meters thick, as revealed by local shallow drill holes. This latter model is not shown in the figure.

The magnetic anomaly (minimum $-300\,nT$) was modelled by Laubersheimer (1997) with the cylinders and it was found that an inversely magnetized body below

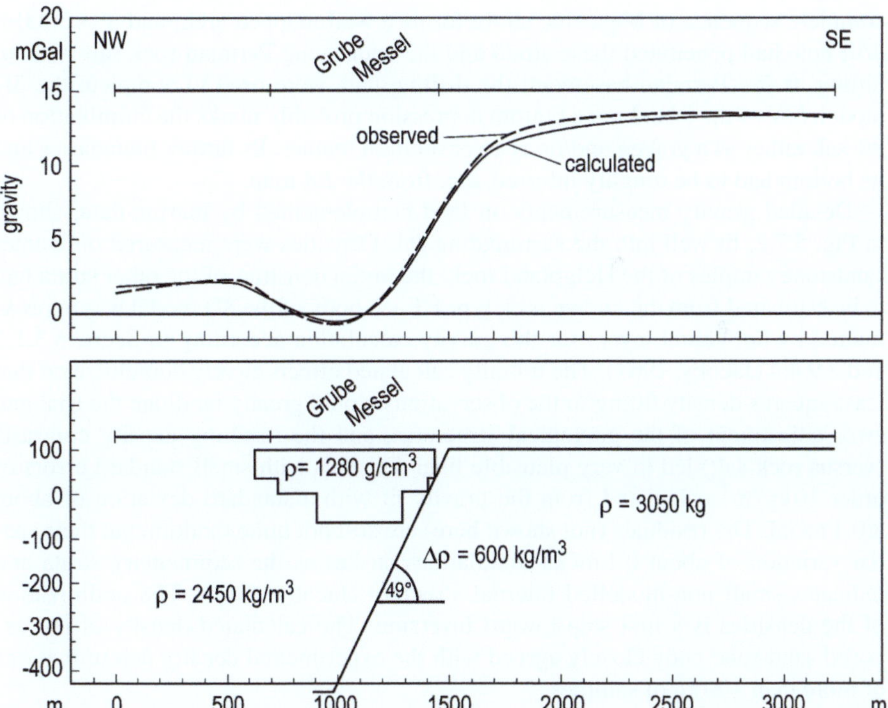

Fig. 6.5.1 A NW-SE section through the Messel Maar crater and across the Messel Fault Zone (MFZ). The crater is represented by three not quite coaxial cylinders (shown in axial sections). The density contrasts were calculated by LSQ adjustment; after Moos (1994)

the sediments would fit the data, but the shallow non-magnetic hole of the crater embedded in positively magnetized country rock could similarly fit the data, if the magnetisation is tapered, for example, by deep weathering of the wall rocks. The models can be tested by deeper drilling. A research well was drilled near the centre (UTM coordinates R31.30, H82.75; in Fig. 6.5.1 at $x \approx 1120$ m) and it did hit the modelled magnetized tuff body at approximately the predicted depth and the diatreme breccia below, the latter at a depth of about 400 m under the surface. Quantitative gravity and magnetic interpretation of the observed anomalies, complemented by geological observations, served well in the investigation of, and essentially demonstrating, the existence of a maar-like crater (Jacoby et al., 2001, 2003).

6.5.2 Salt Structure of Helgoland

For the simple estimates on the salt structure under Helgoland of Sect. 5.7.2 a point or spherical mass was assumed, and arguments for a 2D triangular cross section were given. But enough information exists such that a more quantitative fairly detailed 3D interpretation can be attempted by trial and error, supported by a LSQ density optimization. The island is built of NE dipping Bunter sandstone. An incomplete sequence of Mesozoic strata forms a well mapped arch, and a > 3000 m drill hole had penetrated these strata and the underlying Permian rock salt without hitting its Pre-Permian basement; the drilling data were used to complete the 3D model. SW of the island a sea bottom depression probably marks the culmination of the salt either as a graben and/or as an exsolution feature. Its farther boundaries and its bottom had to be roughly inferred, e.g. from the *BA* map.

Detailed gravity measurements on land complemented by marine data, shown in Fig. 5.7.2, fit well into the surrounding *BA*. Densities were measured on Bunter sandstone samples of the Helgoland rock; the initial densities of the other strata had to be estimated from the known rock types. Each body of the 3D model was approximated by horizontal layers for the gravity calculation according to Sects. 6.3.1.2 and 2.9.4.4 (Jacoby, 1967). The initially calculated effects clearly demonstrated that least-squares density fitting to the observations would greatly facilitate the trial and error adjustment of the geological structures, and the resulting density contrasts (versus rock salt) led to very plausible final densities with small standard errors of order $10 \mathrm{kg/m}^3$ calculated from the gravity fit with a standard deviation of about ± 0.1 mGal. The residuals (not shown here) are still not quite random, but their spatial variation of about 0.1 mGal undulations strikes as the sedimentary strata and indicates small non-modelled internal structure (Jacoby, 1966). The optimization of the densities is a first step toward inversion. The calculated density of the exposed sandstone body closely agreed with the experimental density determinations of more than 100 kg of samples.

The final 3D block model of the arched Mesozoic succession is presented in Fig. 6.5.2 based on the gravity modelling and on the geological surface and drilling data. The succession above the Permian salt is Bunter sandstone (the deepest of

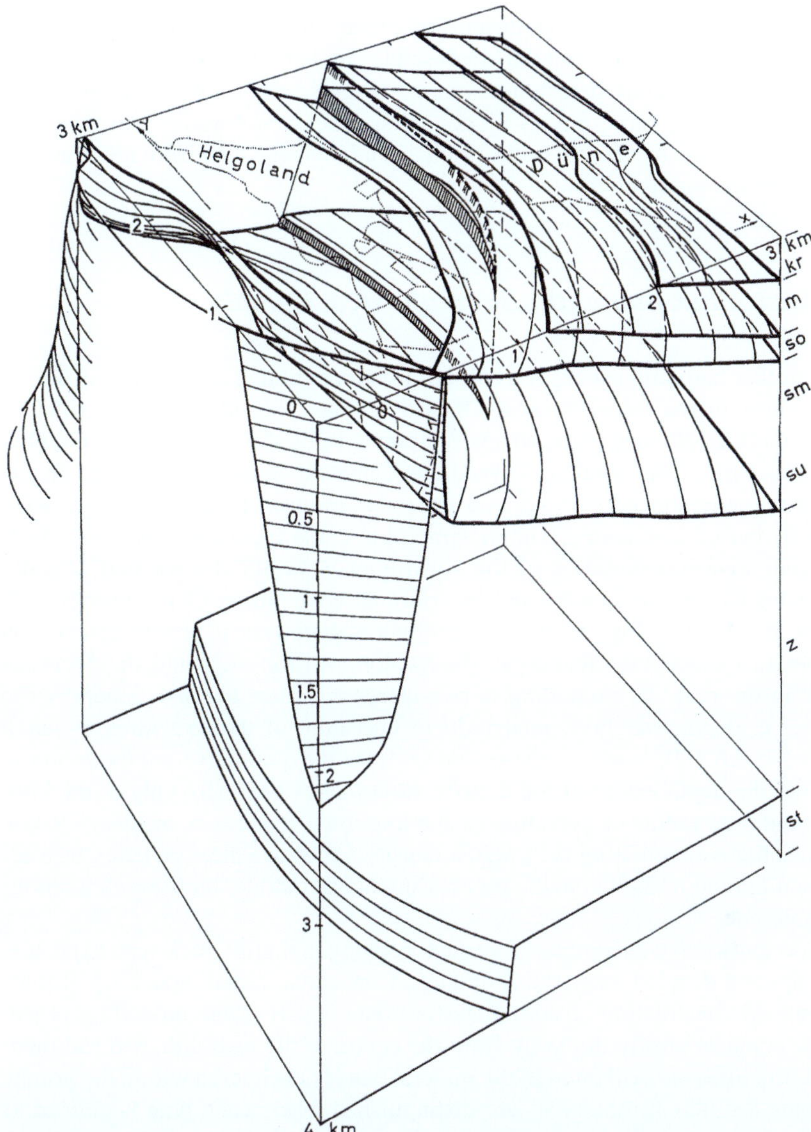

Fig. 6.5.2 3D block diagram of the uplifted supra-salt dome sedimentary structure of the Helgoland area; sub-salt structure hypothetical. Data: surface geology, drill hole data. Legend – kr: Cretaceous; m: Middle Triassic limestone *Muschelkalk*; so, sm, su: Upper, Middle, Lower Bunter sandstone, respectively; z: Permian salt *Zechstein*; Düne: sand dune island; after Jacoby (1966)

which is exposed in the island), "Muschelkalk" (Middle Triassic limestone) and Cretaceous limestone (the latter representing the basement of the separate island of sand dunes, called "Düne"). Upper Triassic and the Jurassic are missing. The model is fairly detailed as adjusted by quantitative trial and error fitting. The one acceptable model found is not unique and the whole model space remains unknown. A hypothetical sub-salt fault and the inaccessible "main fault" SW of Helgoland are uncertain; but Cretaceous limestone exposed at the seafloor SW of Helgoland at the edge of the depression, mostly hidden by sediments and water, proves a large vertical offset versus the Bunter rocks of the island. Seismic data were not available at the time of the interpretation.

6.5.3 Anstruther Batholith: Bancroft Area, Canada

The Anstruther batholith is part of a SW-NE trending granitic arch of gneiss domes and batholiths in the Bancroft area, Ontario, Canada, introduced in Sect. 5.7.3, where a simplified *BA* map of the area was presented (Fig. 5.7.3; W. Jacoby, unpubl., 1970). This part of the southern crystalline Canadian Shield was formed in the Grenville orogeny about 1 Ga ago, and the rocks exposed in the area today were then deeply buried and heated. The metamorphic grade increases westward. Here, a 2D quantitative interpretation of the Anstruther batholith is presented as a detailed geological section constructed by trial and error fitting of the gravity profile (Sect. 6.2.1.2 (1); Fig. 6.5.3); the geological-geophysical profile extends on both sides to include the Glamorgan gneiss dome in the west and the Methuen batholith in the east. 2D modelling is considered adequate because generally the ratio z/r, i.e. depth over horizontal radii of curvature of the structures, is small (Sects. 5.3.1; 5.6.3.2); based on the estimates from the qualitative interpretation in Sect. 5.7.3, the depth extent of the density variations is generally only a few kilometres, while the radius of curvature of the important features is mostly >10km. But in quantitative modelling the gradual density variations must be taken into account from granite to the Grenville metasediments as marble, biotite-rich gneisses and amphibolite.

A priori knowledge encompasses surface geology including rock type exposure, contact dips and detailed magnetic anomalies. Conceptual initial modelling of folds or "downward construction" from the surface data is part of the modelling procedure. The contacts mostly dip away from the centre of the batholith, and rudimentary fold structures are evident as mafic mineral bands or schlieren within the granite body. Some 8 different rock types are distinguished where each type is defined as one uniform unit, though often geometrically disjoint. Gravity effects are calculated with the Talwani method (Sects 2.9.7.4.3 & 2.9.7.4.4). The density contrast between granite and the majority of the wall rocks is initially assumed to be about $\Delta\rho \approx -150$kg/m^3 with variations of up to ±100kg/m^3. The initial density values are adjusted by linear least-squares fitting the observations. The a posteriori absolute densities ρ_k(post) are computed from the adjusted density contrasts $\Delta\rho_k = \rho_k - \rho_o$, the a priori densities ρ_k(pre) (e.g. from measured samples) and the mean reference density $\underline{\rho}_o = \Sigma\rho_{ko}/m$ where m is the number of independent densities and the differences between the $\rho_{ok} = \rho_k$(pre) $+ \Delta\rho_k$(post) result from errors of the ρ_k(pre) and the fact that all are referred to a common background ρ_o (Sect. 6.1.5.1 (2)).

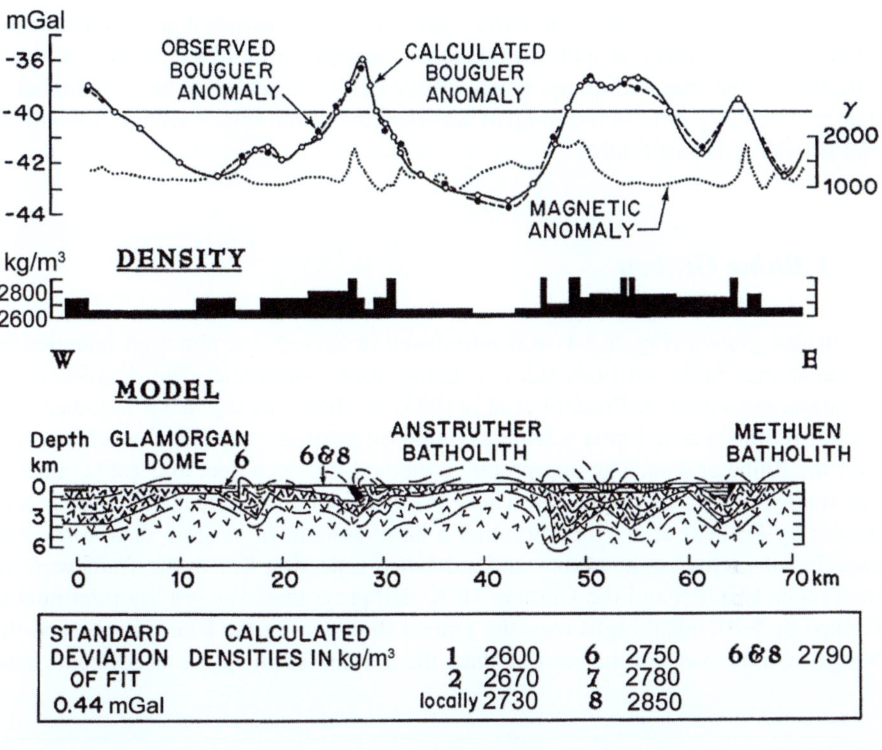

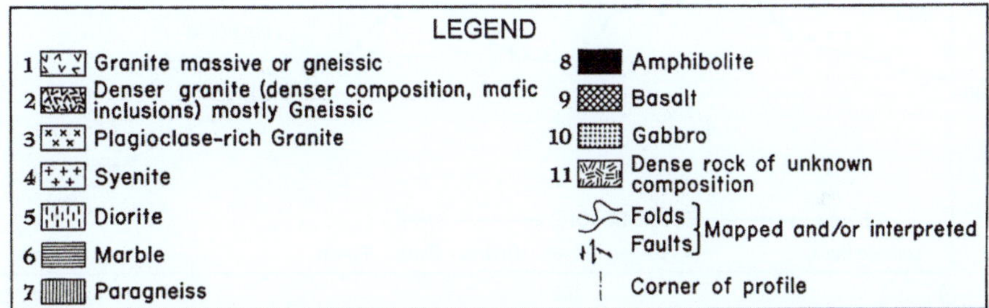

STANDARD DEVIATION OF FIT 0.44 mGal   CALCULATED DENSITIES IN kg/m³

1 2600	6 2750	6&8 2790
2 2670	7 2780	
locally 2730	8 2850	

LEGEND

1	Granite massive or gneissic	8		Amphibolite
2	Denser granite (denser composition, mafic inclusions) mostly Gneissic	9		Basalt
3	Plagioclase-rich Granite	10		Gabbro
4	Syenite	11		Dense rock of unknown composition
5	Diorite			Folds } Mapped and/or interpreted
6	Marble			Faults }
7	Paragneiss			Corner of profile

Fig. 6.5.3 *Upper* crustal section across the Anstruther batholith, Bancroft area, Ontario, Canada. Data: geological maps, densities from rock samples, gravity measurements, partly unpublished; only the rock units 1, 2, 6, 7, 8 of the legend occur in the section shown. The density results have errors of $< 50 \, \text{kg/m}^3$, but their mean relative differences are probably more accurate

The section shown in Fig. 6.5.3 is the result of some 5–10 trial and error iterations. The observations are fitted to within $< 0.5 \, \text{mGal}$. The granitic bodies appear to rise from a deep "layer" and are therefore termed "batholiths"; only in the eastern greenschist terrain, some bodies have the appearance of magmatic intrusions into greeschist facies cover rocks (not in the section). Generally the metasediments appear to be pendants into a continuous "granitic layer", reaching no more than $\sim 2 \, \text{km}$

depth, but the density transition or the "halo" with intermediate densities may reach about 6 km. In agreement with the qualitative interpretation (Sect. 5.7.3), the depth variation of the metasediments is suggestive of the structure to be an undulating density boundary with two dominant wavelengths of the major granite updomings and smaller scales of folding.

6.5.4 Rhine Graben

The Rhine graben (Fig. 5.7.4) was introduced in Sect. 5.7.4; although bounded by lateral master faults on both sides, it is not quite symmetric. The detailed interpretation, presented by Prodehl et al., (1995), is shown in Fig. 6.5.4 together with the *BA* profile which forms a wide asymmetric positive "arch" and a relative low over the Rhine graben. The interpretation method is that of Sect. 6.2.1.2 (1) including studies in 2D, $2\frac{1}{2}$D and 3D. The a priori information comes from references quoted by Prodehl et al., incorporating a large amount of seismic data, refraction experiments, teleseismic studies and a detailed reflection line in the framework of the French ECORS and the German DEKORP programs. The gravity minimum is flanked by 5–10 mGal highs over the graben shoulders of the Black Forest and the Vosges. On the basis of the seismic data, the Tertiary Rhine graben sediments have

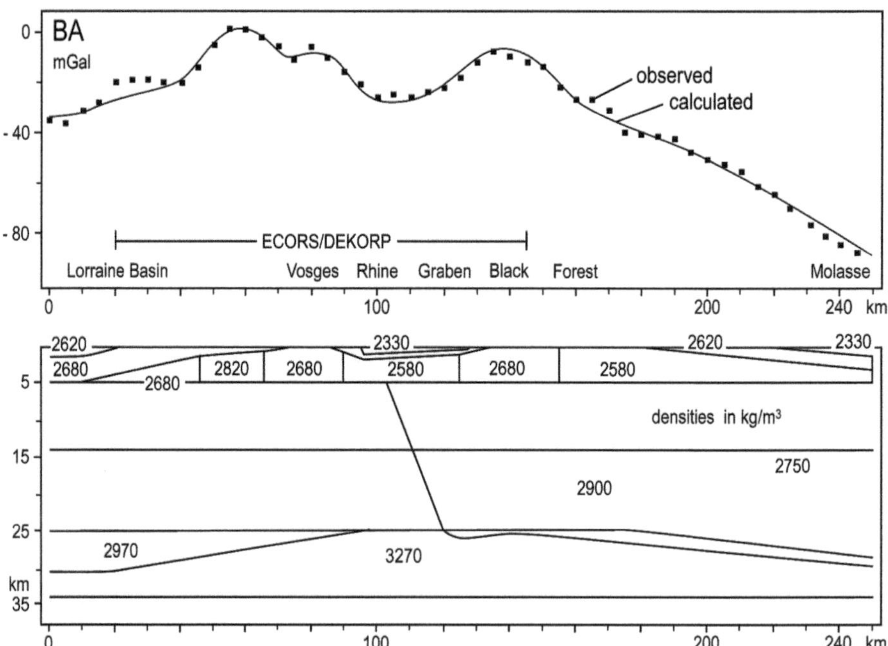

Fig. 6.5.4 Crustal section across Rhine graben at latitude near Strasbourg. Structure based on geological and seismic information (after Prodehl et al., 1995)

a thickness of only 1–1.5 km (instead of the 2.4 km estimated in Sect. 5.7.4 on the basis of a single density contrast of $-200\,kg/m^3$). A stronger density contrast of about $-300\,kg/m^3$ and additional 1 km of pre-rift sediments explain the difference. The uppermost crustal density undulations of about $\pm100\,kg/m^3$ about an average of $2660\,kg/m^3$ are partly based on geological observations, partly guessed to fit the corresponding *BA* undulations.

Outside the graben, the Moho had been estimated in Sect. 5.7.4 from gravity alone to drop 2.5 km to the west and >6 km toward the Alps. It is now (Fig. 6.5.4) taken from the seismic data to drop from 25 km below the graben to 30 km in the west and to 32 km toward the Alps. The main discrepancy with the previous estimate in the west is explained by a hypothetical high-density lowermost crustal body. Generally, this model is founded on more data than that of Sect. 5.7.4. The interpretation does not invoke a low density mantle body below the graben which had been postulated earlier to exist. The interpretation is a case of classical trial and error fitting with a large amount of guiding data.

6.5.5 *The SE Iceland Shelf Edge*

As outlined in Sect. 5.7.5, the gravity anomalies, i.e. *FA*, also *BA* and especially the residual *BA, rBA* (Moho-effect subtracted), all display a dipolar feature at the SE Iceland shelf edge with a high above the shelf break and a low above its bottom. In the semi-quantitative estimate it was treated as an edge effect of a sharp vertical 2D boundary between two adjacent crust-upper mantle structures of balanced vertical density distributions (Fig. 5.7.5, see Sect. 5.6.8). The data sources and geological background information are given by Fedorova et al. (2005). The 2D models presented are adequate since the SE Iceland shelf is very nearly straight. In this chapter, several experimental initial models, based on different geological concepts are investigated with the method outlined above in Sect. 6.2.1.2 (1), and in Sect. 7.4.2 one of these concepts is subjected to a more detailed inversion.

Beside the transition from land to sea, the isostatic edge effect fits the *rBA* data rather convincingly (Sect. 5.7.5). But the separation of effects of the Moho and the edge is not unique. It can be taken as a starting model of a more thorough quantitative interpretation. Several other geological scenarios are sketched in Fig. 6.5.5. A non-isostatic edge (1) is fitted to the *rBA*, and the Moho model and water body are shown only for comparison. (2) Another geological idea is shelf erosion, for example, by glacial abrasion and corresponding uplift of a positive density gradient, evident in an enhanced average crustal density, coupled with sediment deposition at the foot of the shelf slope; the *BA* has been fitted to this model including the Moho. (3) Densification of the upper crust might also result from increased volcanism during a phase of increased plume activity, resulting in forward building of the shelf, again, coupled with light oceanic sediments; here *rBA* was fitted, and a small adjustment of the Moho density contrast was permitted. (4) Finally a non-isostatic or uncompensated edge flexure and rotation is shown, possibly resulting from torques

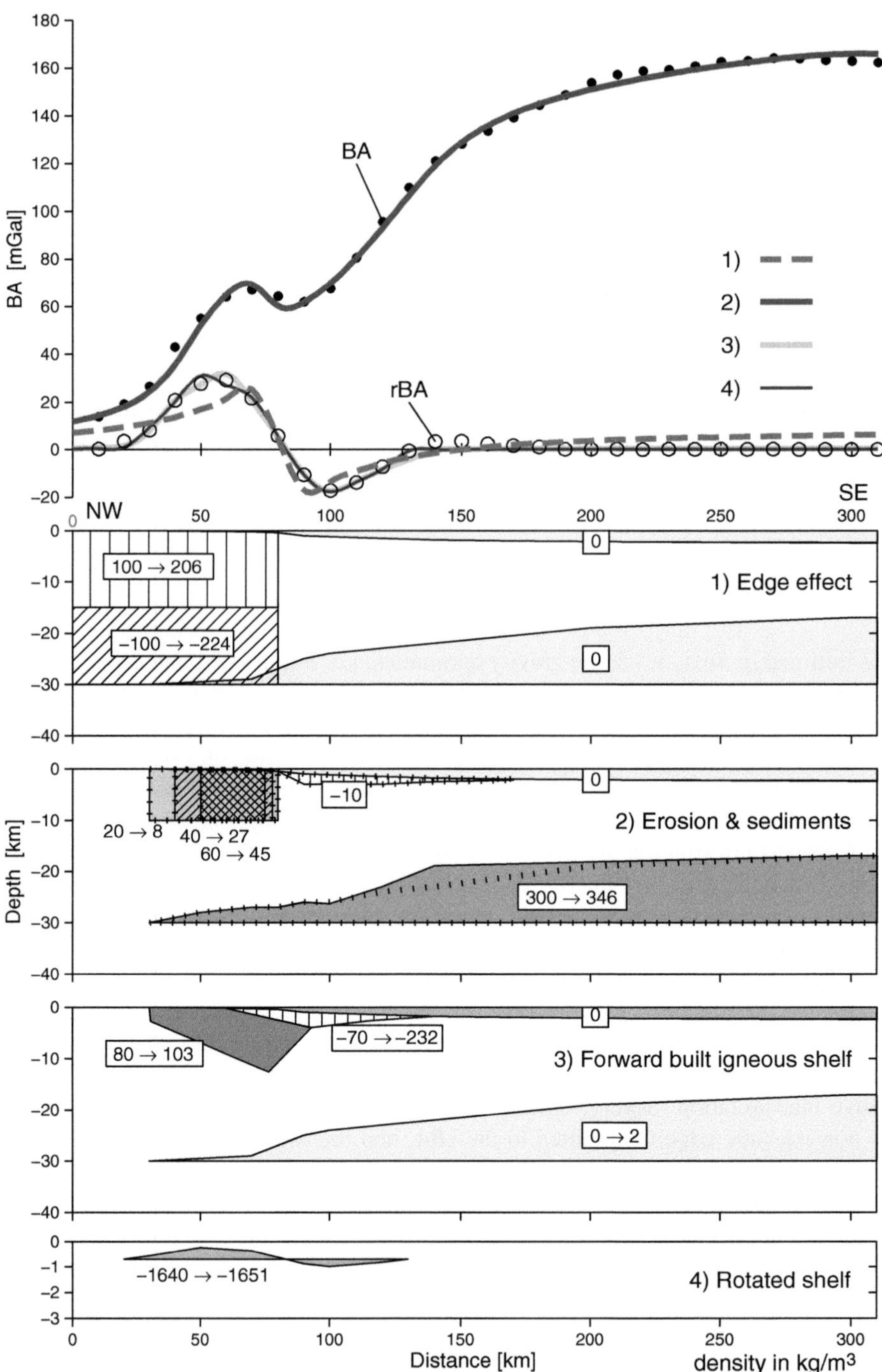

Fig. 6.5.5 (continued)

inherent in the lateral density variation (Jacoby, 1978) and uplifting the outer shelf and slightly depressing the foot of the slope; the *rBA* was fitted. LSQ adjustment of the density contrasts was applied speeding up the trial and error process. These models can be complemented by other ideas of geological evolutionary processes. Traditional trial and error interpretation cannot fully explore the wide range of modelling possibilities and does not solve the non-uniqueness problem. It becomes also obvious that more thorough a priori information is needed.

6.5.6 Spreading Ridges

Before the advent of seafloor spreading, ocean ridges were considered submarine mountain belts with crustal roots, but then it was realized that hot, low-density material rises diapir-like and explains the ridge topography (Jacoby, 1970). In Sect. 5.7.6, a qualitative estimate for Reykjanes Ridge of a density anomaly of $-30\,\mathrm{kg/m^3}$ for an assumed 100 m depth extent was interpreted with a temperature anomaly of about $+300\,\mathrm{K}$. Quantitative gravity modelling (2D: Sect. 6.2.1.2 (1)) has to take into account additional a priori information, mainly data on bathymetry and from seismic studies of crustal structure which shows the Moho to rise towards the ridge axes, instead of dipping with a thickening root. The Reykjanes Ridge was investigated by many workers (see Jacoby et al., 2007, and references quoted there). Here an interpretation of gravity across two ridges is reproduced: the rather slow Mid Atlantic Ridge (MAR: Fig. 6.5.6a), and the fast spreading East Pacific Rise (EPR: Fig. 6.5.6b). The 2D model of a triangular-shaped low-density body of hot rising asthenospheric mantle material from under the diverging cooling and thickening plates was adopted and adjusted by trial and error (Jacoby, 1978, where the available a priori information is quoted). Again, 2D is well justified, although the structures reach more than 50 km depth. Fitted was a modified *BA* reduced for crustal structure, i.e. crust was, so to speak, first "filled up" to mantle density by calculation (Sect. 4.5.3.1). The resulting gravity anomaly should essentially reflect the effect of the asthenospheric wedge.

The shape of the asthenosphere wedge is taken as an equivalent to the density distribution inherent in the isotherms of cooling plates, which should be quite similar in both ridges if normalized to a common spreading rate. It is the physically correct temperature model (McKenzie, 1977, but the asthenospheric wedges is a suitable equivalent. For the MAR with a wedge of 60 km height $\Delta\rho$ was found

---

Fig. 6.5.5 The SE Iceland shelf, compare Fig. 5.7.5. Four different model types portraying different geological processes which might have formed the shelf. (1) Non-isostatic edge fitted to the residual Bouguer anomaly, *rBA*; Moho and water body shown for comparison; (2) fit of *BA*: glacial abrasion causing uplift and enhanced average crustal density, coupled with sediments at the foot of the shelf slope; (3) fit of *rBA*: dense upper crust due to increased volcanism and forward building of shelf, coupled with light slope sediments; (4) fit of *rBA*: fully non-isostatic edge bending or rotation, uplifting the outer shelf and depressing the slope (see text)

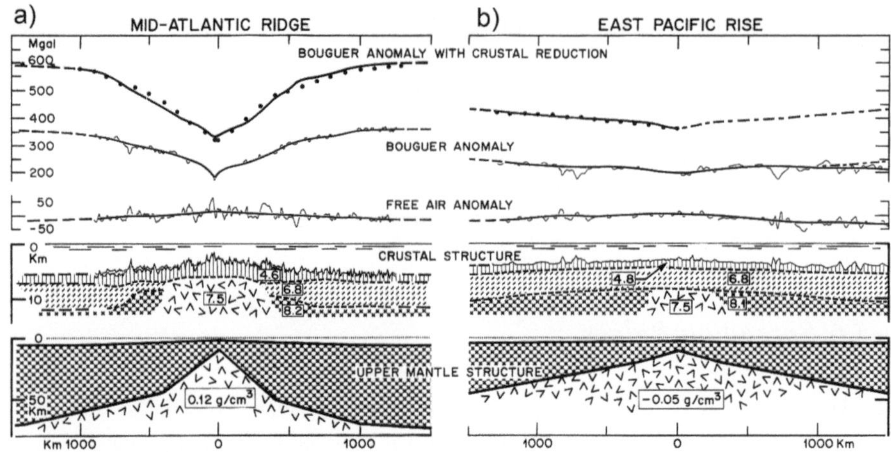

Fig. 6.5.6 Crust-upper mantle sections of spreading ridges, (**a**) Mid Atlantic Ridge (MAR), (**b**) East Pacific Rise (EPR); the sections are based on crustal seismic information (in boxes: seismic P velocities in km/s) and on the concept of thickening lithosphere and rising asthenosphere (after Jacoby, 1975)

to be about $-120\,\mathrm{kg/m^3}$; this is a high value which, for a height of 100 km, would be reduced to about $-70\,\mathrm{kg/m^3}$, still exceeding estimates for the Reykjanes Ridge. For the EPR a much lower density contrast of $-50\,\mathrm{kg/m^3}$ for the central wedge with, however, only 30 km height was obtained (only $-15\,\mathrm{kg/m^3}$ if the height were 100 km).

The discrepancies between the preliminary estimate of Chap. 5 and the present models reflect the limitations of rough estimates, but the differences between the ridges seem substantial enough to be significant. The MAR and EPR are different, for example, in divergence rate, plume occurrence and dynamics. The slow spreading Atlantic is characterized by many near-ridge plumes that inject hot and possibly volatile-rich material into the asthenosphere, thus enhancing the melting and the density deficit, while the fast spreading Pacific is also driven by slab pull such that the asthenospheric upwelling might lag behind. The physically more adequate model of the lateral cooling density anomalies (anomalous isotherms; McKenzie, 1977) is treated by Jacoby & Çavşak, (2005).

6.5.7 Mantle Plumes

Mantle plumes are, by definition, narrow 3D structures rising from great depth, many from the core-mantle boundary at 2900 km depth, but even recent tomographic pictures are not yet focussed enough to be conclusive (Montelli et al., 2004; Nolet et al., 2005). The main problem is that their associated anomalies are hardly known because they cannot be clearly isolated from the effects of other inhomogeneities and at the surface the effects of their deepest parts are too weak and spread out too much. The "plume signal" shown in Fig. 5.7.7 is an average found by deconvolution for a global set of 33 hotspot locations (thought of as delta functions). It is positive at the centre with a negative ring around, indeed suggestive of a plume dipole. In Sect. 5.7.7 it was qualitatively compared with the theoretical isostatic ("Airy") plume "dipole" of Sect. 5.6.9.3.

In the earlier qualitative interpretation attempt (Sect. 5.7.7) it was argued that both the central gravity high and the negative rim are broader than the theoretical effects of the "Airy plume dipoles" of section Sect. 5.6.9.3, and in addition, both are embedded in a wide positive gravity anomaly. Furthermore the central high is much lower than shown in Fig. 5.6.5b. Plume models should be based on reliable tomographic images and/or on dynamic models rendering the dynamic surface uplift associated to positive gravity effects which, however, depends on poorly known rheological parameters. A preliminary 3D model in Cartesian coordinates is taken to fit the gravity "signal" of Fig. 5.7.7. The mass balance between the plume mass deficit and the uplifted mass excess need not be realized in a ductile dynamic system in a continuous process of adjustment. The following model components deviate from the above simple model: (1) the uplifted topography is much lower on average and spread out laterally to, say, 500 km from the centre; (2) since a deep plume itself cannot explain the broad negative rim, a plume head is assumed to extend laterally, or flow out, from its centre, arbitrarily assumed to a distance of $r = 1000$ km, 50 km thick and with a density contrast of only -3 kg/m^3 (but only the product of thickness and density contrast can be significant); (3) the plume itself extends arbitrarily from 100 to 600 km depth with a radius of 100 km and a density contrast also of -3 kg/m^3, however, its influence is negligible (see Sect. 5.6.9.3); (4) an arbitrary "deep" mass has been assumed at 1000 km depth to fit the very broad gravity high which may be related to the plume clustering. It should be emphasized that the model components partly compensate each other in their effects, as inherent in the dipole nature (causing problems), and that any details of the surface-near mass distribution, normally considered in the Bouguer reduction, are ignored here.

A reasonably fitting model (Fig. 6.5.7) has some similarity with the Iceland plume as imaged by seismic tomography (Wolfe et al., 1997; Foulger et al., 2000), except for its additional plume head and the "deep positive mass" which both cannot be "seen" in the seismic images. The numerical values of the model parameters are shown in the figure and given in its caption. The subdued and spread-out uplift and the plume head seem to remain somewhat relevant; any topography caused by increased volcanism is totally ignored. The model is not "mass balanced" if the "deep mass" is neglected.

6.5.8 Tonga-Kermadec Trench, Subduction and Back Arc Basin

A gravity profile across the Kermadec trench, island "arc" and back-arc basin was introduced in Sect. 5.7.8 and shown in Fig. 5.7.8 with a much simplified model. The

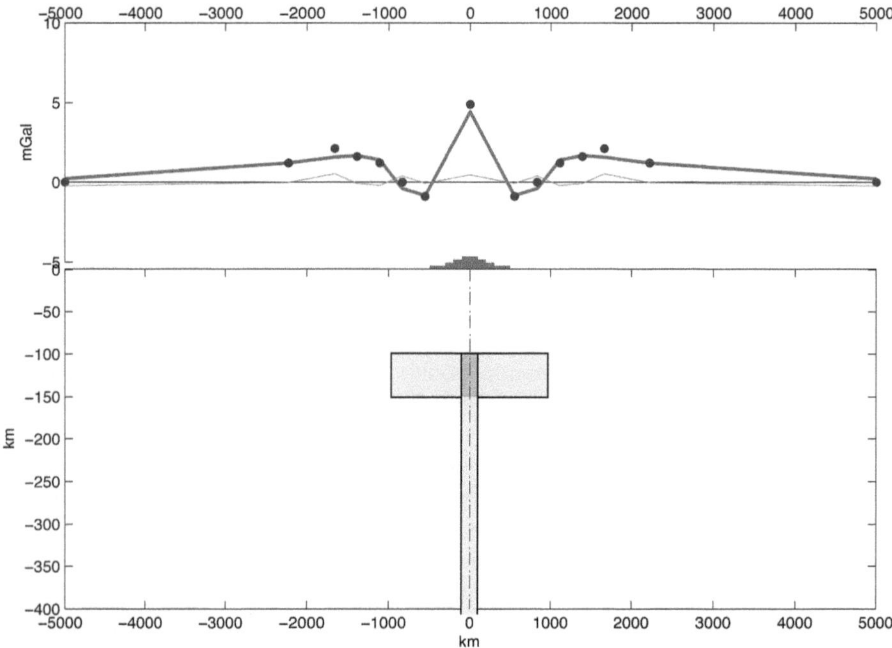

Fig. 6.5.7 Simplistic Cartesian plume model (shown in 10-fold vertical exaggeration to only 400 km depth) which can approximately explain the average plume signal of Fig. 5.7.7 (13 points). The model consists of a vertical circular cylinder of density contrast $\Delta\rho = -3\,\mathrm{kg/m^3}$, radius $r = 100\,\mathrm{km}$, depth extent from $z = 100$–$600\,\mathrm{km}$, which has a head, also cylindrical ($\Delta\rho = -3\,\mathrm{kg/m^3}$, extending laterally to $r = 1000\,\mathrm{km}$ and in depth from $z = 100$–$150\,\mathrm{km}$); where the two bodies overlap, the density contrast is $\Delta\rho = -6\,\mathrm{kg/m^3}$. The uplifted topography (grey, schematic, vertical exaggeration 250 times that of the model), assumed to be radially spread out in conical shape, is modelled as a stack of 4 circular disks ($r = 500, 300, 200, 100\,\mathrm{km}$; each $\Delta z = 15\,\mathrm{m}$ thick and of density $\rho = 2500\,\mathrm{kg/m^3}$). In addition, a "deep" mass (not shown) is modelled as a 2D body of triangular cross section at about $1000\,\mathrm{km}$ depth and of density contrast of about $0.5\,\mathrm{kg/m^3}$; it has no significance other than to fit the broad 2 mGal gravity high around the "mean plume"

descending slab was assumed to be unknown (in the early 1960s) and the positive *FA* and especially the *BA* was qualitatively interpreted by a mass line, directly rendering its depth and the mass excess. The quantitative interpretation includes the a priori knowledge of the subducted slab (Fig. 6.5.8). Also included is the back-arc basin underlain by thin lithosphere and by hot low-density mantle upwelling as well as bathymetry and crustal structure. An early interpretation by a 2D model (Jacoby, 1975), based on references quoted there, serves here as an example for modelling by trial and error (2D approach of Sect. 6.2.1.2 (1)). A satisfactory fit to the available data was achieved. The lower part of Fig. 6.5.8 shows, with no vertical exaggeration, the slab and diapir structure assumed in a section; the top part shows the *FA* and *BA* (Fig. 5.7.8). The slab geometry is constructed from the deep seismicity; the overlying diapir was not geometrically well constrained, but its existence and approximate extent was inferred from strong attenuation (low Q) and

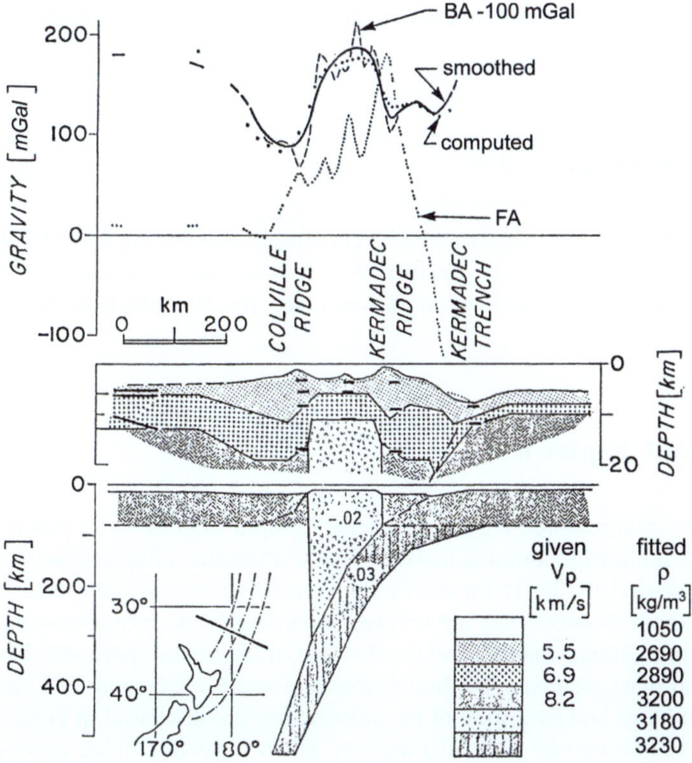

Fig. 6.5.8 Upper mantle and crustal section across the Tonga-Kermadec subduction zone; data basis: slab constructed from intermediate and deep seismicity, lithospheric thickness assumed; low Q, low velocity, low density diapir from seismic data, crust from a few seismic refraction lines (after Jacoby, 1975, and references quoted there)

low seismic velocities. The crust, shown in the middle part of the figure with tenfold vertical exaggeration, is based on seismic refraction experiments.

The model consists of four model units: upper, lower crust, slab and diapir, and the geometry was adjusted in several trial and error steps and alternating linear least-squares fitting to the *BA* for the density contrasts $\Delta\rho$ versus "mantle". The geometrical adjustments in fitting gravity generally amounted to $<4\,$km vertically in the crust and up to 15 km in the slab and diapir. The results fit the observations to about $\pm 5\,$mGal (5% of the *BA* variation) which was considered adequate. From the density contrasts total densities were calculated: 2700, 2900 and 3200 kg/m^3 (with error bounds of <100 kg/m^3) for upper and lower crust and uppermost mantle, respectively; the slab and diapir density contrasts versus mantle came out as +30 and -20 kg/m^3, respectively (error bounds <10 kg/m^3). The example demonstrates that quantitative gravity interpretation with well constrained a priori information can lead to dynamically relevant density models which, in this case, have stood the test of later studies fairly well. The Cartesian 2D geometry of a 500×1000 km^2 section embedded in a spherical earth with a radius of only about ten times the model dimensions causes no gross errors.

6.5.9 Mantle Convection

Mantle convection, qualitatively discussed with very simple estimates in Sect. 5.7.9, is not followed up in this chapter more quantitatively. It would require extensive dynamic convection modelling in spherical coordinates (see, for example, to Schubert & Turcotte, 2001).

6.6 Summary of Chapter 6

Quantitative gravity interpretation requires accurate forward calculation of gravity effects of assumed, arbitrarily complex models. The mathematical expressions derived in Chap. 2 provide the basis for the computational strategies and methods described in the present chapter. They are ordered according to the geometrical aspects of two- and three-dimensionality and the division of the model space into few large or many small mass elements. Methodologies and strategies emphasized and organization, assemblage and execution of the calculations are discussed. It is considered important to concentrate on model aspects which have significant gravity effects and are thus be meaningful. Modelling of subtle details is not given priority, although there are no principle limits to including details. It is a question of economy of quantitative modelling and interpretation.

Available software for 3D, 2D and $2\frac{1}{2}$D gravity (and magnetics) modelling, and examples are quoted in the text. They can be profitably used, if care is taken and the limitations are considered. Application of programs without basic understanding of the underlying principles easily leads to mistakes. Users should do tests. Some aspects of the modelling are illustrated by examples, mostly in 2D which is easier to imagine than 3D (which is also not more quantitative than 2D). Some possibly practicable methods have been described which have never been realized (programmed) and tested but may be developed in the future. Optimization and inversion are the ultimate aim of most gravity studies and the subject of the following and final chapter.

References

Baranov, V.: Sur le calcul de l'influence gravimétrique des structures définies par les isobathes. *Geophys. Prosp.*, 1, 36–43, 1953.

Bosum, W.: Anlage und Interpretation aeromagnetischer Vermessungen im Rahmen der Erzexploration. *Geol. Jahrbuch.*, E20, 63pp., 1981.

Çavşak, H.: Dichtemodell für den mitteleuropäischen Abschnitt der EGT aufgrund der gemeinsamen Inversion von Geoid, Schwere und refraktionsseismisch ermittelter Krustenstruktur. *Ph.D. thesis [unpubl.], Mainz*, 1992.

Fedorova, T., Jacoby, W.R., Wallner, H.: Crust-mantle transition and Moho model for Iceland and surroundings from seismic, topography, and gravity data. *Tectonophysics*, 396, 119–140, 2005.

Foulger, G.R., Pritchard, M.J., Julian, B.R., Evans, J.R., Allen, R.M., Nolet, G., Morgan, W.J., Bergsson, B.H., Erlendsson, P., Jakobsdottir, S., Ragnarsson, S., Stefansson, R., Vogfjord K.: The seismic anomaly beneath Iceland extends down to the mantle transition zone and no deeper. *Geophys. J. Int.*, 142, 1–9, 2000.

Goguel, J.: Calcul de l'attraction d'un polygone horizontal de densité uniforme. *Geophys. Prosp.*, 9, 116–127, 1961.

Hahn, A., Bosum, W.: Geomagnetics – Selected Examples and Case Histories. Geoexploration Monographs 1/10, *Borntraeger, Berlin*, 166pp., 1986.

Götze, H.-J., Lahmeyer, B.: Application of three-dimensional interactive modeling in gravity and magnetics. *Geophysics*, 53, 1096–1108, 1988.

Holstein, H.: Gravimagnetic similarity in anomaly formulas for uniform polyhedra. *Geophysics*, 67, 1126–1138, 2002a.

Holstein, H.: Invariance in gravimagnetric similarity in anomaly formulas for uniform polyhedra. *Geophysics*, 67, 1134–1137, 2002b.

Holstein, H., Schürholz, P., Starr, A.J., Chakraborti, M.: Comparison of gravimetric formulas for uniform polyhedra. *Geophysics*, 64, 1438–1446, 1999.

Jacoby, W.: Schweremessungen auf Helgoland – Auswertung mit Ausgleichsverfahren. *Z. Geophysik*, 32, 340–351, 1966.

Jacoby, W.: Zur Berechnung der Schwerewirkung beliebig geformter Massen mit digitalen Rechenmaschinen. *Z. Geophysik*, 33, 163–166, 1967.

Jacoby, W.R.: Instability in the upper mantle and global plate movements. *J. Geophys. Res.*, 75, 5671–5680, 1970.

Jacoby, W.R.: Velocity-density systematics from seismic and gravity data. *Veröff. Zentr. Inst. Phys. Erde, Nr.* 31, 323–333, 1975.

Jacoby, W.: Lateral density variations and isostasy – or isostasy with balance of torques. *Geoskrifter 10, Aarhus University*, 75–93, 1978.

Jacoby, W.R., Wallner, H., Smilde, P.: Tektonik und Vulkanismus entlang der Messel Störungszone auf dem Sprendlinger Horst: Geophysikalische Ergebnisse. *Z. dt. Geol. Ges.*, 151, 493–510, 2001.

Jacoby, W.R., Mertz, D., Wallner, H.: Students' field research extends knowledge of origin of a UNESCO world heritage site in Germany. *EOS Transactions Am. Geophys. Union.*, 84, 39, 393, 396–397, 2003.

Jacoby, W.R., Çavşak, H.: Inversion of gravity anomalies over spreading oceanic ridges. *J. Geodynamics*, 39, 461–474, 2005.

Jacoby, W., Weigel, W., Fedorova, T.: Crustal structure of Reykjanes Ridge near 62°N, on the basis of seismic refraction and gravity data. *J. Geodynamics*, 43, 55–72, 2007.

Jung, K.: Schwerkraftverfahren in der Angewandten Geophysik. *Akad. Verlagsges. Geest & Portig, Leipzig*, 348pp., 1961 [KJ61].

Laubersheimer, F.: Vermessung magnetischer Anomalien über Ölschieferbecken bei Messel: quantitativer Vergleich verschiedener Vorkommen auf dem Sprendlinger Horst. *Master's thesis [unpubl.], Mainz*, 1997.

McKenzie, D.P.: Surface deformation, gravity anomalies, and convection. *Geophys. J. Roy. Astron. Soc.*, 48, 211–238, 1977.

Montelli, R., Nolet, G., Dahlen, F.A., Masters, G., Engdahl, E.R., Hung, S.H.: Finite-frequency tomography reveals a variety of plumes in the mantle. *Science*, 303, 338–343. PMID 14657505, 2004.

Moos, C.: Gravimetrische Untersuchung zur Frage der Genese von "Messel". *Diploma thesis [unpubl.], Mainz*, 1994.

Nolet, G., Karato, S., Montelli, R.: Plume fluxes from seismic tomography (abstract). *Geophys. Research Abstracts*, 7, 04221, 2005.

Prodehl, C., Mueller, St., Haak, V.: The European Cenozoic rift system. In "Continental Rifts: Evolution, Structure, Tectonics", ed. K.H. Olsen, 133–212, *Elsevier, Amsterdam*, 466pp., 1995.

Schubert, G., Turcotte, D.L.: Mantle Convection in the Earth and Planets. *Cambridge Univ. Press*, 940pp., 2001.

Sebazungu, E.: Investigations of maar diatreme volcanoes by inversion of magnetic and gravity data from the Eifel area in Germany. *PhD thesis [unpubl.], Mainz*, 2005.

Sailhac, P., Gibert, D.: Identification of sources of potential fields with the continuous wavelet transform: Two-dimensional wavelets and multipolar approximations. *J. Geophys. Res.*, 108, B5, 2262, doi:10.1029/2002JB002021, 2003.

Smilde, P.L.: Verwendung und Bewertung von a priori Information bei potentiell singulären Inversionsproblemen am Beispiel der gravimetrischen Bestimmung von Dichteverteilungen. *Deutsche Geodät. Komm., C 490, München*, 1998.

Talwani, M., Ewing, M.: Rapid computation of gravitational attraction of three-dimensional bodies of arbitrary shape. *Geophysics*, 25, 203–225, 1960.

Talwani, M., Worzel, J.L., Landisman, M.: Rapid gravity computations for two-dimensional bodies with application to the Mendocino submarine fracture zone. *J. Geophys. Res.*, 64, 49–59, 1959.

Wolfe, C.J., Bjarnason, I.Th., VanDecar, J.C., Solomon, S.C.: Seismic structure of the Iceland mantle plume. *Nature*, 385, 245–247, 1997.

Chapter 7
Optimization and Inversion

7.1 Introduction

We recapitulate: gravity interpretation aims at models of mass distributions on the basis of observed gravity anomalies or data. "Model" stands for "image" or "virtual representation of reality" in the widest sense with a mixed spectrum of meanings which humans form in their minds for lack of "absolute" knowledge of reality.

Gravitational model building requires the mathematical tools of forward calculating gravity effects of given mass bodies: Newton's law of gravitation (Chap. 2), and strategies of defining acceptable model geometries. The classical strategy is trial and error (Sect. 6.1.2), where models are constructed from the available information, the model effects, δg_m, are calculated, and the differences or residuals, $r = \delta g_o - \delta g_m$ (5.1.5) from the observations, δg_o, will iteratively guide the search. They are reduced by appropriate changes of the model parameters. The procedure is subjective and limited, but so is also the choice of mathematical algorithms. Errors of data and a priori information lead to conflicts, especially if the data sources are of different kinds, but the principal ambiguity makes a priori information mandatory, which depends on the task, in geodesy to efficiently describe the external gravity field, in geophysics to obtain information on geological structures. Here the emphasis is on the geological-geophysical aspects.

Optimization and inversion are envisaged as the essence of gravity interpretation. Both terms have similar meanings and are not always distinct. We understand optimization of model parameters as the effort to reconcile conflicting evidence by minimization of discrepancies by, for example, scaling, en bloc shifting and turning of model bodies (see below). This implies that a problem is overdetermined: more than the minimum information exists required for a solution, thus giving insight into, and increasing, its reliability or accuracy. In view of our principal ignorance, however, more unknowns exist than data, and the relationships between unknowns and observations must be simplified to enable a solution to be obtained. Information, here, is mainly gravity, supplemented by additional observations. Optimization means that all available information is combined into a "best model". The least-squares (LSQ, see below) solution is the best linear and unbiased estimator (BLUE) of the true situation. It is an approximation since the observations and the model relationships

W. Jacoby, P.L. Smilde, *Gravity Interpretation*,
© Springer-Verlag Berlin Heidelberg 2009

represented by the functional matrix A (see below) deviate from the true situation by statistical scatter.

Inversion of anomalies, on the other hand, is defined as the direct mathematical construction of the most acceptable models, which includes also their optimization. Especially interesting variables are introduced in order to determine their values. Inversion is an extension of the so-called direct methods of interpretation (Chaps. 5 and 6) which are based on idealized a priori definitions of mass distributions with parameters directly calculable from a number of certain characteristic data. They include simple estimates (Sect. 5.6) or more sophisticated methodologies (Sects. 6.2 and 6.3). Generally, however, conflicting information exists which gives insight into how much confidence one can have in the results, and the data have uncertainties suggesting weighting schemes and ideas about the "field" or the "anomaly" that is defined by the discrete observations, including the space between them: is it totally undefined or a "smooth" continuation?

Gravity inversion and optimization converge. One may consider inversion as a subclass of general optimization and the variables of inversion are thus a subclass of those of optimization. Generally optimization concerns any quantities supplemented in order to let the observations fit. In contrast, for inversion the chosen quantities are the main interest. Bayesian inversion starts from a priori information with its own error bounds which are treated like the gravity observations. Just like the open question of what gravity teaches about the real mass distribution, inversion is principally under-determined, and variables are introduced in order to find out if they are real, how large and how certain.

In starting with Newton's law of gravitation (Eq. 2.5.5):

$$\delta \mathbf{g} = \int_V d\mathbf{g} = G \int_V \Delta\rho \, (\mathbf{r}/r^3) dV,$$

we note that the fundamental mass element $dm = \Delta\rho dV$ contains density ρ or $\Delta\rho$, and the integral is, in this sense, linear in density, however with an infinite number of infinitesimal mass elements. which cannot be generally resolved for the density distribution. The task can be achieved only if a finite number of mass elements or parameters that describe the mass distribution and sufficient observations are linked by systems of linear equations of the form (see below: Sect. 7.2.1):

$$\underline{\mathbf{A}}\mathbf{x} = \mathbf{y} \text{ or } Ax = y \text{ or } y = A(x) \tag{7.10.1}$$

with the formal (LSQ) solution

$$x = (A^T A)^{-1} A^T y \tag{7.10.2}$$

A or $A(x)$ represents "model relationships", x stands for the m unknown densities and y for the n observations; x and y are described as vectors (usually written as x and y). A and x can be generalized to include other parameters than densities.

Optimization is treated before inversion, and both sections are organized by the logic of equation (7.10.2) from the observations, y, to the model relationships, A, and to the parameters or variables, x. The term "model relationships" is meant to clarify the common term "model" which is used differently in geophysics and in geodesy. In geophysics "model" is usually restricted to the parameters, x, in geodesy "model" usually includes the function or matrix A. Of course, $A(x)$ (or A) and the variables x belong together, but generally both x and A are not known well enough to be absolutely distinguished in a given problem. Uncertain components of A can also be considered variables, x, or they may be considered observations, y, with errors. Components of A that are very certain are considered "conditions", while others are very uncertain model hypotheses or even known to be "wrong" as, for example, intended simplifications of the model relationships. The uncertainty of defining A, x and y results from our ignorance about the real world and specifically about the Earth's interior.

This chapter is largely based on Smilde (1997, 1998), henceforth referred to by PS97, dealing with the use and evaluation of a priori information for potentially singular inversion problems and the determination of density distributions by gravity. Important texts for the general theory of geophysical inversion problems are Menke (1984) and Tarantola (2005).

7.2 Optimization

Optimization is a very general task in every field of human activity, especially in science. In gravity interpretation, it is the attempt to apply potential theory and optimally map observations into density models. Redundant data involve discrepancies which indicate errors. Here, *errors* are understood as differences from the true values which, however, are usually unknown (note that the "statistical error", below, has a different meaning). Optimization is error minimization, based on error theory of measurements and theoretical conditions (as the sum of angles in a plane triangle).

The notions of *error* and *uncertainty, precision* and *accuracy, probability* or *likelihood, reliability, stability* and *resolution* may need some clarification. "*Uncertainty*" is often synonymously used as "*error*" (or statistical error), that refers to the "scatter" of the observations or determinations of a variable. The "*standard error*" is also a statistical error which will be defined below. "*Precision*", the opposite of the statistical error, means smallness of the statistical scatter or close repeatability which need not be centered at the true value if the observations are affected by *systematic errors* (first definition). "*Accuracy*" of a result, on the other hand, expresses the *likelihood* of being close to the true value, and in this sense, it is the opposite to "error" (first definition). *Reliability* has about the same meaning as accuracy and implies that the given value can be trusted within certain limits, i.e., can be tested by other means. "*Probability*" or "*likelihood*" of approaching a true value are usually difficult to estimate and thus subject to personal judgment; both terms are synonymous and a high probability is the opposite of large uncertainty. The *stability* of a result expresses its insensitivity to data errors and to changes in the initial assumptions, for example, of iteration in the case of a non-linear problem. *Resolution* describes the independence of variables or of distinguishing results to be unique and really distinct. It is also related to the discrimination of neighboring results and their significance.

7.2.1 Theory

The theory is largely based on Eq. (7.10.1), beginning with observations, y, constraining the model relationships, A, with their variables, x. The basis of the model relationships is Newton's law of gravitation (Eq. 2.5.5)

$$\boldsymbol{\delta g} = \int_V d\boldsymbol{g} = G \int_V \Delta\rho\,(\boldsymbol{r}/r^3)\mathrm{d}V = -\boldsymbol{\nabla}\delta U$$

or for the potential:

$$\delta U = -G \int_V \Delta\rho/r\,\mathrm{d}V \tag{7.2.1}$$

The integrand is the density distribution $\Delta\rho(x,y,z)$ in the source volume, V, and the kernel is $\boldsymbol{r}/r^3$, or for U, $1/r$; $\boldsymbol{r}$ is the distance vector from a volume element $\mathrm{d}V$ to P (at the origin 0). These integral equations are examples of Fredholm's integral equation of first kind (Arfken 1985, p. 865) which has the form

$$g(u) = \int_a^b K(u,v)\rho(v)\mathrm{d}v \tag{7.2.2}$$

with $g(u)$ and $\rho(v)$ at the coordinates u and v, respectively, and the kernel $K(u,v)$ in the form of Newton's law. It has generally no unique solution except in special cases. Methods leading to solutions are called "regularization" narrowing or removing the non-uniqueness and the instability at least mathematically. It is assumed in $\underline{A}x = y$ or $Ax = y$ (Eq. 7.1.1) that the unknown density distribution can be described by only m parameters, x, and that n observed gravity values, y, are given. A is the operator matrix based on (Eq. 2.5.5) that projects the density model onto the observations. The non-uniqueness is represented by the nullspace $Ax_o = 0$ which can be added to (Eq. 7.1.1) without changing y, as the external gravity effects are zero. A "nullspace" is defined by models generating no external field, for example, defined by the differences between any two fitting models. Say, $[\rho(x,y,z)]$ describes such a model in geometrical space and $[\rho_i(x,y,z)]$ and $[\rho_k(x,y,z)]$ have the same external gravity effect, $\delta g^i(x,y) = \delta g^k(x,y)$, then the model $[\rho_i(x,y,z) - \rho_k(x,y,z)]$ generates $\delta g^{i-k}(x,y) = 0$.

The matrix A is in this case singular and thus the inverse operator A^{-1} does not exist which otherwise would render a unique solution. Any of the solutions $x + x_o$ must be judged by a priori external criteria and may be found either relevant or more or less acceptable or unacceptable. If all solutions except one are unacceptable, the problem is uniquely solvable.

Those external criteria could be expressed as obeying the auxiliary conditions, $Bx = z$, and could be extracted from (Eq. 7.2.2) by writing

$$\begin{aligned} Ax &= y \\ Bx &= z \end{aligned} \rightarrow \begin{bmatrix} A \\ B \end{bmatrix} x = \begin{bmatrix} y \\ z \end{bmatrix} \tag{7.2.3}$$

If only one solution exists, then the inverse of the combined matrix exists and A has then been successfully regularized.

7.2.1.1 Observations, Model Relationships, Variables

Observations and ideas are the basis of the image we have of the world. Observations and experiments are intimately entangled with questions, theoretical relationships and models. Observations in geodesy and geophysics are primarily of geometrical and gravitational nature, but in a broad sense, include other physical quantities and pre-existing knowledge about an object of study. Even theoretical relationships can be considereds observations. In gravity, observations of various kinds at locations, (x_i, y_i, z_i), complemented by extra-gravity information, are applied to the question of the inaccessible underground density distribution. Observation are always affected by errors, and there are never sufficient observations.

Optimization is defined here as optimally solving the situation when observations are conflicting. When are they conflicting?

- When they have too large scatter or contain blunders or outliers (in case of observations of the same kind or of different kinds).
- When their theoretical relationship is wrongly defined, mostly in the case of observations of different kinds or because the theory is really wrong, or only too much simplified.

These two types of conflicts might be actually equivalent.

In geodesy, a theoretical relationship between the observations is sometimes called a "model", sometimes a "condition", especially when this relationship is very certainly known, as a mathematically derived equality or inequality. In geophysics, the relationship between the observations is usually given by an "image" or model of the hypothetical underground gravitating density distribution. Its description contains optimizable parameters or variables. Additional conditions are needed because of the ambiguity, such as a priori density bounds, idealizations and simplifications of geometry, for example, "smoothness" (see Chap. 5).

A fundamental condition for parameter optimization is that a solution exists and that the parameters are sensitive to the observations, i.e. that they are resolvable. Irresolvable quantities are, for example, laterally constant mass distributions as Bouguer plates and confocal spheres which generate laterally constant effects which cannot be separated. Mass distributions should be laterally variable. In areas with constant observed gravity, a priori information on variable mass "anomalies" permits to derive tractable models by geological gravity reductions (Sect. 4.5.3.3). However, in that case the a priori information cannot be optimized by gravity.

Often the theoretical relationship between the observations is known up to some quantities with poorly known values. In this case optimization is the selection of the values of these variables such as to reduce and to minimize the conflicts, for example, by scaling and en bloc shifts of model bodies. But usually some additional types of conflicts between observations will remain after optimally adjusting the variables. In the case of inversion, however, variables (with usually well defined physical meanings) are introduced with the purpose of determining their unknown values (see Sect. 7.2.2.2).

If error-free data and/or model relationships are in conflict, no solution exists. On the other hand, models can fit noisy data approximately (unless constructed such as to nonsensically fit the errors). If gross and systematic errors are eliminated, the probability distributions of the random errors must be carefully assessed. Weights will be applied accordingly, large ones in the case of narrow error probability distributions and vice versa. Optimization thus means adjusting the models, to be as compatible as possible with all available information, or the least severe conflict with any specified constraint or best compromise or trade-off between the different data errors. This non-trivial task must be defined.

7.2.1.2 Optimization Criteria, Probability, Likelihood, Accuracy, Reliability, Uncertainty, Errors

Conflicts can arise only if redundant observations exist, say n, more than needed to solve for m unknowns; there are then $n - m$ redundant observations. Solving conflicts between observations in the optimal sense requires criteria, which define what a solution is optimal. Optimization criteria depend on what is expected of the most probable solution. Types of criteria are:

– Geometrically minimal "distance" between the given observation values and their "deconflicted" (optimized, adjusted) values.
– Probabilistic criteria take the statistical properties of the observations into account. The real observations are assumed to be scattered realizations of hypothetical, conflict-free observable quantities. The "deconflicted" optimal solution consists of calculated values for these quantities, which would be theoretically realized with the highest probability or likelihood by the given observations, if their probability distribution were known.

For more than one observation, "distance" can be defined in the usual Euclidian way as the square root of the sum of the squared individual distances (errors) or the L_2-norm. That establishes the least-squares (LSQ) method. More generally the individual distances might be weighted appropriately. "Distance" might be defined differently: e.g. by the sum of absolute individual errors ($\Sigma |r_i| = $ min!, L_1-norm) which is less sensitive to the largest discrepancies. Higher norms emphasize larger discrepancies; L_∞ will correct the largest discrepancy only. If the statistical properties are given by the Gaussian probability distribution (see below), the solution is the same as for the geometrically motivated least squares method. Assigning different accuracies, or better: standard deviations, to individual observations is equivalent to individually weighting the distance components. "Least-squares" methods are most popular for error minimization and parameter optimization. If the errors are

Gauss-distributed, LSQ renders the optimal solution. If not, optimal solutions are possible but not necessarily so. A qualified decision which norm is to be chosen has to be taken depending on the basis of the nature of the observations and the given error statistics.

The least-squares principle is illustrated by the case of n observations, $y_i = y_{obs,i}$ of a single unknown variable x ($m = 1$). A theoretical deduction is given below (Sect. 7.2.1.2.1). C.F. Gauss (1777–1855) assumed that the individual measurement errors are stochastic, i.e., statistically independent. For arbitrary y and $r_i = y_i - y$, the estimated variance, $S(y) = \Sigma r_i^2/(n-1)$ has a parabolic shape, if plotted versus y, with its minimum at $y = x = \Sigma y_i/n$ ($x = \Sigma w_i y_i/\Sigma w_i$, if weighted with w_i); x is the arithmetic mean of the observations and the solution of minimization. $\sqrt{\Sigma r_i^2}$ represents the minimum "distance" between the observational values. The frequency distribution of the residuals, r, is maximum at, and symmetric about, the mean, x, forming a bell-shaped curve. Plotted versus a general coordinate y, the error frequency or Gaussian error distribution is given by $f(y) = 1/[s(2\pi)^{1/2}]\exp[-y^2/(2s^2)]$, characterized by the standard deviation $s = S^{1/2}$ as a measure of the width of the scatter, and the uncertainty of the mean is $s_m = s/n^{1/2}$. It must, however, not be forgotten that S can generally be only roughly estimated on the basis of limited observations.

The least-squares principle is extendable analogously to $m > 1$ for models defined by m variables ($m > n$) and $m - n$ redundant observations. An ensemble of n observations, δg_i at the station coordinates $x_i = (x_i, y_i, z_i)$, $i = 1$, n, may represent a "gravity anomaly" which may be described as $\delta g(x_i)$ along a 1D profile (x) or $\delta g(x_i, y_i)$ in a 2D (x, y) array on a map. For modelling, δg_i is taken to be the effect, $f(x_i, p)$, of a density model defined by physical and geometrical parameters $p = (p_1, p_2, \ldots p_m)$ in (x, y, z, ρ) space. The parameters are the variables p ($= x$ in the sense of Eq. (7.10.1)) of the optimization or inversion problem; $f(x_i, p)$ is linear in density but generally non-linear in the coordinates. Solving for p, is a minimum problem analogous to the one-variable case and is treated later. For optimization of non-linear cases, an initial or a priori estimate p_0 of the parameter vector is assumed to exist, and the optimal p is to be determined from the n observations. Admittedly, in view of our ignorance, such problems are only artificially over-determined ($n > m$) by restricting m with appropriate simplifications, while in reality it is always under-determined ($m > n$).

Minima $S(p)$ and $S(p_1, p_2)$ for one or two variables p_k are easily imagined graphically as a depression of a line or a surface, but for more parameters the $S(p)$ form hyper-surfaces in the p-coordinate system. $S = \Sigma r_i^2/(n-m)$ and $r_i = \delta g_i - f_i(p)$. Since the surface S depends on all residuals r_i, all model effects, $\delta g_i = \delta g(x_i)$, at the observation points, $P_i = P(x_i)$, are functions of the model parameters p_k. The derivatives $\partial r_i/\partial p_k = \partial(\delta g_{obs,i} - f(x_i, p))/\partial p_k = J_{ik}$ are a measure of how the residuals would respond to parameter changes. The whole n × m Jacobian matrix $\{J_{ik}\}$ ($i = 1$, n; $k = 1$, m) thus forms the basis of a minimum problem for the S surface, the solution of which gives the minimum directly in the linear case, and the direction of the search by iteration in the non-linear case. The J_{ik} elements, by their size (steepness of slope), can also be envisioned as sensitivities of the effects against parameter changes, and the Jacobian can be interpreted as "sensitivity matrix". Densities in fixed geometry present a linear problem. For non-linear problems of geometry the hyper-surface may become complicated with many local minima. Several strategies exist for the search of the minimum parameter set p. Improvement need not occur at every point x_i nor with every parameter p_k, and the numerical procedure may pass over the S-minimum and miss it.

7.2.1.2.1 Theory of the Least-Squares Method

A theoretical derivation of the least squares principle is based on the condition that the observations y correspond to a certain expected relationship such that no discrepancies exist in this respect between them, expressed by the condition equation

$$Uy = 0 \tag{7.2.4}$$

where the matrix U describes the condition or conditions and where the uncertainty is quantified by the covariance matrix C_u; the equation can be represented by the conditional probability density distribution $\phi(Uy \mid y)$, common to all observations y. Uy is the deviation from the assumed condition, and $\phi(y \mid \tilde{y})$ is the probability density of y given an actual observation $\tilde{y}$ with the covariance matrix C_y. The effective a posteriori probability density is then under both assumptions

$$\phi(\widehat{Uy},\hat{y} \mid \tilde{y}) = \phi(Uy \mid y) \cdot \phi(y \mid \tilde{y}) \tag{7.2.5}$$

where "a posteriori" is symbolized by "^". The relationship between these three probability densities is illustrated in Fig. 7.2.1.

With the symbolic expression $\phi \propto N(a, B)$ (saying: ϕ has a normal distribution with the expectation a and the covariance matrix B), the assumption of normal distributions of the two probability densities is

$$\phi(y \mid \tilde{y}) \propto N(\tilde{y},C_y) = \alpha_1 \exp\left(-{}^1\!/_2 (y-\tilde{y})^T C_y^{-1}(y-\tilde{y})\right)$$
$$\phi(Uy \mid y) \propto N(0,C_U) = \alpha_2 \exp\left(-{}^1\!/_2 (Uy)^T C_U^{-1}(Uy)\right) \tag{7.2.6}$$

The a posteriori probability density is (Menke 1984; Tarantola 2005):

$$\phi\left(\widehat{Uy},\hat{y} \mid \tilde{y}\right) = \alpha_3 \exp\left(-{}^1\!/_2 [(y-\tilde{y})^T C_y^{-1}(y-\tilde{y}) + (Uy)^T C_U^{-1}(Uy)]\right)$$
$$= \alpha_4 \exp\left(-{}^1\!/_2 (y-\hat{y})^T C_{\hat{y}}^{-1}(y-\hat{y})\right) \tag{7.2.7}$$

with

$$\begin{cases} \hat{y} = \tilde{y} - \left(U^T C_U^{-1} U + C_y^{-1}\right)^{-1} U^T C_U^{-1} U\tilde{y} \\ C_{\hat{y}} = \left(U^T C_y^{-1} U + C_y^{-1}\right)^{-1} \end{cases} \tag{7.2.8}$$

or equivalently

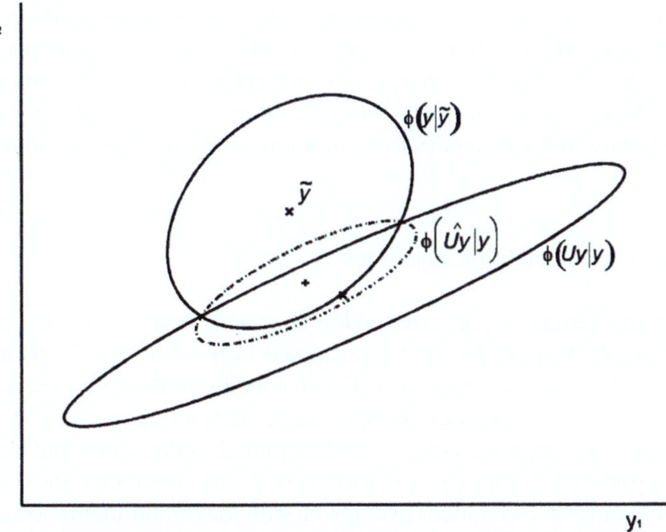

Fig. 7.2.1 Illustration of the a priori and a posteriori probability densities for 2 observations and a connecting theory or condition; within the ellipses the probability density exceeds a certain value. $\phi(y|\tilde{y})$ is the probability density of y, given by two observations $\tilde{y}_1$, $\tilde{y}_2$; $\phi(Uy|\tilde{y})$ is the probability density of the condition $Uy = 0$ to be true for y; $\phi(\widehat{Uy}, \hat{y}|\tilde{y})$ is the a posteriori probability density for $Uy = 0$ to be satisfied by the observations $\tilde{y}$

$$\begin{cases} \hat{y} = \tilde{y} - C_y U^T (U C_y U^T + C_U)^{-1} U \tilde{y} \\ C_{\hat{y}} = C_y - C_y U^T (U C_y U^T + C_U)^{-1} U C_y \end{cases} \tag{7.2.9}$$

where $\hat{y}$ maximizes the probability density and it is the desired maximum likelihood estimator which minimizes its variance, i.e. the squared term in (Eq. 7.2.7):

$$(y - \tilde{y})^T C_y^{-1} (y - \tilde{y}) + (Uy)^T C_U^{-1} (Uy) = \text{minimal} \tag{7.2.10}$$

Equations (7.2.8) and (7.2.9) determine a solution when there are discrepancies among the data, i.e., when the real observations $\tilde{y}$, and conditions are incompatible among themselves or with the data. That is the natural situation in research and exploration into the unknown.

In the special case $C_y = I$ (i.e., all uncertainties of the observations are equal and independent) and a solution vector y that satisfies the conditions, $Uy = 0$, the solution y and the observed vector $\tilde{y}$ have the minimal "distance" in the n-dimensional observation space, if for y the least-squares solution $\hat{y}$ is chosen. The distance is equal to the (equally weighted) L_2 norm of a vector

$$(y - \tilde{y})^T (y - \tilde{y}) = \|(y - \tilde{y})\|_2 = \left(\sum_n (y - \tilde{y})^2 \right)^{1/2} \tag{7.2.11}$$

where $\| \ldots \|_2$ refers to the L_2 norm.

The maximum likelihood solution is identical to the least-squares solution in the case of normal distributions. If $C_y \neq I$, the inverse covariance matrices in (Eq. 7.2.10) modify the definition of distance by weighting the components (to take different units or accuracies of the y-components into account). If $Uy \neq 0$, its components contribute to the distances to be minimized and are thus minimized simultaneously.

7.2.1.2.2 Regularization of Condition Equations

If, as usual, the condition equations are assumed to be exactly correct, then $C_U = 0$ and C_U^{-1} is not defined. A solution can be calculated only with (Eq. 7.2.9) and $(UC_yU^T)^{-1}$ must not be singular. Therefore fewer conditions than observations must exist and the conditions must not contradict each other, i.e. they must be independent; the equation system must not be overdetermined. If these preconditions are not satisfied, a covariance matrix $C_U \neq 0$ will "soften" the conditions; they will not be met exactly but only on (weighted) average or with maximum likelihood under the assumption of the normal distribution. This means that the conditions are considered uncertain, implying that the condition equations are regularized.

7.2.1.2.3 Partitioned Observations

There are special cases of the condition equation (7.2.4) where part of its observations $\tilde{y}$ are very inaccurately known. Then the observations can be partitioned into two sets, x (inaccurately known) and y (well known), respectively, which are not correlated, and we can write:

$$
\begin{cases}
Uy = 0 \\[2em]
C_y,\, C_U
\end{cases}
\Rightarrow
\begin{cases}
[A - I] \begin{bmatrix} x \\ y \end{bmatrix} = 0 \\[2em]
C_{xy} = \begin{bmatrix} C_x 0 \\ 0 C_y \end{bmatrix},\, C_A
\end{cases}
\tag{7.2.12}
$$

where C_A is the limited probability with which the partial matrix A is correct. The solution for the partial set x, after substituting it into (Eq. 7.2.9), is

$$
\begin{cases}
\hat{x} = \tilde{x} - C_x A^T \left[A C_x A^T + C_y + C_A \right]^{-1} [A\tilde{x} - \tilde{y}] \\[1em]
C_{\hat{x}} = C_x - C_x A^T \left[A C_x A^T + C_y + C_A \right]^{-1} A C_x
\end{cases}
\tag{7.2.13}
$$

or substituted into (7.2.8):

$$
\begin{cases}
\hat{x} = \tilde{x} - \left[A^T [C_y + C_A]^{-1} A + C_x^{-1} \right]^{-1} A^T [C_y + C_A]^{-1} [\tilde{y} - A\tilde{x}] \\[1em]
C_{\hat{x}} = \left[A^T [C_y + C_A]^{-1} A + C_x^{-1} \right]^{-1}
\end{cases}
\tag{7.2.14}
$$

The observation equation, which is the basis of the solution equation (7.2.14) corresponding to the condition equation (7.2.12), is:

$$\begin{cases} \begin{bmatrix} A \\ I \end{bmatrix} x = \begin{bmatrix} y \\ x \end{bmatrix} \\ \\ C_y, C_x \end{cases} \tag{7.2.15}$$

If the partial set x is totally uncertain, then $C_x = diag(\infty)$, hence $C_x^{-1} = 0$ and $\tilde{x}$ is arbitrary, e.g. $\tilde{x} = 0$. If, moreover, $C_A = 0$, then (Eq. 7.2.15) corresponds to the least-squares solution of the observation equations $Ax = y$ with x as the unknown variables:

$$\hat{x} = (A^T C_y^{-1} A)^{-1} A^T C_y^{-1} \tilde{y} \tag{7.2.16}$$

This means that the variable x can be considered as an observation of which one knows nothing, which seldom is quite true. The essence of this is: if unknowns "run away" from reasonable values, their uncertainty C_x must be reduced from $\begin{bmatrix} \infty & 0 \\ 0 & \infty \end{bmatrix}$ to finite values requiring a priori information.

7.2.1.2.4   Adjustment in Phases

The variables x and their covariance matrix need not be arbitrarily guessed as can be shown by the so-called adjustment in phases. Given the observation equation $Ax = y$, with $C_x = diag(\infty)$ and $\tilde{x} = 0$, the observations y are divided into two groups (assumed uncorrelated for simplicity, but not necessarily)

$$\begin{bmatrix} A_1 \\ A_2 \end{bmatrix} x = \begin{bmatrix} y_1 \\ y_2 \end{bmatrix} \tag{7.2.17}$$

which becomes

$$\begin{cases} \hat{x} = \left[[A_1\ A_2]^T (C_{y_1 A_1} + C_{y_2 A_2})^{-1} \begin{bmatrix} A_1 \\ A_2 \end{bmatrix} \right]^{-1} [A_1\ A_2]^T (C_{y_1 A_1} + C_{y_2 A_2})^{-1} \begin{bmatrix} \tilde{y}_1 \\ \tilde{y}_2 \end{bmatrix} \\ \\ C_{\hat{x}} = \left[[A_1\ A_2]^T (C_{y_1 A_1} + C_{y_2 A_2})^{-1} \begin{bmatrix} A_1 \\ A_2 \end{bmatrix} \right]^{-1} \end{cases} \tag{7.2.18}$$

which is equivalent to:

$$\begin{cases} \hat{x}_{12} = \left[A_2^T C_{y_2 A_2}^{-1} A_2 + C_{\hat{x}1}^{-1} \right]^{-1} \left[A_2^T C_{y_2 A_2}^{-1} \tilde{y}_2 + C_{\hat{x}1}^{-1} \hat{x}_1 \right] \\ \\ C_{\hat{x}12} = \left[A_2^T C_{y_2 A_2}^{-1} A_2 + C_{\hat{x}1}^{-1} \right]^{-1} \end{cases} \tag{7.2.19}$$

with

$$\begin{cases} \hat{x}_1 = \left[A_1^T C_{y_1 A_1}^{-1} A_1\right]^{-1} A_1^T C_{y_1 A_1}^{-1} \tilde{y}_1 \\ C_{\hat{x}_1} = \left[A_1^T C_{y_1 A_1}^{-1} A_1\right]^{-1} \end{cases} \qquad (7.2.20)$$

The latter equation corresponds to the adjustment with only the observations $\tilde{y}_1$ in the first phase; in the second phase of Eq. (7.2.19) adjustment is carried out with only the observations $\tilde{y}_2$, but taking into account the result $\hat{x}_1$ and its a posteriori covariance matrix $C_{\hat{x}1}$. The solution is the same as that obtained in one go.

In the special case

$$\begin{bmatrix} A \\ I \end{bmatrix} x = \begin{bmatrix} y \\ x \end{bmatrix} \Leftrightarrow \begin{bmatrix} A \\ I \end{bmatrix} (\tilde{x} + \delta x) = \begin{bmatrix} y \\ x \end{bmatrix} \Leftrightarrow \begin{bmatrix} A \\ I \end{bmatrix} \delta x = \begin{bmatrix} y - A\tilde{x} \\ x - \tilde{x} \end{bmatrix} \qquad (7.2.21)$$

with $\tilde{x} = \hat{x}_1$ and $C_x = C_{\hat{x}_1}$, the result of the second phase is:

$$\begin{cases} \hat{x} = \tilde{x} + \delta x = \tilde{x} + \left[A^T C_{yA}^{-1} A + C_x^{-1}\right]^{-1} A^T C_y^{-1} [\tilde{y} - A\tilde{x}] \\ C_{\hat{x}} = \left[A^T C_{yA}^{-1} A + C_x^{-1}\right]^{-1} \end{cases} \qquad (7.2.14a)$$

which is again Eq. (7.2.14), i.e. the well-known equation for the least-squares solution of observation equations with additional so-called a priori information $\tilde{x}$ and C_x. So adding a-priori information corresponds to solving the problem with previously available data, which could also have been solved in one single step. The a-priori information can be interpreted as the summarized results of some previous observations y_1.

7.2.1.2.5 Combined Covariance Matrices

The a priori covariance matrices C_y and C_A always occur together in observation equations and are usually not considered separate. Comparison of Eqs. (7.2.8) and (7.2.14) shows that the uncertainty of the condition equation, C_U, or of the observation equation $(C_y + C_A)$ implies the uncertainty of the respective right-hand sides. The two covariance matrices are thus combined to $C_{yA} = C_y + C_A$; when it is known, the uncertainty of the kernel A must be taken into account together with that of the observations; the uncertainty values of A have thus also been called "parametrization errors". An example in gravity optimization is the real small-scale scatter due to neglected shallow small-scale mass anomalies which cannot principally be explained by larger or deeper density models.

The main principle is that model simplifications (also intended ones) must be accounted for by increasing the uncertainty of the observations $y(C_y)$.

7.2.1.2.6 Non-Trivial Nullspace of A

If the inverse covariance matrix $C_x^{-1} = 0$, then in (Eq. 7.2.14), $A^T (C_y + C_A)^{-1} = A^T C_{yA}{}^{-1} A$; its existence follows from an analysis of the generalized eigen-vectors of A:

$$\begin{cases} Av_j = \sigma_j w_j \\ A^T w_i = \sigma_i v_i \end{cases} \quad \text{or} \quad \begin{cases} = \bar{A}V = WS \\ = \bar{A}^T W = VS \end{cases} \tag{7.2.22}$$

$\bar{A}$ equals A with zeros added to complete A to a square matrix, which in principle does not change the analysis. $S = diag(\sigma_{i \leq p} \neq 0, \ \sigma_{i>p} = 0)$ are the eigen-values, also called "singular values", ordered by size, with a number of p values $\neq 0$; V and W are matrices with the corresponding eigen-vectors as columns, whence follows (dropping the bar above A) the definition (symbolized by: "$=!$"):

$$\begin{cases} A^T AV = VSS \text{ DEF } VS^2 \\ AA^T W = WSS \text{ DEF } WS^2 \end{cases} \tag{7.2.23}$$

If $Ax_0 = 0$ and $x_0 \neq 0$, A has a non-trivial nullspace, i.e., does not contain zeros, and (7.2.23) implies at least one eigen-value to be zero. This means: S^{-2} does not exist, nor exist the following derived matrices: $(A^T A)^{-1} = VS^{-2}V^T$, and $(A^T C_{yA}{}^{-1} A)^{-1} = ((A^T C_{yA}{}^{1/2})(C_{yA}{}^{1/2}A))^{-1} = (A_c{}^T A_c)^{-1} = V_c S_c{}^{-2} V_c{}^T$ (since C_{yA} DEF $C_{yA}{}^{1/2} C_{yA}{}^{1/2}$ is positive definite, hence not singular, and $A^T A$ and $C_{yA}{}^{1/2}A$ DEF A_c (7.2.18) have the same nullspace as A). The matrix rows of V corresponding to the zero eigen-values span the nullspace and are its axes.

The existence of a non-trivial nullspace of A is equivalent with a singularity of $(A^T C_{yA}{}^{-1} A)^{-1}$. The matrix is the a posteriori covariance matrix with V_c as singular values and $S_c{}^{-2}$ as eigen-vectors. $S_c{}^{-2}$ represents the a posteriori variances of the linear combinations of the variables in the columns of V_c. These two matrices define the m-dimensional standard hyperellipsoid, i.e., multidimensional standard deviations. The hyperellipsoid axes are along the eigen-vectors V_c with variances $S_c{}^{-2}$, in the respective directions according to the respective eigen-vectors. If the covariance matrix is singular, at least one element of S_c is zero, the corresponding standard deviation is infinite and the standard ellipsoid degenerates in this direction to an infinite cylinder; no respective solution is preferred. A corresponding regularization is necessary. If A is not to be changed (by some implicit side condition) then C_x^{-1} must be chosen other than the zero matrix, i.e. a finite uncertainty around an a priori estimate must be assumed. The covariance for such a regularized solution is calculated from:

$$\begin{cases} \begin{bmatrix} A \\ I \end{bmatrix} x = \begin{bmatrix} y \\ x \end{bmatrix} \\ C_{xy} = \begin{bmatrix} C_y & 0 \\ 0 & C_x \end{bmatrix} \end{cases} = \begin{cases} \begin{bmatrix} A_c \\ I \end{bmatrix} C_x^{1/2} x_c = \begin{bmatrix} y_c \\ C_x^{1/2} x_c \end{bmatrix} , \\ C_{yc} x_c = I \end{cases} \text{ with } \begin{cases} A_c = C_y^{-1/2} A \\ y_c = C_y^{-1/2} y \\ x_c = C_y^{-1/2} x \end{cases} \Rightarrow$$

$$([A^T I])C_y^{-1}\begin{bmatrix} A \\ I \end{bmatrix} + C_x^{-1})^{-1} = \left(C_x^{1/2} [A_c^T I] \begin{bmatrix} A_c \\ I \end{bmatrix} C_x^{1/2} \right)^{-1} =$$

$$= C_x^{1/2} V_c (S_c^2 + I)^{-1} V_c^T C_x^{1/2} \text{ with } (A_c^T A_c)^{-1} = V_c S_c^{-2} V_c^T \qquad (7.2.24)$$

In contrast to (7.2.23) with $V_c S_c^{-2} V_c^{\mathrm{T}}$, the matrix $(S_c^2 + I)$ to be inverted does not contain zeros any more and the inverse is uniquely defined; the a posteriori standard deviation is limited, even if S_c contains zeros or small values.

The quintessence is that adding a priori information means removing dimensions from the nullspace.

7.2.1.2.7 Numerical Instabilities

If the overdetermined Eq. (7.2.16) is analytically stable and regular, it can still happen that it becomes numerically unstable or singular, if the variables have widely differing values and physical units. If the orders of magnitude are known they can be normalized to order 1 by letting $x_{norm} = C_{norm} x$, and the equation $A_{norm} x_{norm} = y$ is to be solved with $A_{norm} = A C_{norm}^{-1}$. With this normalized kernel A_{norm} substituted into (7.2.16), the solution x_{norm} determines the original variables as $x = C_{norm}^{-1} x_{norm}$. Within numerical accuracy, no numerical instability or singularity should occur.

If small rounding errors have a strong influence on the calculation of the inverse, then the problem is probably physically unstable or singular, not just numerically, and needs additional physical information. If the regularized system of equations

$$\begin{bmatrix} A \\ I \end{bmatrix} x = \begin{bmatrix} y \\ x \end{bmatrix} \qquad (7.2.25)$$

is to be solved, then for the normalization matrix C_{norm}, the "square root matrix" $C_x^{1/2}$ of the a priori covariance matrix $C_x = C_x^{1/2} C_x^{1/2}$ can be taken (e.g. the Choleski factor matrix or ES, where E = matrix of the eigen-vectors of C_x and S = diagonal matrix of its square roots; Tarantola 2005), and the a priori covariance matrix for the normalized variables x_{norm} will be the unit matrix. In many cases it will suffice to choose a diagonal matrix for the normalization matrix containing the assumed standard deviations, if the range of x-values is approximately known, as usual in the case of variables with a clear physical meaning. In other cases, as with unconstrained variables or the coefficients of ordinary polynomials or of harmonic expansions, the needed a priori knowledge is usually missing and regularization is not so easy, but that is generally so for variables with unclear physical meaning. Only mathematical regularization can help here removing singular values as described in Sect. 7.2.1.2.6.

7.2.1.2.8 Summary

The essential message of this section, 7.2.1, on the theory of optimization is that LSQ is based on mathematical principles which guide the search for the most reliable density distributions of geological relevance by fitting the parameters or variables of appropriate a priori models to gravity observations. Appropriate means capability of leading to stable solutions of the fundamental system of regularized linear equations (7.10.1). Important methods and problems of regularization have been discussed theoretically for a number of typical situations.

In the context of inversion of model variables, it is likely that the need for regularization arises when the only source of information on the variables x is in the observations y (Sect. 7.2.1.2.3), for example, when they have no physical meaning and predictable value as the average reading of a relative gravimeter. If the meaning of a variable is clear it usually also implies an estimate of its value and its uncertainty. Such a priori information renders the problem regular and unique and is not an unwanted influencing of the results. It rather comes from earlier adjustment phases (Sect. 7.2.1.2.4). Observations might be lost or forgotten, but the results (estimates and covariance matrices) carry the same information as the original data. Such an earlier phase is the experience of expert scientists mostly reflecting the former results accurately enough as implicit a priori information. However, the solution should be stable, i.e. not too sensitive against variations of the a priori information, as it is also with the pure condition equation. If the solution is unstable regardless, it generally means that the new information of the current phase has no strong weight, i.e. it is not very important relative to the a priori information.

The application of the theoretical aspects to practical problems of gravity model optimization will now be considered.

7.2.2 Practice

Optimization begins with the questions to be answered by gravity interpretation, guiding the planning and the actual measurements, reductions and the analysis of the data (Chaps. 3, 4). The aims determine also the design of appropriate models (Chaps. 5, 6), which should describe the anticipated geological situation. This requires geological knowledge and imagination, but the researcher should be always open to the unexpected.

A compromise is to be attempted between accurate and detailed forward modelling and optimal fitting of the observations and a priori information. Practical optimization is hardly imaginable without efficient computer routines (as INVERT, PS97, attached to this book on a CD). The computational demands are high, and the programs must be applied with mathematical understanding of their limitations. The basic principles and possibilities are explained below. In any optimization procedure choices must be made of specific algorithms, solution methodology, strategies of linearization and iteration, etc., and there is ample space for experimentation.

The system of linear equations to be solved for optimization is $y = Ax$ or $y = A(x)$, (Eq. 7.1.1). First, the observations or data, y, (see Chaps. 3, 4) are discussed, then the model relationships, A, and finally the variables or parameters, x, in connection with the optimization and inversion requirements.

7.2.2.1 Observations

7.2.2.1.1 Types, Spatial Distribution

The following types of observations can be distinguished:

- originally measured values as the classical "reading",
- measurements combined into a single "observed value" as the arithmetic mean (possibly as noted in a field book or done by internal processing in modern instruments),
- reduced, filtered, corrected values,
- interpolated and extrapolated values.

"Observations" may be the original numbers archived in a field book, on tape or in digital storage and may result from some "automatic" processing and even reduction. For example, absolute gravity meters average many "drops" and record the mean values and standard deviations. It is a case of adjustment in phases (Sect. 7.2.1.2.4).

All observable and derivable gravity field quantities (Sect. 2.7.4) can play the role of observations: gravity itself, so-called gravity gradients, components of the gravity gradient tensor (Sect. 2.8), sea surface topography as a measure of potential anomalies or geoid undulations and some of the other second derivatives of the potential. Observations are supplied by pendulums (Sect. 3.2.2), gravimeters and related instruments (Sects. 3.2.3, 3.2.4, 3.2.5, 3.2.6, 3.2.7), gradiometers and the torsion balance (Sect. 3.2.9), satellite altimetry (Sect. 3.2.8) and by astrometry and levelling instruments (plumb-line deflection). There is no sharp boundary between observed and derived quantities. But as a general rule, it is better to interpret observed quantities before too much data processing to avoid the additional uncertainties. Other than gravity observations, as seismic and magnetic, may be included. Gravity interpretation, model optimization and inversion depend on gravity data *and* a priori information (non-gravity).

Data preparation may also involve their reduction in number. Data not relevant for a problem are sensibly removed. (1) There may be too many for practical handling; an example is interpolation between data points, but care must be taken of how to interpolate, e.g. topography and maintaining theoretical relationships as the Laplace equation. (2) If model components are excluded, data correlating with them would affect the remaining model in the solution and should be removed. Into this category belong the gravimeter drift correction if no temporal gravity variation is modelled. The standard reductions (Chap. 4) belong here if no complete Earth model is included, and, if only a local anomaly is modelled, also the removal of a regional field (Sect. 4.7.7) described by unconstrained variables on the basis of the data themselves. Instead, more a priori information can be included into the optimization.

The spatial layout of the observations in the field (see Sect. 3.4) will be designed in view of the target and the a priori knowledge about it, but depends also on logistics, economic constraints, etc. The station configuration should be dense, where the most important field variations are expected for the problem solution. This may become known only during a survey.

In the case of unfavourable data distributions, surprises may happen in the optimization process. The recognition and characterization of "the anomaly" or the "field", on the basis of limited discrete observations is an implicit initial step of interpretation (Sect. 1.4). A similar argument pertains to the background or reference level of the observations relevant for total mass estimates (Sect. 2.7.6). What is an anomaly if only few discrete observations exist and some plausible shape of it is anticipated? Is it permitted to complement the data, for example, by gridding? Or should the interpretation be strictly limited to the original observations? The consequences of data manipulations must be explored by tests and comparisons. For example, mathematical field analysis as filtering may be facilitated by gridding it, but it may turn out that a model body is moved unrealistically by the minimum condition in the optimization, which is influenced by gridding or can be prevented by stronger constraints (see 5.1.5). While gridding can be arbitrarily densified, the a posteriori standard deviation is formally reduced which must be prevented by increasing the a priori standard deviation (Sect. 7.2.1.2.5):

Data point distribution determines parts of a model to be more or less resolvable. Stations may be set up along customary 1D profiles or on 2D maps and sometimes in 3D distributions (buildings, underground works, boreholes, airborne surveys in combination with the surface; see Chap. 3). 1D profiles (usually not exactly straight) hardly allow the detection of density variations sideways from the profile and thus limit modelling essentially to 2D density sections, allowed only if it is known that no significant lateral density variations exist. $2\,^{1}/_{2}$ D models (Sects. 2.9.8; 6.2.3) are appropriate only if the lateral density variation can be estimated.

The optimized density distribution will not be affected critically if the reductions are smaller than the a posteriori accuracy of the data, as usually the case. The mean Bouguer density or the standard vertical gradient (Sects. 4.5.2 and 4.8) locally deviates from the standard values due to the geological situation. Density is never constant and varies inside the topographic mass systematically with topography (dense rock may also resist erosion), and the reductions with locally derived, not completely certain quantities introduce uncertainty or conflicts. Anomalies result from observations and standard reductions with clearly defined standard ("normal") values (see discussion of the fundamental properties of anomalies in Sects. 4.1 and 4.3), and possible differences of the local values or uncertain reduction parameters can be included into the parametrization and optimization as additional model components.

Averaging and smoothing the original data can reduce potential conflicts beforehand, but also introduce unexpected new conflicts! Combining observations will make them statistically dependent. The statistical properties of each observation type will be different. Theoretical relationships might exist between certain types of observations and can be anticipated, but unknown relationships between other types cannot. If the spatial spectra of the field are insufficiently known, for example, due to too sparse measurements, interpolation will overlook high-frequency components and suffer from aliasing.

7.2.2.1.2 Accuracy Estimates, Outliers, Correlation

The a priori standard deviations are estimated from the observation and reduction errors (see Sects. 3.1; 4.8). Experimental accuracy estimates can be derived from repeated measurements or between different types of measurements (see Sects. 3.6.1; 3.8.3). But when the sample size is not very large, the measurement process has to be reviewed which is necessary in almost all cases. The values found for the measurement accuracy must be considered rough estimates. Statistical correlations between observations (see Sect. 4.7.5) must be considered for a correct relative weighting of individual observations. If this is difficult, optimization of the Tikhonov regularization Parameter (Sect. 7.2.2.3.3) should be considered. The accuracy of the data or their likely errors are very important for optimization of models, because fitting the model effects to the data cannot constrain the optimization better than the data errors allow.

The norm chosen affects the error estimates. The L_2 norm is rather sensitive to data that deviate by large discrepancies. The L_1 norm is more robust, but the solution algorithms are more complex and evaluating the solution is more difficult. It is possible to combine the advantages of both methods (e.g. Nolet, 1987) by an iterative procedure where the a posteriori residuals of an L_2 solution optimally adjust the a priori standard deviation of the data at the next iteration step such that effectively the L_1 norm is optimized.

Statistically distinguishing outliers from statistical scatter, as by data snooping (see Sects. 3.8.1; 4.7.2), is important but difficult. Outliers are defined by being incompatible with the optimization criteria; for example, in case of the geometrical distance minimization the relevance of outlier size is not proportional to the defined distance function. In the case of the probabilistic method, the statistical properties of outliers do not correspond to the assumed statistics of the normally scattered observations. Correlations between observations must be considered for a correct statistical description and therefore usually influences the relative weighting of observations. But especially interpolation and filtering will introduce additional correlations which should not be ignored. Estimating correlation from repeated measurements, while combining different observations with complex theoretical relationships, is difficult. The correlation introduced by the theoretical relationship must be either removed or be taken into account in the optimization process (full covariance matrices C_y).

7.2.2.2 Model Relationships

"Model relationship" (see Sect. 7.1) has a broader and more general meaning than the usual term "model", expressing the uncertainty of how to formulate the basic problem. It is based on the system of linear equations $Ax = y$ or $A(x) = y$ (Eq. 7.1.1) for the unknown variables x on the basis of the observations y and the model relations $A(x)$ or A. There is no unambiguous "natural" prescription for the definition of x, y and A; they are all intimately entangled and mutually linked and cannot be strictly kept apart. Usually several possibilities exist as a matter of philosophy and the choice of A and x and their separate treatment are somewhat arbitrary. In other words, nature does not "know" such distinctions of roles or absolute definitions. Usually A incorporates physical laws, such as Newtonian gravitation or potential, but a priori knowledge or assumptions may be built into Ax beforehand with affects on the variables x and the observations y. For example, numerical constants, if accurately known, may be included into the physical laws A or into the observations, y, but also, if uncertain, into the variables, x. Some relationships and/or variables are hypothetical. To make the problem tractable, some choices may be necessary, for example, idealizations and simplifications of the model relationships, though objectively "wrong". Balancing the different requirements is the art of modelling.

Model relationships include mathematical conditions, for example, of equality or inequality and rankings of size and importance of individual model components. A given model has different parts which are mutually related, not independent, in other words, in most cases the different components are not orthogonal. Adding or leaving out one component will influence the results of optimization for the others, unless the models are described by orthogonal functions, as for example Fourier series, Chebyshev polynomials and spherical harmonics.

Two different modelling philosophies are typically held by researchers: either aiming at a detailed description of geological structures or emphasizing principal questions (Sect. 6.1.3). For modelling and optimization, the former aim requires detailed a priori information such as good seismic images, which in special cases, can be improved by gravity optimization, for example, in seismic "blind spots". However, the danger is to become merely descriptive. The latter more principal approach is rather exploratory and has more options of parametrization beyond the mere description of structures. The limitations in the numbers of optimizable (or invertible) variables is a further argument for this approach, a topic discussed as reparametrization (Sect. 7.2.2.3.1.1).

7.2.2.2.1 Alternatives

Distinctions between different model relationships are investigated by hypothesis testing or simultaneous solution of corresponding parameters. It should encompass the possibilities and the essentials of geological or physical reality. Mostly various alternative mathematical descriptions exist which each has its advantages and disadvantages, as adaptability to the complexities of nature and adequacy or inadequacy and accuracy or uncertainty in relation to the data. Different mathematical descriptions might have different problematic ranges (as, e.g., map projections with their own unprojectable areas (poles) and characteristic "errors" in projectable locations: shape conserving conformal projections versus area conserving authalic projections). Complex geometries always have alternative possibilities to be constructed. For simpler geometries parametrization may be carried out (1) with "large" rectangular prisms by defining their corner coordinates (Sects. 2.9.6.1; 2.9.7.4; 6.3.1), but inflexible in adapting to more complex shapes or (2) by fine point grids (Sect. 6.3.2.2

(1)) which, however, beside possibly suffering from discretization errors, cannot be stably optimized mathematically: small changes in observation values or model parameters result in very different optimized results which must be avoided, for example by reparametrization (Sect.7.2.2.3.1.1). Grids can be favourably oriented in space by coordinates rotation (see integration, Sect. 2.9). Parametrization can be done in various coordinates, as Cartesian, cylinder and spherical, or in different domains, as space or functional domains (power series, harmonic series, spherical harmonics, wavelets). The Gauss-Legendre quadrature in spherical coordinates of Sect. 2.10.7.3 is an example. As each parametrization is best adaptable for certain aspects of optimization, tests and comparisons are recommended.

7.2.2.2.2 Complexity, Accuracy, Conditions

Ideally the information about the a priori parameter values and the a priori standard errors is reliable and independent. The optimization procedures should then find the best compromise for the a posteriori parameters, but this does not happen automatically due to incomplete, inaccurate and/or contradictory knowledge, inadequate parametrization and error statistics. Often parameter uncertainties can only be estimated which is difficult when the relationships between data and side conditions are not obvious. Thus, complex model relationships may include observations with unclear statistical properties, which require analysis before optimization and much work must be done before optimization. If this total state of knowledge is exploited by relating real a priori probability distributions to all data and side conditions, reliable a posteriori probability distributions may result, but the solution may offer little new information.

When model relationships are not very accurately known they have themselves the character of optimizable variables. They may be considered observables with given or assumed error bounds as discussed above. Depending on the details or vagueness of geological a priori information, the model relationships may be extended by additional conditions and relations between separate model units (illustrated by the case of the SE Iceland shelf: Sect. 7.4.2).

Conditions can be interpreted as model relationships themselves. For example, mathematical conditions as equality $(=)$, inequality or exclusion $(\neq)$ or comparison $(<, >)$ usually have the highest priority. Examples for equality are repeated measurements of the same quantity (length of a rod, gravity value at a station, however, after subtraction of tidal and other temporal effects) or the sum of angles in a triangle; radii are always > 0 (although formally negative values may appear in a calculation with a special meaning), so are absolute densities. Or geological bodies do not occur above the Earth's surface: $z > 0$ (z pointing down and $z = 0$ at the surface). These mathematical conditions, though partly based on physics are stronger than physical relationships as a priori density bounds or practical conditions to be met by model design, as idealizations and simplifications of geometry, for example, smoothness (Sects. 7.2.2.3.4, 7.2.2.3.5).

7.2.2.3 Variables

Once the model relationships have been defined, the variables or model parameters can be calculated or optimized. In the case of linear problems the calculation is in one step and in non-linear problems it usually involves iteration. In geological modelling, the initial or a priori assumed model parameters (Chaps. 5 and 6) are usually somewhat hypothetical density distributions in 3D geometry (x, y, z), often simplified in 2D (x, z). They depend on the task and should be suited to the geological structures in question.

7.2.2.3.1 Types, Parametrization

Models are described by numbers with given meanings, and so are the gravity effects at certain locations (coordinates). It must be defined which model characteristics are to be optimized (see Sect. 6.1.3). Parameters or variables can be assumed either free, fixed, bounded by a priori standard deviations or limited by inequality conditions and they may be subject to a priori mutual dependencies or correlations by a pre-determined correlation matrix or controlled by pre-set condition equations.

7.2.2.3.1.1 Reparametrization

In most cases, different kinds of parametrization are possible for a given problem (see Sect. 6.1.4). Among the arguments for the choice to be made for a particular parametrization, three are most important: completeness, compactness and numerical efficiency. Not excluding each other, these conditions may, however, be in mutual conflict. Simple large mass units are well suited to optimization within their principal limitations, as they cannot fit observed small-scale variations, or they may be parameterized by fine grids of mass points, each with 4 parameters (coordinates, mass), which have less geometrical limitation but very many parameters. Fine grids of finite mass elements approach the properties of Fredholm's integral equation with infinitesimal mass elements. Such "complete" parametrizations with large numbers of small mass elements, while principally linear functions of the element densities, loose the linearity by the necessary regularizations in the form of additional conditions. Reparametrization can reduce the degrees of freedom by combining many of them to fewer adjustable parameters. It is meant to optimally use the given a priori information by limiting the variables with conditions as additional observation equations, e.g., smoothing or density contrast boundaries subjected to optimization of fewer parameters. One might define, for example, a grid of points and parameterize their relative masses (or densities) by a rather *smooth function* of the coordinates, say a Fourier series of low order which complies with the a priori information on limitations to the density variation and possibly on existing density gradation. Or a volume of fixed shape may define a *density contrast* which differs from that outside; the boundaries may be defined by conditions of equality or inequality (such as if $x \geq c$, $\rho = \rho_1$, else $\rho = \rho_2$), by barrier or by penalty functions (Sect. 7.2.2.3.5). The smaller number of variables, the smoother the model effects which affect the noise level and the standard deviation of the observations must be adapted to this type of parametrization error (Sect. 7.2.1.2.5).

7.2.2.3.1.2 Parametrization with Only Trivial Nullspace

Parametrizations with no formal ambiguity are represented by one or very few suitable uniform model bodies with density contrast boundaries. A list of simple geometries which can be defined to approximate geological structures rather naturally may include in **3D**:

- point masses or spheres (with 4 parameters each: mass m, coordinates x,y,z); if on sufficiently fine grids, point masses approximate integration over infinitesimal mass elements by summation over finite mass elements (see (Eq. 2.6.1), Sects. 5.6.3.1, 6.3.3.1 (3), 6.3.3.2 (1)));
- polyhedra or closed triangulated body surfaces with n corner points ($3n + 1$ parameters) (see Sects. 2.9.6.2, 6.3.1.2 (1));
- rectangular prisms, i.e. vertical columns with rectangular horizontal bottom and top planes, usually in regular grids based on digital terrain models (DTM) with heights (and densities) at n corner points or centres; the top planes might be determined by averaging from the values at the bordering corners ($3n$ parameters: n densities, $2n$ values of z, reduced to n, if the bottom z-value is fixed, plus possibly 3 parameters for origin and orientation of the DTM) (see Sect. 2.9.6.1: (Eqs. 2.9.41, 2.9.42, 2.9.43) with arrangements: Sects. 6.3.1.1 (2), (4), (6); 6.3.1.2 (5); 6.3.1.2 (3), 6.3.2.2 (2)));
- horizontal disks or layers of polygonal circumference and thickness $d = z_{top} - z_{bottom}$, which is a generalization of the above prism ($2n + 3$ parameters: density, z_{top}, z_{bottom}, $2n$ corner x, y coordinates) (see Sect. 2.9.4.2: (Eq. 2.9.30); Sects. 6.3.2.1 (2); 6.3.2.2 (3)).

2D:

- horizontal infinite mass lines or cylinders, e.g., in a regular grid in the vertical x, z plane (for n lines $3n$ parameters: n line masses $\rho^+ = \rho\pi R^2$ (either ρ^+ or ρ or R), possibly plus 4 parameters for origin x, y, z and azimuth); such line masses, like point masses, in sufficiently fine grids approximate integration by summation (see Sect. 2.9.3.1: (Eq. 2.9.7); Sects. 2.9.7.1.1: (Eq. 2.9.49); Sects. 5.6.3.2; 6.2.1.1 (1); 6.2.21);
- horizontally infinite beams with polygonal cross section in the vertical plane, called "Talwani beams" ($2n$ parameters: corner point coordinates x, y, plus possibly 4 parameters for density, azimuth and origin x_o, y_o; see Sects. 2.9.7.4.3, 2.9.7.4.4: (Eqs. 2.9.50, 2.9.51, 2.9.65); Sects. 6.2.2.1 (6); 6.2.1.2 (1)).

The list is not complete, and other kinds of parametrization can be combined or new ones can be invented. Modelling of undulated seismic boundaries may aim at optimizing the corresponding density contrasts either constant or characterized by a simple gradient. Surface boundaries of geological maps may be continued downward and borehole data can render a priori information exploited by such parametrizations, also well suited for detailed inversion. While gravitation is a linear function of density, geometry and density always form products, and for large changes of geometry, a solution ansatz for non-linear equations will be required.

Small-scale density anomalies as occurring at body edges may be suitably parameterized. Even if the ambiguity seems to disappear, such parametrizations are never quite realistic, as they will not represent all possibilities existing in nature. The ambiguity is reduced seemingly by combining small and large-scale structures or by basic simple shapes of uniform mass elements assembled irregularly or in regular grids as suited to the a priori information. Possibilities exist to design parametrizations with formally no ambiguity (and such with merely trivial ambiguity; null-masses or null-densities can always be added) which have advantages and disadvantages.

7.2.2.3.1.3 *"Complete" Smooth Parametrizations*

Famous examples of "smooth" functions with no ambiguity are the Fourier series and the spherical harmonics (Sects. 2.10.5, 2.10.7). The smoothness is determined by the maximum degree and order assumed for fitting the given discrete data. Downward continuation of Fourier-transformed observations (see Sect. 2.10.5.3; Eqs. 2.10.14, 2.10.15), coupled to the equivalent stratum (Sect. 2.7.3) renders the density unambiguously under the condition that the intermediate space or depth range is empty ($\rho = 0$) or has a constant density ($\rho = $ const). With $b_{nm} = $ gravity coefficients:

$$\begin{cases} \rho^*(x,y,z) & = (1/NM) \sum_{n=0}^{N-1} \sum_{m=0}^{M-1} x_{nm}(z) e^{-2\pi i(nx/N+my/M)} \\ x_{nm}(z) & = (2\pi G)^{-1} b_{nm} \exp\left(-z\left(n^2+m^2\right)^{1/2}\right) \end{cases} \quad (7.2.26)$$

The spherical harmonics equally render a unique solution under the equivalent conditions:

$$\rho(\phi,\lambda,r) = \sum_{n=0}^{N} r^n \sum_{m=0}^{n} \left(x^c_{nm} \cos m\lambda + x^s_{nm} \sin m\lambda\right) P_{nm}(\cos\phi) \quad (7.2.27)$$

and

$$x^c_{nm}, x^s_{nm} = \left[(2n+1)(2n+3)/4\pi GR^{2n+3}\right] b^c_{nm}, b^s_{nm} \quad (7.2.28)$$

7.2.2.3.1.4 *Transformation of Variables*

Reparametrization can be achieved by transforming the parametrization into any other more suitable form, for example, from simple "macroscopic shapes" (polyhedra or specialized prisms or disks) to sets of many small "microscopic" elements (6.2.2.1, 6.3.2.1). It may be advantageous, on the basis of a given fine grid, to decide if a mass element belongs into one or another body, depending on the definition of a boundary. The effect of the microscopic mass elements is calculated separately and the speed of computation may be increased by tabulation and interpolation routines. The reparametrization can also be carried out in the opposite direction, since relatively simple large volumes of uniform density defined by few variables should be better optimizable. Typical shapes include cylindrical bodies approximated by polygonal outline and vertical, horizontal or oblique orientation. The relief of density boundaries can be described by digital terrain models. Definition and estimation of a priori values and standard deviations should permit easy input. Most of the usual parametrizations are combinations of simpler ones (polyhedra composed of elementary tetrahedra, DTM composed of rectangular prisms, polygonal 2D Talwani beams composed of elementary steps, etc.). The horizontal disk may approximate a vertical circular cylinder, and the circle can be generalized to ellipses or cycloids or trochoids capable of describing horizontally complex geological structures with only a limited number of adjustable parameters (Appendix A1). A further generalization may be an arbitrary orientation of the various above bodies, involving coordinate transformations (Sect. 2.4.3.1).

7.2.2.3.2 A-Priori Information, Correlations, Conditions

A priori information of some kind is always included in gravity modelling and optimization, even if implicit and invisible for the inexperienced worker, e.g., the damping by the factor e^{-z} in Eq. (7.2.26). A priori information ranges from ideas and hypotheses to detailed structures as supplied by other methods of geology and geophysics (Chaps. 5, 6). The most important rule for estimating the standard deviations of the a priori variables is the so-called *two-sigma rule*, that is to adjust the a priori standard deviations of different parameters (e.g. density and depth of a body) by first asking, which deviations would be unacceptable independently, say, at the 95% level defining a range of $\pm\,2\sigma$ (e.g. $100\,\mathrm{kg/m^3}$ and $10\,\mathrm{m}$) and then to assume half these values as suitable standard deviations, σ ($50\,\mathrm{kg/m^3}$, $5\,\mathrm{m}$).

Correlations can be investigated by calculating the a priori covariance matrix (Sect. 7.2.2.3.8) of the regularized variables, or if a priori standard deviations of the variables are given, a correlation matrix expresses that the chosen variables are not formulated as being independent from each other (non-orthogonality). In case of geometrically or temporally ordered variables, a correlation length, d_0, can be estimated within which a variable should not change more than a certain fraction of its standard deviation from its neighbours. The correlation coefficient r_{ij} can be estimated from the "distance" d_{ij} between the variables i and j with a correlation function, for example, $r_{ij} = \exp(-(d_{ij}/d_0)^p)$ or $r_{ij} = 1/(1 + (d_{ij}/d_0)^p)$, where p is a suitably chosen constant.

"Correlation" can be treated by the above covariance and correlation analyses. On the other hand, if obvious in advance, correlations can be included in the problem formulation and hence in optimization. This reduces the number of variables stabilizing the optimization. Variables with correlated gravity effects are, for example, the density and the volume of a body (mass = density × volume), or coordinates of mutual boundary elements between contiguous model volumes, if shifted, change both volumes equally but with opposite sign.

7.2.2.3.3 Tikhonov Regularization

Tikhonov regularization is a method of determining the mutual relative influence of the data versus that of the given a priori information in an optimization problem. The standard form is (Tikhonov & Arsenin, 1977):

$$(Ax - \tilde{y})^{\mathrm{T}}(Ax - \tilde{y}) + \lambda^2 (x - \tilde{x})^{\mathrm{T}}(x - \tilde{x}) = \|Ax - y\|^2 + \lambda^2 \|x - \tilde{x}\|^2 = \min \quad (7.2.29)$$

or generalized:

$$(Ax - \tilde{y})^{\mathrm{T}}(Ax - \tilde{y}) + \lambda^2 (Bx - \tilde{z})^{\mathrm{T}}(Bx - \tilde{z}) = \|Ax - \tilde{y}\|^2 + \lambda^2 \|Bx - \hat{z}\|^2 = \min \quad (7.2.30)$$

The data are divided according to (Eq. 7.2.3), and the covariance matrices are unit matrices, multiplied by the scalar λ^2. The generalization to full covariance matrices can be transferred into the form (see Sect. 7.2.1.2.6):

$$(Ax - y)^{\mathrm{T}} C_{yA}{}^{-1}(Ax - y) + \lambda^2 (Bx - z)^{\mathrm{T}} C_{zB}{}^{-1}(Bx - z) =$$
$$(A'x - y')^{\mathrm{T}}(A'x - y') + \lambda^2 (B'x - z)^{\mathrm{T}}(B'x - z) \quad (7.2.31)$$
$$\text{with } A' = C_{yA}{}^{-1/2}A, \ y' = C_{yA}{}^{-1/2}y, \ B' = C_{zB}{}^{-1/2}B, \ z' = C_{zB}{}^{-1/2}z$$

where, for example, $C^{1/2} = ES$ and E is an orthonormal matrix of eigen-vectors (columns) of the positive definite covariance matrix $C = C^{1/2}C^{1/2}$; S is the corresponding diagonal matrix of the square roots of the eigen-values.

The standard form is the least-squares minimizing criterion with the solution (Eq. 7.2.14). In the Tikhonov regularization it is essential that λ^2 is a free optimizable parameter. A balance is intended between adjusting to the new data, y, and to the old data, i.e. the a priori information or the side conditions. If λ^2 is included in C_{zB} it represents the certainty of the a priori, and when $1/\lambda^2$ is included in C_{yA} it quantifies implicitly the parametrization error C_A (see Sect. 7.2.1.2.5). In the quite common case that the weighting of the a priori information relative to the data is much more difficult than the determination of the covariance matrices, especially C_z or B', then optimization of λ^2 is recommendable. If, on the other hand, the relative mutual weighting of the individual pieces of a priori information is uncertain, then λ^2 is only one among many free parameters. In this case optimizing only λ^2 will most likely not lead to the optimal result. An important example are densely gridded observations dominating the data relative to the a priori information on the variables. Adapting λ^2 corresponds to finding a better mutual relative weighting while leaving the data grid unchanged.

The regularization parameter λ^2 can well illustrate the so-called trade-off or exchangeability of new data and a priori information. Plotted in double-logarithmic scale, the residual terms in (Eq. 7.2.30), $\|Ax - \tilde{y}\|$ and $\|Bx - \tilde{z}\|$, as function of the regularization parameter λ^2, represent in most cases an L-shaped curve (Fig. 7.2.2). A small λ^2 fits the new data well but permits a large deviation from the a priori information, while a large λ^2 leads to results which mostly reflect the a priori information and practically neglect the data. If both data sets are consistent, the choice of

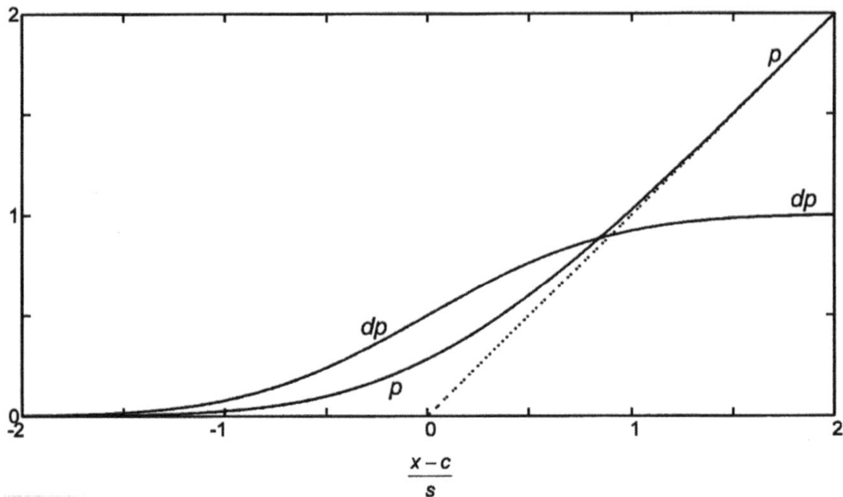

Fig. 7.2.2 The L-curve shows, when the Tikhonov regularization parameter λ is varied from small (*upper left*) to large (*lower right*), on the abscissa: how the fit to the new data varies and, on the ordinate: how the fit to the a priori information (or old data) varies, both together demonstrating the trade-off

λ^2 is irrelevant. In the case of discrepancies, the optimization of (Eq. 7.2.1–2) with λ^2 as a free parameter results in a solution which is equally fitted to the data and to the a priori information. The implicit assumption is that the whole new data set has the same weight as the total previous information. Such an assumption is sensible if the old information consists of expected a priori values of individual model parameters and the new information consists of many repeated or mutually checking measurements, which might gain a large weight dominating the previous information. A cumbersome analysis of the whole data material and its statistical relevance is then replaced by the optimization of the regularization parameter. Often the λ^2 value at the "bend" of the L-curve is taken as the optimum from whereon one residual term strongly deteriorates while the other improves only slightly. Under certain circumstances an additional upper bound may be set for one or the other residual term, or for both, such that the regularization parameter may be further limited. The "edgehog" procedure (Jackson, 1973) is based on such assumptions.

The Tikhonov regularization invites us to think about what can be said a priori about the variables x. If all variables are independent and contained in the a priori information, a hypothetical singularity is prevented from the very start, for as formulated below in Sect. 7.3.2.3.2, these variables predetermine the deepest point of the variance topography; the condition $Ax = y$ will only shift it, if it contains better information. The generalized form of the Tikhonov regularization leaves the choice of the B-matrix and the a priori information z open. Regularization with the Sobolev norm and with inequality conditions represents special cases which require an optimization of the regularization parameter.

7.2.2.3.4 Smoothness or Sobolev Condition

Smoothness conditions are frequently needed in optimization and inversion problems, e.g. to control the data scatter relative to suitable model dimensions. Fredholm's integral equation (7.2.2) in the form of the Newton integral (Eq. 2.5.5) is a distance-dependent smoothing operator. It causes very small-scale far-away density variations to become unobservable. They are thus part of the nullspace and must be eliminated by regularization. The inverse operator roughens the observations and translates every small-scale random scatter into a strong variation of the variables.

A smoothness condition can be defined, if a partial set of the variables x are considered values of the same function at their positions ξ, or according to Menke (1984), if they have a natural order, as for example, density as a function of space coordinates or temperature as a function of time, where the function $x(\xi)$ is the smoother, the smaller the derivatives $\delta^n x/\delta \xi^n$. Such a condition is:

$$(B_{S_2^N}x - \tilde{z})^{\mathrm{T}}C_{zb}^{-1}(B_{S_2^N}x - \tilde{z}) = (\alpha_o x(\xi) + \sum_n \alpha_n \delta^n x/\delta \xi^n - \tilde{0})^{\mathrm{T}}C_{zB}^{-1}$$

$$(\alpha_o x(\xi) + \sum_n \alpha_n \delta^n x/\delta \xi^n - \tilde{0}) \tag{7.2.32}$$

For equidistant ξ_i, it is, with the constants a_0 and a_1:

$$B_{S_2^1}x = \alpha_o x(\xi) + \alpha_1 \delta x(\xi)/\delta \xi = \alpha_o x(\xi_i) + \alpha_1(x(\xi_{i+1}) - x(\xi_i))/\Delta \xi$$

$$= \alpha_o diag(x) + \alpha_1/\Delta \xi \begin{bmatrix} -1 & 1 & 0 & \dots & 0 \\ 0 & -1 & 1 & \dots & \\ \dots & \dots & \dots & & 0 \\ 0 & & \dots & 0 & -1 & 1 \end{bmatrix} \tag{7.2.33}$$

or

$$B_{S_2^2}x = B_{S21}x + \alpha_2 \delta^2 x(\xi)/\delta \xi^2$$

$$= B_{S21}x + \alpha_2((x(\xi_{i+1}) - x(\xi_i)) - (x(\xi_i) - x(\xi_{i-1})))/\Delta \xi^2$$

$$= B_{S21}x + \alpha_2/\Delta \xi^2 \begin{bmatrix} 1 & -2 & 1 & 0 & \dots & 0 \\ 0 & 1 & -2 & 1 & & \\ & & \dots & \dots & & 0 \\ 0 & \dots & & 0 & 1 & -2 & 1 \end{bmatrix} \tag{7.2.34}$$

The pure Sobolev L_2-norm S_2^{N} lets the minimum condition extend to the Nth derivative, hence $\alpha_n = 1$ for all $n = 1, N$. In practice, every α_n can be set 0 or 1 just as needed for the smoothness wanted.

7.2.2.3.5 Inequality Condition (Exclusion Condition)

Certain ranges of parameter values are usually excluded, e.g., by the given a priori information or the physical situation, such as the space above the Earth's surface.

Exclusion can be achieved by inequality conditions. Inequality conditions as $x \leq c$, $x \geq c$ or $c_1 \leq x \leq c_2$ provide a priori conditions only in a limited sense. They correspond to an infinite uncertainty, i.e. $C_x = diag(\infty)$, as long as the condition is satisfied, but as the solution is about to transgress the boundary, they correspond to absolute fixing the variable at the value of c, i.e. $C_x = 0$. A possible ambiguity will no longer extend from $-\infty$ to $+\infty$, but will be limited by inequality conditions to the permitted region. It may, however, happen that no solution exists if the conditions exclude each other forming an overdetermined system of condition equations which itself must be regularized (Sect. 7.2.1.2.2).

The usual methods of solving equation systems with inequality conditions are

- search methods, as the simplex method and the gradient method which check the side condition for each iteration;
- reparametrization of the variables (applied in Sect. 7.2.2.3.1.2) by the introduction of the $N(0,1)$ normal-distributed variable $x' = \log((x \pm c)/(x_{o \pm} c))/s$ (instead of the ordinary variable x) with the back-transformation $x = (x_o \pm c)\exp(sx') \pm (-c)$ and the parameters $x_o(\approx x)$ and $s\ (>> 1)$, free to choose; it leads to a solution for x which obeys the condition $x < c$ (upper sign) or $x > c$ (lower sign) (compare Tarantola, 2005); the implicit probability density of x is thus condensed toward the boundary and dilated in the opposite direction;
- penalty and barrier methods add to the function to be minimized a term which increases toward the boundary:

$$(y - Ax)^{\mathrm{T}} C_{yA}^{-1} (y - Ax) + \lambda^2 \Pi_{...}(x, c, \ldots) = \min, \text{ with}$$

$$\Pi_{\text{barr}}(x, c, s) = -\ln((x \pm c)/s + 1) \text{ or } \Pi_{\text{pen}}(x, c, s, r) = \max(0, ((x \pm c)/s)^r)$$

$$(7.2.35)$$

The barrier Π_{barr} increases to infinity at the boundary c, such that it cannot ever be reached, but increases before reaching c holding back the solution at a distance within the permitted region. In contrast, the penalty Π_{pen} rises only within the forbidden region such that the solution will pass c until the penalty term reaches an equilibrium with the rest of the minimum function.

Inequality conditions of physically meaningful parameters (e.g. radius > 0, 10^4 kg/m^3 > density > 0, depth > 0, etc.) are idealized probability densities which are, indeed, small near the boundaries and are thus reasonably represented by barrier methods. The penalty method does not totally exclude the forbidden region, but the adjustment leads to a weighted mean. It may not be critical that some conditions are not met, since a completely satisfactory solution is no possible in this case. By trespassing in the forbidden region, the result may show which conditions might be incompatible.

An example of a *penalty* function, increasing the resistance with the approach of the boundary and further within the region $x > c$, is the following $p(x')$ with $x' = (x - c)/s$, c and s being constants, and the $+$ and $-$ sign for the two conditions $x < c$ and $x > c$, respectively (Eq. 7.2.35):

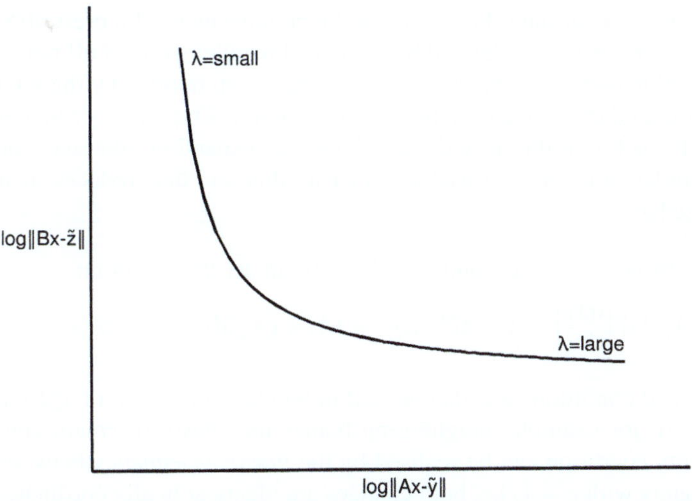

Fig. 7.2.3 The penalty function p and its derivative dp for an upper limit c; it rises toward a linear asymptote $(\mathrm{d}p(x') \approx \text{constant})$, and the transition length is determined by s $(x' = (x-c)/s)$

$$
\left\{
\begin{array}{l}
p(x') = {}^{1}/_{2}(x' + x'\mathrm{erf}(x') + \pi^{-1/2}\exp(-x'^{2})) \\[4pt]
dp(x') = {}^{1}/_{2} \pm {}^{1}/_{2}\mathrm{erf}(x') \\[4pt]
\begin{bmatrix} A \\ dp(\pm x')/(\pm s) \end{bmatrix} dx = \begin{bmatrix} \mathbf{y} - \mathscr{A}(x) \\ p(\pm x') \end{bmatrix}
\end{array}
\right.
\qquad (7.2.36)
$$

The function $p(x')$ (Fig. 7.2.3) is linear and nearly zero in the permitted region except near $x = c$. In the "forbidden" region, it rises toward a linear asymptote $(dp(x') \approx \text{const})$, and the transition length is determined by the value of s. This definition guarantees that the linearized function intersects zero at or near c, and if a current solution $x' > 2s$, $p(x')$, the penalty function takes on the role of an observation $\tilde{x} = c$ with a standard deviation of $1s$.

A disadvantage of the penalty and barrier methods is that in most cases assumptions are violated, either that of the non-linearity of the system of equations to be solved or that of normal distributions of the data (or a priori information). Generally it must be decided if the assumptions inside the permitted region are to be maintained while violated only near the boundary, or if a potentially forbidden solution is to be prevented from reaching the boundary largely preserving the assumptions but already sensing the influence still within the permitted region and trespassing the boundary strongly impeded.

7.2.2.3.6 Linearly Related Variables

Reparametrization can also exploit *linear dependencies* between variables, i.e. a change Δp_2 of parameter 2 is linearly related to the change Δp_1 of parameter

1: $\Delta p_2 = c\,\Delta p_1$, c being a constant. For example, block movement of a prescribed volume of constant point masses is defined by equal and parallel shifts of all points in that volume. Or all masses, i.e., the density as the mass sum divided by the total volume, may be permitted to change each by the same amount. Or point coordinates may change linearly such that the prescribed volume can expand or contract. The fact that scaled parallel shift relates variables to each other and thus reduces their numbers is expressed as:

$$\mathscr{A}(x_1,x_2) = y \text{ with } x_1 = \tilde{x}_1 + dx_1, \text{ and } x_2 = \tilde{x}_2 + dx_2 \text{ under the condition}$$

$$dx_2 = c\,dx_1 \Rightarrow [A_1\ A_2]\begin{bmatrix} dx_1 \\ dx_2 \end{bmatrix} = y - A(\tilde{x}_1,\tilde{x}_2) \Rightarrow (A_1 + cA_2)dx_1 = y - \mathscr{A}(\tilde{x}_1,\tilde{x}_2)$$

Linear relations and conditions are also needed in regularizations with large but simple geometries. If, for example, neighboring bodies may have to remain connected, the contiguity condition can be realised by the *linear dependency* between the affected parameters with $c = 1$, i.e. both changes are identical in all coordinates of the connected points. This reduces the number independent variables, too. Similarly, such bodies can also be scaled and shifted by the same kind of linear relations. Furthermore, rigid rotations can be provided by the proper relations between coordinate changes in the optimization process.

7.2.2.3.7 Lanczos Inverse

The Lanczos inverse can be regarded a special case of the Backus-Gilbert approach (see Sect. 7.3.1.2.2). The generalized eigen-value analysis (Sect. 7.2.1.2.6) leads to the so-called *singular value decomposition* (SVD) of A:

$$A = WSV^{\mathrm{T}} \tag{7.2.37}$$

The columns of V are an orthonormal basis for the model space of A, i.e. for the variables x; the columns of W are an orthonormal basis for the projection space of A, i.e. for the observations y. The vectors of V belonging to the generalized singular values $S_i = 0$, constitute the model nullspace, i.e. the linear combinations of variables which have no effect on the observations. The corresponding vectors of W constitute the partial space of the discrepancies of the observations, i.e. the linear combination of observations which cannot be generated by the variables. For the non-regularized observation equation, $Ax = y$, follows from (Eq. 7.2.23) that S contains the singular values and V the eigen-vectors of the a posteriori covariance matrix of the variables $(A^{\mathrm{T}}A)^{-1}$.

The so-called Lanczos inverse (Lanczos, 1961; Jackson, 1972) of A is defined as $VS_L^{-1}W^T$, where $S_L = diag(\sigma_{i\leq p} > 0,\ \infty_i > p)$ or $S_L^{-1} = diag(1/\sigma_{i\leq p} > 0,\ 0_{i>p})$. Since S_L consists only of elements > 0, the inverse exists always and projects nothing into the nullspace; the solution has a component of length zero in the direction of the nullspace vector. To be a sensible solution technique, the variables must be deviations dx from the estimates x, as is the case with solutions according to the minimum norm or the Sobolev norm. Thus, the implicit regularization or a priori information is that the solution should contain only components that can be projected by A. In the case of gravity inversion, it means a restriction to gravimetrically relevant density distributions.

If no a priori information or sensible estimates of x exist, then this solution is the most reasonable one; it says nothing about components of which nothing can be said. A nearly singular matrix A, and thus an unstable solution x, is characterized by $\sigma_i \ll 1$ or $1/\sigma_i \gg 1$. The instability can be avoided by damping the solution through forcing the $1/\sigma_i$ values to be small by the damping parameter α_i, such that $S_D^{-1} = diag(\alpha_i/\sigma_i)$ with $1 \geq \alpha_i \geq \alpha_{i+1} \geq \alpha_p \geq 0$ (*damped singular value decomposition*, DSVD); when $\alpha_{i \leq j} = 1$ and $\alpha_{i > j} = 0$, we have the so-called *truncated singular value decomposition*, TSVD (see Hansen, 1993). DSVD with $\alpha_i = \sigma_i^2/(\sigma_i^2 + \lambda^2)$ corresponds to a Tikhonov regularization with the regularization parameter λ^2. The TSVD has the solution $\hat{x} = VS_L^{-1}W^T y$ with $y = Ax_y = WS_L V^T x_y$. From this follows for the second criterion of Sect. 7.3.1.2.2 that $\hat{x} = x_r = R_y y = VS_L^{-1}W^T W S_L V^T x_y = x_y$ and hence $R_x = I$. This regularization is optimal according to the Backus-Gilbert method, and the parametrization always leads to a solution, but since the decomposition is usually unknown beforehand, the only known fact is that some badly determined linear combinations of the variables have been removed from the solution, but it is not known a priori which solution is preferred. It is possible to control this by considering the SVD of the regularized problem $A_{Bc}(dx_c) = dy_{yc}$, instead of the original problem $A(dx) = y - Ax$:

$$\begin{cases} \begin{bmatrix} A \\ B \end{bmatrix} dx = \begin{bmatrix} y - Ax \\ z - Bx \end{bmatrix} \qquad A_{Bc}dx_c = dy_{zc} \\[2em] \qquad\qquad \Rightarrow \\[1em] C_{yz} = \begin{bmatrix} C_y & 0 \\ 0 & C_z \end{bmatrix}, C_{dx} \quad C_{yz_c} = I, C_{dx_c} = I \end{cases} \quad \text{with} \quad \begin{cases} A_{Bc} = \begin{bmatrix} C_y^{-1/2} & 0 \\ 0 & C_z^{-1/2} \end{bmatrix} \cdot \begin{bmatrix} A \\ B \end{bmatrix} C_{dx}^{1/2} \\[2em] dy_{zc} = \begin{bmatrix} C_y^{-1/2} & 0 \\ 0 & C_z^{-1/2} \end{bmatrix} \begin{bmatrix} y - Ax \\ z - Bx \end{bmatrix} \\[2em] dx_c = C_{dx}^{-1/2}dx \end{cases}$$

$$(7.2.38)$$

If the side conditions $Bx = z$ do regularize the problem completely, the solution is the same as that with the Tikhonov regularization. If the side conditions do not suffice and unexpectedly a nullspace remains, the Lanczos inverse or the TSVD/DSVD, as a "black box", provides a unique solution which is as close as possible to the estimate x, in the sense of a minimizing norm with respect to C_{dx}. The problem has been recently investigated also by Sebazungu (2005).

7.2.2.3.8 Covariance Matrix

A general analysis of linear relationships between variables is furnished by the a priori covariance matrix (see below Sect. 7.3.2.2.4) of the regularized variables or, if the standard deviations are estimated independently from it, a correlation matrix.

Such a matrix can be calculated directly in most cases (see Sect. 7.2.2.3.2). Another possibility exists, if a side condition, $Bx = z$, relates several or all variables to each other. It can be optimized in a first-phase adjustment, based on the a priori $\tilde{x}$, their standard deviations S_x, and the uncertainty S_z, within which the side condition is to be satisfied. Values of x corresponding to the condition equation are estimated with $S_x^2 B^T (BS_x^2 B^T + S_z^2)(B\tilde{x} - \tilde{z})$: the covariance matrix follows from

$$S_x^2 - S_x^2 B^T (BS_x^2 B^T + S_z^2)BS_x^2.$$

If the standard deviation of z is chosen zero, side conditions could possibly be defined to be in mutual conflict. The changes of the variables may also satisfy the side condition $B\,\delta x = 0$, which automatically holds for the a priori information but is used only to determine the covariance matrix of the variable changes as in $S_x^2 - S_x^2 B^T (BS_x^2 B^T + S_z^2)BS_x^2$, above. But this requires the conditions to be linear, or after each solution of the basic equation the covariance matrix must be recalculated.

7.2.2.3.9 Accuracy

Reliable a priori values of the chosen parameter values and their standard errors are essential for the whole complex of the model relationships (Sect. 7.2.2.2.2), but knowledge is usually incomplete, inaccurate and contradictory and the parametrization may be questionable. The parameter uncertainties themselves can often be only vaguely guessed. The input of the a priori information may be tedious, and proper weighting of auxiliary conditions is often difficult. All the a priori choices affect the optimization results, partly intended, partly uncontrollably. Ambiguity can be cancelled or reduced only in the mathematical *solution*, but if too much is prescribed a priori, the solution will no longer provide new information from the data. Most of these problems are solvable by the 2σ rule (see below), especially relevant to inversion.

7.2.2.3.10 Application of Regularization Methods

Regularization must be balanced between too little and too much which is not easy. Over-regulating returns only the input with the solutions, and under-regulating leads to unstable solutions. A stable and reliable solution rarely offers new unexpected details which can be gained only with the risk of ambiguity and instability. The trade-off between solvability and stability can be tested by the Tikhonov regularization parameter λ^2, controlling the relative weighting of data and a priori information as illustrated by the L-curve (Sect. 7.2.2.3.3; Fig. 7.2.2). The task of inversion is to find the best compromise between the influences of the data and of the side conditions. Properly weighted standard deviations may be problematic, even if the Tikhonov parameter is optimized automatically. It is advisable to adjust the a priori standard deviations of different parameters by applying the *two-sigma rule* of Sect. 7.2.2.3.2.

The definition of a minimization condition for the parameters with the Tikhonov regularization is facilitated by giving the parameters a form in which the conditions are easily formulated or deduced. For the latter, the parameters should have approximately a normal distribution. The Sobolev norm (Sect. 7.2.2.3.4) assumes that the parameters are chosen to have a "natural" order.

Real or realistic a priori probability density distributions of all data and side conditions calculated *before* the solution is calculated reflect the total state of knowledge expressed by the a posteriori probability distributions. If serious discrepancies have arisen, desirable additional observations or observational methods may be suggested or rethinking of the problem. Alternatively, one might begin with minimal a priori information, analyze the solutions by comparison with the data and side conditions for unwanted properties and iteratively add more and more a priori information until only acceptable solutions are generated taking all information into account. If resolution is insufficient new observational methods or ideas should be considered. In this case, the main work is done *after* each mathematical solution. The "nullspace shuttle" of Deal & Nolet (1996) is an example of this approach, modifying an initial solution, determined by the Lanczos inverse according to the auxiliary conditions; however, only that part of the modified intermediate solution is accepted into the final solution that does not contradict the data, i.e., it lies in the nullspace of the original problem.

The first strategy requires a vision of all aspects of the problem and compression of partial aspects of information, to keep the data quantity manageable and to allow proper weighting and compliance of correlations between them. It is often a numerical problem, but it permits a better analysis of the result in its various aspects and employs the capabilities of inversion in the literal way. The iterative strategy derives partial results rather separately. This is no inversion any more but rather forward modelling with the disadvantages of both methods, intended inversion combined with incomplete possibilities to incorporate all information in a way the researchers are not aware of. The influence of the neglected information remains unclear. In practice, the choice will often lie somewhere between these extremes depending on the human and numerical possibilities.

The Bayesian approach is to include non-gravity a priori information, not merely as initial assumptions (as would be needed for any iterative scheme), but as data on some or all model parameters p_k, $k = 1$, m, with estimated error distributions. They are treated like data in the inversion-optimization process and their residuals, $r_k = p_{ok} - p_k$, are included in the minimization of the variance S. This permits the combination of data sets of different nature and different physical units, non-dimensionalized or normalized by their individual estimated standard deviations σ_k (see Sect. 7.2.1.2) and could be further developed to joint inversion of original observations. If the a priori σ_k values are unknown, they have to be very carefully estimated, but their sizes have only second-order effects on the solution, where the a priori parameter values have first-order effects. Effects of changes in the relative sizes of the σ_k on the results of fitting should be investigated. They should be trusted only if well established or if they have no strong influence on the results. Assuming normal error distributions permits the application of efficient LSQ-oriented algorithms, easily rendering a posteriori

standard deviations. The justification of the statistical assumptions must be checked
a posteriori. If normal distributions are well established (combination of normally
distributed errors leads to collective normal distributions), the least-squares method
renders the most likely parameter values, the largest probability densities and the
smallest variances (by definition, see Rao, 1973).

Superposition of many processes of observation and data reduction, each with its
normal distribution, leads again to normal distributions, such that the variance is a
good measure of the scatter or error distribution.

Suspicion against bias and preconceived (Bayesian) ideas has led to supposedly
unbiased inversion programs fitting the gravity data to any desired degree and producing apparently accurate images of underground density distributions. However,
implicit assumptions are easily overlooked. Close fitting of the data and attractive
models deceive the inexperienced researcher about their uncertainties. "Unbiased"
results are questionable if in conflict with reliable a priori information. Hence, exclusively forward modelling programs are still much in use which accurately calculate
gravity effects but cannot easily be made to fit the observations and do not systematically investigate the full range of the a priori knowledge with its uncertainty
limits. Therefore a compromise is needed between accurate detailed forward modelling and optimal fitting of the observations using any valid a priori information
and illuminating the range of possibilities.

7.3 Inversion

Inversion has the aim to determine geologically and geodynamically important density distributions inside the Earth which cannot be obtained directly from samples.
One of the first steps of inversion is the recognition and characterization of "anomalies" which suggest certain mass distributions. While optimization requires an initial
approximation, inversion should find the solution, for example for certain variables
of special interest, directly from the data. The solution should be optimal. However
in potential fields a priori search ideas are mandatory.

Gravity fields or anomalies are estimated from discrete data (Sect. 1.4) where
"field" usually means "continuous variation of a quantity". Model effects are usually quantified as continua (outside the source volumes), but calculations are carried
out usually with discrete numbers at a limited set of observation points. Different
components of a density model are not "orthogonal": they influence each other mutually in the inversion and optimization and affect all variables. Data distributions
influence the ultimate inversion results by the minimum condition over the whole
data set (see 5.1.5), sometimes seriously. The solution to the inverse problem lies in
its regularization by the available a priori information and adequate parametrization.
In Bayesian inversion the parameters are accompanied by their uncertainties. It may
be extended to adjustment in phases (Sect. 7.2.1.2.4) or even joint inversion which
requires the primary observations (e.g. seismic travel times, etc.) and their respective theories to be included. Gravity is the subject of this treatise, and all other data

are regarded a priori information helping to regularize the inversion problem. For a seismologist the situation would be reversed.

In practice the range of acceptable models can by narrowed by a priori knowledge, in exceptional cases to zero (see Sects. 7.2.2.3.1, 2.7.3), but only as a mathematical possibility. The uncertainties in the parameters of a priori models and in the observations will leave room for many acceptable models, characterized by the a posteriori parameter values and their uncertainties. Search for acceptable models resembles trial and error, and optimization may be guided, e.g., by the Gaussian method (Sect. 7.2.1.2). If that way the main minimum is missed, systematic or random search and genetic model variations (see below: Sect. 7.3.3) are possibilities. The different strategies can be combined offering the possibility to choose the best.

7.3.1 Overview

Optimizing algorithms are at the heart of inversion. They can be chosen such as to suit the task at hand and are available in the literature on numerical mathematics. Practical inversion methodology is essentially the same as for optimization. Theoretical aspects specific to inversion problems are treated in this section. Emphasis is on regularization, parametrization and evaluation of the results, the calculation of a priori and a posteriori covariance, correlation and resolution matrices etc. Direct methods (Sect. 7.3.2) solve the normal equations, $Ax = y$ (Eq. 7.1.1), by matrix inversion, applied to linear and non-linear problems. The latter usually require linearization and iteration, the calculation of the Jacobians and evaluation of convergence properties and abort criteria. Iterative algorithms as conjugate gradients, LSQR (Paige & Saunders, 1982) etc. search a solution by iteration, without explicitly calculating $(A^{\mathrm{T}} A)^{-1}$ or even A^{-1} along deterministic mathematical rules like line minimization, steepest descent routines or conjugate gradients. Evolutionary and similar algorithms exploit chance and large numbers of tests, realizable by efficient computers (see Sect. 7.3.3). The assessment of the results needs criteria and permits a judgment about the accuracy, reliability and resolution. The theoretical concepts are illustrated by two case studies (Sect. 7.4) and in several tasks.

7.3.1.1 General Remarks: The Situation

For inversion, important questions are:

o Which data have the greatest influence on the solutions, which ones have little influence?
o What kind of additional information will optimally lead to acceptable solutions?
o What is the cost-benefit relation between data gathering and knowledge gain?
o What can be sensibly gained from a solution and what not?
o How can a density distribution be derived from the available data, with emphasis on gravity, which corresponds to all known aspects, adds new aspects and avoids conflicts?

○ Can observation methods or instruments be suggested that most completely and
   accurately determine a density distribution?

Explicit and implicit ways to find answers are discussed.

The gravity inverse problem is an ill-posed problem, since no que solutions exist (Hadamard, 1953). Conditions $Bx = z$ were treated for optimization (Sects. 7.2.1,
7.2.2.3.8). A principal possibility is to exclude the nullspace by defining the problem
with a parametrization that cannot represent a non-trivial nullspace (Sect. 7.2.1.2.6).
In gravity, a finite number of point masses leaves no non-trivial nullspace but their
effects will not generally fit the data. A most complete parametrization, providing
free variables for many if not all aspects of the theory, should be chosen in order
not to exclude possible solutions and to fit the observations optimally within the statistical scatter. Examples in gravity inversion are fine grids of small mass elements
(see Sects. 6.2.2, 6.3.2) or functional series to high order (see 2.95–8). An arbitrarily large, but still limited number of mass points (mass > 0) at different locations is
principally determinable, but only under the fictitious condition of infinite and infinitely accurate data and infinitely accurate arithmetic. Generally such parametrizations must, however, be strongly regularized to permit numerically stable solutions.
Another kind of conditions are exploited with bodies extending from the surface
to large depth; this permits sort of extrapolation to depths, where the gravity effects are very small. Singular value decomposition (SVD) is helpful for unfavorable
parametrizations (see Sect. 7.3.3), because it drops the insignificant small singular values, and thus defines a much more compact parametrization which can be
extracted from the original one. Removal of insignificant undeterminable variables
or avoiding singularity by properly designing density models (e.g. using orthogonal functions, as Fourier series) are examples of successful regularization. These
methods circumvent nullspaces mathematically, leaving the physical and geological realization to the interpreter and can thus be called "extraction of a preferred
solution out of many" achieved with the aid of side conditions.

7.3.1.2 Requirements, Strategies

7.3.1.2.1 A Priori Information

A priori information is strictly required by the condition that gravity inversion has
to be formally overdetermined, to be solvable. A priori information provides regularization. Generally information contains internal conflicts which reveal the uncertainty of a problem, and in this sense inversion is a subclass or subset of general optimization (as expressed in Sect. 7.1). But inversion goes also beyond optimization
as the direct construction of the most acceptable models by introducing interesting
unknown variables in order to determine their values.

Several general methods of regularization exist which do not represent explicit
geometrical or physical a priori information, but are a more general type of a priori
information. Examples are conditions of smoothness and exclusion (Sects. 7.2.2.3.4
and 7.2.2.3.5), the Lanczos inverse (Sect. 7.2.2.3.7), discussed above, and the

balancing of the influence of the data and a priori information by the Backus & Gilbert approach, below.

7.3.1.2.2 Backus-Gilbert Method

The Backus-Gilbert method (Backus & Gilbert, 1968, Sabatier, 1977) tries to formulate how to best regularize inversion problems in a general way or to show which solution should be preferred: the one which satisfies the condition of the minimum variance or that which corresponds to the a priori "reality". The aim is not the best fit of the model to the data, but the best-possible reconstruction of "reality", taking into account that reality must be deduced from the data y, but can be represented in the data only partially. The nullspace is not represented. Moreover, the data are uncertain. "Reality" is thus understood as only the facts, x_y, which are described by the (consistent or fitted) observations, y. If the solution must be regularized, the reconstruction x_r will be influenced and distorted by the a priori information relative to the facts x_y : $x_r = R_x x_y$ and ideally $R_x = I$. The projection of x_y onto the observations is obviously $A x_y = y$, and the reconstruction cannot be done other than via the data: $x_r = R_y y = R_y A x_y = R_x x_y$. What is desired is a reconstruction matrix R_y which satisfies the conditions:

(1) The uncertainty $R_y^T A^T C_{yA} A R_y$ of the individual reconstructed x_r which follows from the uncertainty of the data and the uncertainties of the projection A (expressed by C_{yA}) is to be minimal.
(2) The reconstruction x_r is to be as similar as possible to the situation x_y following from the observations: $R_x = R_y A$ should ideally equal the identity matrix I.

These conditions contradict each other in most cases, as experience shows. For example, a detailed density distribution (many parameters) can fit the data very well, while it will, however, have large uncertainties. In contrast, a large-scale mean density can be determined very accurately, however, with a poor fit to the detailed data. One has to choose (1) the relative weights of the two conditions, corresponding to the Tikhonov regularization parameter λ, and (2) the definition of similarity with the identity matrix. The matrix R_x must then be calculated from the two conditions revealing the "best" reconstruction $x_r = R_y y$. The second condition can be of help in choosing the regularization method and as a criterion in evaluating a solution (see Sect. 7.3.2.2.8).

The analysis can be carried out with the direct methods for linear (Sect. 7.3.2.1) or non-linear (Sect. 7.3.2.3) problems or by iterative methods (Sect. 7.3.3.1). Only for linear or mildly non-linear problems the simple matrix algorithms are appropriate.

7.3.1.2.3 Summary of Possibilities to Define A Priori Information in a Gravity

Computational tools (such as INVERT, PS97, attached to this book on a disk) should provide for all the possibilities discussed above. Parameters or variables have to be defined either free, fixed, bounded by a priori standard deviations and option-ally limited by pre-set conditions and mutual dependencies or by correlations (as a correlation matrix or condition equations). Parametrizations calculable by internal or external modules ought to permit principal modelling geometries to be defined adaptable to given a priori information, complemented by the possibility to import externally calculated model effects. The relative weighting of the parameters and the data are preferentially optimizable with the Tikhonov regularization parame-ter (Sect. 7.2.2.3.3). Linear or non-linear or conditions should be definable, and it should be possible to place starting parameter values near suitable solutions since otherwise the iteration may immediately "jump" into "nonsensical" or "forbidden" "terrain". Last but not least, the inversion tool must provide all the a posteriori in-formation needed to evaluate the results.

7.3.2 Direct Algorithms

Methods are called "direct" if they solve, mostly by matrix inversion, the normal equations (7.2.14a) $x = (A^T A)^{-1} A^T y$. In linear problems the unknowns are rendered in one step. Mildly non-linear problems are most easily linearized (if possible) and then iterated, each step solving (Eq. 7.1.2) and updating the solution until an abort criterion is satisfied. The elements of A are the Jacobians, J_{ik}, i.e. the derivatives of all model effects, $\partial \delta g(x_i)/\partial p_k$, (Sect. 7.2.1.2).

7.3.2.1 Linear Problems

If the underlying functional relationship between the variables x and the observa-tions y is linear, the solution (Sect. 7.1.2) renders optimized variables and the inver-sion of the data in one step. Newton's law is linear in density (also in geometrical scale). Hence, theoretically determining density (or scale) is a linear problem, if the geometry of the geological bodies is known or given. Generally, however, ge-ometry is unknown, and large point grids, without further regularization, will not automatically lead to the correct stable solution, not even approximately. Rather the non-linear question of geometrical distribution remains open.

Cases of linear inversion do occur in practice, when the geometry of a structure is well determined, for example by seismic methods, or if not so accurately deter-mined, at least as an experimental step. Linear inversion for density was demon-strated in some case histories (Sects. 6.5.2; 6.5.3; 6.5.8). Another linear case, where the geometry is assumed known, is the optimization of the Bouguer density and vertical gravity gradient (Sects. 4.5.3.1; 4.5.2); this is briefly described below as a variant of the Nettleton method (Sects. 3.6.3.6) which exploits the relation between the Bouguer anomaly and topography, generally in graphical approximate form.

7.3.2.1.1 Principles, Methods, Tasks

Solving linear equations, including large sets of them, is a common task of numerical mathematics and will thus be treated only briefly. The special case of separating or isolating correlated effects in gravity data sets by sequentially solving linear equations is treated thereafter.

7.3.2.1.1.1 Solving Linear Equations

The classical Gauss-Newton method for solving systems of linear equations seems the most common tool for practical inversion, capable of optimizing models or model parameters as discussed above. It has proven fully adequate as it can always find a minimum if it exists (see Sect. 7.3.2.3.2).

Many methods are available for matrix inversion. The pivoting-free *Choleski decomposition* is advantageous for systems of equations of moderate size and saves memory. The normal matrix is decomposed into an upper and a lower triangular matrix which are transposed matrices, saving storage space; the method is numerically stable without pivoting (reordering of the matrix elements). For very large systems of equations it is time-consuming and memory demanding (as is the Newton method generally if sparseness of the matrices is not exploited). The situation is characterized by the number of parameters which usually must be limited through suitable simplifications. If the aim is mainly to evaluate the results carefully, compact representations are required anyhow. Under these circumstances, the calculation of the model effects usually takes more time than the matrix inversion. If larger problems are to be routinely solved, other algorithms must be considered, e.g., iterative ones without matrix inversion or LU decomposition (Gollub & Van Loan, 1989, Chap. 4) which decomposes the normal matrix similarly and lends itself very well for exploiting sparse matrix techniques more economically in terms of computer time. Singular Value Decomposition (SVD) solves the model equations on the basis of a generalized inverse without calculating the normal matrix whose condition number equals the square of that of the original model equations which are thus easier to solve accurately. SVD is suited also for singular, nearly singular and numerically unstable problems. Several implementations of these algorithms are widely used.

7.3.2.1.1.2 Separating Correlated Effects in Data by Sequentially Solving Linear Equations

The task of isolating the effects of the vertical gradient used in the height reduction, $-\delta g_\mathrm{h}$, and the Bouguer density assumed in the Bouguer and terrain reduction, $-\delta g_\mathrm{B}$, is discussed here first, independently from the solution algorithm applied. Removing their mutual linear relations and those with the observations δg, before solving for their values, is a refinement of the traditional Nettleton method, the results of which are affected by such relations (Jacoby, 1966). Calculating first the linear relationships (regression coefficients) of δg_B with δg, δg_h with δg and δg_B with δg_h and subtracting them from δg, better fitting values are obtained for the Bouguer density and the vertical gravity gradient by solving observation equations of the residual gravity plus residual height and mass effects. The unknown Bouguer anomaly from deeper sources can be approximated, e.g., by a polynomial fit. Steep slopes of topography are advantageous for stable solutions; indeed, such geometry approaches the 3D situations discussed in Sect. 3.6.3.5. The application to Helgoland (Sect. 3.6.3.6) proved rather convincing. Of course, there may exist more complex relationships, not eliminated by linear correlations.

7.3.2.2 Assessment of Solution (Residuals, Covariance, RMSE, F-Test, Standard Deviations, Correlation, Resolution)

When a solution has been found by regularization and the adjustment of the variables to be unique and stable and a density distribution has been calculated for further research, it is important to first evaluate the quality of the solution. The quality involves the internal consistency between the data and side conditions, the adequacy of the parametrization, the uncertainties of the variables, the necessity of data and side conditions, the novelty of the solution or the lack of it, and the suitability of the solution method for the problem at hand.

For this, several characteristics or criteria can be obtained from the optimization and inversion process itself. A complex problem will, however, not easily reduce to a few criteria. It is important, as a kind of contrast enhancement, to define them such that they will take on simple values such as 0 or ± 1 or $\gg 1$ or $\ll -1$, if significant, and intermediate values would indicate insignificance. Hence, the properties of solutions and general characteristics should be compared to each other in the form of differences or ratios to arrive at normalized criteria independent from the problem at hand. In the Bayesian approach (7.2.2.3.10) all a priori and a posteriori information, including the basic equations, are usually described by normal distributions, where the most important results are the expected values and their uncertainties. The evaluation of the solution will thus be their comparison with suitably chosen quantities, indicating that the *expectation* either

- corresponds with the criterion or
- deviates from the criterion within acceptable limits or
- deviates strongly or is incompatible with the criterion;

or the *uncertainty* is either

- much narrower than the criterion value or
- it is comparable with it or
- it is much broader than it.

Figure 7.3.1 illustrates all 9 combinations of the above possibilities and interpretations. The actual meaning will depend on the given case and the purpose of the comparison. For example, only solutions of a pre-set quality may be acceptable, or the computational method may be appropriate only within certain limits of linearity, etc. In gravity, a priori limits (e.g. of density) may be decisive, and it will be

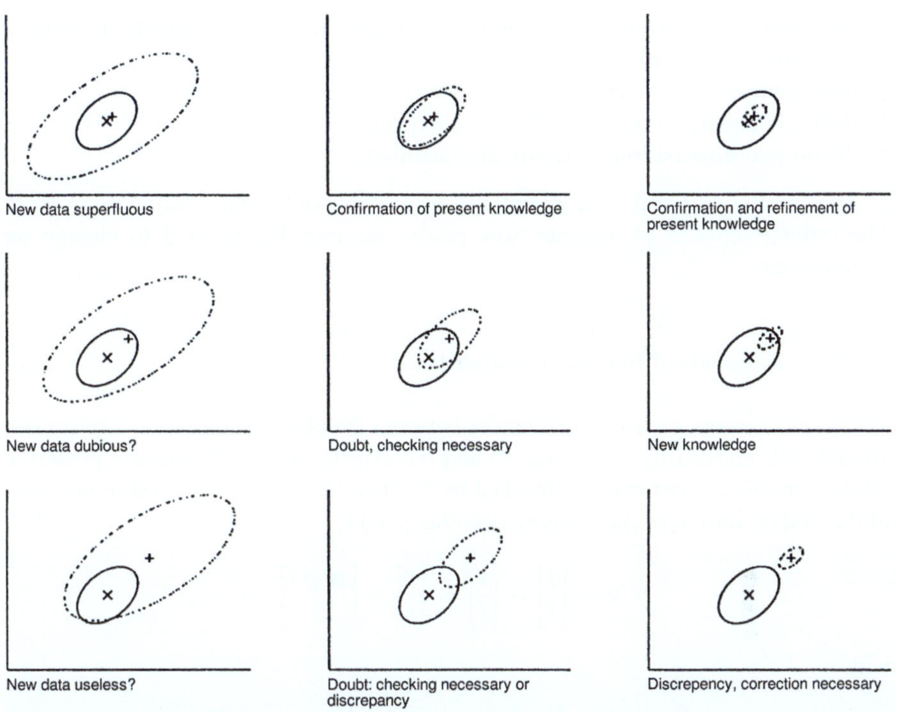

Fig. 7.3.1 Graphic illustration and interpretation of the 9 different situations in comparing the expectation and the uncertainty of a solution based on "new data" with a criterion which represents the a priori or previous knowledge

important to decide whether new information has been gained or something already known has been only confirmed.

The criteria may refer to single parameters, to selected combinations or to the whole set. Single parameters can be judged generally only if they have physical meaning or can be related to a meaningful interpretation. In these cases, the evaluation will be simple. In some cases, however, the parameters have a meaning only in combination, for example, the coefficients of a polynomial alone cannot be judged, but the functional values can (see Sect. 2.10.3). Criteria encompassing all variables will permit the most compact assessment of the quality of a solution in a problem-independent form because they are defined in the same coordinate system as the solution itself.

7.3.2.2.1 The Expectation: The Residuals

Evaluation implies judging probabilities. Probabilities can be estimated correctly only if data and a priori information comply with the following conditions:

1. the standard deviations and covariance matrices are chosen correctly to describe the statistical uncertainty;
2. they are free of gross errors;
3. the basic equation $Ax = y$ (7.10.1) is adequate;
4. the normal distribution is statistically justified.

The conditions 1 and 4 are essentially equivalent, and 2 and 3 follow from them. The order is chosen to indicate how easily one may be inclined to change the assumptions.

7.3.2.2.2 Residuals of Individual Variables

Residuals or errors e are the differences between the observations (also a priori side conditions), marked by a tilde (as $\tilde{y}$) and the effects of the optimized a posteriori model variables or parameters, marked by "^" (as "$\hat{x}$"); "error" is used in the sense of the first definition of the introduction (Sect. 7.1):

$$e = \begin{bmatrix} \tilde{y} \\ \tilde{x} \end{bmatrix} - \begin{bmatrix} \hat{y} \\ \hat{x} \end{bmatrix} = \begin{bmatrix} \tilde{y} \\ \tilde{x} \end{bmatrix} - \begin{bmatrix} \mathscr{A}(\hat{x}) \\ \hat{x} \end{bmatrix}$$

with

$$\begin{cases} \hat{x} = \tilde{x} + \left(A^\mathrm{T} C_{yA}{}^{-1} A + C_x{}^{-1} \right)^{-1} A^\mathrm{T} C_{yA}{}^{-1} \tilde{y}) \\ C_e = C_{yA} - A \left(A^\mathrm{T} C_{yA}{}^{-1} A + C_x{}^{-1} \right)^{-1} A^\mathrm{T} \end{cases} \tag{7.3.1}$$

These residuals are $N(0, C_e)$ normal-distributed if data and a priori information have a normal distribution $N((y,x),(C_{yA},C_x))$. By comparison with this probability distribution the validity of the assumptions can be tested. Generally the individual residuals are judged in view of their significance for the problem to be solved.

7.3.2.2.3 Root Mean Square Error (RMSE) and F-Test

Some simple numbers can be provided to judge the results of any kind of inversion problem. The *shifting variate* is the length of the error vector or the weighted sum of n squared $N(0, I)$ normally distributed variables, according to the L_2 norm that is minimized

$$e^\mathrm{T} \begin{bmatrix} C_{yA} & 0 \\ 0 & C_x \end{bmatrix}^{-1} e = \varepsilon^2 = \min \tag{7.3.2}$$

If r linear combinations of m variables are not regularized, only $(n-r)$ error components (degrees of freedom) contribute to the error vector, and the shifting variate is, by definition, χ^2_{n-r}-distributed with the expectation $n-r$. Under the assumption that the data follow $N((y,x),C_{yA},C_x)$, the probability that $\varepsilon^2 \geq \varepsilon_c{}^2$ is equal to the percentile $\chi_{n-r}{}^2(\varepsilon_c{}^2)$. The quantity $\varepsilon^2/(n-r)$ is the weighted mean squared residual. If the covariances are valid up to a common variance factor σ^2, derived from a sample of size k, the F-statistic $\varepsilon^2/((n-r)\sigma^2)$ has $F_{n-r,k}$-distribution: it compares the squared length of the residual vector with the expected length, where both are uncertain. If the chosen σ^2 is valid, as though taken from a quasi-infinite sample, the F-statistic is $F_{n-r,\infty}$ distributed, equivalent to $\chi^2_{n-1,k}/(n-r)$. For suitably scaled covariance matrices, $\sigma^2 = 1$, such that the F-statistic equals $\varepsilon^2/(n-r)$. It is a value estimated for the covariance factor of the data (a priori information). If the a posteriori standard deviations, $\hat{\sigma}_i$, are multiplied with this factor, as it appears to describe the given uncertainty better, the variables thus connected with the fitted standard deviations, are no longer normally distributed, but have the Student's t-distribution, $e_i(n-r)^{1/2}/(\varepsilon \cdot \hat{\sigma}_i) \propto t_{n-r}$.

If the F-statistic ≈ 1, all the above assumptions appear to be valid.

If the F-statistic $\gg 1$, the variance factor should be multiplied with it to bring it up to 1 without changing the solution and the unweighted residuals, suggesting that the assumptions are not valid within the standard deviation multiplied by $\varepsilon^2/(n-r)$.

If the F-statistic $\ll 1$, the solution fits better than expected, and the basic equation (or model) could possibly be further specified. If the assumptions hold, two different solutions with an F-statistic $\ll 1$ are statistically indistinguishable, and one can conclude that the difference between both solutions lies effectively in the nullspace of the basic equation.

Shifting variate and F-statistic provide a joint evaluation of the data and the a priori information.

The weighted Root Mean Square Error (wRMSE) permits independent evaluation of either form of information. For the data and for the variables it is defined differently by:

$$\left[(\tilde{y} - \mathscr{A}(\hat{x}))^{\mathrm{T}} C_y^{-1} (\tilde{y} - \mathscr{A}(\hat{x}))/n\right]^{1/2}$$

and

$$\left[(\tilde{x} - \hat{x})^{\mathrm{T}} C_{\tilde{x}}^{-1} (\tilde{x} - \hat{x})/m\right]^{1/2}, \text{ respectively} \qquad (7.3.3)$$

They are ordinary averages disregarding the degrees of freedom and have no F-distribution, but the user can thus judge the quality of the data fit relative to the fit of the a priori information. They are plotted against each other in double-logarithmic scale in the L-curve of the Tikhonov regularization (Sect. 7.2.2.3.3; Fig. 7.2.2). The statistical quantities wRMSE of the variables and of the data, the F-statistics and the coordinates of the Tikhonov L-curve of the current solution provide a means to assess the quality of the complete (multidimensional) solution by three simple numbers.

7.3.2.2.4 The A Posteriori Covariance Matrix

The evaluation of the covariance matrix C concentrates on combination of two variables by their correlation or larger groups of variables by the eigen-value/eigen-vector analysis. A covariance matrix C without normalization can rarely be interpreted as such because its elements often differ in physical meaning and units.

Therefore the matrix C is decomposed into the dimensionalized a posteriori matrix $diag(\sigma)$ of the standard deviations, where $\sigma_i = C_{ii}^{1/2}$, and the non-dimensional correlation matrix R; this is achieved by the decomposition $C = diag(\sigma)\ R\ diag(\sigma)$, where $\sigma_i^2 = C_{ii}$ and $R = diag(1/\sigma)\ C\ diag(1/\sigma)$; see Sects. 7.3.2.2.7, 7.3.2.2.10.

7.3.2.2.5 The Standard Deviation σ

The a posteriori standard deviation permits an evaluation of the individual variables, which is only useful, as mentioned, if they are physically meaningful. In evaluating the a posteriori standard deviation, one asks:

(1) Is it below a certain physically realistic criterion value σ_{crit}?
(2) Or, for individually regularized variables: has σ significantly decreased relative to the a priori value σ_{prior}?

The answer should be $\sigma/\sigma_{crit} < 1$, or $\ll 1$ and ideally should approach 0, but not exactly zero (the covariance matrix would then be singular).

For σ_{crit} the maximum acceptable uncertainty of the result may be chosen, such that $\sigma/\sigma_{crit} > 1$ suggests that the solution is effectively singular and urgently needs regularization. In extreme cases, the variable may be removed from the inversion. Or it can indicate that the uncertainty exceeds a given range of interpretability. Or the range of linearity of the observation equations may be exceeded and must be regarded with caution. In general, σ_{crit} is given in conjunction with an expected value of the variable, and it might be sensible to add these quantities as observations and standard deviations directly to the system of equations as a measure of regularization. If $\sigma/\sigma_{crit} \approx 1$, σ has not decreased significantly and thus none of the data was able to pin down this variable and the regularization was really necessary. The result $\sigma/\sigma_{crit} \approx 0$ indicates that the data alone fix the solution and regularization is unnecessary. However, a precise solution (with a small standard deviation) can be wrong, and the regularization may be needed to prevent this, i.e. it may be important for the problem at hand, even if superfluous for the solution process.

7.3.2.2.6 The Correlation Matrix

The correlations R_{ij} measure how closely two variables, x_i and x_j, are grouped around that straight line which represents with the smallest residual variance a linear relation between them if infinitely many samples of normally distributed data would be considered. But the R_{ij} are estimated only from a few scattered data. As the covariance matrix is positive definite, $R_{ii} = 1$ on the diagonal expresses complete self-dependence. For all others $(i \neq j): -1 < R_{ij} < +1$. R_{ij} near ± 1 indicates a strong dependence suggesting that one of the variables is unnecessary as it can be calculated from the other and carries no new information. This can be exploited for regularization, and a significant section of the regression line with the least scatter could be chosen to be represented by a point x_i, x_j, such that $x_j = bx_i$. The slope b of the straight line is

$$b = 2\sigma_i\sigma_j R_{ij}/\max\left[\left(\sigma_i{}^2 - \sigma_j{}^2\right) \pm \left(\sigma_i{}^4 + \sigma_j{}^4 + 2\sigma_i{}^2\sigma_j{}^2\left(2R_{ij}{}^2 - 1\right)\right)\right]$$
$$\approx \mathrm{sign}(R_{ij})\sigma_j/\sigma_i, |R_{ij}| \to 1 \qquad (7.3.4)$$

This follows from the fact that each of the partial matrices Cij

$$\begin{bmatrix} \sigma_i{}^2 & \sigma_i\sigma_j R_{ij} \\ \sigma_i\sigma_j R_{ij} & \sigma_j{}^2 \end{bmatrix}$$

contains the coefficients which describe an ellipse: $x^\mathrm{T}C_{ij}x = \sigma_i^2 x_i{}^2 + 2\sigma_i\sigma_j R_{ij}x_i x_j + \sigma_j^2 x_j^2 = 1$. The two half axes λ_{max} and λ_{min} with the directions v_{max} and v_{min} result from the two solutions of the eigen-value problem:

$$C_\lambda v = \begin{bmatrix} \sigma_i{}^2 - \lambda & \sigma_i\sigma_j R_{ij} \\ \sigma_i\sigma_j R_{ij} & \sigma_j{}^2 - \lambda \end{bmatrix} v = 0.$$

Non-trivial solutions exist only if $|C_\lambda| = 0$, hence, for the half-lengths and the directions of the axes, respectively:

$$\lambda = e^{1/2}\left(\left(\sigma_i{}^2 + \sigma_j{}^2\right) \pm \left(\sigma_i{}^4 + \sigma_j{}^4 + 2\sigma_i{}^2\sigma_j{}^2\left(2R_{ij}{}^2 - 1\right)\right)\right),$$
$$v = \begin{bmatrix} \lambda - \sigma_j{}^2 \\ \sigma_i\sigma_j R_{ij} \end{bmatrix} \qquad (7.3.5)$$

The slope b follows from the ratio of the second over the first component of the large semi-axis v_{max}, which itself is calculated with $\lambda = \lambda_{max}$. This standard ellipse is the two-dimensional equivalent of the standard deviation and describes how the variables are grouped. A very narrow and inclined ellipse corresponds to $R_{ij} \approx \pm 1$, a circular or axis-parallel one corresponds to $R_{ij} \approx 0$. The principal axes lie in the direction of the eigen-vectors of C_{ij} and are interpreted as the linear combinations of the two variables with the largest and smallest standard deviations, represented by the lengths of the principal axes, which are given by the square roots of the of the eigen-values.

7.3.2.2.7 Eigen-Value/Eigen-Vector Analysis of the A Posteriori Covariance Matrix and the Criterion Covariance Matrix

Generally the covariance matrix is too large and complicated to be easily interpreted. The axes of a hyperellipsoid are simpler to imagine, and the classical standard ellipse can be generalized to an m-dimensional standard hyperellipsoid describing the collective grouping of m variables. The connection between the covariance matrix C, the square roots of the m eigen-values $\zeta_i{}^2$, i.e. the singular values (always positive, as C is positive definite) in the diagonal matrix $S = diag(\zeta_1 \geq \zeta_2 \geq \ldots \geq \zeta_m)$ and the respective eigen-vectors e_i in the columns of matrix E (orthonormal, i.e. $EE = 1$) is:

$$Ce_i = \zeta_i{}^2 e_i, \quad i = 1, \mathrm{m} \Rightarrow CE = ESS \Rightarrow C = ESSE^\mathrm{T} \qquad (7.3.6)$$

The square roots of the eigen-values, so-called "singular values" are interpreted as the standard deviations of the linear combinations of the variables which are defined in the corresponding eigen-vectors. If the vector variables x are normally distributed with the expectations μ and the covariance matrix C, i.e., $x \propto N(\mu, C)$, x can be transformed via the eigen-vectors:

$$E^{\mathrm{T}}x \propto N\left(E^{\mathrm{T}}\mu, E^{\mathrm{T}}CE\right) = N\left(E^{\mathrm{T}}\mu, E^{\mathrm{T}}(ESSE^{\mathrm{T}})E\right), \text{ and hence } E^{\mathrm{T}}x \propto N\left(E^{\mathrm{T}}\mu, S^2\right).$$

If linear combinations in the eigen-vectors occur which have a physical meaning, their standard deviation can be compared to a criterion value defined for the combination, e.g. two densities occurring as the sum and the difference.

An *example*: of m variables $x_{\mathrm{k}=1,\mathrm{m}}$, the first half of the variables ($\mathrm{k}=1,\mathrm{m}/2$) describe the depth of a layer top and the second half of the variables ($\mathrm{k}=\mathrm{m}/2+1,\mathrm{m}$) describe the bottom. The components of the eigen-vectors are thus defined in relation to the base $[x_{\mathrm{k}=1}\ldots x_{\mathrm{k}=\mathrm{m}/2}\ \ x_{\mathrm{k}=\mathrm{m}/2+1},\ \mathrm{m}\ldots x_{\mathrm{k}=\mathrm{m}}]^{\mathrm{T}}$. The eigen-vectors $\left[\frac{1}{\mathrm{m}/2}\ldots\frac{1}{\mathrm{m}/2}0\ldots0\right]^{\mathrm{T}}$, $\left[0\ldots0\frac{1}{\mathrm{m}/2}\ldots\frac{1}{\mathrm{m}/2}\right]^{\mathrm{T}}$ and $\left[\frac{1}{\mathrm{m}}\ldots\frac{1}{\mathrm{m}}\frac{1}{\mathrm{m}}\ldots\frac{1}{\mathrm{m}}\right]^{\mathrm{T}}$ correspond to the mean depths of the top, the bottom and the whole layer, respectively. The standard deviations of these quantities are the corresponding singular values.

If variables have different physical units, the singular values and eigen-vectors cannot be evaluated immediately. Take, for example, the components a and b of an eigen-vector to have the units "m" and "kg/m^3"; the linear combination would be $a[\mathrm{m}] + b[\mathrm{kg/m}^3]$ which is physical nonsense. Even non-dimensionalization (e.g. division by 1 m and 1 kg/m^3) would leave singular values which, if ordered by size, strongly depend on the units chosen.

An *example*: assume the variables of a whole lithosphere section to be (1) the density anomalies, known a priori to about $\pm 0.5\,\mathrm{g/cm}^3$ and (2) depths with an uncertainty of $\pm 30,000\,\mathrm{m}$; both are uncorrelated. After adjustment the standard deviations are, e.g., $0.1\,\mathrm{g/cm}^3$ and $2000\,\mathrm{m}$ with a correlation coefficient of 0.8. The singular values of the a posteriori covariance matrix are thus $2000.00\ldots$ for the eigen-vector $[0.00\ldots0.99\ldots]$, representing essentially only the depth accuracy, and $0.06\ldots$ for the eigen-vector $[0.99\ldots0.00\ldots]$, describing essentially only the density accuracy. From this, one would deduce that the depth uncertainty has decreased (about correctly) with a factor of 15, and the density uncertainty by a factor of 10 (instead of 5). The considerable correlation is totally invisible because the eigenvectors are essentially parallel to the coordinate axes. If, however, the coordinate system is scaled with the estimated uncertainties, i.e., the density values are divided by 0.5 and the depth units by 30,000, then the singular values are $0.20\ldots$ and $0.038\ldots$ for the eigen-vectors $[0.96\ldots0.26\ldots]$ and $[-0.26\ldots0.96\ldots]$, respectively. Now the strong correlation is recognizable from the inclined orientation and the different size of the eigenvectors and the improvement relative to the assumed uncertainties is directly indicated by the singular values.

To solve this problem the original variables x can be subjected to a coordinate transformation T to the new $x_c = Tx$, where the transformed a posteriori covariance matrix:

$$T^{-\mathrm{T}}CT^{-1} \tag{7.3.7}$$

represents the standard hyperellipsoid $x^{\mathrm{T}}Cx = x_c{}^{\mathrm{T}}T^{-\mathrm{T}}CT^{-1}x_c = 1$ in the new coordinates. The correlation matrix $R = diag(1/\sigma)\, C\, diag(1/\sigma)$ can thus be interpreted as the standard hyperellipsoid in the coordinate system of the a posteriori standard deviations of the variables.

To determine whether or not the accuracy of the solution corresponds to some given criterion, for meaningful comparison first a *criterion covariance matrix* C_c must be defined by a suitable transformation of the a posteriori covariance matrix. The matrix C_c is usually calculated by one of the following procedures:

1. The criterion is given by the standard deviations $S = diag(s_{\mathrm{i}=1,\mathrm{m}})$ of certain ortho-normalized linear combinations L of the variables, i.e. the transformation matrix $T = C_c^{1/2} = SL^{\mathrm{T}}$ or $T^{-1} = C_c^{-1/2} = LS^{-\mathrm{T}}$ is known (the *criterion standard deviations*, as computational quantities, symbolized by s_{i}).
2. The criterion is given by the standard deviations $S = diag(s_{\mathrm{i}=1,\mathrm{m}})$ of the variables and their correlation matrix R, and the transformation matrix must be calculated from the square root matrix of R : $T = C_c^{1/2} = SR^{1/2}$ or $T^{-1} = C_c^{-1/2} = R^{-1/2}\, S^{-1}$.
3. The criterion must be deduced from conditions between the variables: $Bx = e$ with $x \propto N(0, C_{xB})$ and $e \propto N(0,\ C_e)$; the covariance matrix of the variables which correspond to this condition is, after (Eq. 7.2.9), $C_c = C_{xB} - C_{xB}(BC_{xB}B^{\mathrm{T}} + C_e)^{-1}BC_{xB}$ from which follows, as specified: $T = C_c^{1/2}$ or $T^{-1} = C_c^{-1/2}$.

If a singular value of the transformed covariance matrix (Eq. 7.3.7) equals 1, the standard deviation in the direction of the eigen-vector is equal to that of the criterion. If it is greater, the aim is missed. If it is smaller, the expectations are surpassed.

The transformations easiest to interpret are, beside the trivial transformations I, pure scaling of the original coordinates (without rotation), used also above. Especially for variables with unique physical meaning, one should consider re-normalizing them with the a priori standard deviations $S = diag(s_{\mathrm{i}=1,\mathrm{m}})$, which in this case also play the role of *criterion standard deviations* s_{i} (see above):

$$S^{-1}C_xS^{-1}. \tag{7.3.8}$$

Then the singular values and the components of the eigen-vectors have units most easily interpreted. It is a compromise between maintaining the original physical coordinate system and the subjective weighting of the results.

If the criterion matrix is equal to the a priori covariance matrix, C_x, and if $C_x^{1/2}$ represents an interpretable transformation, in the best case, for example, if C_x is a diagonal matrix, then with (Eqs. 7.2.13, 7.2.14), the transformed a posteriori covariance matrix is:

$$C_x^{-1/2}C_{\hat{x}}\left(C^{-1/2}\right)^T = C_x^{-1/2}\left(A^{\mathrm{T}}\right)C_{yA}A + C_x^{-1})^{-1}\left(C_x^{-1/2}\right)^T$$

$$= I - C_x^{-1/2}A^{\mathrm{T}}\left(AC_xA^{\mathrm{T}} + C_{yA}\right)^{-1}A\left(C_x^{-1/2}\right)^T$$

The central expression is positive definite and the second term of the right-hand side expression is positive semi-definite, i.e. positive definite when $A^T A$ is not singular. Furthermore, for the singular values of the matrix sum $(I - C)$, $eig(I - C) = 1 - eig(C)$; hence, the eigen-values (the squares of the singular values in this case) of the transformed a posteriori covariance matrix must be >0 and ≤ 1. Values close to 1 indicate that the accuracy of the corresponding linear combinations of the variables has not improved; values near 0 mean that the eigen-vector is well resolved. If an eigen-value of the left-hand expression is equal to 1, then the corresponding eigen-value of the second term of the right-hand expression is 0. It means that $A^T A$ is singular and that the corresponding eigen-vector lies in the nullspace of A. This way, the nullspace of the non-regularized problem can be re-established from a fully regularized solution. So one is able to determine which combination of variables would be badly resolved without a priori information although a stable solution is achieved by including it. This way, the a priori information that is not verified by observations can be recognized.

In most cases, not all variables will be regularized with a priori information, especially if criteria can hardly be formulated (see Sect. 7.2.1.2.7). For such variables the following may be done:

1. they are removed from the a posteriori covariance matrix and do no longer appear in the eigen-vectors;
2. let the respective diagonal elements $C_x^{-1/2} = 0$, implying an infinite a priori standard deviation. Then the eigen-values are zero for eigen-vectors in the directions of these variables and appear thus as infinitely improved;
3. let, for these variables, $C_x^{-1/2} = 1$ (or another arbitrary value) in order to exhibit possible interactions with other variables.

7.3.2.2.8 The Resolution Matrix

"Resolving" different variables means to be able to distinguish, to disentangle their values. What is meant here is the distinction of what the observations tell about the solution in comparison to the a priori information. The resolution matrix R_x is central to the Backus-Gilbert method of regularization (Sect. 7.3.1.2.2) and is defined by $x_r = R_y y = R_y A x_y = R_x x_y$. From this and (Eq. 7.2.13) is derived:

$$\begin{cases} x_r - \tilde{x} = C_x A^T \left(A C_x A^T + C_{yA} \right)^{-1} A \left(x_y - \tilde{x} \right) = R_x \left(x_y - \tilde{x} \right) \\ C_{\hat{x}} = C_x - C_x A^T \left(A C_x A^T + C_{yA} \right)^{-1} A C_x \end{cases} \tag{7.3.9}$$

from which follows:

$$R_x = I - C_{\hat{x}} C_x^{-1} \tag{7.3.10}$$

The most important properties of R_x are:

1. R_x describes how strongly the solution x_y which exactly corresponds to the consistent or adjusted observations is distorted by the regularization. In the ideal case, if $R_x = I$, the regularized solution is not distorted, i.e., it is perfectly resolved by definition.
2. If C_x is arbitrary, the ideal situation exists only if $C_{\hat{x}} = 0$, i.e. the a posteriori covariance matrix should be the null matrix. Of course, it is desirable that the scatter of the adjusted variables is zero.
3. If $C_x = diag(\infty)$, $R_x = I$. According to the definition, the trivial conclusion is that without regularization there is no influence of regularization on the solution.
4. If it turns out that $R_x = I$ in spite of regularization, then it follows from $x_r - \tilde{x} = R_x(x_y - \tilde{x})$, that the a priori information has no influence on the results.
5. Data errors propagate, even in the ideal case $R_x = I$, according to the adjustment method R_y, into the solution x_r.
6. If $R_x \neq I$ and the data are in conflict with the a priori information, then the discrepancies will generally spread across several variables although there may be only one conflicting variable.
7. The element $R_x(i,j)$ (line i, column j) describes the change (relative to the unregularized solution) of variable i, caused by a discrepancy of one unit in the variable j between data and a priori information. In other words: if the data try to shift variable j by one unit to a value contradicting the a priori information, then (also or in its place) variable i will change by $R_x(i,j)$ units to better fit the data. If the diagonal element $R_x(i,i) < 1$, the required change of the variable i is incomplete, and $R_x(i,i) < 0$ expresses that the feedback of regularization reverses the sign of the change required by the data.
8. R_x is seldom symmetric. $R_x(i,j)$ and $R_x(j,i)$ will have reciprocal physical units, and the influence of a strongly regularized or fixed variable will spread over less regularized variables but not vice versa.
9. If the units of the variables are changed such that they exactly agree with the a priori information, i.e. on both x_r and x_y, the transformation $C_x^{1/2}$ is applied, then
$$x_r = (I - C_{\hat{x}}C_x^{-1})x_y \Rightarrow C_x^{-1/2}x_r = C_x^{-1/2}(I - C_{\hat{x}}C_x^{-1})C_x^{1/2}C_x^{-1/2}x_y \Rightarrow x_r^{\mathrm{T}} = (I - C_x^{-1/2}C_{\hat{x}}C_x^{-1/2})x_y^{\mathrm{T}}$$ and hence $R_x' = (I - C_x^{-1/2}C_{\hat{x}}C_x^{-1/2})$. By this transformation the matrix has become symmetric remaining positive definite. However, the $R_x(i,j)$ in transformed units tell us little more about their meanings. Alternatively, as in (Eq. 7.3.8), the a priori $S = diag(\sigma_{i=1,m})$ can be chosen for the transformation matrix, again as *criterion standard deviation* (Sect. 7.3.2.2.7):

$$R_x'' = (I - S^{-1}C_{\hat{x}}C_x S^{-1}) = S^{-1}(I - C_{\hat{x}}C_x^{-1})S \qquad (7.3.11)$$

which is a similarity transformation. $R_x''(i,j)$ describes by how many standard deviations variable i changes when variable j was to be changed by one, if not prevented by the a priori information or the regularization. The matrix elements are non-dimensional, and in the special case of $C_x = S^2$, R_x'' is even symmetric.

The practical application of this discussion will be illustrated in the case history of Meerfeld Maar (Sect. 7.4.1).

7.3.2.2.9 Eigen-Value/Eigen-Vector Analysis of the Resolution Matrix

The interpretation of the eigen-values and eigen-vectors of the resolution matrix is related to the standard hyperellipsoids and criterion covariance matrices, although resolution matrices are generally not symmetric and therefore not describing hyperellipsoids. The criterion covariance matrix implies the question of whether the a posteriori uncertainty of an arbitrary linear combination $L^\mathrm{T}x$ is smaller than the criterion. The uncertainty is defined by the norm $(L^\mathrm{T}x)^\mathrm{T}C(L^\mathrm{T}x)$ of the vector $(L^\mathrm{T}x)$ relative to a covariance matrix C, or LCL^T, as only corresponding linear combinations are of interest. With the a posteriori covariance matrix $C_{\hat{x}}$ and the criterion covariance matrix C_c, the question is equivalent to:

$$LC_{\hat{x}}L^\mathrm{T} \le L\,C_cL^\mathrm{T} \Rightarrow C_{\hat{x}}L^\mathrm{T} = \lambda C_cL^\mathrm{T} \Rightarrow C_c^{-1}C_{\hat{x}}L^\mathrm{T} = \lambda L \Rightarrow C_{\hat{x}}C_c^{-1}L = \lambda L \text{ with } \lambda \le 1.$$

The first equation expresses that the hyperellipsoid described by $C_{\hat{x}}$ should always be smaller than that defined by C_c. If the subsequent equation is solved for all linear combinations, the eigen-values and eigen-vectors of $C_{\hat{x}}C_c^{-1}$ are obtained which represent extrema and saddle points of the size ratio of the hyperellipsoids. Since $C_{\hat{x}}C_c^{-1}$ is not symmetric, it does not describe a hyperellipsoid, and the directions of the extrema are usually non-orthogonal (Fig. 7.3.2) as the principal axes of a hyperellipsoid would be. Orthogonalization (as described above) by the transformation matrix $C_x^{-1/2}$ carries no real advantage since it is no longer clear in which coordinates the principal axes would be defined.

From (Eq. 7.3.10) follows that the resolution matrix, similar to the covariance matrix, makes a comparison, however, not with the uncertainty, but with its reduction:

$$C_c^{-1}C_{\hat{x}}x = \lambda x \Rightarrow (I - C_{\hat{x}}C_x^{-1})x = \lambda x.$$

The eigen-values are between 0 and 1. Near 0 means minimal reduction of the uncertainty, not better than the criterion; near 1 means maximal uncertainty reduction with hardly any effect from the regularization. Eigen-vectors with very small eigen-values represent linear combinations of variables which are strongly influenced by

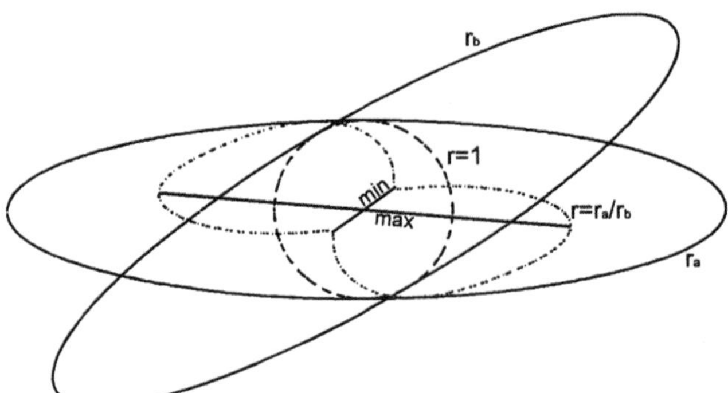

Fig. 7.3.2 The ratio of the sizes of two concentric ellipses and directions of its extrema (as for $C_x \cdot C_c^{-1}$ with identical centre) in polar coordinates

the a priori information. Eigen-values ≈ 1 show a strong uncertainty decrease, i.e. nearly exclusively by the data.

Also here it is reasonable for the analysis of the eigen-vectors, to apply a similarity transformation with the a priori standard deviations such that the size of their components will show whether they are large relative to the criterion (Sects. 7.3.2.2.7, 7.3.2.2.8). The interpretation of the resolution matrix and its eigen-values shows a close relationship with the concept of the criterion matrices applied in the geodetic coordinate determination (e.g. Baarda, 1973).

7.3.2.2.10 Comparison of the Eigen-Value/Eigen-Vector Analysis of the Covariance and Resolution Matrices

In Sect. 7.3.2.2.7 the eigen-values and eigen-vectors of the a posteriori covariance matrix, normalized with the criterion a priori standard deviations, were calculated: $S^{-1}C_xS^{-1}$ (Eq. 7.3.8) and in Sect. 7.3.2.2.9 the eigen-values and eigen-vectors of the resolution matrix (as similarity-transformed also with the a priori standard deviations) were derived: $R_x'' = S^{-1}(I - C_{\hat{x}}C_x^{-1})S$ (Eq. 7.3.11). What is their connection? It is simple if the a priori covariance matrix is equal to the a priori standard deviation squared: $C_x^{-1} = S^{-2}$; then the two matrices are complementary to 1, and the eigen-vectors are identical.

If unregularized variables exist, their a posteriori standard deviation will be limited and probably correlated with other variables, hence they will be represented in several eigen-vectors of (Eq. 7.3.8). Their resolution, however, appears perfect relative to their infinite a priori standard deviation, and for the respective eigen-vector, the equation (7.3.11) will have an eigen-value of 1.

If the a priori covariance matrix assumes correlations between variables, the eigen-value/eigen-vector analysis of (Eq. 7.3.11) emphasizes linear combinations of variables which would have large uncertainty without regularization. Strongly regularized linear combinations appear in (Eq. 7.3.8) among the eigen-vectors with small uncertainty, as intended by the regularization, while by (Eq. 7.3.11) they are counted as poorly resolved.

Generally however, the eigen-vectors of (Eqs. 7.3.8 and 7.3.11) differ because of completely unregularized variables (which cannot show up in Eq. 7.3.11) and correlations. Though many eigen-vectors will be equal, if regularization was only weakly applied, especially for the unequal ones it is of interest, whether a small uncertainty, i.e. a small eigen-value in (Eq. 7.3.8) is connected also with good resolution and whether this would be due only to a priori information or also to the data. If in (Eq. 7.3.11) no similar vector occurs, the question is difficult to answer. Similar to the case of a posteriori standard deviations of individual variables, the ratio of a posteriori over a priori uncertainty must be calculated for the linear combination considered. If, instead of S, the individual eigen-vectors of (Eq. 7.3.8) are chosen for the transformation matrix, $S^{-1}C_{\hat{x}}S^{-1} = ED(ED)^{\mathrm{T}}$ with D being a diagonal matrix of the singular values and E being an orthogonal matrix of the eigen-vectors, the resolution in the respective direction is

$$I - DEC_x^{-1}E^{\mathrm{T}}D \qquad\qquad (7.3.12)$$

An eigen-vector of (Eq. 7.3.8) with small uncertainty *and* poor resolution appears certain only due to the a priori information. Generally, however, small uncertainty will be associated with good resolution.

7.3.2.2.11 Residuals of the Eigen-Vectors of the A Posteriori Covariance

Whether or not a linear combination of variables with small a posteriori uncertainties lies at the right place, is indicated by the residual of the corresponding eigen-vector of the a posteriori covariance matrix. If, as in Sect. 7.3.2.2.10, $S^{-1}C_{\hat{x}}S^{-1} = ED(ED)^{-1}$ (D: diagonal matrix of the singular values, E: matrix of the eigen-vectors), then the weighted residuals of these eigen-vectors are $(e^{\mathrm{T}}EC_x^{-1}E^{\mathrm{T}}e)^{1/2}$. Poorly resolved but strongly regularized eigen-vectors (Eq. 7.3.12) are expected to have small residuals; but if large, non-linearity of the basic equation is suggested which may be ambiguous near the optimum, this ambiguity does not necessarily reach up to the location of the a priori information, so that the solution might be unique again if such a priori information is defined. In the case of only slightly regularized eigen-vectors with very small residuals the a priori information has already predicted the information contents of the new data. In contrast, large residuals of well determined variables suggest discrepancies between them.

7.3.2.2.12 Application: Design of New Observational Methods

Finally, a practical application of the eigen-value/eigen-vector analysis is to reveal deficits in the existing data or data types and to suggest which ones should be obtained by new measurements or new methods. It poses a design problem for new measuring instruments or methods. If several eigen-vectors are poorly resolved, it may suffice to measure a linear combination of them, and mutually independent measuring procedures can be combined arbitrarily to arrive at a useful design. The a priori information used will influence the design of new methods by adding data, which may result from real observations. For example, seismically determined models supply indirect density information and direct geometrical information on contrast surfaces, and they could be incorporated in a joint inversion procedure by including refraction and reflection laws, ray tracing and seismic velocity-density relations.

Most observation methods will solve non-linear combinations of the variables. Thus the search must be for new methods which provide the wanted eigen-vectors after linearization. The design of observation methodology is itself an ill-posed inverse problem (Penrose, 1991), where the eigen-vectors can help, but where no unique solution can be defined. No algorithm can be proposed, only experienced insight and ideas may serve as regularization.

7.3.2.2.13 Summary: Evaluation Criteria for Gravity Inversion

The a posteriori information for a solution permits a well educated evaluation of the results, their reliability and accuracy, in view of the geological, geodynamic or exploration aims of interpretation, especially when the model situation is complex. The evaluation criteria relevant to gravity inversion have been described. The criterion standard deviations, written as a diagonal matrix, are the basis for transformations (Sects. 7.3.2.2.7 and 7.3.2.2.9, Eqs. (7.3.8, 7.3.11)). They can be chosen such that the quantities expressed as numerical values in this system (standard deviations and eigen-vectors) are easily interpreted. Usually the criterion standard deviations are the a priori standard deviations. For the individual variables the next quantities are evaluated in their original physical units:

- the values of the solution,
- the residuals and
- the a posteriori standard deviations.

Statistical units (unit-less) of interest are:

- wRMSE of the variables;
- wRMSE of the data;
- the F-statistic;
- the coordinates of the Tikhonov L-curve of the current solution.

In relation with the *a posteriori covariance matrix* the results to be assessed are:

- the a posteriori standard deviation relative to the criterion a posteriori standard deviation, equivalent to (1 − resolution), for the individual variables;
- the correlation coefficients
- the eigen-values of the a posteriori covariance matrix in the coordinate system of the criterion standard deviations;
- the corresponding eigen-vectors, expressed in the units of the criterion standard deviations;
- the residuals of the eigen-vectors, expressed in the units of the criterion standard deviations;
- the resolution of the eigen-vectors (1 − a posteriori/a priori standard deviation of the eigen-vectors);
- the linearity factors for the current solution ± one eigen-vector, i.e., the change of the shifting variate (Sect. 7.3.2.2.3) when an eigen-vector is added to, or subtracted from, the current solution.

In relation with the *resolution matrix* interesting quantities are:

- the resolution matrix itself, normalized with the criterion standard deviations;
- the eigen-values of the resolution matrix;
- the corresponding eigen-vectors, expressed in the units of the criterion standard deviations.

In addition:

- the second derivatives to assess the non-linearity (Sect. 7.3.2.3.2)

7.3.2.3 Mildly Non-linear Problems

Mildly non-linear problems are problems which can be linearized such that iterative solutions are stably possible. In the present context a problem is defined to be linearizable, if within about a 2σ interval of all observations and all estimated variables the model relationship is sufficiently linear that the iteration will converge to only one solution. This solution should be within the same two standard deviations also the optimum of the original non-linear problem. Such a situation will, of course, depend on the model relationships and the variables of the specific problem.

7.3.2.3.1 Principles, Problems: Linearization, Initial Values, the Damping Parameter

The *linearization of the basic equation*, i.e. the calculation of the Jacobian matrix of the derivatives $\partial r_i/\partial p_k = \partial(\delta g_{obs,i} - f(x_i,p))/\partial p_k = J_{ik}$, can be obtained by analytical or numerical differentiation. In the last case, the differentiation distance can be chosen proportional to the estimated a posteriori standard deviations, i.e. the inverse diagonal elements of $A^T C_{yA}^{-1}A + C_x^{-1}$, or to the a priori (and the normalizing) standard deviations. If the basic equation can be considered linear only around the minimum, the former option guarantees that the search is done only there. The proportionality factor may be chosen in the order of 0.1. Caution is necessary because of possible problematic interference between the discretization of the variables and numerical differentiation. A discretization of the parameters should not depend on their values, and in any case, the discretization distance should be much smaller than the differentiation distance. Otherwise the derivatives (finite difference coefficients) can be seriously in error. In critical situations, tests with different discretizations are necessary.

7.3.2.3.2 Expansion of A(x) to Second-Order Terms

To assess problems of the linearization, the expansion of $A(x)$ to second-order terms is considered.

The least-squares solution of the observation equation $Ax = y$ with x as the unknown variable renders the expression (Sect. 7.2.16) that minimizes the expression (see Sect. 7.2.1.2.3):

$$\varphi(x) = (y - \mathscr{A}(x))^T C_{yA}^{-1}(y - \mathscr{A}(x)) \tag{7.3.13}$$

A is the linearized version of $A(x)$. The Taylor expansion till second-order terms is:

$$\varphi(x) \approx \varphi(x_i) + \partial\varphi(x_i)/\partial x^T \Delta x + \tfrac{1}{2}\Delta x^T \partial^2\varphi(x_i)/\partial x\partial x^T \Delta x \tag{7.3.14}$$

where

$$\partial \varphi(x_i)/\partial x^T = -2\partial \mathscr{A}^T/\partial x^T C_{yA}{}^{-1}(y - \mathscr{A}(x_i))$$

$$\partial^2 \varphi(x_i)/\partial x \partial x^T = 2\partial \mathscr{A}^T/\partial x C_{yA}{}^{-1}\partial \mathscr{A}/\partial x^T - \partial^2 \mathscr{A}^T/\partial x \partial x^T C_{yA}{}^{-1}(y - \mathscr{A}(x_i))$$

$$(7.3.15)$$

At the minimum, the gradient $\boldsymbol{\nabla}\varphi(x) = 0$:

$$\boldsymbol{\nabla}\varphi(x) = \partial \varphi(x_i)/\partial x - \partial^2 \varphi(x_i)/\partial x \partial x^T \Delta x = 0 \Rightarrow$$

$$\Delta x = \partial \varphi(x_i)/\partial x (\partial^2 \varphi(x_i)/\partial x \partial x^T)^{-1} \Rightarrow \qquad (7.3.16)$$

$$\Delta x = \partial \mathscr{A}^T/\partial x^T C_{yA}{}^{-1}(y - \mathscr{A}(x_i))$$

$$\times \left(\partial \mathscr{A}^T/\partial x C_{yA}{}^{-1}\partial \mathscr{A}/\partial x^T - \partial^2 \mathscr{A}^T/\partial x \partial x^T C_{yA}{}^{-1}(y - A(x_i)) \right)^{-1}$$

For a linear kernel, $\partial \mathscr{A}/\partial x^T = A$ and $\partial^2 \mathscr{A}^T/\partial x \partial x^T = 0$. This confirms the equality with $\hat{x} = (A^T C_y^{-1} A)^{-1} A^T C_y^{-1} y$ (see Sect. 7.2.1.2.3, Eq. (7.2.16)). If in an iterative method, starting from a current solution $\varphi(x_i)$ not the absolute minimum directly, but only a better solution $\varphi(x_{i+1}) < \varphi(x_i)$ is aimed at it suffices to step a small distance $\Delta x_{i,i+1} = x_{i+1-x_i}$ in a linear relation with $\boldsymbol{\Phi}$ the opposite direction of the local gradient: $\boldsymbol{\Phi} = F\nabla\varphi = F\delta\varphi(x_i)/\delta x$, where F is a square matrix, such that:

$$\varphi(x_{i+1}) = \varphi(x_i) + \delta\varphi(x_i)/\delta x^T \Delta x_{i+1,i} = \varphi(x_i) - \delta\varphi(x_i)/\delta x^T F \delta\varphi(x_i)/\delta x \quad (7.3.17)$$

The condition $\varphi(x_{i+1}) < \varphi(x_i)$ is always met when $\delta\varphi(x_i)/\delta x^T F \delta\varphi(x_i)/\delta x > 0$ which corresponds to the definition of F being positive definite. This conforms to a Newton-Raphson method for determining the root of a gradient function $\nabla\varphi$. An arbitrary positive definite matrix F differs from the special choice $F = F_{opt} = (\delta \mathscr{A}^T/\delta x C_{yA}^{-1}\delta \mathscr{A}/\delta x^T - \delta^2 \mathscr{A}^T/\delta x \delta x^T C_{yA}^{-1}(y - \mathscr{A}(x_i)))^{-1}$ in that the former renders only a slight improvement and the latter renders the optimal solution. In other words, in the case of using F, the lowest point is looked for by stepping downslope in the steepest direction, while in the case of F_{opt} the curvature as the linear or non-linear "topography" is used to find the minimum in one step. To this end a hyperparaboloid with the same curvature is placed at the current solution and renders with (Eq. 7.3.16) the coordinates of its minimum. If $\mathscr{A}(x)$ is linear, the solution is correct, if not, non-linear iteration should be continued. The following considerations are important:

- The Newton method finds roots of the gradients, but for non-linear $\mathscr{A}(x)$ they may be also maxima or saddle points, which however can be checked;
- Approximations of the function $\varphi(x)$ by the hyperparaboloid may be locally poor and may not place the new solution near the (nearest) minimum;
- The a posteriori covariance matrix of the variables $C_x = F^{-1}$ is now influenced by the second derivative of $\mathscr{A}(x)$: $-\partial^2 \mathscr{A}^T/\partial x \partial x^T C_{yA}^{-1}(y - \mathscr{A}(x_i)) \neq 0$. The uncertainty is corrected upward or downward relative to the linearized version depending on the curvature of $\varphi(x)$. Close to a minimum the predictions are improved. But if maxima exist, there must also be inflexion points with zero

curvature, and F does not remain positive definite. Hence at such points the method fails to decide whether to go to the minimum or the maximum introducing again a singularity. Moreover, the second derivative is usually not considered, because (1) it is often quite small, (2) its effect is only to find the minimum a bit faster and (3) its calculation is tedious.

Consequently, when a current approximate solution x_i lies near a desired minimum, the solution method for weakly non-linear basic equations should make the step Δx so small that it will not reach another minimum. Such a damping of the iterated solution is in effect a regularization with the side condition $\Delta x = 0$; its co-variance matrix $C_{\Delta x}$ defines the local search or damping radius. The regularized and damped basic equation for iteration i is thus:

$$\begin{cases} \begin{bmatrix} \partial A/\partial x \\ I \\ I \end{bmatrix} \Delta x = \begin{bmatrix} y - A(x_i) \\ x - x_i \\ 0 \end{bmatrix} \\[6pt] C_{yA,x,\Delta x} = \begin{bmatrix} C_{yA} & 0 & 0 \\ 0 & C_x & 0 \\ 0 & 0 & C_{\Delta x} \end{bmatrix} \end{cases} \qquad (7.3.18)$$

with the solution:

$$\begin{cases} x_{i+1} = x_i + \left[\partial A^T/\partial x^T C_{yA}^{-1} \left[\partial A^T/\partial x \right] + C_x^{-1} + C_{\Delta x}^{-1} \right]^{-1} \\ \quad \left[\partial A/\partial x \right]^T C_{yA}^{-1} \left[y - A[x_i] \right] + C_x^{-1} \left[x - x_i \right] \right] \\ C_{xi+1} = \left[\partial A^T/\partial x^T C_{yA}^{-1} \left[\partial A^T/\partial x \right] + C_x^{-1} \right]^{-1} \end{cases} \qquad (7.3.19)$$

The "regularization observation" $(x - x_i)$ pulls the solution at every step back to the a priori information, but the "damping observation" $\Delta x = 0$ is the same in all steps. When the solution has converged to the minimum, Δx is necessarily zero and damping has become superfluous. Therefore $C_{\Delta x}^{-1}$ is used only to determine the solution path; it does not appear in the a posteriori covariance of the converged solution, for the statistics should be independent from the path.

7.3.2.3.3 Iterative Solution

For the mildly non-linear case, the Gauss-Newton algorithm was extended to the Levenberg-Marquardt algorithm (Levenberg, 1944; Marquardt, 1963, Press et al., 1992). It is applied to the initial values of the variables (Sect. 7.3.2.3.4) and regularizes, i.e. damps, the change of the solution from step to step. To keep the damping in accordance with the a priori information, as discussed below (Sect. 7.3.2.3.2), the covariance matrix of the parameter changes per iteration step, $C_{\Delta x}$ of Eq. (7.3.19), can be chosen proportional to the a priori standard deviations $S_x = diag(\sigma_{i=1,m})$, i.e. λS_x. Since the free variables $(\sigma = \infty)$ cannot be damped this way, a "normalizing standard deviation" must be prescribed to them in order to define the proportionality factor of damping which should reflect the expected range of values, but the choice is not critical.

In the case of a linear basic equation, the optimal solution lies at the bottom of a parabolic valley, the walls of which rise accordingly if moving normal to the standard ellipsoid, i.e. along the eigen-vectors of the a posteriori covariance matrix. From the minimum, the topography rises by one unit if one moves by a unit of a singular value along the eigen-vector. This is because the singular values of the a posteriori covariance matrix indicate how large the uncertainty is in some particular linear combination which is also connected with a data uncertainty of one standard deviation. One standard deviation of the data from the optimal solution changes the shifting variate by one unit, as given by the slope of the above topography. In case of a non-linear basic equation the slope can be higher or lower. An increase of the shifting variate by < 1, when the solution is changed by one singular value unit in the direction of the eigen-vector, indicates that the effective a posteriori standard deviation is larger than calculated. Obviously the data are not fitted that much worse than should be expected and the non-linearity results in a larger acceptable solution range than calculated. A decrease (negative increase) indicates that one has gone downslope – by accident? – and the optimum solution has not yet been reached. Very large values indicate a deterioration of the solution by more than one standard deviation. The estimate of the a posteriori standard deviation has to be reduced accordingly, and the solution range is smaller than deduced from the linear model. The approach is related to the so-called "edgehog" procedure (Jackson, 1973).

7.3.2.3.4 Initial Values of Variables

Initial values of the variables are needed to start the iteration, but in order to converge, they must lie in the "environment" of the solution looked for. Moreover, in the Bayesian approach they should represent the best a priori information. Initial values are not always easy to choose if the a priori information is vague and the problem may be more strongly non-linear than anticipated (Sect. 7.3.2.3.2). There is the danger that the solution does not reach the main minimum. Especially problematic starting points or locations are where the φ-topography is extreme, i.e., either flat or highly curved, even a singularity. It is therefore important to start at a favourable point, for example, in a region of negative curvature around the expected minimum. Thus the choice of the starting point is part of the regularization.

7.3.2.3.5 Damping Parameter λ

The damping parameter λ serves to affect the speed of convergence. A small λ leads to fast convergence when the basic equation is approximately linear. Large λ values are needed when convergence deteriorates. λ can be recalculated at each iteration step. Near the minimum, the error of fit should decrease at each iteration and λ can be set to zero. If not, then one might just step back and increase the damping until a better solution is found. Then the damping may be decreased until it is smaller than some fraction of the a priori standard deviation (e.g. 1/100). If the solution still improves, no damping need be imposed: $\lambda = 0$. Some optimizing algorithms for the damping parameter employ a systematic search as *back tracking* (Press et al., 1992), but no clear rule seems to exist and a slow change of λ usually suffices. If the problem, in spite of regularization, is singular, a solution can be guaranteed by letting $\lambda > 0$. Only when the largest of the estimated a posteriori standard deviations (which are the inverse diagonal elements of $A^T C_{yA}^{-1} A + C_x^{-1}$) exceeds some multiple of its a priori value (e.g. 100), singularity is indicated.

7.3.2.3.6 Abort Criteria (χ^2-Change, Model Differences, Damping)

The damping parameter renders also the abort criterion if the improvement remains consecutively below some small value (e.g. 0.01 for several steps). In this situation, also the change δx_i from step i to step $i+1$ should be considered in relation to the covariance matrix:

$$(\delta x_i)^{\mathrm{T}} C_{\hat{x}}^{-1} (\delta x_i), \tag{7.3.20}$$

in order to ensure that the change of the solution is really small relative to what the data can resolve. If a small deterioration occurs in the iteration (e.g. a change of the F-statistic by < 0.01 and of the normalized solution vector by < 0.1), the solution cannot be accepted, but it is tested if the change occurs within the region of the previous minimum or is on the way to another. Such measures are necessary in view of non-linearity and imprecise arithmetic, for example, from numerical differentiation. If both relaxing the damping and strong tightening do not result in a significant change of the solution, it can be assumed that the minimum has indeed been found.

7.3.3 Other Algorithms

Some model relations are so strongly non-linear that linearization will generally fail to lead to stable solutions. In such cases, other strategies alternative to the Newton method guided by the derivatives of basic functions may be more successful, such as iterative algorithms, systematic or random (Monte Carlo) grid search and variants of genetic or evolutionary methods of search (Goldberg, 1989; Holland, 1992). They search the solutions by random variations of the parameters and selecting only acceptable solutions or by rejecting unacceptable ones.

Consider the non-linear basic equation $A(x) = y$ which cannot be easily linearized to $Ax = y$. Fortunately, ordinary problems of gravity inversion are often reasonably linearizable in the environment of interest, or they can be formulated this way. But problems arise if F (Eq. 7.3.17) is singular and the topography has no curvature and no minimum in a certain direction; if F is nearly singular with a topography resembling a very flat bowl, and its deepest point is difficult to find; like a ball the solution can roll back and forth and come to a stop arbitrarily by the slightest influence, it is unstable. Regularization bends the rims of the bowl up to better define a minimum (illustrated in Fig. 7.3.3). If regularization is carried out by replacing $A^T C_{yA}^{-1} A$ with $A^T C_{yA}^{-1} A + C_x^{-1}$, the solution space is concentrated into the region of the a priori information x_i. If $C_x = diag(\sigma^2)$ with small σ^2, the solution will move across an only small distance down the local gradient toward the optimal solution.

In the case of *strongly non-linear basic equations* one should use other algorithms, because regularization with C_x will not suffice to obtain a unique result. For example, a systematic search on a narrow grid of parameter "points" (which would require many attempts) or a random search by a Monte Carlo method (see below). Or imagine a topography characterized by many minima and maxima, but nevertheless generally descending toward the principal minimum (e.g. a density distribution consisting of many shallow small bodies and a large deep one). A Monte Carlo survey may then first identify a low region or one with frequent lows, on which the search can then be concentrated. Evolution algorithms (see below) generate from previous accidental good solutions new solutions with smaller modifications, from which the best are chosen for further improvement (e.g. by a systematic adjustment). Also the SIMPLEX method (Nelder & Mead, 1965) concentrates on new solutions near the best previous one, but only in a descent direction determined from a set of solutions. It works well where the basic equation is relatively smooth, but can be rather easily trapped in a local minimum. The method does also not calculate the lowest point but

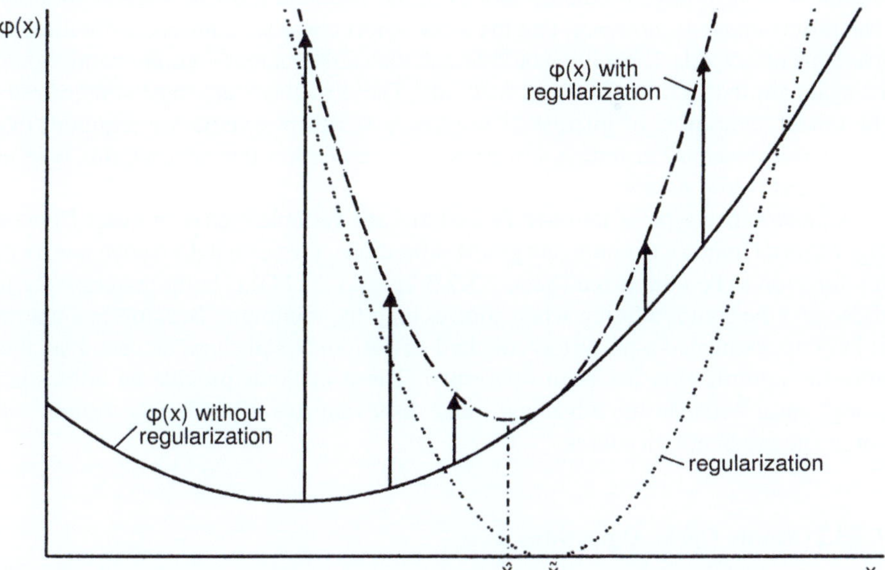

Fig. 7.3.3 Deformation of the function to be minimized by the a priori information

an arbitrary point lower than the current solution, nor does it calculate the gradient, i.e. the maximum slope.

7.3.3.1 Iterative Algorithms

Iterative algorithms of matrix inversion as opposed to direct algorithms (Sect. 7.3.2.1.1.1) try to gather explicitly or implicitly information on the matrix inverse by an iterative procedure. The Levenberg-Marquardt algorithm (Sect. 7.3.2.3.1) also solves iteratively, but in that case during each iteration step the matrix inversion itself is usually still performed by a direct method (LU, Choleski, etc.). The rationale of matrix inversion by iterative algorithms is to speed up the computation and reduce memory consumption relative to direct algorithms. For a matrix of size $N \times N$ the time requirement of direct algorithms usually scales with N^3 and the memory consumption with N^2. In contrast, many iterative algorithms even never build up the inverse matrix, but only search for the optimal solution, requiring memory scaling with only N. And for non-linear problems, in each iteration step not the optimal solution of the linearized problem is searched for, but only a better one. This simplified optimization step saves computation time, since in subsequent iteration steps the linearization and therefore the achievable optimum the will change anyhow.

A well known type of iterative algorithms is the conjugate-gradients method. It is a minimization algorithm, which searches for the minimum by stepping iteratively more or less directly downhill. Especially the LSQR flavour of this type of algorithms, which solves the least-squares problem, could be considered for gravity inversion of very large problems. Not even the Jacobian must be stored explicitly. The disadvantage is, however, that the a-posteriori covariance matrix is not directly available afterwards. Even if it would be calculated the memory requirements would be again similar to those of direct methods. These methods are interesting, if only the model solution is of interest. If solution assessment criteria are required, like a-posteriori standard deviation, covariances or resolution information, this type of algorithms is less suitable.

An alternative type of iterative procedures are variable metric or quasi-Newton algorithms. Contrary to conjugate gradient methods, the second derivative matrix of the function to be minimized (Sect. 7.3.2.3.2; Eq. (7.3.14)) is built up iteratively in order find the solution faster, when approaching the minimum. Because this matrix is built up, estimated a-posteriori standard deviation, covariances, etc. are available after the optimization has been performed. These methods provide an interesting compromise between the advantages and disadvantages of direct algorithms and conjugate gradient procedures.

7.3.3.2 Monte Carlo Algorithms

Monte Carlo search routines work with random numbers or random number generators, i.e. they "throw dice". Starting from am initial vector x_o of the variables or parameters, random changes $\boldsymbol{\delta x}$ are added, where for each parameter δx_k samples are generated from a Gaussian distribution with the standard deviation σ_k, assumed a priori for x_k. A large sample of solutions, $\boldsymbol{y} = A(\boldsymbol{x}_0 + \boldsymbol{\delta x})$, is then calculated and analyzed statistically. Outliers with respect to the a priori frame of knowledge can be removed immediately. The rest is then plotted for single variables or couples of them, or general frequency distributions are calculated for multi-dimensional solutions and the maximum of the distribution may be taken as the best solution plus or minus some measure of its standard error.

7.3.3.3 Genetic and Evolutionary Algorithms

Genetic algorithms are search techniques to find exact or approximate solutions to optimization and represent a particular class of evolutionary algorithms inspired by evolutionary biology. The technique somewhat simulates biological evolution by combining the previously best solutions with the hope that the offspring may be even better. The term "evolution" is not quite appropriate, because the algorithms aim at solutions in the vicinity of the a priori information while biological evolution has no aim (or is not teleological). Populations consist of abstract representations in the form of arrays of bits or values of variables (*chromosomes, genotype*) of candidate solutions (*individuals, phenotypes*). Evolution happens in generations. The advantages are that the approach can produce reasonable solutions even of strongly non-linear problems with many minima and possibly with singularities in some regions and they can easily escape from local minima in which a guided solution may have been trapped. The disadvantage is that a statistical quality evaluation of the solutions is quite difficult.

The basic scheme of the algorithm is the following:

(1) A starting model is constructed as a population of randomly modified individuals centred on the a priori information. Its parameters are varied within assumed statistical distributions, usually normal distributions with given standard deviations. A generation of solutions (*individuals*) is thus produced.
(2) The solutions are ordered according to the F-statistic (F suggesting *fitness*). The fitness function measures the quality of a solution.
(3) The worst *individuals* are rejected or removed.
(4) They are replaced by new solutions (*children*) which are arbitrary (chance) combinations of the parameter values (*genes*) taken (*inherited*) from the remaining individuals (*parents*). This recombination is called "*cross-over*".
(5) The parameter values of the new solutions are slightly changed at random (*mutation*). The equally inherited *mutation factor* determines how strong the average mutation is. The mutation factor is different for each variable and is itself subject to mutation.
(6) Individuals that are too similar to each other are removed, where the similarity is defined by the F-statistic and by (Eq. 7.3.20). This prevents *inbreeding* and encourages co-existence of several *niche populations*.

(7) The so-called slack variables (variables of no physical importance for the problem at hand, like the mean of relative gravity readings) are, at each step, optimized by the Newton method because evolution does not work for them and would inhibit the search for a minimum.

(8) The abort criterion is examined and, if not yet met, the procedure is repeated from point (2) (*new generation*).

Some characteristics of the algorithm are:

- Each *individual* may have its own parameter sequence. In the cross-over, neighbouring parameter values are likely to remain together, such that positive or fitting combinations tend to be handed down, but the sequence does mutate occasionally as well, so that better ones have a chance to be found.
- An individual exists only for a certain number of iterations and can extend its existence only by producing better offspring. Accordingly, *successful* individuals (with small F-statistics) produce more offspring than less successful ones do.
- Evolutionary algorithms more or less play the game of genetic evolution by random mutations and recombinations accompanied by rejection of the unfit. In contrast to biological evolution that has no aim and measures success with the ability survive, gravity inversion generally aims at fitting some a priori information and success is measured also by fitting it.
- Performance can be quite varied, and diversity is important such that it should be enforced, by emphasizing mutations and bringing in *random immigrants*. Generally, genetic algorithms are good at finding reasonable global solutions, but the absolute optimum is better found by classical matrix inversion, once the general region has been found. It is thus effective to combine the different methods.

For more details and possibilities of genetic algorithms see Frazer & Burnell (1970), Goldberg (1989), Mitchell (1998), Fogel (2006). Various problems and variants of the method are discussed in Wikipedia (http://en.wikipedia.org/wiki/Genetic_algorithm).

7.4 Case Studies

Inversion with emphasis on the above principles is illustrated with two case studies, the Meerfeld Maar (Sect. 7.4.1) and the SE Iceland shelf edge (Sect. 7.4.2). The Meerfeld Maar in the Eifel, Germany, was introduced in Sect. 5.7.1. The SE Iceland shelf (Sects. 5.7.5, 6.5.5) is an exploratory case. The two examples differ in the amount and nature of the a priori information and in the approach to inversion. In the first case, the analysis of statistical characteristics and evaluation play a more important role than the geological aspects, treated in Sects. 5.7 and 6.5. But since the Bayesian approach (Sect. 7.2.2.3.10) intimately links the geological a priori information to the statistical inversion problem, the two sides cannot be really separated. In the second case the situation is characterized by very vague a priori information essentially consisting of several speculative geological ideas, four of which were presented in Sect. 6.5.5. Here, only one of these is investigated, and the inversion strategy involves trial and error with different model types, including schematic ones.

The program INVERT was applied. The concepts of regularization as described were tested in practice and strengths and weaknesses became evident. An important aspect was structuring the abstract concepts into a comprehensible format that was to help the users in a reasonably short learning period to understand the many evaluation criteria and to arrive at better and more useful solutions in less time. Some geological assumptions are simplified, the basic principle may, however, be explained better, by limiting the spectrum of the interacting pieces of information from the start.

7.4.1 Meerfeld Maar

7.4.1.1 The Geological Framework

The a priori information includes the volcanic processes forming a maar and the rocks within which it originated. The Quaternary West-Eifel volcanic field is characterized by many localised bowl-shaped depressions ("craters"), many of the younger ones filled party with lakes (from which the term "maar" originates).

The depressions result from phreatomagmatic steam explosions and sagging of country rock into the reaction chamber between magma and water during rather short eruption cycles (Lorentz, 1986; Meyer, 1988). Magma ascends through faults and fractures into which water flows from above in sufficient quantities to generate large amounts of steam. Most maars occur in former river valleys. Triggering of steam explosions depends on pressure being $< 25\,\mathrm{b}$ and on shock waves. The explosions fracture the cooling juvenile magma and the country rock. A mixture of fragments is ejected and deposited as a tuff wall around the crater. It ranges in size from fine-grained ash and lapilli to bombs, xenoliths (solid already in the ascending magma) as well as country rock which usually dominates in volume. After an explosion, water flows again into the deepest parts of the reaction chamber and may lead to the next steam explosion, such that the fragmentation front steps downward, and the rock is loosened and destabilized above the chamber, sagging down in blocks. Material falling back and cleared out during the eruptive cycles partly falls back forming a cone-shaped volume of loose material, until the supply of magma and/or water comes to an end, and either more lava rises to the surface or the pipe being left filled with fragmented material. The results are heterogeneous structures, reflecting the different genetic processes. The deeper parts of fragmented rock are called "diatreme", upon which more or less volcanic material of the final phases may be deposited as welded tuff or even built-up scoria cones. The left-over depression is subsequently more or less filled by ground water and the former river, forming a lake and a sediment body of eroded tuff and country rock with added muddy organic material. The shape of the sediment body will be that of a bowl widening upward if the depression is successively filled, but the sediment body may narrow upward, if higher-density screen from the slopes dominates the filling process.

Larger maars, as the Meerfeld Maar, have the following characteristics.

The surface is strongly modified by later sedimentation and/or erosion; with time the maar lakes disappear.

The surface dimensions are obscured and often revealed by geophysical signatures of gravity and magnetic anomalies.

The maximum depths of the diatremes seems to be 2000–2500 m.

The upper quarter of a diatreme seems to have rather shallow wall slopes, of to about 35°, which depends on the texture of the country rock (Stachel & Büchel, 1989).

According to Lorentz (1986) the original depth over width ratio of the diatreme is about 1:1 (Fig. 7.4.1).

In comparison, the deeply eroded and deeply mined diatremes of similar origin in the Kimberley region of South Africa (diamond mines in the kimberlite pipes) have usually very steep walls sloping 79–85° in their lower parts (Hawthorne, 1975), but less, higher up (about 60°).

The maar craters have an original width over depth ratio from 1:3 to 1:7 (Lorentz, 1986), mostly 1:5 (Stachel & Büchel, 1989).

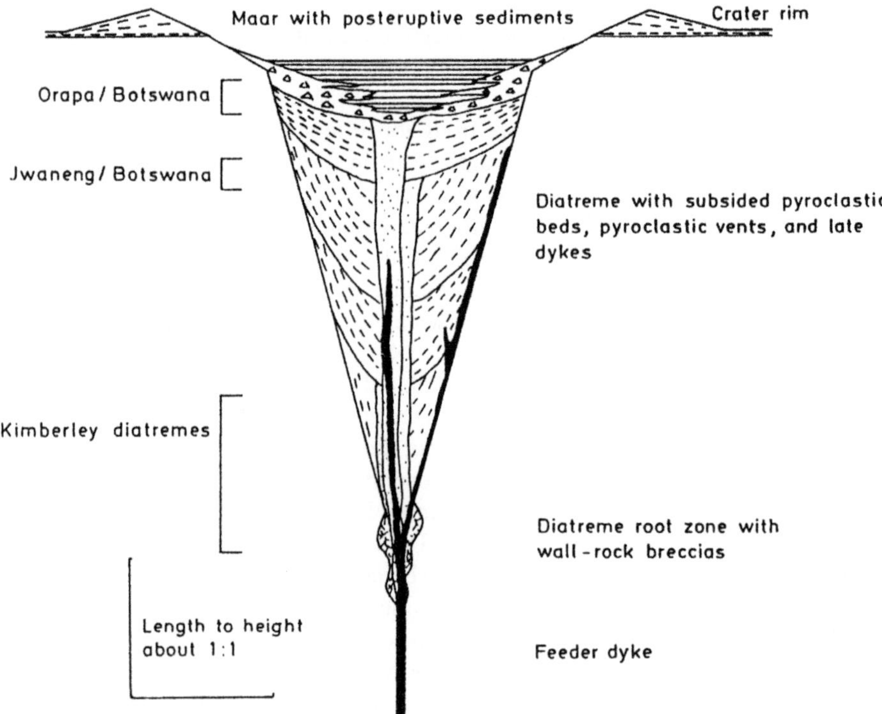

Fig. 7.4.1 Schematic model of a maar diatreme with post-eruptive crater sediments and dykes after Lorenz (1986). The model is based on kimberlite diatremes in Botswana and the Kimberly district of South Africa. Dimensions of dykes and pyroclastic vents (*dotted*) are exaggerated

The original maximum slope of the maar craters is probably $< 35°$ (Stachel & Büchel, 1989).

The Meerfeld Maar (Fig. 7.4.2) originated about 29 000 years ago (Lorentz & Büchel, 1980) within the Devonian rocks of the West Eifel. Its rim has a diameter of about 1700 m and the depth of the morphological depression is presently about 160 m. Geoelectric resistivity measurements revealed a thickness of post-volcanic sediments of about 190 m. These relations correspond to a width:depth ratio of 1:5, and geological observations suggest only minor erosion of the original surface. Considerable relicts of the tuff wall still exist at the NW, N and NE flanks, supported by strong magnetic anomalies. In the NE they reach 15 m thickness (Meyer, 1988). To the W and SE direction spread-out tuff deposits occur which come possibly from eruptive centres at the corresponding sides of the crater edge (see Fig. 7.4.2; Hunsche, 1973). The diatreme surface diameter is about 1200 m, as deduced from magnetic data (Fig. 7.4.2; after Büchel, unpubl., 1990). The plan form is nearly circular, possibly slightly elongated in the NNW-SSE direction. If the diameter-depth ratio is 1:1, the diatreme should extend to 1100–1700 m depth. The maar lake that originally filled the whole crater has been partly filled by a sediment fan from the Meerbach stream on the S side. Directly around the south shore there is a wet region. The village Meerfeld to the SW is situated on the fan which partly consists of back-transported tuffs. On the W, N and E sides, the magnetically identified diatreme edge very closely coincides with the edge of the sediments which abut against exposed Devonian rocks only a few tens of meters further out. Above the crater rim, faults exist which seem related to the depression.

7.4.1.2 Rock Densities

The densities of the various rock types are known within certain error bounds. The Devonian marlstones and sandstones have a density of about $2650\,\mathrm{kg/m}^3$ (Mertes, 1983), but the average porosity of 5% reduces it to $2500\,\mathrm{kg/m}^3$ (Henk, 1984), confirmed on hand specimens, possibly with a bit higher density (G. Büchel, pers. communication). The tuffs consist of 75% country rock and 25% juvenile components (including bombs) with a porosity of about 30%. If dry, the mean volume density is estimated about $2000\,\mathrm{kg/m}^3$ (Mertes, 1983; Henk, 1984; Stachel & Büchel, 1989), rising to $2200\,\mathrm{kg/m}^3$ when water-saturated. The lake sediments, if consisting of sand, silt or clay have densities of $1800–2150\,\mathrm{kg/m}^3$ (Stachel & Büchel, 1989). Peat and mud have densities near $1000\,\mathrm{kg/m}^3$. The sample to sample scatter is occasionally large, but partly an artefact of sample handling. Within the rock formation relative standard errors are estimated to be of the order of 5%.

7.4.1.3 Gravity Data

Gravity data were collected in 1990 during a geophysical field course by Johannes Gutenberg University Mainz, and in 1992 data gaps were filled and some points

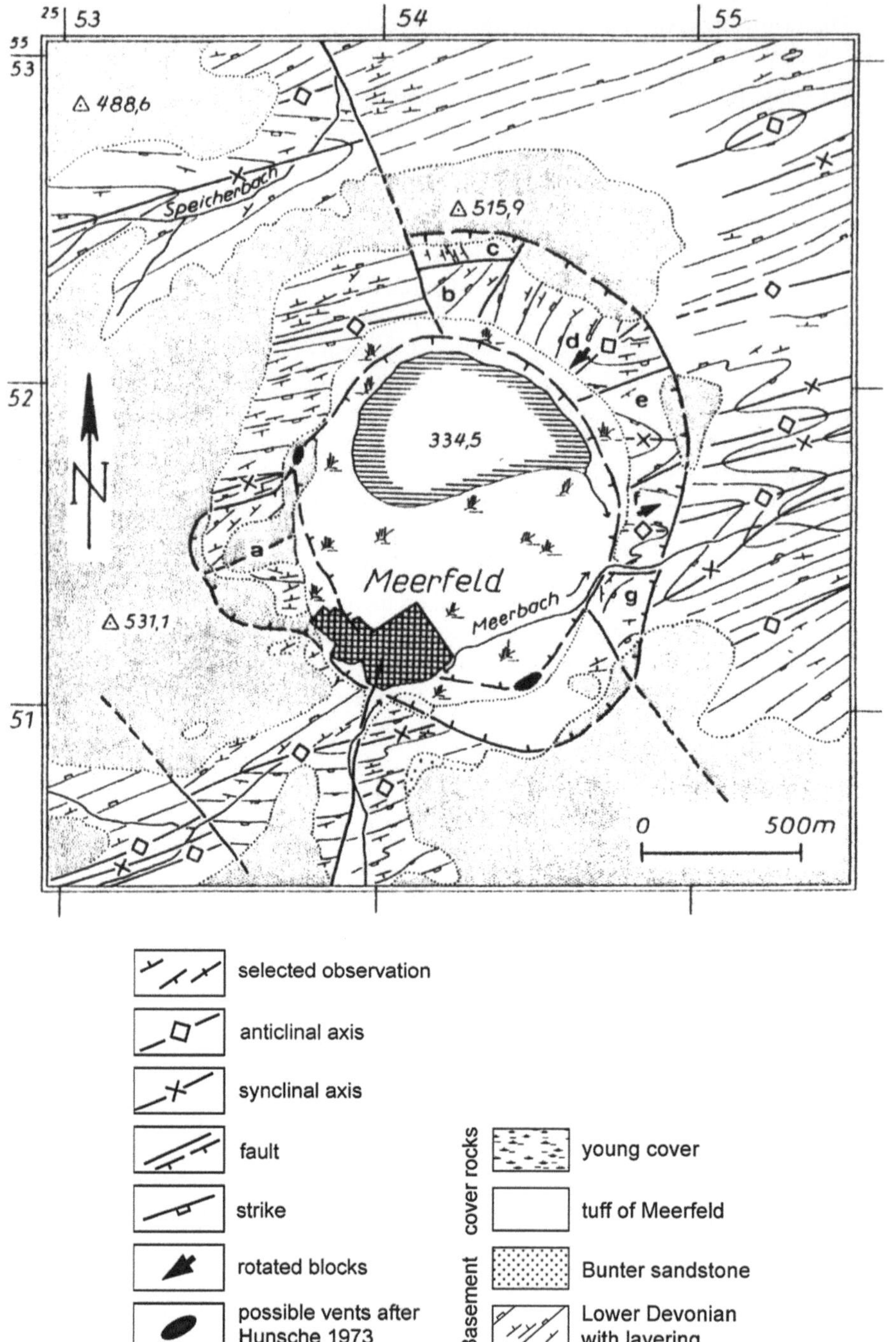

Fig. 7.4.2 Geological sketch map of the Meerfeld Maar after Meyer (1988). Devonian basement shown by strike lines of folded strata, fold axes, and NNW-SSE trending pre-maar faults; crater bottom: lake and alluvium; normal faults at crater walls, arrows: block rotation; rim of diatreme mostly inferred; tuff cover (white with dotted boundaries); possible eruptive centres (*black*)

were checked. 122 stations are distributed along roads and partly in fields in a roughly $2 \times 2\,\mathrm{km}^2$ area with the highest station density at the bottom of the maar, less on the slopes and on the upper plateau (see Fig. 5.7.1c). A Sodin Geodetic W410T gravimeter was used. The 3D station coordinates were surveyed with a TOPCON ET2 total station (theodolite with electronic distance measurement). The survey network was adjusted and tested for accuracy in connected loops. The relative errors of the coordinates were estimated to be generally better than 5 cm. The program GRAVI (Smilde, 1995; see Sect. 3.3.2) was applied to calculate the gravimeter drift and the normal, height and Bouguer reductions (see Sects. 4.5.1, 4.5.2, 4.5.3). The tidal reduction (Sect. 4.2) was implicitly included in the drift correction. The terrain reduction (Sect. 4.5.3.2) was calculated on the basis of the $40 \times 40\,\mathrm{m}^2$ digital terrain model of the Landesvermessungsamt Rheinland-Pfalz (State Geodetic Survey of Rhineland-Palatine), Koblenz.

A conservative estimate of the errors of the reduced gravity anomalies is about 0.06 mGal, which is mainly caused by the uncertain instrument drift correction. The Bouguer density used in the reductions (Bouguer plate and terrain) was $2550\,\mathrm{kg/m}^3$.

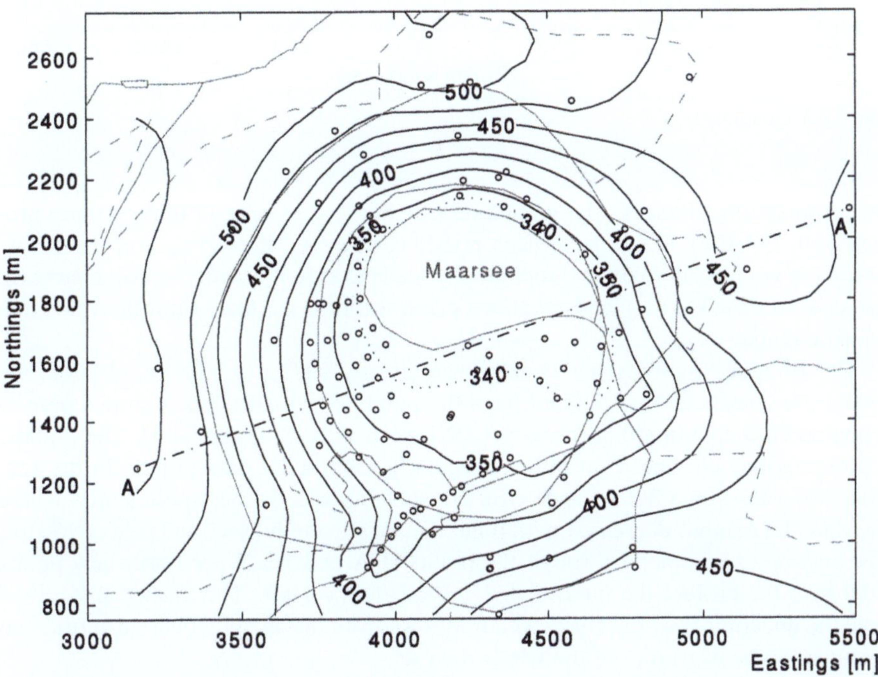

Fig. 7.4.3 *Top*: contour lines (25 m) of elevation of the Meerfeld Maar, constructed only from the observed station elevations. *Bottom*: WSW-ENE section A-A′ through the Meerfeld Maar (approximately along the Variscan strike); the circles show point values in a strip; their sizes decrease linearly with distance from 0 to 150 m; the *solid line* describes the weighted interpolation, and the *dashed lines* show its uncertainty taken from the spatial covariance function of the data (see also Fig. 5.7.1c)

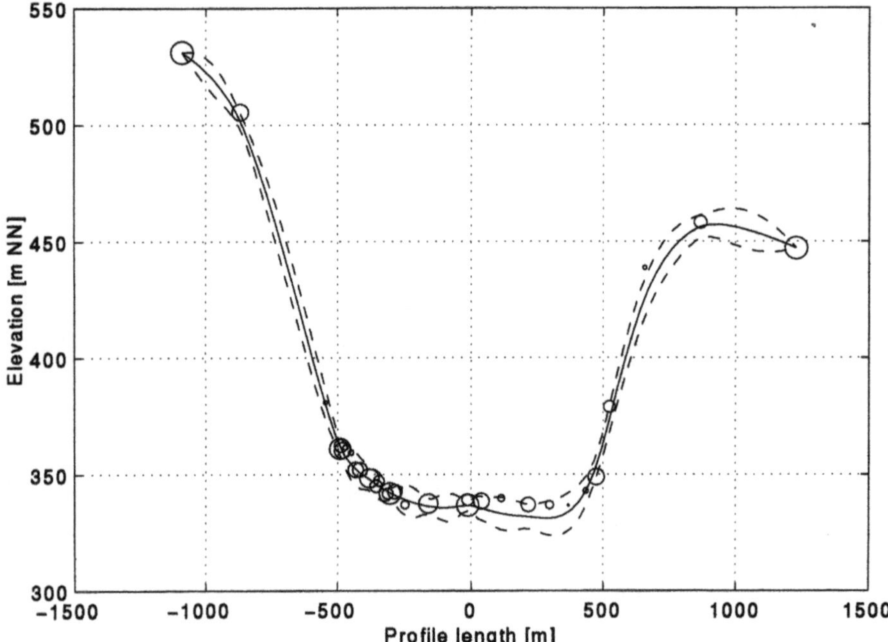

Fig. 7.4.3 (continued)

Its optimization with a Nettleton method (see Sect. 3.6.3.6) or in the inversion process with INVERT would have been possible, however, the strong correlation between the gravity anomaly and topography leads to systematically wrong results (in this case to $2840\,\mathrm{kg/m^3}$), and reliable a priori information must limit the value to a realistic range.

The Bouguer anomaly has been shown already in Fig. 5.7.1c. In addition, the topography has been interpolated from the point coordinates and is shown here as a contoured map and along the WSW-ENE section A-A$'$ (Fig. 7.4.3). The graphic representations are based on rather irregularly distributed data points. In the sections, the values in a 300 m wide strip are shown projected perpendicularly, where the size of a symbol decreases with point distance from the section (zero at 150 m). The section goes exactly through the points at A and A$'$. Where only few points exist near the profile, the interpolated curve is less certain. The dashed lines illustrate the uncertainty of the curve which follows from the spatial covariance function (assumed to be isotropic) of the whole data set.

The interpolated values must be considered with caution, since the location of possible disturbing bodies affects the correlation function locally. Hence, the assumption of a uniform and isotropic covariance function may have an unwanted effect on the determination of the target density distribution.

Some characteristics of the Bouguer anomalies are pointed out.

- Outside the maar crater, a weak "regional" trend is observed, rising from NNW to SSE. It is so weak that its effect will be averaged out when it is neglected in the modelling, since the data are distributed quite regularly around the crater.
- The minimum is remarkably peaked, which indicates a very shallow density anomaly. Lake sediments narrowing upward are a possible source, however, very soft soil in this region (making gravity measurements difficult) close to the lakeshore suggests groundwater saturation.
- Outside the crater rim gravity slightly decreases outward on the WSW to NW or N side. A possible source are the tuffs deposited there. However such an effect is not observed on the tuff deposits to the NE and SSE.
- Unfortunately, the transition from the steep gravity increase inside the crater rim to the flat, "undisturbed" field outside is not well defined because of an insufficient number of stations. A reliable definition of the deeper structures requires a better data basis especially in this region.

7.4.1.4 Parametrization of the A Priori Information

The aim of gravity inversion is in this case to determine the density distribution describing the diatreme, i.e., the shape and the density contrast with the country rock. Internal density variations in the diatreme and also in the country rock, for example, by possible alteration are of interest as well, but are probably not resolvable.

Accordingly, the parametrization to be designed for the density distribution will not permit a fit to the data with a better accuracy than $0.1\,$mGal, as standard deviation. For a normal distribution of the data, this means, if all other assumptions are met, that about 95% of the a posteriori residuals should be $< 0.2\,$mGal, or that the residuals should be $< 0.2\,$mGal with a probability of 95%.

7.4.1.4.1 Bouguer Density

As discussed above (Sect. 7.4.3.1.2), the Bouguer and terrain reduction was carried out with the density $2550\,$kg$/$m^3, which has removed the largest part of the gravity variations related to the different station elevations. This implies that for all gravity anomalies not yet explained, suitable source bodies must be defined. The reference density is thus defined for the body contrasts, but since it is not certain that $2550\,$kg$/$m^3 is representative for the country rock (not explicitly modelled), the Bouguer density could be adjusted together with the other density contrasts, especially if their absolute values are known. Since they are assumed as density contrasts relative to the assumed absolute Bouguer density, a strong negative correlation between these is expected and should be defined a priori. For the start, $2550\,$kg$/$m^3 is assumed error-free, but the assumption might be dropped at a later stage.

7.4.1.4.2 The Diatreme

The conical shape suggests a parametrization with conical circular disks, i.e. horizontal conic sections. Ellipticity seems negligible in view of the a priori information and its very small surface effect for the deeper parts. The most important aspect is the radius, i.e. its variation with depth. The magnetic edge anomalies define the upper radius quite reliably, and information on the diatreme shape from kimberlite pipes suggest an initial assumption on the slope of the pipe walls. Circular vertical cylinder disks are simpler than conical ones and to avoid non-negligible parametrization errors, a sufficiently fine division is necessary, which may become coarser with depth. Near the surface, however, the effect jumps at the edges, such that here a conical disk was assumed. It was defined by an upper and a lower radius and a triangulated side. This avoids a strong gradient of the effect at the edge, which is (1) not observed (maybe because of the insufficient point density, but it would, nevertheless, influence the solution, because it would tend to place the edge between points), and (2) the steep gradient surrounding a flatter region represents a strong non-linearity in the relation between radius and density. The gravity effect is nearly zero outside the disk and jumps rather abruptly to the Bouguer plate effect when moving onto the disk. Gradient-based optimizing procedures has problems of finding the minimum which might be found only by trial and error. The problem is less severe, when the gradient decreases and extends over a wider zone. This is not necessary for the deeper disks with vertical walls. The interpretation of the conical side of the top disk may be literally such a shape, or it may involve mixing of material and density gradation, and may finally be taken as a model error. Another possibility is to model such a density gradation by a cylindrical disk with a density contrast decreasing toward the edge. This can be realised by nesting many disks with increasing radii and correspondingly smaller density contrasts, such that by addition at the centre the total contrast is established (see Sect. 6.1.5.1 (3.2); Fig. 6.1.1d). In this way also a broader gentle gradient is generated.

The parameters of the cylindrical circular disks (see Sect. 7.2.2.3.1.2) are: density, depths of top and bottom, centre coordinates and radius. The conical disk has two radii, at the top and bottom. The circles were approximated by 18 equal straight sections (polygon), which causes a negligible error relative to the station distances and the other generalizing assumptions (e.g. the circular shape).

At the surface the radius is known, but at depth there is better a priori information on the slopes of the diatreme walls than on the radii which vary more strongly. Nevertheless, it appears easier to reparametrize the radii and link them to the slope as a function of depth. Fitting the depths as a function of the radius is another possibility, but depth determination is more strongly non-linear and more problematic than the non-linearity of radius determination. Unfortunately a symmetric variation of the slope transforms into an asymmetric variation of the radii, but by a fine division of the diatreme into thin disks, the problems more or less disappear (as shown below). Moreover, the uncertainty of the slope is anyway not likely to be symmetric. The a priori information on the slope is reparametrized to become a priori information on the individual radii by translating the shape of the diatreme, based on the slopes, into radii at selected depths. The standard deviation σ_r of the radii r is chosen to be large, because the individual radii are hardly known: $\sigma_r = 200$ m if $r > 200$ m, σ_r is otherwise smaller, down to 100 m at the bottom. Since the uncertainty of the disk radii is asymmetric, the inequality condition (Fig. 7.2.3; see Sect. 7.2.2.3.5) was imposed: $r > 0$ with a transition range of 5 m (Eq. (7.2.10), with $c = 0$, $s = -5$ in $x' = (x - c)/s$, where "minus" signifies a lower bound). The assumed standard deviations, combined with the disk intervals of 25–250 m and a maximum radius of 500 m imply a very large range of angles, possibly leading to a singularity, certainly too unstable and rarely acceptable solutions. The individual disks would be varied in size so much that finally each individual disc would function to fit a small subgroup of data points.

Such behaviour is suppressed by condition equations which connect the radius change of each disk with that of its neighbour. For an upper disk i (depth z^i, radius r^i) and a lower disk $i + 1$ (z^{i+1}, r^{i+1}) the slope $\alpha^{i,i+1}$ follows from $\tan \alpha^{i,i+1} = (z^{i+1} - z^i)/(r^{i+1} - r^i) = \Delta z^i / \Delta r^i$. Hence the radius difference is $\Delta r^i = \Delta z^i / \tan \alpha^{i,i+1}$. For a variation of α between α_{min} and α_{max}, the variation of the radius differences is $\sigma_{\Delta ri} \approx (\Delta z^i / \tan \alpha_{max} - \Delta z^i / \tan \alpha_{min})/4 (2\sigma$ rule). If the initially assumed model radii describe the slope correctly: $(r_0^{i+1} - r_0^i) = \Delta r_0^i = (\Delta z^i / \tan \alpha^i + \Delta z^i / \tan \alpha^{i+1})/2$, for arbitrary radii $r^i = r_0^i + dr^i$ we have $\Delta r^i \propto N(\Delta r_0^i, \sigma_{\Delta ri})$ and $\Delta dr^i \propto N(0, \sigma_{\Delta ri})$, and for the radius changes relative to the initial model the condition equations are:

$$
\begin{cases}
Ux = 0 \\
C_x, C_U
\end{cases}
\Rightarrow
\begin{cases}
\begin{bmatrix}
1 & -1 & 0 & & 0 \\
0 & 1 & -1 & & \\
\cdots & \cdots & \cdots & \cdots & 0 \\
0 & \cdots & 0 & 1 & -1
\end{bmatrix} dr = 0 \\
S_r = diag(\sigma_r), \ S_{\Delta r} = diag(\sigma_{\Delta r})
\end{cases}
\tag{7.4.1}
$$

where σ_r is the standard deviation of the individual radii (here 200 m). From these condition equations, the covariance matrix of the radii, or more exactly, of the deviations from the radii of the initial model, can be calculated with:

$$
C_{dr} = S_r - S_r U (U S_r U^T + S_{\Delta r})^{-1} U^T S_r
\tag{7.4.2}
$$

(see Eq. (7.2.9) and Sect. 7.2.2.3.8). If this covariance matrix of the radii is included as a priori information, the radii will closely follow the condition of limited slope angles for neighbouring disks in the solution, if the components of $S_{\Delta r}$ are chosen small. With increasing distance between disks the radius can be fitted to the data with less and less constraints provided by the condition. A common change by, say, 200 m would be hardly prevented. The standard deviations calculated with the diagonal elements of the covariance matrix, will be smaller than 200 m, because each individual radius is also constrained by the angle condition. Since the uncertainty of the a priori values of the slope angle decreases with depth, the a priori uncertainty of the radius variation will also have to decrease with depth.

Table 7.4.1 lists the assumed initial model data (shown together with the results, below, in Fig. 7.4.6a). The elevation above mean sea level and the depth below the surface at 336 m elevation are shown. The top radius of 550 m and the slope angles (with maximum and minimum value) at the given depths represent the a priori information. Columns with the position in the last row empty define the relation between a disk and the next subjacent one. The ΔR per 100 m values follow from the slope angles and are the basis, together with the top 550 m, for calculating the deeper radii. From the minimal and maximal slope angle follows $\sigma_{\Delta r}$. Negative maximum slope angles (of the lowermost disks) mean the diatreme could widen downward. Table 7.4.2 presents, in its first columns, the uncertainties of the individual disks.

The density contrast of the disks relative to the country rock is limited narrowly to the anticipated range of the contrast between tuff and the Devonian, i.e. $2050\text{--}2550 = -500 \pm 50\,\text{kg/m}^3$, thus the density contrast should be between 400 and $600\,\text{kg/m}^3$ at 95% probability, or a discrepancy of 0.1 mGal (the assumed modelling accuracy) violates the observations as much as a deviation of $50\,\text{kg/m}^3$ violates the assumed a priori density contrast. Density could be assumed as a known or unknown function of depth, however, since probably both densities, of diatreme and Devonian, increase with depth, the contrast may be close to constant and is assumed as such for simplicity.

The disk centres are coupled to each other: all lie directly above each other. The a priori information does suggest that a diatreme does not blast its way down along a winding path and usually produces a roughly circular cross section (supported by the results, see below). The standard deviation of the centre coordinates is set 100 m in X and Y, and the location was set at the BA minimum at $X = 4252$ and $Y = 1647$ m (Easting: [25]54252, Northing: [55]51647). Although this implies an unwanted correlation between the gravity data and the a priori information, not arranged for in the inversion program, the assumed uncertainty of 100 m will leave any correlation unnoticeable, at least if the variables behave not too non-linearly. Moreover, the geological a priori information suggests the same values within the uncertainty.

A further possible parametrization error of modelling with plane horizontal disks is that not all stations lie exactly on the surface of the uppermost disk. The question is how to take the mass in between into account. The following possibilities exist.

(1) A body can be defined by a DTM (digital terrain model) of the plane upper disk surface, the terrain and the prolongation of the conical disk side. This is a tedious procedure which cannot be built into the optimization process without further approximations.
(2) The upper disk surface is placed directly below the lowest station and the intervening mass is neglected. As seen in Fig. 7.4.3, the topography above the assumed diatreme in the profile range ±500 m is relatively flat and the intervening volume is small. Moreover, all stations outside the diatreme do lie above its upper boundary, such that its gravity effect is correctly calculated.

Table 7.4.1 A priori data on depths and radii of the 19 model disks. The upper two lines refer to the uppermost conical disk (see text)

	Z-lower [m NN]	Z-upper [m NN]	D-lower [m]	D-upper [m]	ΔD [m]	D-center [m]	ΔD-center [m]	Dip angle α [°]	α max. [°]	α min. [°]	Radius R [m]	ΔR per 100 m [m]	$\cup(\Delta R)$ [m]
1	336	336	0	0		0	13	35	26	65	550	143	10
2	324	324	13	13		13	6	36	25	60	534	137	5
3	311	324	25	13	13	19	19	37	27	53	526	134	11
4	286	311	50	25	25	38	25	39	29	54	500	126	13
5	261	286	75	50	25	63	25	41	31	56	469	116	12
6	236	261	100	75	25	88	25	43	34	58	440	107	11
7	211	236	125	100	25	113	38	46	35	61	413	98	16
8	161	211	175	125	50	150	50	49	39	63	377	87	18
9	111	161	225	175	50	200	50	54	44	66	333	74	15
10	61	111	275	225	50	250	50	58	49	70	296	62	13
11	11	61	325	275	50	300	50	63	55	73	266	51	10
12	−39	11	375	325	50	350	50	68	60	76	240	41	8
13	−89	−39	425	375	50	400	50	72	67	79	220	32	6
14	−139	−89	475	425	50	450	63	77	71	83	204	23	7
15	−214	−139	550	475	75	513	100	77	71	84	189	23	12
16	−339	−214	675	550	125	613	163	77	71	83	166	23	19
17	−539	−339	875	675	200	775	225	77	71	83	129	23	26
18	−789	−539	1125	875	250	1000	250	85	78	−88	77	9	29
19	−1039	−789	1375	1125	250	1250	313	89	84	−86	55	2	30
20	−1414	−1039	1750	1375	375	1563					49		

Table 7.4.2 A priori and a posteriori results for the model parameters (explanations see text)

Variable	A-priori	σ_{prior}	σ_{corr}	σ_{norm}	Inequality	Result	Difference	σ_{post}
r1 [m]	550	200	48.6	200		587	37.0	10.3
r2 [m]	534	200	47.7	200		574	39.7	7.0
r3 [m]	526	200	47.5	200		566	40.1	6.5
r4 [m]	501	200	46.6	200		530	28.6	5.8
r5 [m]	469	200	45.5	200		454	−15.5	7.8
r6 [m]	440	200	44.7	200		399	−40.7	9.3
r7 [m]	414	200	44.1	200		358	−55.7	11.3
r8 [m]	377	200	43.0	200		297	−79.7	15.1
r9 [m]	333	200	41.7	200		234	−98.5	19.3
r10 [m]	297	200	40.8	200		189	−107.8	20.6
r11 [m]	266	200	40.1	200		153	−113.4	21.0
r12 [m]	240	200	39.8	200		124	−115.7	21.1
r13 [m]	220	200	39.7	200		103	−116.7	21.2
r14 [m]	204	200	39.6	200		87	−116.9	21.3
r15 [m]	189	189	39.6	189		72	−116.9	21.4
r16 [m]	166	166	39.6	166		51	−114.7	21.6
r17 [m]	129	129	40.2	129		25	−103.7	21.6
r18 [m]	77	100	41.6	100	$> 0 \pm 5$	1.0	−76.0	(2.57) 2.00
r19 [m]	55	100	44.8	100	$> 0 \pm 5$	0.2	−54.8	(2.10) 1.64
r20 [m]	50	100	50.2	100	$> 0 \pm 5$	0.1	−49.9	(2.05) 1.60
x [m]	4252	100	100	100		4246	−6.0	1.9
y [m]	1647	100	100	100		1628	−19.1	2.1
d [kg/m^3]	−500	50	50	50		−620	−120	250
c [mGal]	0	∞	∞	1000		0.49	0.49	0.02

(3) The elevation of the stations is placed at the uppermost disk surface. For stations
directly above the diatreme which actually lie higher, the disk depths could be
interpreted as though defined relative to the local terrain; the disks are, so to
speak, fitted to the terrain. This is acceptable in the case of disks close to the
stations and not too steep terrain; but for more distant parts of the anomalous
body errors incur. Fortunately these are small due to the distance and have less
influence.

Figure 7.4.4 shows the calculated gravity effect of the diatreme which results
from shifting the stations from the terrain level to 0.5 m above the uppermost disk
level. The total effect nowhere reaches 0.15 mGal, i.e. it is of the order of the assumed standard deviation of the gravity data and the modelling. Outside the crater
the original elevation would be undoubtedly better, and the largest error occurs at
the edge of the diatreme and on the sediment fan in the SW part of the maar. The
difference effect demonstrates how the gradient of the gravity effect increases closer
to the edge of the anomalous body. Above the diatreme, consideration of the correct
elevation is insignificant because the effect of the upper disk approximates that of a
Bouguer slab, which does not depend on elevation above the disk.

Thus, for the inversion, the stations are left at the actual terrain elevation. Possible
corrections could be estimated on the basis of the figure. For comparison, the final

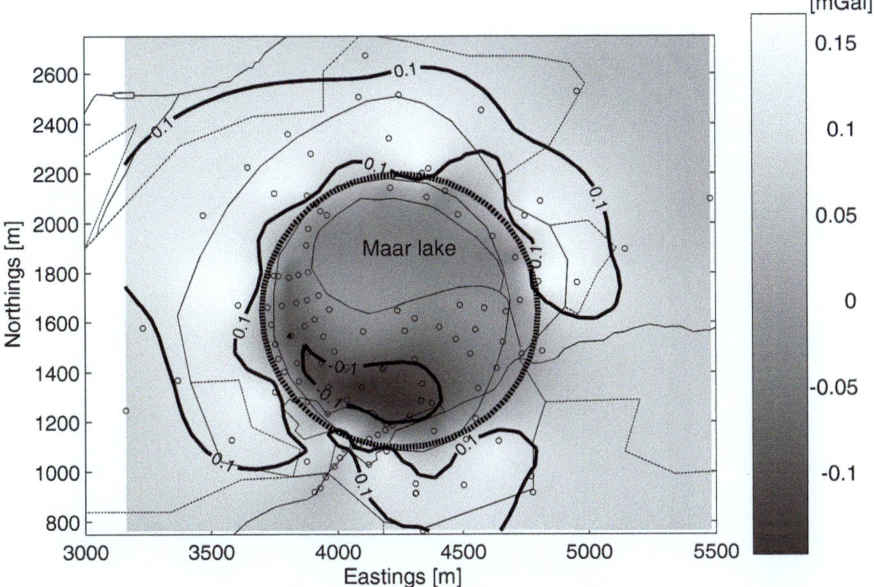

Fig. 7.4.4 Contours and shading of the change of the gravity effect when the stations are shifted down from their observation position to 0.5 m above the upper model disk

gravity residuals (observed minus final model effects) are shown in Fig. 7.4.5. For the discussion see below (Sect. 7.4.1.5.2).

7.4.1.4.3 The Lake Sediments

Since the lake sediments lie above the diatreme, their gravity effects are correlated with the diatreme effects and hence they mutually affect their density results. It would therefore be desirable to determine the sediment density distribution in magnitude and geometry. The a priori information on density is rather reliable and could be easily optimized because their effect are closely linearly related the sediment thickness, but this is hardly known. The body is thus not modelled initially.

If the sediments are distributed with some similarity or correlation with the diatreme, they would probably decrease the diatreme density and affect its shape in the solution. If the sediments occur only locally or irregularly, they will be evident in the corresponding residuals.

7.4.1.4.4 The Tuffs Outside the Crater

The tuffs could have a similar systematic effect as the sediments, but only few measurements have been made over them. Their influence on the whole inversion will

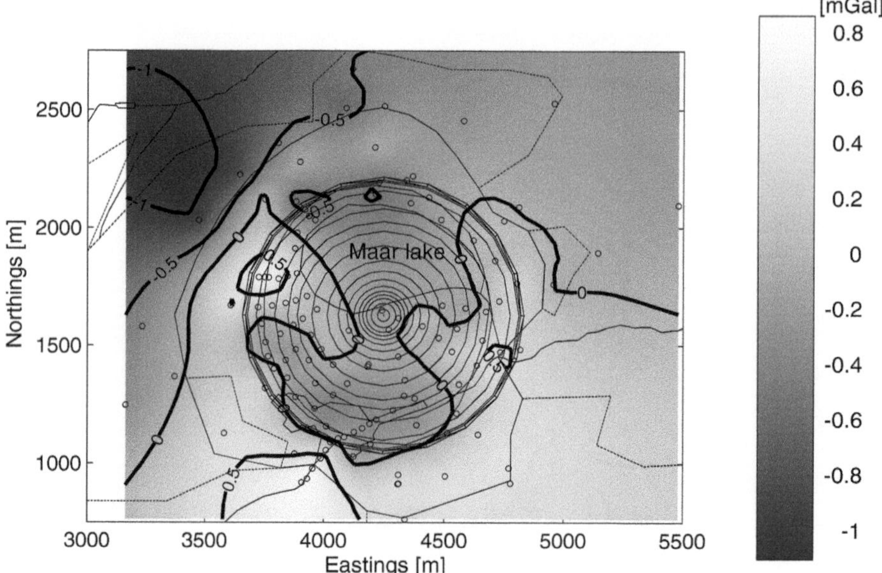

Fig. 7.4.5 Contoured and shaded residuals (observed – calculated) after the inversion; the a posteriori disk radii are also shown

thus be small. According to geological mapping they are not regularly distributed, i.e. they show poorly correlated behaviour with the circular diatreme and will not influence much the inversion results for the diatreme. An improved model may take the tuffs into account, for example, by a circular or elliptical ring, which can be realised by two disks of opposite density contrast (outer disk: $\Delta\rho$ (tuff versus Devonian); inner "compensating disk: $\Delta\rho$ (Devonian versus tuff)). Or some more detailed information might be modelled by a DTM.

7.4.1.4.5 The Regional Trend

If the stations are not very asymmetrically distributed around the crater, the regional trend may be neglected in an optimal fit of the data. The effect of the trend on an essentially radially symmetric model field would largely cancel itself. The neglect will increase the data standard deviations and the F-statistics. In an improved model a linear trend (like an inclined plane) could be defined with two additional variables which would be included in the optimization as slack variables, such that the gradient (magnitude and direction) would be free in the solution. A higher-order polynomial is hardly justified by the data, and due to the small number of stations outside the crater the corresponding trend surface would tend to "partly explain" the gravity minimum caused by the diatreme. In such a case a strong regularization would be necessary in a way limiting the parameters to describe only the trend surface and to separate it from the diatreme effect.

7.4.1.4.6 The Additive Constant

An additive constant in the fit of the model effects to the observations is needed in the optimizing task, because the reference of the Bouguer anomalies and the effects calculated with an arbitrary reference density must be correctly referenced to each other. The constant is a slack variable.

7.4.1.4.7 Summary

The parametrization has resulted in 24 variables to be optimized:

> 20 radii, 2 for the conical disk and 18 for the cylindrical circular disks, henceforth referred to by **r1–r20**, from top to bottom;
> the two common centre coordinates, called **x** and **y**;
> the density contrast of the diatreme relative to the Devonian, called ρ;
> the additive constant, called **c**.

Table 7.4.2 summarizes for the variables the a priori initial values and the conditions. Column 3 contains the standard deviations of the individual variables, as defined before the calculation of the correlation matrix of the radii. Column 4 contains the standard deviations after the calculation with the conditions (the diagonal elements of C_x in Eq. (7.2.1 – 2)). Column 5 contains the normalizing standard deviations, used in presenting the results: the residuals, the a posteriori standard deviations, the eigen-vectors and the eigen-values of the a posteriori covariance matrix (see Sect. 7.3.2.2.10). Column 6 shows, for the lowermost 3 disks, the inequality conditions assumed.

7.4.1.5 Results

7.4.1.5.1 Geology

Figure 7.4.6b shows the geologically relevant aspect of the results which is also presented numerically in the last three columns of Table 7.4.2. The most conspicuous characteristics are the much shallower slope angle in the upper part of the diatreme and the reduction of the maximum depth to 600–700 m below sea level or about 1000 m below the surface, not 1700 m as in the a priori model. The radii of the three lowermost disks are all < 1 m, suggesting that they were constrained by the inequality condition; otherwise they would have turned negative (in the INVERT implementation, the gravity effect would have also changed sign to warrant continuity and uniqueness). In Fig. 7.4.7, the initial and final forms of the diatreme demonstrate that between 600 and 800 m depth the shape was largely maintained while the radius was strongly reduced. The radii of the two upper disks have slightly increased beyond the initial value of 550 m; at places it is wider by about 100 m than the magnetically observed diatreme edge. The density contrast of the diatreme fill became

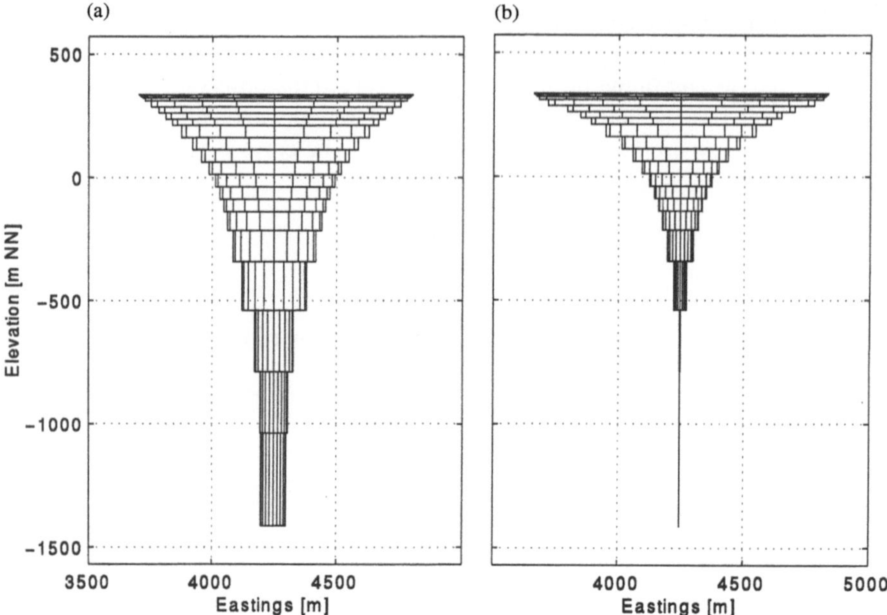

Fig. 7.4.6 Side views of the diatreme model: (**a**) initial a priori model (density contrast $-500\,\mathrm{kg/m^3}$) and (**b**) final result of the inversion (density contrast $-620\,\mathrm{kg/m^3}$, the vertical axis has shifted 6 m to the W and 19 m to the S). The top disk is conical and all others are vertical cylinders; the disk parametrization is by polygons, the edges of which are also shown

more negative, i.e. -620 instead of $-500\,\mathrm{kg/m^3}$. This may have the same cause as that of the increase of the upper disks. The centre coordinates were only insignificantly changed. Figure 7.4.5 shows a map of the residuals (observed − modelled) and Fig. 7.4.8 shows four profiles with symbols identical to those of Figs. 5.7.1c and 7.4.3.

7.4.1.5.2 Gravity Anomaly

- Especially in the NW, the negative residuals seem to be strongly correlated with the non-modelled tuffs at their transition to their Devonian basement at the crater edge;
- in the SSW of the crater lies a negative anomaly which is probably related to the non-modelled stream sediment fan;
- the negative anomaly to the N of the lake and the positive one to the W side cannot be explained without better knowledge of the local geology; perhaps the latter might be connected with western eruptive centre of tuff (see Fig. 7.4.2) suspected by Hunsche (1973); however, no such anomaly is seen near the other similarly suspected centre to the S;

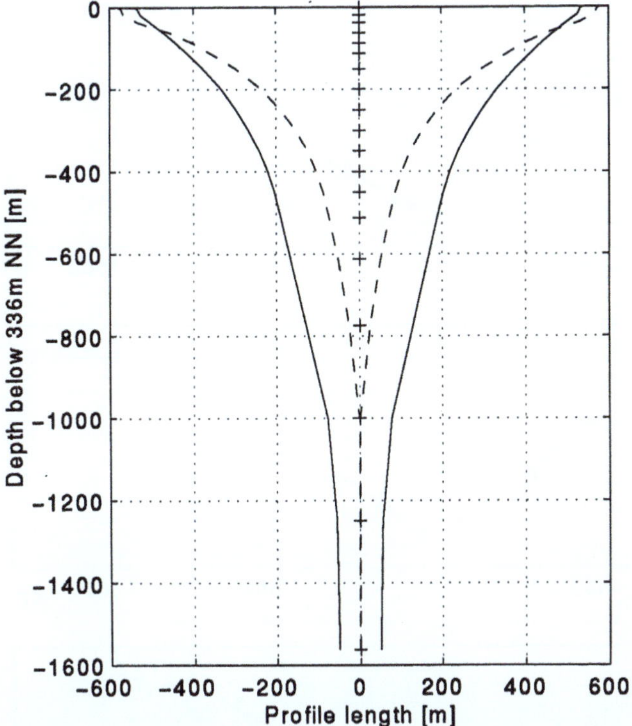

Fig. 7.4.7 Comparison of the a priori model (*solid line*) and the result (*dashed line*); the crosses on the axes are the disk centres and the lines are the envelops

- the sharp *BA* peak at the crater centre could not quite be fitted, but the residuals here are not large (-0.14 mGal) by comparison with the residuals generally;
- no uniform correlation of the residuals can be recognized with the circular structure, only a correlation with change of sign in the SSW-NNE direction and a positive WNW-ESE gravity ridge with, however, the most negative residuals in the "far" WNW (tuffs).

7.4.1.5.3 Comprehensive Evaluation of the Residuals

In summary, the diatreme model deviates from the a priori information in some significant aspects but it is an acceptable possibility, especially in view of its lack in separating the sediments from the diatreme. The fit of the gravity data is a little less convincing, as evident in the amplitudes of many of the residuals. This situation is quantitatively demonstrated by the following numbers:

- wRMSE of the observations: 3.81,
- wRMSE of the variables: 1.25
- F-statistics: 14.98

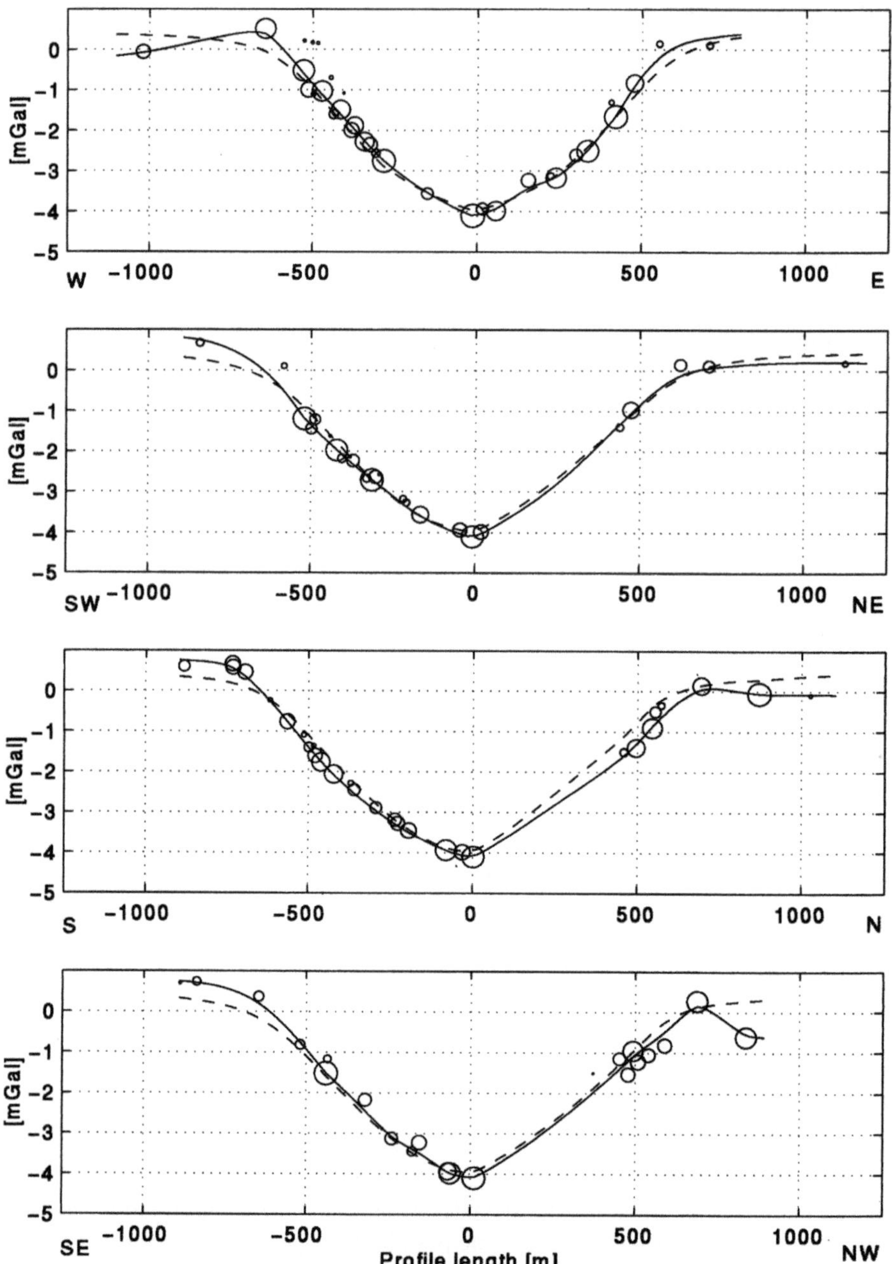

Fig. 7.4.8 Four gravity profiles across the Meerfeld Maar, W-E, SW-NE, S-N, SE-NW, comparing the observed *BA* (*solid line*) and modelled gravity effects (*dashed line*). The observations are shown by circles and their size symbolizes the station distance from the profile (largest to smallest corresponds to 0–150 m, respectively). The positive anomalies near inflow of the stream Meerbach and its exit are probably explained by the deep cuts into the Devonian basement rocks, since neither the topography of the DTM of the state geodetic survey nor the assumed Bouguer density will be perfectly correct

The uncertainty of the observations (including the expected modelling errors) was exceeded by nearly a factor of 4 on average, and the uncertainty of the variables (including the a priori information on the slope of the diatreme) by $1\frac{1}{4}$. Probably the standard deviation of the observations was chosen too small and could be adjusted. Optimizing the Tikhonov regularization parameters suggests an increase of the present standard deviations by a factor of 3–4, but then the diatreme structure would hardly be resolved by the gravity data. If in an improved model the gravity trend and the occurrence of the tuffs outside the crater are taken into account, the diatreme would again be resolvable. For the purpose of this example, this simple solution, although not fitting the gravity data extremely well, was accepted, because it exhibits a better resolution of the model parameters.

Summarizing: the statement that some assumptions about the model and about the statistical distribution of the data or a priori information are not true, is wrong with a probability of $< 5\%$, because the quantile $F_{121,\infty;95\%} = 1.22$ is much smaller than the F-statistic (see Sect. 7.3.2.2.7).

7.4.1.5.4 Stability of the Iteration

In Fig. 7.4.9 the history of the variables and some evaluation parameters during the 40 iterations is plotted. The change of values of the variables **r1–r20** and **x, y, d, c** is shown normalized with the final a posteriori standard deviation. The maximum change thus means that a variable in one step has increased or decreased by (more than) one a posteriori standard deviation (upper or lower bound, respectively); **r2–r4** have initially grown, **r6–r20** have shrunk, as evident in the figure.

The line marked by **Chi** represents the change of the shifting variate (Sect. 7.3.2.2.3), i.e. the squared length change of the residual vector of the observations and variables jointly, normalized with the a priori standard deviations of the observations or of the a priori covariance matrix of the variables. If **Chi** remains within its bounds, the squared length change of the residual vector is smaller than about one a priori standard deviation squared. Given the 122 observations and 24 variables, of which 23 have an a priori value (as have the observed values), there are 121 degrees of freedom. Hence the bounds indicate a change of the F-statistics of $\pm 1/121 \approx \pm 0.008$.

The line **Xi** represents the length change of the vector of the variables jointly, normalized with σ_{norm}, i.e. the normalizing standard deviation (see Table 7.4.2). **Xo** shows the same vector, normalized with the estimated a posteriori standard deviations after each iteration. The line **lam** shows, in a logarithmic scale, the damping factor λ used, where the bounds represent $10^{\pm 5}$. Due to the non-linear behaviour of the uppermost disk, the damping was limited to a minimum 0.1, such that the variables were set to preserve their values of the previous iteration within a maximum of 10 times their normalizing standard deviations. When **Chi** grew larger, the damping was also increased.

The shifting variate changes after about 20 iterations at each step less than a standard deviation, but occasionally deteriorates, in which case the damping must be

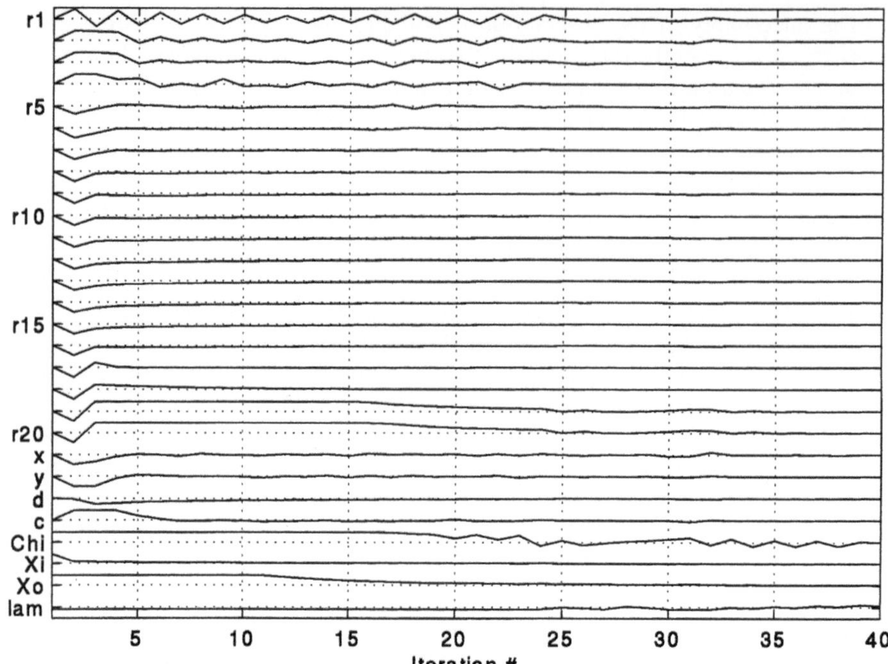

Fig. 7.4.9 Variation of the variables and iteration parameters during the iteration history, shown in order to demonstrate the stability or instability of the iteration (see text)

increased. The variables themselves, however, change after iteration 25 much less than one a posteriori standard deviation, although the damping allows for at least that freedom. The instability of the shifting variate during the last 10 iterations thus indicates a relatively high sensitivity of the data values to minimal changes of the variables. The cause is the mentioned steep gradient generated by the upper disks. The radii of the upper disks seem therefore most difficult to determine without the increased damping from iteration 25 onward. But since the changes always alternate sign the total error will even after many iterations be smaller than one a posteriori standard deviation.

The centre coordinates **x, y** approach their final values already after 5 iterations. The longest lasting influence of the instability originates from the radii **r19** and **r20** of the two lowermost disks which affect also the neighbouring disks and the density contrast. In the first iteration they jumped into the negative region and needed many iterations to find the non-linear equilibrium between the pressure of the penalty function toward positive values and the common pressure of the data and the rest of the a priori information toward negative values. The changes are small relative to the a priori standard deviations, but the penalty function is included in the calculation of the a posteriori standard deviations, such that their values are quite small and the changes relative to them are large, as long as the solution lies within, or very close to, the forbidden region.

On the whole, the solution seems to reliably represent a local minimum. This is a necessary basis for quantitatively evaluating the criteria which were discussed in Sect. 7.3.2.2.

7.4.1.5.5 A posteriori Standard Deviations of the Radii

Figure 7.4.10 presents the disk thicknesses (middle) and the standard deviations of the radii (left) and radius differences (right), plotted versus depth in logarithmic scale. The standard deviations of the individual radii significantly decrease (a posteriori) when the gravity data have also to be explained. The a priori decrease of the standard deviations with depth mainly results from the slope angle condition for the pipe walls. The inversion reduces the radius standard deviations by 80% to about 100 m depth; below 200 m depth, the reduction is nearly constant at 50%; the lowermost disks are practically fixed by the inequality condition. It is difficult to estimate how much an asymmetric probability distribution (which cannot be realised in the solution method) would affect the uncertainly of solely permitted radius values. On the whole, the gain of information from the gravity data seems to be considerable for all individual radii.

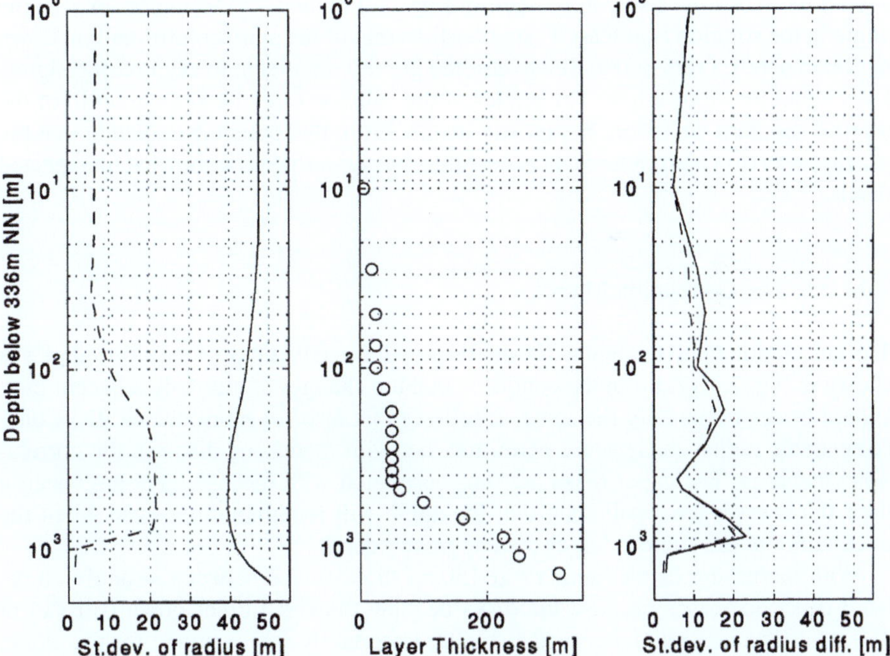

Fig. 7.4.10 A priori (*solid line*) and a posteriori (*dashed line*) values of three quantities versus (logarithmic) depth characterizing the diatreme model; *left*: standard deviation of the disk radii; *right*: standard deviation of the radius differences; *middle*: disk thicknesses which affect the standard deviations

The right-hand picture shows the standard deviations of the radius differences from disk to disk, which are closely connected with the slope of the wall. They were calculated by continuation of the variances and covariances by the neighbouring radii. For this quantity the information gain is much smaller than for the individual radii. The gravity data do not seem to affect neighbouring disk radii differently, except between 20 and 90 m and between 130 and 270 m, and the reduction of the uncertainty is maximally 30%.

The relative approach of both curves (left-hand picture) between 100 and 400 m depth is caused partly by the thickness increase of the disks, which is not proportional to their gravity effects. The Backus-Gilbert regularization method (Sect. 7.3.1.2.2) has, among others, the aim to avoid such effects, by composing the variables such that they have as uniform variance reduction as possible (i.e. such a resolution). This approach was not implemented in INVERT, but an iterative optimization would be possible on the basis of the information presented.

The radii were introduced to define a simple connection between the a priori information on the slope angle and the gravity effect. Therefore also the uncertainty of the slope angle should be considered. This is done by calculating the connection between the disk thicknesses with the radii and the standard deviations of their differences. Figure 7.4.11 shows the results of this calculation for the initial (left) and final (right) models. The centre line shows for the result that the slope angle should be $< 30°$, as also can be seen in Figs. 7.4.6b and 7. The uncertainty of the angle in the solution is at least $5°$ smaller than that of the a priori information, down to 200 m depth. From 300 m downward the gravity data seem to say nothing significant about the slope angle. The angles of the solution often lie at the margin of the a priori standard deviation, but mostly inside. From this one may conclude that the gravity inversion has rendered an acceptable, but nevertheless somewhat unexpected result.

7.4.1.5.6 The Correlation Matrix

The correlation matrix calculated from the condition equations of the slope angles is shown in Fig. 7.4.12. From the condition that the changes of mutually adjacent radii should be approximately the same, it follows that also for more distant disks only comparable radius changes are permitted, but with increasing distance the correlation decreases. The upper disks are thus correlated with more neighbours because their thicknesses are small such that the angle will remain similar even when the uncertainty of the angle is larger than at greater depth.

With increasing depth (until **r9** at 150 m) the distance between strongly correlated disks decreases because the disks become thicker; further down (till **r14** at 450 m) it increases again because a smaller uncertainty was assumed for the angle. The lowermost disks are hardly correlated, because their uncertainty is increased and their thickness is also larger.

The a posteriori correlation matrix (Fig. 7.4.13) shows a much reduced correlation for all disks above about **r11** (at 300 m depth) relative to the a priori correlation.

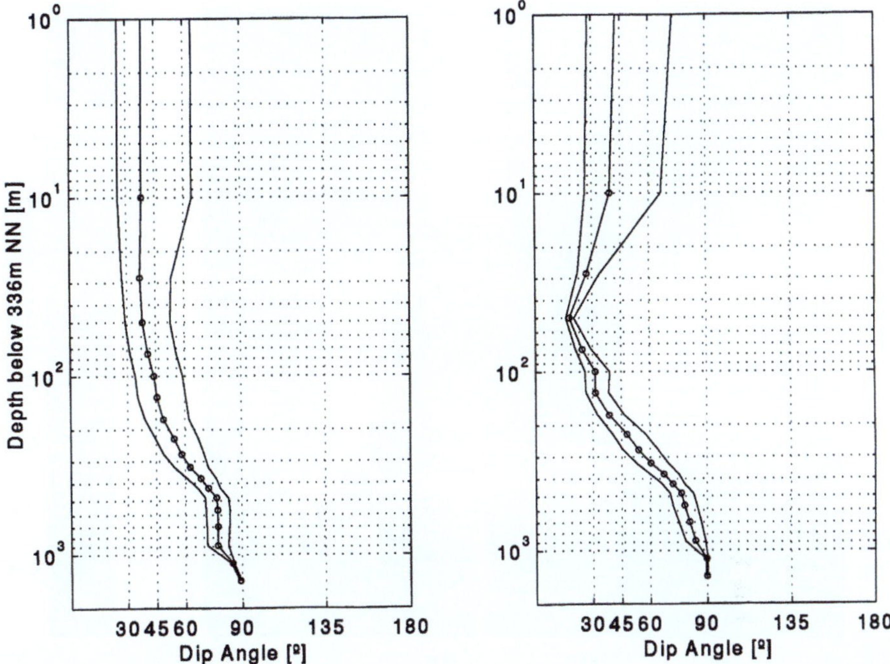

Fig. 7.4.11 Wall slope angles and their error bounds versus (logarithmic) depth, *left*: a priori, *right* a posteriori

The gravity data seem to tell much more than the a priori information does. However, they do not say the same as the a priori information, as follows from the reduced slope angles (see Fig. 7.4.11). Radius **r4** (25–50 m) and **r10–r17** (250–800 m) are negatively correlated with the first three disk radii: the effect of the upper more extended bodies can be compensated by the lower more compact bodies. A downward continuation of the gravity field (Sect. 2.10.5.3, Eq. (7.2.26)) would project a wide component only to greater depth. The disks **r11–r16** (300–600 m) are mutually correlated, but still as much as by the a priori information, which again supports the poor resolution by the gravity data in this depth range. The lowermost disks are effectively uncorrelated, because they are constrained only by the inequality condition.

The density contrast is negatively correlated with the upper disk radii and positively with the lower ones. The means that if the density contrast is less negative than found by the present solution, then the radii of the upper disks would have to decrease and those of the lower disks should increase. Since the density contrast of the solution is remarkably negative (-620 instead of a priori $-500\,\mathrm{kg/m}^3$), it might be a bit more realistic that the density contrast may be somewhat weaker (perhaps less influenced by the neglected lake sediments) and the diatreme walls would then also be less curved, which, too, would correspond better to the a priori information.

The centre coordinates are not appreciably correlated with any other variables.

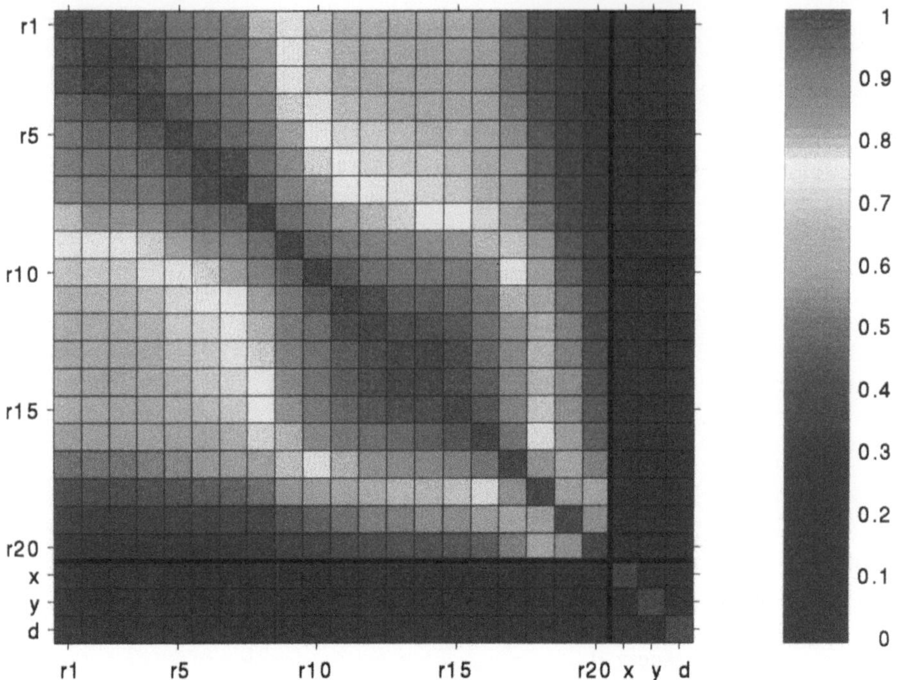

Fig. 7.4.12 A priori correlation matrix of the variables with the angle condition (note the grey scale!)

The additive constant is positively correlated with nearly all radii and with the density contrast, as would be expected. But only to a minor degree, therefore no serious possibility of a mutual influence exists, and the values of the more interesting variables are not rendered doubtful by this.

7.4.1.5.7 Eigen-Vectors of the Correlation Matrix

In Fig. 7.4.14 the eigen-vectors (centre) and singular values (square roots of eigen-values) (left) of the a posteriori covariance matrix are shown together with the resolution and residuals (also left) and the linearity factors (right). The singular values are expressed relative to the normalizing standard deviation of Table 7.4.2 (see (Eq. 7.3.8) and Sect. 7.3.2.2.10). The lines show the relative contribution of the individual variables to the eigen-vector, normalized with the largest amplitude. For example, the density contrast **d** in the first (uppermost) eigenvector is plotted at the upper bound, with also the **r10–r15**; it means that they contribute the same to the eigen-vector (in relation to their respective normalizing standard deviations: 200 m for **r10–r14**: 189 m for **r15** and: 50 kg/m^3 for **d**). Eigen-vectors represent only directions, hence the vertical scales of the individual lines are arbitrary and vary from

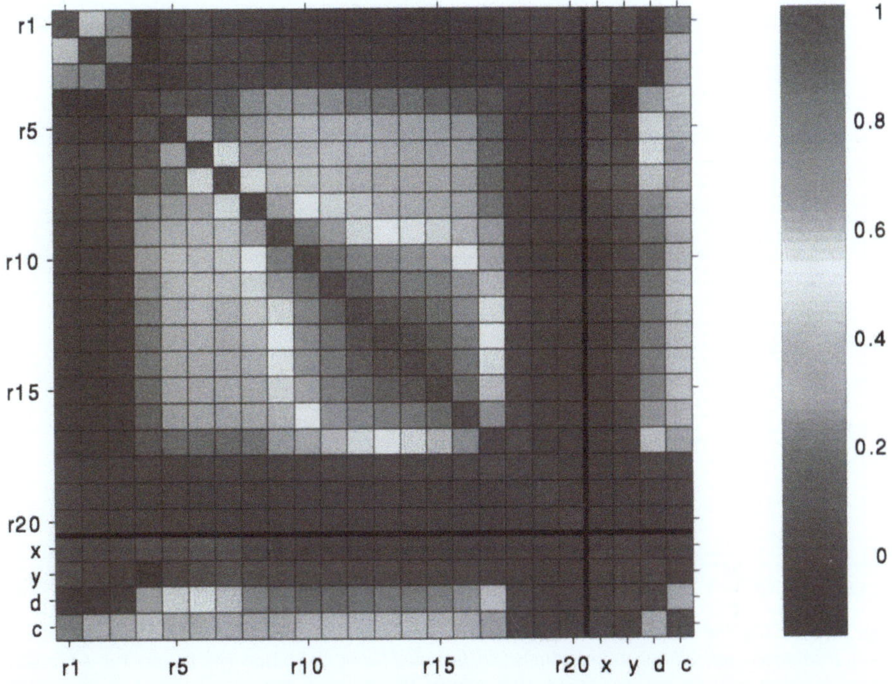

Fig. 7.4.13 A posteriori correlation matrix of the variables (note the grey scale!)

line to line. This is graphically necessary to make the largest (most important) components obvious and to help recognizing the components of eigen-vectors.

The singular values (circles in left part of Fig. 7.4.14) show how large the a posteriori standard deviation of the respective eigen-vector is relative to its normalizing standard deviation. The normalizing standard deviation is, for all variables, equal to the a priori standard deviation, except for the additive constant c ($\sigma_{\mathrm{norm}} = 1000$, $\sigma_{\mathrm{prior}} = \infty$). For the radii, it is the standard deviation which had been taken for calculating the covariance matrix of the radii from the conditions concerning the slope angle. The singular values represent the extreme values of reduction of the uncertainty relative to these standard deviations.

The maximum reduction occurred for the slack variable c (the last eigen-vector); $s_{\mathrm{norm}} = 1000$ has effectively no significance. The two eigen-vectors shown above c represent the lowermost disks that were "frozen" with an uncertainty reduced to about 1/100 of originally 100 m. The other eigen-vectors show a reduction between 1/100 and 1/10 of the uncertainty, with the exception of the first eigen-vector which has almost not improved. This is the eigen-vector corresponding to the correlation between the density contrast and the curvature of the diatreme wall, mentioned in the previous section. Simultaneously decreasing the density contrast and the curvature is hardly constrained, neither by the gravity data nor by the a priori information on the slope angle. The curvature has the same standard deviation where these data

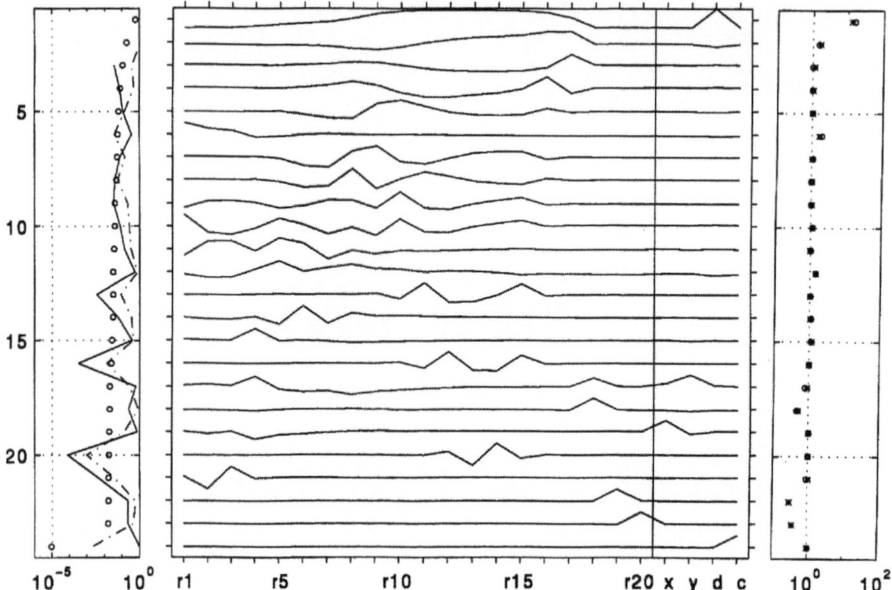

Fig. 7.4.14 Singular values of the a posteriori covariance matrix (*middle*); *left*: eigen-values (*circles*), resolution (*solid line*), and residuals (*dash-dotted line*); *right*: linearity factors for +1*eigen-vector (*circles*) and −1*eigen-vector (*asterisks*); for the units see text

are not taken into account. As it turned out later, many more eigen-vectors are only poorly resolved by the gravity data, though better by the conditions related to the slope angle. The first eigen-vector is, however, not determined well by these conditions either, because it represents the longest wavelength in the change of the radius differences. Two adjacent radii nearly perfectly obey the angle condition, but across many disks the condition has little influence. Generally it is noted that the long-wavelength components are less well determined than the short wavelengths, especially at greater depths.

The strong uncertainty of the first two eigen-vectors is somewhat relaxed by the linearity factors (right-hand side of figure). They show how much the square root of the shifting variate grows, when the solution changes by one a posteriori standard deviation (equal to the singular value) in direction of the corresponding (normalized) eigen-vector. For linear basic equations this length change should be 1 (see Sect. 7.3.2.3.3). For the first eigen-vector this value is about 16, and to a first-order approximation it can be assumed that the a posteriori uncertainty is $< 1/16$, instead of 1, because a change by one a priori standard deviation of this linear combination of variables apparently would cause a change of 16 (not 1) times the a priori standard deviation of the linear combination of the observations. The non-linearity is, plausibly, linked with the functional connection between density contrast and disk radius. If both change as defined in the eigen-vector, the gravity effect does not change by adding the individual components, as is the case with the radii mutually, but by multiplying the density contrast by the radius. The non-linear effect of the upper disks is recognizable in eigen-vector #6, but it is not very strongly visible. The penalty function even weakens the gradient of the minimizing function, such that the change of the solution by one a posteriori standard deviation worsens the fit to the data even less, than is the case in the linearized equation.

The line at the left edge of Fig. 7.4.14 connects the points that represent the resolution of the eigen-vectors (Eq. 7.3.12), in this case referring also to the a priori information on the slope angle. The eigen-vectors, though having small eigen-values, i.e. being well determined, are just badly resolved. So, they are well determined only by the a priori information on the angle. Here it is evident that the short-wavelength variations of the radius differences are poorly resolved by the gravity data. The upper two eigen-vectors have resolution values smaller than 10^{-6} and were not plotted.

The residuals of the eigen-vectors are represented by the dash-dotted line at the left edge of Fig. 7.4.14, again relative to the normalizing standard deviations (see Sect. 7.3.2.2.11). Only the residuals of the first two eigen-vectors are essentially 1, which means that the a priori information does not have to be impaired significantly by fitting the data. Most new aspects of the solution are obviously represented by the long-wavelength aspects. Generally the resolution and the residuals correlate, though the well resolved linear combinations of the variables do not fully agree with the a priori information, for good resolution does not correspond to negligible residuals (with the exception of the additive constant), but non-resolved combinations of the variables are indeed kept at the initial values, because small resolution is coupled with small residuals (which is not always the case in non-linear problems).

7.4.1.5.8 Eigen-Vectors of the Resolution Matrix

The resolution of the diatreme density distribution by the gravity data is specified yet more accurately by the singular values and eigen-vectors of the resolution matrix (Fig. 7.4.15) than by the resolution of the eigen-vectors of the a posteriori covariance matrix. The upper singular values (1–10) and eigen-vectors of the resolution matrix demonstrate the poor resolution of the short-wavelength radius changes; the larger the wavelength, the better is it resolved. As from eigen-vector 15 or at the latest 18 (depending on the demands on the gravity data by the modeller), the resolution is effectively equal to 1, i.e. optimal. Such a high resolution is shown also by the additive constant c and the centre coordinates x and y. Eigen-vector 21 describes a general increase of the radii together with a decrease of (already negative) density contrast value. Since each of these contributions changes gravity in the same direction, the linear combination is well resolved. Eigen-vector 20 is more or less the opposite of the first eigen-vector in Fig. 7.4.14. They are, in any case, orthogonal relative to each other, because the components of the radii have the opposite sign. Eigen-vectors 19, 18 and 14 again indicate that the radii of the upper 11 disks (to 300 m depth) can be resolved by the gravity data in a broader scale and at greater depths. In the case of very short wavelengths the a priori information determines the shape of the diatreme.

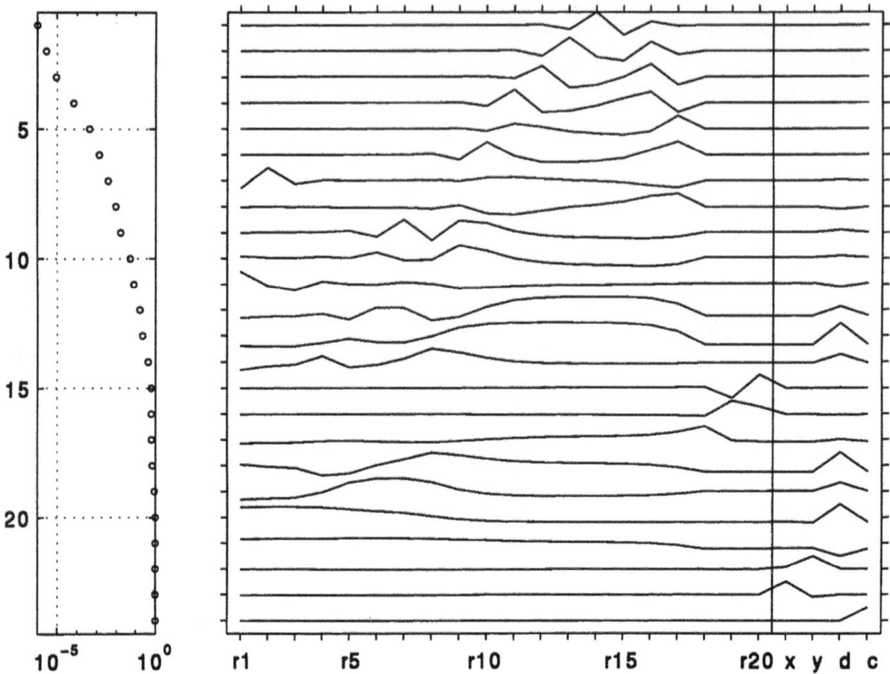

Fig. 7.4.15 Singular values (*left*) and eigen-vectors (*right*) of the resolution matrix

7.4.1.5.9 The Resolution Matrix

Figure 7.4.16 shows the resolution matrix which presents information different from the previous figures. While large and small values of elements on the main diagonal do mean well and poorly resolved variables, respectively, the emphasis of this picture is the influence which the a priori information on a variable has on the other variables (see Sect. 7.3.2.2.8). The influence is, again, expressed relative to the normalizing standard deviation. The clearest example is the row for the density contrast **d**. When radius **r10** increases by one normalizing standard deviation (200 m), because the gravity data demand it but the a priori standard deviation opposes it, the density contrast **d**, instead, becomes smaller (more negative) by about 0.8 times its normalizing standard deviation $(50 \, kg/m^3)$. In contrast, a prevented increase (becoming more positive) of the density contrast by $50 \, kg/m^3$ results in a less than 0.1×200 m radius shrinking of disk **r10**, as can be seen in column **d**. Below the diagonal, larger values appear than above it, implying that a priori information limiting the radii of the upper disks will influence the radii of the lower disks, while such an influence of the lower disks on the upper disks is very weak. Disks **r10–r17** (250 to 800 m depth) appear to be affected only jointly by the gravity data, for individual disks are hardly affected by the data, as evident from the small values on the diagonal, while the neighbours react equally as seen in the uniform magnitude

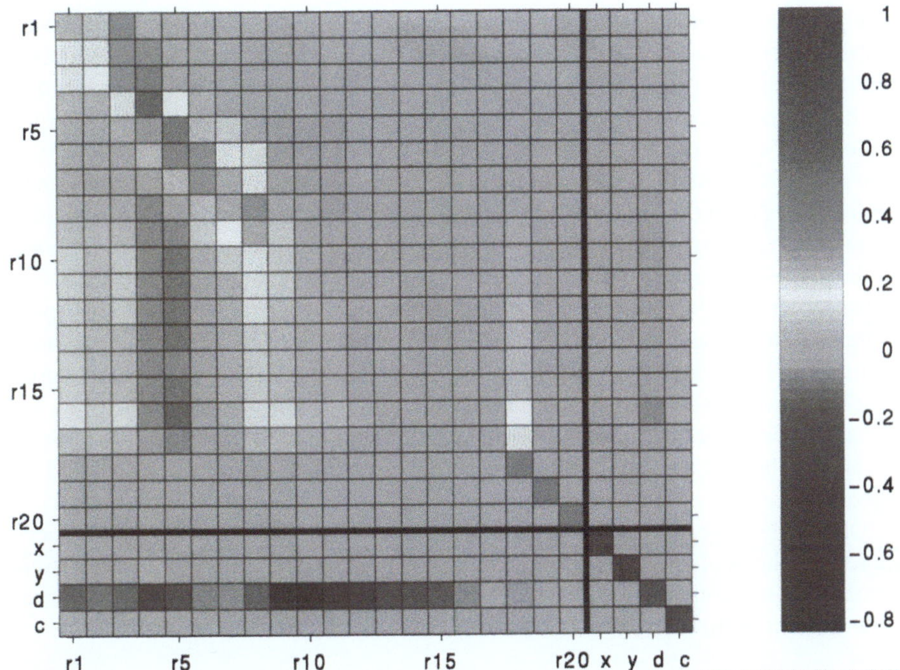

Fig. 7.4.16 The resolution matrix

of the values in this region. The three lowermost disks show null resolution since
they are almost completely determined by the penalty function (see Sect. 7.4.1.5.7),
restraining them from becoming negative.

7.4.1.6 Concluding Remarks on the Example Meerfeld Maar

The Meerfeld Maar is an example of a nearly singular gravity inversion, mainly because the size and the density contrast of the diatreme are highly correlated with
each other. Therefore much a priori information was collected which can help in
solving the problem. Since it is also a tutorial example, some of the available a
priori information was consciously dropped which hampers the geological interpretation. This actually demonstrates how important it is to include as much a priori
information into the modelling as possible.

The resulting model has a few unexpected aspects, especially the low slope angle
of the upper part of the diatreme and its reduced depth extent. These aspects appear
to only weakly contradict the known facts. The eigen-values and eigen-vectors of the
resolution matrix demonstrate that one needs to put much effort into the definition
of the a priori information concerning the small- to medium-scale depth range at a
depth of 300–600 m, because in this region the gravity data provide little information

and determine at most just a mean radius. Evaluation of the a posteriori covariance matrix clearly demonstrates that the joint optimization of the gravity data and the a priori information improves knowledge. The a priori information is concentrated on the deep small-scale part, while the gravity data emphasize the shallower or larger-scale part. Further modelling should concentrate more carefully on a priori information and gravity data that tell something about the upper 200 or 300 m depth range. This follows from the largest eigen-vector which shows that the density distribution in this depth range can be fully compensated by a modified density contrast and radii in the diatreme range of 300–800 m depth. An incorporation of the lake sediments between 40 and 190 m, as determined from geo-electric data by G. Büchel (pers. comm., 1996), would strongly affect the results concerning the slope and curvature of the diatreme walls and, hence, the depth extent and density contrast at depth. Since the respective information seems to be missing in the present gravity data and in the a priori information used, new measurements should concentrate on these aspects, and the first step suggested is areal gravimetric surveying at and beyond the crater rim.

Obviously, the case of the Meerfeld Maar emphasizes the methodological aspects more than the geological questions. Since, however, geological or related questions are the driving motivation of gravity inversion, the discussion also showed what can be done next in this direction.

7.4.2 SE Iceland Shelf: Edge Effect

7.4.2.1 Principal Ideas

The second case history differs from the above in important aspects. It is the conspicuous dipolar gravity anomaly at the shelf edge which calls for an investigation (Sect. 5.7.5), and solid a priori information on the causative structures is essentially missing. The relationship with Iceland, one of the most prominent hotspots on Earth, and its structure and evolution are questions and only vague a priori information: is the gravity anomaly explained by the edge effect between two different types of oceanic versus Icelandic crust, or does it reveal a continental splinter under the shelf? Fedorova et al. (2005) speculated about such a splinter of the Greenland shelf split off and carried east when the spreading ridge jumped to the west from the Norwegian basin, some 26 Ma ago, to initiate the Kolbeinsey ridge north of Iceland. The spreading history at the latitudes of Iceland was complex and it is possible that a Greenland splinter was carried to the east, to form a more or less continuous band from the SE Iceland shelf at about 63 °N to Jan Mayen at 71 °N. The Jan Mayen Ridge has been investigated by refraction seismics and inferred to have a continental basement (Kodaira et al., 1998). The a priori information has the nature of a general concept which must be considered as a whole with large uncertainty of the individual parameters describing the geometry and densities. The hypothesis of the structure being a continuation from, or a least related to, the Jan Mayen Ridge can be taken as a priori information to be tested by inversion – similar to the Meerfeld Maar where the South African kimberlite pipes played such a role in the inversion (Sect. 7.4.1).

The situation calls for an exploratory strategy of parametrization, especially the two-sigma rule (Sect. 7.2.2.3.2). The individual parameter errors may be chosen too wide and then narrowed, and condition or correlation matrices may be assumed a priori. This section will follow such a strategy: testing the shelf edge effect, a continental splinter and the conditions imposed by the corresponding structures.

The SE Iceland shelf was introduced in Sect. 5.6.8 emphasizing the conspicu-ous dipolar gravity anomaly accompanying the edge and steep slope of the Iceland plateau down to the ocean basin. Figure 5.7.5 presents an averaged NNW-SSE pro-file across the 50 km wide shelf with $x = 0$ at 64 °N intersecting the coast at about 63.7 °N, 17.5 °W (inset for location) and shows the topography and bathymetry, the free air anomaly (*FA*), the Bouguer anomaly (*BA*) and the residual *BA* (*rBA*) ob-tained by subtracting an estimated Moho effect from the *BA*. The data errors have been estimated from the averaging two neighbouring profiles, each based on spa-tially distributed gravity and topography data in compartments of a 20 km strip and they range from 1 to 6 mGal and average about 2–3 mGal. Parametrization errors of the same order were added to take the short-wavelength data scatter into account.

All three different kinds of gravity profiles display the dipolar anomaly near the shelf edge. It is most pronounced in the *FA* with a 60 mGal asymmetric peak above the shelf break, a steep drop to the bottom of the plateau slope and a more gentle -10 mGal asymmetric gravity low at the foot of the slope. The dipolar shape is not completely removed by the Bouguer reduction, but it appears more rounded and shifted landward (about 10 km for the peak and 20 km for the low), superimposed on the general oceanward rise of the *BA* to 160 mGal, reflecting the crustal thinning. With its effect estimated and removed, the *rBA* isolates a dipolar feature with a 30 mGal peak, shifted 20 km farther landward from the *FA* peak, and the 20 mGal low shifted 10–20 km landward from the *FA* low. The $\sim$ 20 km distance between the *rBA* and the *FA* dipolar features seems significant.

As the Moho rise toward the ocean basin is not constrained by detailed seis-mic a priori information, the *rBA* is only preliminary. In Sect. 5.7.5 a qualitative interpretation was offered for it as an edge effect, i.e. a vertical edge of two lay-ers on top of each other, each 15 ± 5 km thick with lateral density contrasts of $+144 \, \text{kg/m}^3$ overlying $-144 \, \text{kg/m}^3$ density contrast at depth. Furthermore, the *rBA*, in Sect. 5.7.5, indicates that the double layer is probably laterally limited landward. In Sect. 6.5.5 three more models of different geological nature were investigated (Fig. 6.5.5), but only the edge or dipole model is followed up here by inversion (with the program INVERT) because it is attractive as an idea and would test the above hypothesis of a continental splinter. The other geological ideas, not individ-ually treated, are closely related to the edge effect and will be linked with it below. Details can be worked out as Task 7.5.

As a first experiment, the *rBA* is inverted in order to test this conceptual model of a laterally limited vertical mass dipole. A very simple model of two rectangular 2D blocks (coupled adjustable density contrasts $+/-100$ with each an uncertainty of $\pm 100\,\mathrm{kg/m^3}$) was assumed as a priori information with fixed depth extent (0–15 and 15–30 km); the horizontal limits of the upper and lower blocks were assumed at $x = 40$ and 80 km, on the basis of the *rBA*, and were left adjustable ($\pm 20\,\mathrm{km}$). The a priori standard errors of density contrast can be considered rough estimates of 2σ representing for each body a range from 0 to + or $-200\,\mathrm{kg/m^3}$ and the mutual difference of $400\,\mathrm{kg/m^3}$ as an extreme. Fixing the depths (a priori error ± 0) is justified only by the preliminary nature of the test. The result of this inversion is shown in Fig. 7.4.17. It confirms the viability of such a structure in a very crude form. The calculated densities are (in the case shown) $+83$ and $-83\,\mathrm{kg/m^3}$; interestingly, the gravity data shift the lower block some 20–30 km seaward below the shelf edge. The *rBA* is not considered further because of the problem of the poorly known Moho assumed.

The rapid transition from the Iceland shelf to the ocean basin itself generates an edge effect of a gravity dipole signal (Sect. 5.6.8) by the shallow "negative top" of the water mass versus upper crust and the "positive bottom" of the mantle versus lower Icelandic crust. It is attempted to separate the two "dipole" effects by inversion. Thus the next step is to include the Moho in the inversion of the *FA* and the

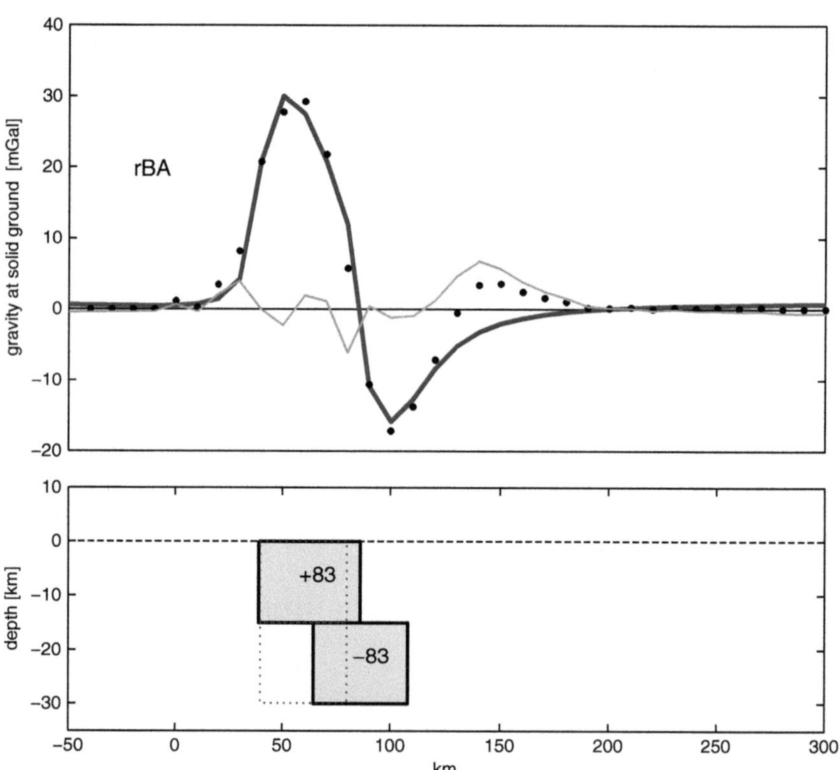

Fig. 7.4.17 The SE Iceland shelf: preliminary inversion of the *rBA* with an idealized crustal dipole model

BA. They contain all the effects of the internal crust and mantle structure with the transition from the Iceland shelf to the deep sea, while the effect of topography and bathymetry is treated differently in the two types of gravity anomaly. The *FA* contains the unreduced effects of topography and bathymetry which must be modelled, here in an idealized 2D manner, while the *BA* results from detailed Bouguer and terrain reductions. It is of interest to investigate the influence of this difference.

7.4.2.2 A Priori Information on the Crustal and Mantle Structure

The a priori information as discussed above, includes (1) a concept of the origin and nature of the shelf, including the seismically determined crustal structure of the Jan Mayen Ridge (whose relevance is, however, hypothetical and was therefore included only towards the end), (2) the morphology, (3) the background knowledge of oceanic and Iceland crust with data from further inland, and (4) strong localized magnetic anomalies accompanying the shelf edge. No detailed seismic data are available in the region proper, except further inland under the Vatnajökull ice cap, and the situation is similar in the ocean basin. Only general a priori knowledge of the ocean crust exists. Its average thickness may be thicker than normal under the anomalously shallow North Atlantic basins, affected by the Iceland plume. Crustal thickness of Iceland is debated and is today believed to reach a maximum of about 40 km under NW Vatnajökull (Darbyshire et al., 2000), where earlier "thin crust models" (15–20 km) were based on partly the same seismic data, but on a different interpretation. The lower crust in the "thick crust model" has a very small density contrast against the material below (about $100 \, kg/m^3$; Menke, 1999; Kaban et al., 2002; Fedorova et al., 2005) and may be considered a transient layer intermediate between mantle and crust. Under the circumstances, the gravity data may have a stronger weight than the a priori information in the inversion.

Under the circumstances, local isostasy of the shelf region may be taken as an initial a priori assumption to be tested, implying the Moho to be an expanded mirror image of topography/bathymetry (called "isostatic Moho" or Moho model (2)).

Another possibility is to combine the seismic data with the topography/ bathymetry and gravity data in the whole region by regression analysis, as presented by Fedorova et al. (2005) on the basis of the "thick crust model". From the regionalized experimental regression coefficients, Moho values were calculated on a 25 km grid of topography and gravity points, and a new Moho map was plotted on this basis. Lack of detailed knowledge of internal crustal structure and the statistical noise suggest that the individual point errors were about ±3 km. The map is only approximate. If taken as the best available a priori information on the Moho, a profile across the shelf can be constructed (called "a priori Moho" or Moho model (3)). The map of Fedorova et al. shows a band of thickened crust (to > 30 km) accompanying the SE shelf and continuing to the N along the western margin of the Norwegian basin to the Jan Mayen Ridge (as mentioned above) which has been suggested, on the basis of a seismic experiment (Kodaira et al., 1998) to contain continental rocks split off the Greenland margin when the Kolbeinsey ridge was initiated. In the seismic section the continental basement appears to be $\sim 15\,$km thick and $\sim 60\,$km wide and layered (pre-rift sediments, "continental upper crust" and "continental lower crust"), and overlain by a much thinner basalt cover. The eastern edge towards the Norwegian basin was not defined by the seismic experiment, and to the west the upper continental layers appear stretched, perhaps some $100\,$km. In contrast, a simplified uniform body is envisaged for the SE Iceland shelf with a similar shape and of similar dimensions, but a thin NW-ward continuation under Iceland is not considered, nor would it be revealed by gravity.

For the densities, values between 2600 and $2900\,\mathrm{kg/m^3}$ may be regarded appropriate a priori limits, implying $\sigma \approx 300/4$ or $75\,\mathrm{kg/m^3}$. The density contrast of sea water $\Delta\rho_\mathrm{w} = \rho_\mathrm{w} - \rho_\mathrm{uc}$ against uppermost crustal rocks is about $1600 \pm 100\,\mathrm{kg/m^3}$. The depth-averaged Moho density contrast (15–40 km), $\Delta\rho_\mathrm{m} = \rho_\mathrm{m} - \rho_\mathrm{lc}$, is assumed $160 \pm 100\,\mathrm{kg/m^3}$, which is low, but supported by regional isostasy. In Sect. 5.7.5 values of 300 and $100\,\mathrm{kg/m^3}$ were considered with the lower value more appropriate for the depths from 20 to $40\,$km. On the very large scale of the North Atlantic, isostasy is probably also disturbed by the Iceland mantle plume, but the related deeper density distribution has only gently varying gravity effects along the (radial) profile and may be neglected here.

The marine sediments at and below the Iceland plateau slope are approximately known (see Fedorova et al., 2005). To avoid associated effects in the inversions, the gravity anomalies were "sediment-corrected", i.e., the compaction of the sediment layer to basement density by about 10% was estimated and added to the water depth, and also taken into account in the locally isostatic Moho depth.

7.4.2.3 Parametrization Strategy

Two-dimensional parametrization is sufficient, since the radius of curvature of the shelf edge between $14°$ and $21°\,$W is $\sim 300\,$km, compared to model depths to $<40\,$km (Sect. 5.3.1, Fig. 5.3.1) and the errors relative to 3D modelling are small.

The uncertain situation of the a priori information motivates experimentation with several parameterizations of different geological nature (Fig. 6.5.5) with the common feature of a shallow landward density excess and a corresponding deep crustal density deficit (Sect. 6.5.5). Before the crust-internal "mass dipole model" is treated, a homogeneous crust is investigated to clarify the exclusive shelf edge effect with three different parameterizations of the Moho (Fig. 7.4.18): (1) an idealized ramp-like isostatic transition from the shelf to the ocean basin (drop: $2\,$km, width: $20\,$km, oceanward density contrast $\Delta\rho_\mathrm{w}$ a priori assumed $-1600 \pm 100\,\mathrm{kg/m^3}$, a posteriori $-1586\,\mathrm{kg/m^3}$) coupled with an isostatically balanced $20\,$km thick mantle layer between 10 and $30\,$km depth and a density contrast $\Delta\rho_\mathrm{m} = 160 \to 159\,\mathrm{kg/m^3}$, (2) the locally isostatic model of observed topography-bathymetry and a Moho with $-\Delta\rho_\mathrm{w}/\Delta\rho_\mathrm{m} \approx 10$ and its top at about $10\,$km, and (3) the a priori model or section through the crust map by Fedorova et al. (2005), which shows, with a $\pm3\,$km uncertainty, thickened crust below the shelf edge (maximum $30\,$km), from where the Moho rises to both sides, into the ocean basin gradually to $13\,$km and toward NNW

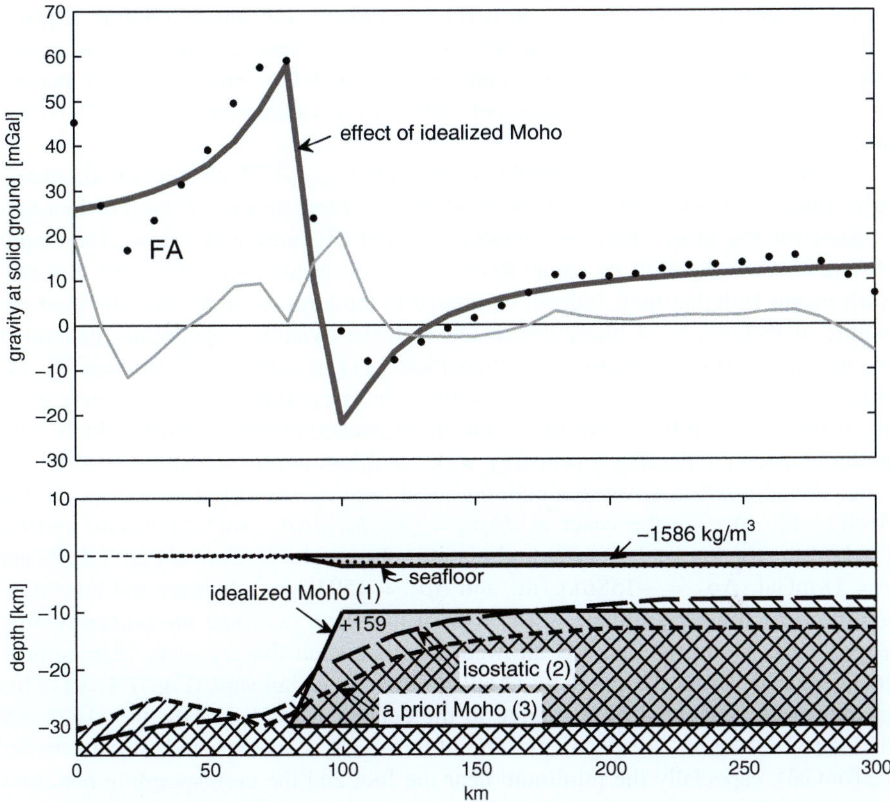

Fig. 7.4.18 Experimental density models of the ocean basin and the Moho (with uniform crust) for the SE Iceland shelf and transition to the ocean basin: (1) local isostasy; steep idealized slope of 2 km across 20 km ($\Delta\rho_{\text{sea}} = -1600\,\text{kg/m}^3$) and Moho rising from 10 to 30 km ($\Delta\rho_{\text{Moho}} = 160\,\text{kg/m}^3$); an idealized ocean basin belongs to this (*solid line*); (2) local isostasy, Moho as mirror image of the observed topography/bathymetry, expansion factor $\Delta\rho_{\text{sea}}/\rho_{\text{Moho}}$; the observed bathymetry is shown by a *dotted line*; (3) Moho based on a priori information ($\Delta\rho_{\text{Moho}} = 160\,\text{kg/m}^3$), not strictly isostatic, but with the same observed bathymetry

to 25 km depth close to the Iceland coast from where it drops to 40 km depth below NW Vatnajökull (not strictly isostatic).

7.4.2.4 Inversion of the FA

7.4.2.4.1 The Iceland-Ocean Transition

The observed *FA* shows distinct dipole signal which, as argued above, may be a composite of two similar superimposed signals with a relatively small lateral offset. It is difficult to separate the sources and the inversion may become unstable, based on gravity alone, unless properly regularized, i.e. unless reliable a priori information prescribes such a dipole. In the present case the inversion is essentially a tool for testing hypotheses and their plausibility. Since the shelf margin is the principal geological feature and a strong FA accompanies it, the FA is first investigated.

Surprisingly, the simplest model (1), shown in Fig. 7.4.18, produces without any adjustment a gravity effect (not shown) which is more similar to the FA than calculated for the more "realistic" models (2) and (3) presented below. The major differences are the shift of some 20 mGal and the peaked extrema. The constant shift means little but may, indeed, be related to the large-scale positive FA over the whole of Iceland, most likely connected with the dynamic topography generated by the Iceland mantle plume. Inverting model (1) as a priori information by adjusting only the additive constant c and the density contrasts leads to a fairly good fit of the FA (± 7 mGal), except for the more peaked extrema, and residuals with patterns mainly reflecting a probably wider and less abrupt transition. If the densities are adjusted in a way maintaining local isostasy (by correlating $d\Delta\rho_m$ at the Moho with $d\Delta\rho_w$ of the water as $d\Delta\rho_m = (h_w/h_m)d\Delta\rho_w$, with h_w (2 km) and h_m (20 km) the thicknesses of water and mantle), the results (shown in Fig. 7.4.18) are $c = 18$ mGal, $\Delta\rho_w = -1586$ kg/m^3 and $\Delta\rho_m = 159$ kg/m^3. In a second inversion, additionally, the x coordinate of the foot of the slope (x_w) and the corresponding edge of the Moho rise (x_m) is allowed to adjust such that $x_m = x_w$. The adjustment, indeed, shifts the point by about 8 ± 0.4 km oceanward (Fig. 7.4.19). This change is accompanied by a change in the densities $\Delta\rho_w$ and $\Delta\rho_m$ to -1623 ± 9 and $+162 \pm 1$ kg/m^3, respectively. The fit to the FA is, as expected, slightly better (± 6 mGal), especially the minimum near the foot and the corresponding residuals are smaller. In both above inversions, the residuals landward from the slope are wavy and reminiscent of rBA, however of smaller amplitude (< 10 mGal).

The gravity effect of model (2) does not fit the FA (Fig. 7.4.20). Though of little meaning, the constant shift has the wrong sign (opposite to above) and stems from the greater thickness of the assumed mantle layer, defined by the Moho maximum depth of 40 km below central Iceland. It appears related to the assumed 160 kg/m^3 density contrast extending to 40 km depth. More critically, the dipolar double amplitude is only about one half of that observed (70 mGal). Doubling the amplitude could be achieved by scaling density with a factor of 2, but it would contradict the a priori information. An inversion of the FA (to ± 10 mGal, F-statistics: 108) with a priori Moho model (2) and only the constant c and the isostatic densities adjustable, renders the densities $\Delta\rho_w$ and $\Delta\rho_m - 1665$ and $+166$ kg/m^3, respectively. Beside the density scaling, the abruptness of the transition affects the amplitude. It would have to be less smooth than taken from the morphology, hence would also deviate from local isostasy. Part of the geological smoothing may be by sediments below the slope, say, between $x = 90$ and 100 km; a maximum sediment thickness of 1 km and a density contrast of, say, $\Delta\rho = -240$ kg/m^3 would give plausible -10 mGal. However, introduction of a corresponding sediment correction (i.e. at most 1 km of sediments compacted to upper crust density) has only an insignificant effect on the results (which is also true for the other models). Of course, beside the conflict with

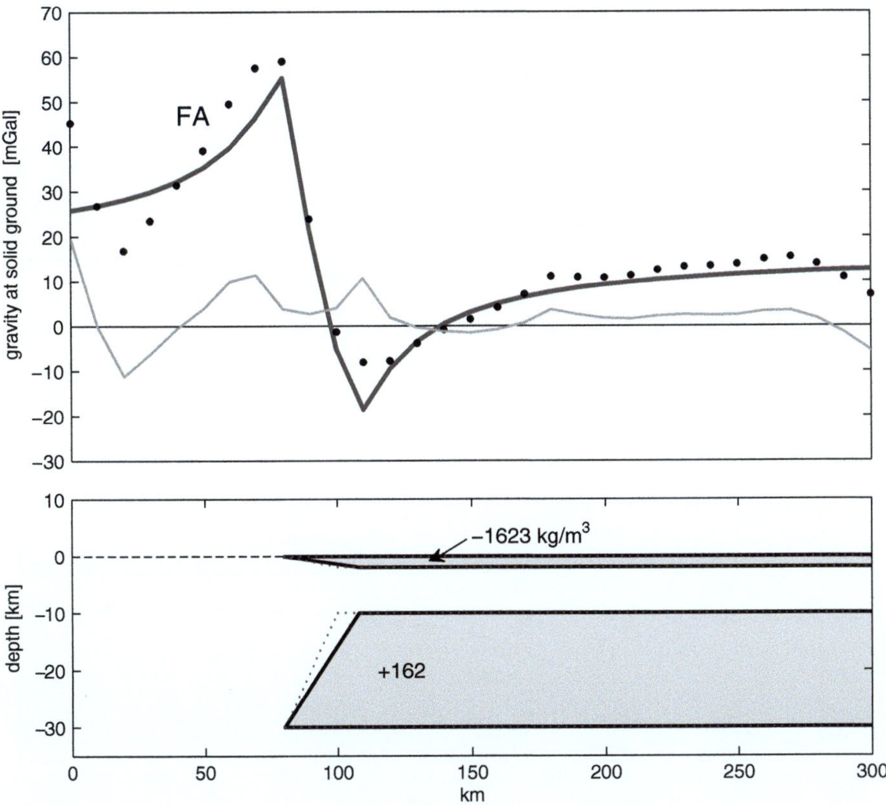

Fig. 7.4.19 Idealized model (1) as a priori information for the inversion of the *FA*, with adjustment of the densities $\Delta\rho_w$ and $\Delta\rho_m$, maintaining isostasy, and of the x coordinate of the foot of the slope and the corresponding upper Moho edge

gravity, local one-to-one isostasy at 10–20 km scale is unlikely. This can be studied by an inversion of the *FA* and permitting also the Moho morphology to adjust, starting with model (2). The results are not shown, but a strong Moho undulation (amplitude ±10 km; $\Delta\rho_w - 1604$, $\Delta\rho_m + 160\,\mathrm{kg/m^3}$) superimposed on the a priori Moho is calculated for which no independent information exists. A high is found below the shelf break between two lows, one near the coast line of Iceland, the other at the shelf foot. Nevertheless, the *FA* residuals are still large.

The fit of the effects of model (3) to the *FA*, with only the constant c adjustable (−10 mGal), is worse than in the above cases. This is not surprising and the large residuals (to 30 mGal) are negatively correlated with the a priori Moho, accordingly. When the density adjustment is included, the fit is still very poor (wRMSE: ±15 mGal, F-statistics: 262, Fig. 7.4.21), and the results are $\Delta\rho_w$, $\Delta\rho_m =$ $-1437, +144\,\mathrm{kg/m^3}$, respectively. These values are questionable, but strongly depend on the constraints imposed on them. The reason is the deep Moho under central

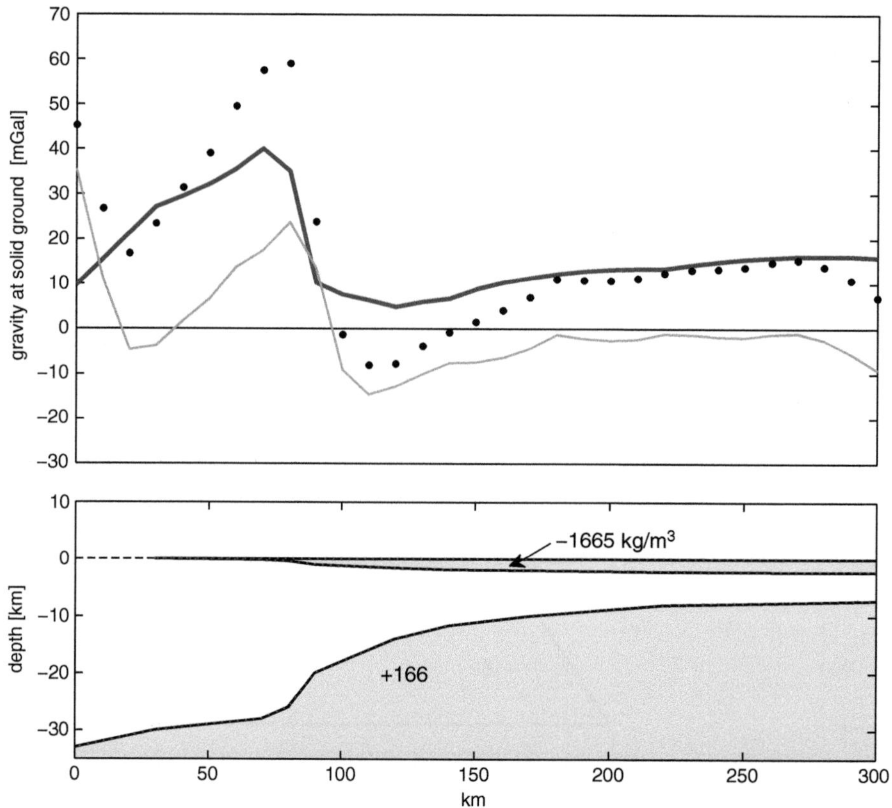

Fig. 7.4.20 Fitting the *FA* with the isostatic Moho model (2) defined as an expanded mirror image of the bathymetry; the inversion shown by adjusting $\Delta\rho_w$ and $\Delta\rho_m$ maintaining isostasy, results in $\Delta\rho_w = -1665$ and $\Delta\rho_m = +166\,\text{kg}/\text{m}^3$)

Iceland included also in model (3), so that the gravity data require a low Moho density contrast $\Delta\rho_m$ (to which $\Delta\rho_w$ is assumed correlated to maintain isostasy).

The above results are not supported by a priori information. Hence, with a homogeneous crust, no acceptable fit to the *FA* is obtained. If, for a test, the Moho is allowed to adjust within ± 5km, it is calculated very similar to that obtained for Moho model (2) as starting point: $\Delta\rho_w$, $\Delta\rho_m = -1607, +161\,\text{kg}/\text{m}^3$, respectively. Both Moho models differ only in the a priori assumptions which, in Bayesian inversion, are also fitted, but the data dominate the Moho depth variation calculated to explain the *FA* edge anomaly amplitude. As a strong Moho undulation is not likely, the arguments for a crustal inhomogeneity as a dipolar crustal structure (Sects. 5.7.5, 6.5.5; Fig. 6.5.5 (1)) are more convincing with a buried continental sliver in the lower shelf crust (Sect. 7.4.2.2).

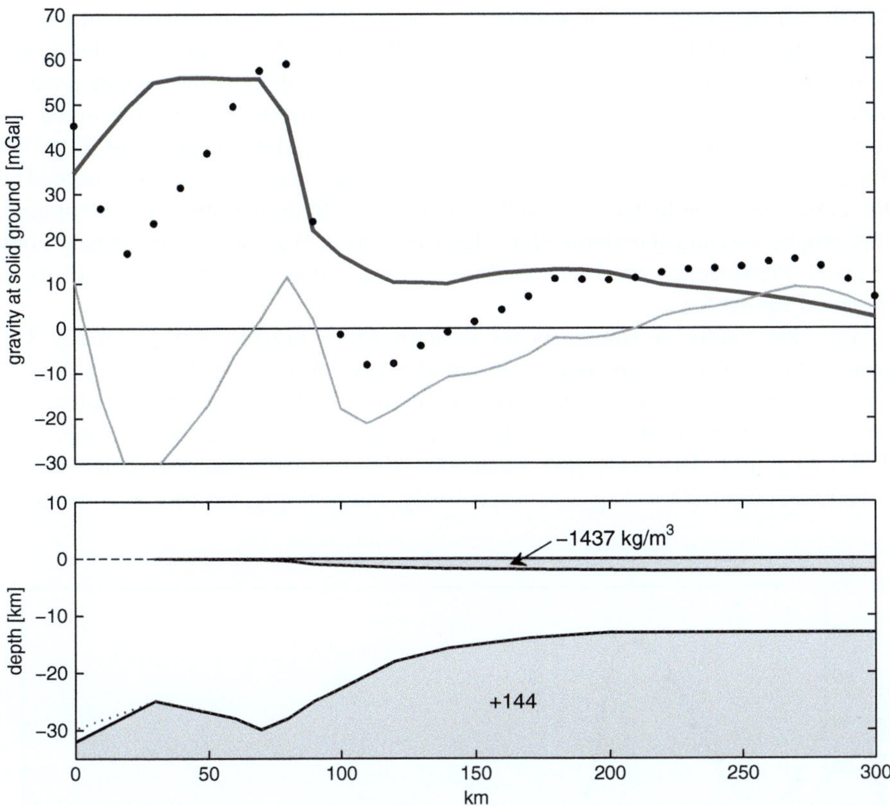

Fig. 7.4.21 Fitting the *FA* with the a priori Moho model (3) by adjusting $\Delta\rho_w$ and $\Delta\rho_m$ roughly maintaining isostasy. The resulting density contrasts (-1437 and $+144\,\mathrm{kg/m^3}$) are rather low

7.4.2.4.2 A Crustal Mass Dipole Added

The vertical mass dipole is added to Moho models (2) and (3). Since no concrete a priori information exists on its geometry, the parametrization begins with rectangular 2D block form, $15 \times 40\,\mathrm{km^2}$ in section, similar to Fig. 7.4.17. Adaptation of the bottom to each Moho model will follow. The densities and the lateral boundaries are assumed adjustable, followed by a more general geometrical adjustment. The a priori lateral density contrasts are assumed, as for the test with the *rBA*, to be $+/-100\,\mathrm{kg/m^3}$ with the same uncertainty. The fit (not shown) to the gravity data was immediately improved and the adjusted density contrasts of the initial model came out to $+/-124\,\mathrm{kg/m^3}$, implying absolute densities of, say, $2700{-}2800\,\mathrm{kg/m^3}$, embedded in 2600 and $2800{-}2900\,\mathrm{kg/m^3}$ material, respectively from top to bottom. A general vertical density increase with z was thus taken into account.

Refinement of the oversimplified initial model is a necessary, very important step. It includes adaptation of the bottom of the lower block to conform with the a priori

Moho and parametrizing the geometry to better fit the *FA*. This led to an experimental rhomboidal shape of the lower crustal body of $\sim 15\,$km vertical and $\sim 60\,$km horizontal dimensions, buried under a roughly 15 km thick denser upper crust of similar shape and 200 m water. Ideas on how a continental splinter is formed are too vague for a priori information. If this basalt and the dense upper crust of the SE shelf are equivalent, one might argue that the difference in thickness is related to the closeness to the Iceland plume. Furthermore, to fit the *FA* observations on land, topography was added underneath the land stations, and a shallow low-density body of loose Pleistocene gravel and sand was added optionally (turning out not to be essential for the fit).

First the results are shown for the isostatic Moho model (2) (Fig. 7.4.22). The lower crustal body comes out much smaller and more triangular with a density contrast $-85 \pm 10\,$kg/m^3. The upper crust body is inclined oceanward, but shifted from the a priori position by $< 5\,$km oceanward (density contrast 116 ± 7, core $174 \pm 8\,$kg/m^3). The correlated water and Moho density contrasts come out as -1600 ± 20 and $160 \pm 2\,$kg/m^3, respectively. Interestingly, the inversion modifies

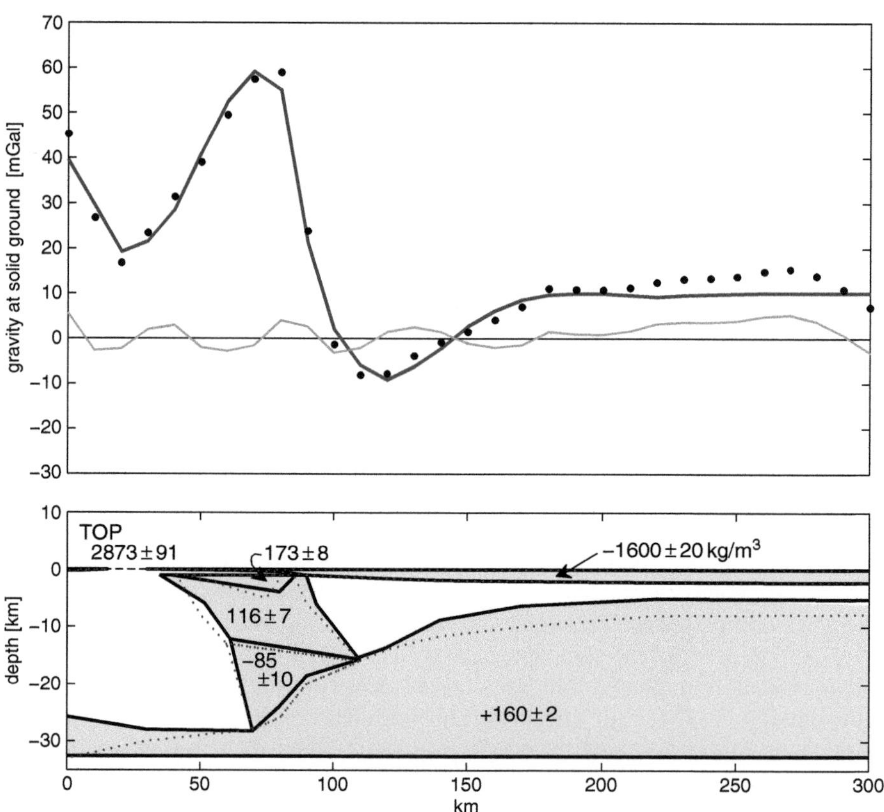

Fig. 7.4.22 Fitting the *FA* with the a priori Moho model (2) and the shelf crustal mass dipole added

the Moho model (2) in the direction of model (3) with a slightly thickened crust under the shelf. Topography under the land stations is very roughly approximated (with a density of $2873 \pm 91 \, kg/m^3$).

The inversion result for the somewhat more realistic a priori Moho (3), with the mass dipole of rhomboid bodies added, is presented in Fig. 7.4.23. The gravity fit is similar; the F-statistics is about 4, the wRMSE (see 7.3.2.2.4) is 1.3 mGal for the observations and 1.6 times the a priori standard deviations of the variables which, under the circumstances, is not very significant. The Moho density contrast is, again, calculated to be $160 \pm 1 \, kg/m^3$, assumed correlated to the water density contrast of $-1601 \pm 10 \, kg/m^3$. The denser upper body shown $(+128 \pm 7 \, kg/m^3)$ is inclined oceanward with a core of increased density (adding up to $+192 \pm 8 \, kg/m^3$). The rather big a posteriori low-density lower crust $(\Delta\rho : -120 \pm 8 \, kg/m^3)$ still resembles an inclined rhomboid similar to the continental crustal body suggested for the Jan Mayen Ridge by the seismic study of Kodaira et al. (1998). The land topography and the *sander* body are constrained by only very few stations and are calculated to have densities of 2180 ± 336 and $-184 \pm 42 \, kg/m^3$, respectively.

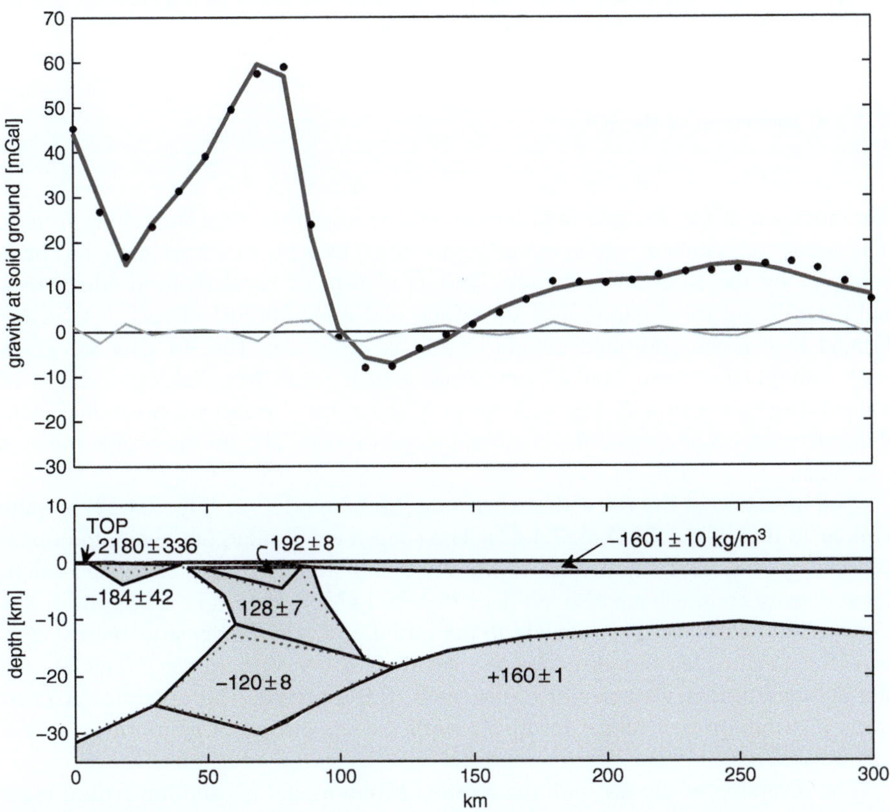

Fig. 7.4.23 Fitting the *FA* with the a priori Moho model (3) and the shelf crustal mass dipole added

Two aspects are emphasized. (1) One is the problem of the high (negative) correlation between the effects of the upper and lower crustal body, rendering the inversion unstable if not strongly regularized. Consequently small variations of the geometrical a priori parameters easily lead to large changes in the a posteriori density contrasts of the order of <100 to $>300\,\text{kg/m}^3$ (corresponding to about 2σ). The gravity data, by themselves, cannot resolve the details of the volume and shape of the bodies nor the densities individually, as volume and density are also highly correlated. (2) The principal result of the high/low density vertical dipole, however, is very robust. It does not depend on details of the a priori information. If the densities are loosely constrained, the amplitude of the variation from top to bottom tends to increase, but it leads to only an insignificantly better fit of the data. The constraints imposed on the model are justified only on the basis of the whole geological concept. Details of density and geometry are uncertain, but comparison of the two Moho versions suggests, that the a priori, not strictly isostatic version with a Moho depression under the shelf, more than the strictly isostatic Moho version, favours a low density, possibly continental root. However, importantly the lower body is not resolved by gravity alone. If, however, assumed a priori, it is supported by the inversion. This is not an absolute proof for its existence, but proof or verification cannot principally be expected from gravity in this case.

7.4.2.5 Inversion of the BA

Also the *BA* is inverted with the same a priori information in order to investigate the influence of the Bouguer and terrain reductions on the final modelling results. The ocean mass deficit, which is well constrained by a priori information, has been removed by the Bouguer reduction. This is a form of regularization which may help stabilizing the separation of the Moho and crust-internal effects. It is investigated how much additional regularization is necessary. The *BA* data are given with individual station standard deviations which result from taking averages of values distributed in a 20 km wide strip. The largest a priori observational standard errors σ occur around the shelf and slope (4–6 mGal), but the profile mean is $\pm 2\,\text{mGal}$.

The inversion of the *BA* with the isostatic Moho model (2) (Fig. 7.4.24) is quite similar to that of the *FA* (Fig. 7.4.22). The shapes are similar, but again, the upper crust body is shifted slightly landward and has a slightly bigger core. The calculated density contrasts are also similar (98 ± 7, 147 ± 8 and $-111 \pm 9\,\text{kg/m}^3$). The a posteriori Moho fits more closely to the initial one, and the density contrast $\Delta\rho_\text{m}$ is $155 \pm 4\,\text{kg/m}^3$. In this inversion the wRMSE is 0.9 a priori, i.e. $\pm 2\,\text{mGal}$, and the assumed initial variables are fitted well (0.9 a priori). The upper/lower crust mass distribution is a robust result but with the same kind of restrictions as discussed above.

The inversion of the *BA* with the a priori Moho model (3) and the crustal mass dipole is, in addition, constrained by condition matrices concerning the rhomboid shape (and the density of the core of the upper crustal body). In contrast to rigidly

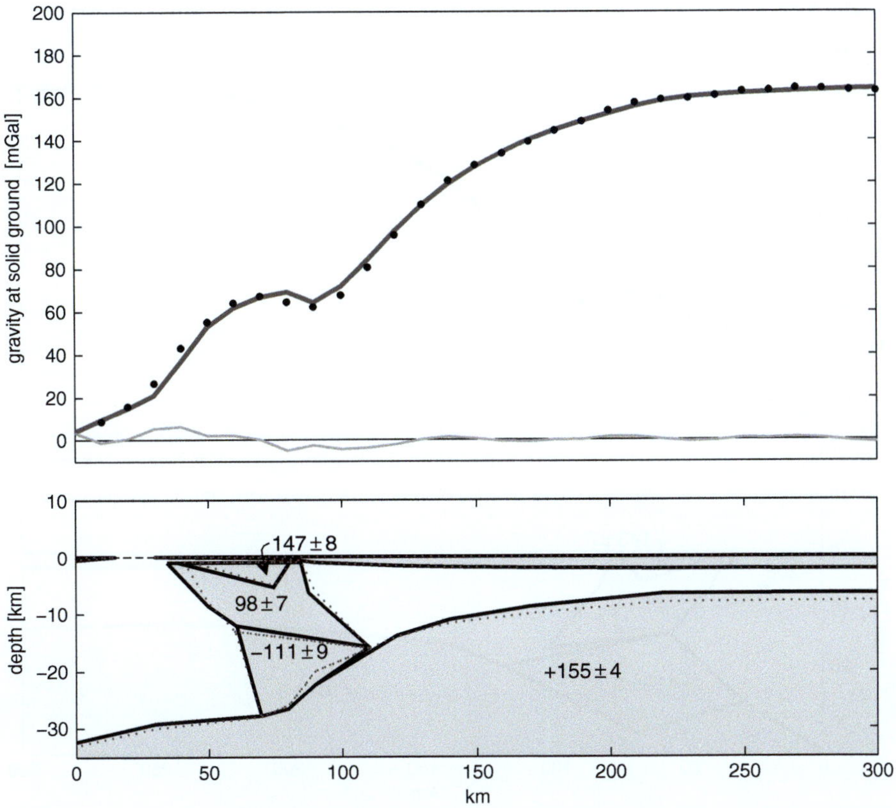

Fig. 7.4.24 Fitting the *BA* with the a priori Moho model (2) and the shelf crustal mass dipole added

fixing relationships between parameters, condition matrices permit them to be constrained more loosely, within specified a priori standard deviations replacing the fixed correlation between the variables. As a brief illustration, take for example, two variables x_1 and x_2 which are a priori expected to change approximately the same way: $\Delta x_1 \approx \Delta x_2$; then the condition can be expressed as $\Delta x_1 - \Delta x_2 = 0 \pm s_x$, i.e., relaxed by permitting an a priori standard error s_x.

The results are shown in Fig. 7.4.25. The wRMSE of the observations is 0.5 times the a priori values ($\sim 1\,\mathrm{mGal}$), the corresponding value for the model is equal to the a priori assumptions of the parameters and the F-statistics is about 1.1. Remember that a priori parameter values and their standard deviations are individually only poorly constrained. The characteristic features required by the *BA* data in the shelf region (where $\sigma_{\mathrm{data}} \approx \pm 5\,\mathrm{mGal}$) are similar to the results of inverting the *FA* (Fig. 7.4.23), but the initial model was slightly modified (the local Moho depression defined by only 3 points which were given the nominal a priori $\pm 5\,\mathrm{km}$ standard deviation). In this solution, the upper body is shifted landward by $\sim 6\,\mathrm{km}$, and the lower body is shifted down by $\sim 1.5\,\mathrm{km}$. The Moho density contrast is $+164 \pm 3\,\mathrm{kg/m^3}$. The upper body density contrast is $95 \pm 6\,\mathrm{kg/m^3}$ with core of $143 \pm 7\,\mathrm{kg/m^3}$, and the

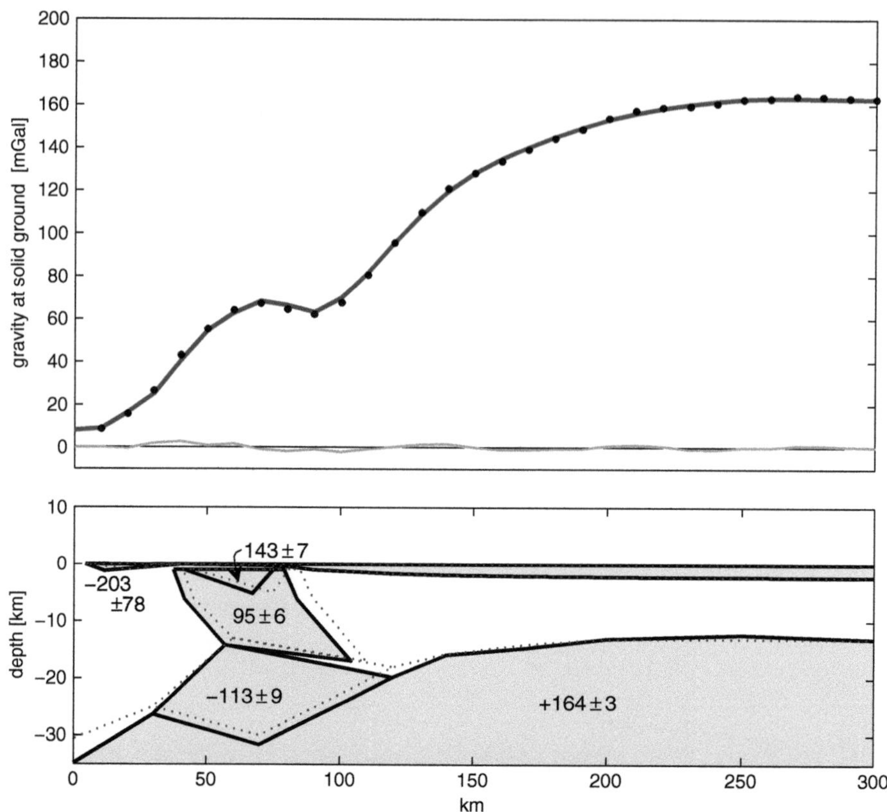

Fig. 7.4.25 Fitting the *BA* with the a priori Moho model (3) and the shelf crustal mass dipole added

lower crustal body has a density contrast $\Delta\rho$ of $-113 \pm 9\,\mathrm{kg/m^3}$. The *sander* body near the coast was included and calculated to be only about 1 km thick with a density contrast of $-203 \pm 78\,\mathrm{kg/m^3}$. Only by tightly ($\pm 10\,\mathrm{kg/m^3}$) constraining the density contrasts (especially for the lower body), volume and shape of the model can be determined with the individual coordinates are only loosely constrained (e.g. $\pm 10\,\mathrm{km}$). The principal dipole nature is, however, again a robust result. All the differences between the models of Figs. 7.4.23 (*FA*) and 7.4.25 (*BA*) are well within the uncertainties of the a priori information.

7.4.2.6 Discussion of Errors and Conclusions

The *rBA, FA* and *BA* all are characterized by a conspicuous dipolar anomaly from which the investigation starts by suggesting itself a somewhat dipolar mass distribution of crustal dimensions. The various inversion attempts point to a vertical dipole in the shelf crust, and the conclusion is difficult to escape that the lower crust has an anomalously low density complemented by a relatively dense upper crust. This speculative interpretation involving the hypothesis of a lower crust of continental origin is supported by the inversion results. High uplift during the early splitting phase of the Greenland shelf led to stronger erosion removing the less dense uppermost rock (comparable to model 2 of Fig. 6.5.5) and the early magmas may have been more mafic than ordinary oceanic and Icelandic basalts (comparable to model 3 of Fig. 6.5.5). Large magnetic anomalies near the shelf edge may support these ideas. Comparison to the seismic model of Jan Mayen Ridge by Kodaira et al. (1998) reveals surprising similarities of structure which may suggest – cautiously – that the continental splinter hypothesis and the so-called "a priori Moho" are reasonable hypotheses, though not proven by independent evidence. There is one large difference between the SE Iceland shelf and Jan Mayen Ridge: the magmatic cover of the continental block appears much thicker on the former than on the latter. Indeed, this seems plausible in view of the proximity to the Iceland plume at the SE shelf and the greater distance around Jan Mayen.

Gravity has the power to render a hypothesis more or less likely. In the present case, the hypothesis of a "light" lower crust and dense upper crust is suggested intuitively by a positive gravity anomaly above a locally depressed Moho. It is strengthened by the fact that even for the "isostatic Moho" without a local depression the hypothesis is supported by the inversion.

Relative to the inversion situation of the Meerfeld Maar (Sect. 7.4.1), beside the geological situation and scale, the a priori information on the structure and the expected dipolar mass distributions is vague. It leads to an intrinsically singular and unstable inversion aggravated by the two features in close proximity that are genetically related to each other: a vertical mass dipole consisting of a light continental splinter from Greenland in the lower crust under a denser magmatic upper crust just where the crust sharply changes thickness from Iceland to the ocean basin. Near-singularity is demonstrated by the sensitivity of the a posteriori variables to small changes in the regularization with locally large gravity residuals. Nevertheless, the a priori geological guiding principle proved acceptable.

The reason for the large errors and deviations in some of the series of inversions (some of which are reported) lies in the uncertain a priori information on the individual parameters. The a priori information on any single parameter has large error bounds. The a priori information is realistic only concerning the whole geological situation which is thus the main a priori constraint. Note that most individual parameters have the potential of large trade-offs. As argued above, the resulting nearly singular situation requires either stronger regularization concerning the parameters than justified by their a priori knowledge, or alternatively, the whole concept must be constrained by assuming condition or correlation matrices, in order to satisfy the whole geological situation, for example, the tentative shape of the source bodies. This was tested in the last inversion shown in Fig. 7.4.25. Confidence in the principal result comes from its consistency and robustness, independent from details of the Moho and of the body shapes.

How convincing is this in view of the other model types of Sect. 6.5.5 (Fig. 6.5.5)? These were characterized by densification through top erosion (2), dense volcanics

(3) and block rotations (4). Relatively small adjustments to their model parameters can satisfy the gravity data equally well. Possibly a valid argument is that all these models are non-isostatic, but the dimensions of $< 100\,$km render this argument weak. The models (2) and (3) have much similarity with the dipole mass model, but missing the weakly resolved low-density lower crustal block. Since it generates the negative gravity side lobes, models (2) and (3) must include another component generating such an effect: mainly the sediments below the plateau slope which must be more effective (density contrast times thickness) than already taken into account. This would be a case of regional mass compensation or isostasy, which is also a realistic concept. It is an interesting demonstration of the difficulty, if not impossibility, of discriminating local and regional isostasy merely on the basis of gravity.

The present example is a case where the gravity data are of greater importance than the a priori information on shape, volume and density contrasts. Missing constraints on the individual model parameters do not prevent a general model type to be modelled by inversion. The analysis shows that specific additional (mostly seismic) a priori information is needed on the sediments and on the crustal structure of the SE Iceland shelf and also on the evolution of the North Atlantic. This demonstrates the value of gravity inversion in specifying questions for future research.

The choice of the SE Iceland shelf as a demonstration of gravity inversion has been made for its tutorial value, not because of a final proven result. In fact, the result is still debatable. However, this is not different in essentially all cases of gravity inversion.

7.5 Outlook

Independent a priori information reduces or even completely solves the ambiguity problem of potential field interpretation. Optimization, for example, by least-squares fitting in the case of inconsistent data, requires implicit and explicit a priori information. Moreover, extensive a priori information is desirable to avoid unacceptable or unusable solutions.

The structure of the a priori information determines how it is transferred into a mathematical formulation of the regularization. Mostly this leads to non-linear optimizing problems. How strongly the non-linearity affects the stability of a solution can be tested by experimenting or by including the higher derivatives of the basic equations (which may be cumbersome). In practice, regularization is necessary to prevent instability by keeping the number of variables acceptably small.

The distribution of the observations relative to the a priori models has an effect on the optimization or inversion results. If not sufficiently constrained, a model body may be moved far from the a priori expectation if the data are better satisfied that way. Sufficiently strong regularization is then required. It is also the question of what is an anomaly if "too few" discrete observations exist relative to an anticipated anomaly. Interpolation on a grid then represents a form of regularization.

The two case histories represent two different situations, one with considerably detailed a priori information (of which some was not even used), the other with general but rather vague a priori information on an ensemble of variables, but not much on them individually. The strategies chosen were thus different.

Comparison of the deviations in the values of the solution with the data and the a priori information enables an evaluation of a current solution to be made, especially if the existing a priori information is explicit and detailed. For a finer evaluation of the potential solution space (especially in the first example), however, a comparison of the a posteriori covariance matrix of the variables is very important with various suitably chosen criterion covariance matrices. Especially the normalizing standard deviations, mostly chosen to be equal to the a priori standard deviations, and the a priori covariance matrices have been emphasized (Sects. 7.2.2.3.8, 7.3.2.2). The former tell us something about the extent of the solution space under the assumption that it is adequate, and in a clearer form than possible considering only the a posteriori covariance matrix. The latter forms the basis for analysing the a priori information more thoroughly and determining its influence on the solution, relative to the gravity data. The solution method can be optimized with the aid of these evaluation criteria, because they tell which a priori information is superfluous and which properties are important and, if not yet used, should be used. For this purpose the meaning of the elements of the correlation and resolution matrices and their eigenvalues and eigen-vectors should be intensively considered, as demonstrated in the case of Sect. 7.4.1.

In the second case (Sect. 7.4.2) this kind of analysis reveals that parts of the model depend completely on the a priori information, here especially the lower body of low density. Only if it is assumed that the a priori concept is correct, can gravity inversion tell something about the shape and density of the whole vertical mass dipole. This is not too surprising since in many such situations gravity alone cannot solve the problem of the source distribution. The chosen approach to modelling the dipolar anomaly was exploratory, beginning with a primitive mass dipole. After a principal fit of the gravity anomaly, a rhomboidal cross section, matching the a priori Moho at the bottom, was tested with some freedom of adjustment to fit the *BA* or *FA*. The cross section was optimized in shape and orientation. This led to a rhomboidal low-density lower crust (of supposedly continental origin) under 200 m of water and a denser upper crust of similar shape and dimensions (supposedly of volcanic origin). The seismically derived cross section of the Jan Mayen Ridge, buried under 1 km of water and only some 2 km of basalt and post-rift sediments, was initially not taken as a priori information for the construction of the a priori model, but the similarity lends support to the preferred a posteriori model as the result of inversion.

For a situation characterized by so little a priori knowledge, experimentation by trial and error is a reasonable way to find or define a priori information, e.g. on acceptable basic shapes. If a typical characteristic shape is suggested or found, it may be preserved in principle during the inversion, for example, roughly rhomboidal, by some form of condition (or covariance) matrix which would impose the condition that neighbouring coordinates (points) would be shifted in a similar fashion. In the case of the SE Iceland shelf, this is a plausible, though geologically not very strong condition. At the experimental stage it produces a fairly stable type of solution. This was the case in the present modelling of the crustal shelf structure.

Practical applications require numerical tools as the program INVERT which was developed in connection with a number of problems of various geological nature. It became evident that even strongly singular problems can be solved such that no contradiction to established facts arises. Nevertheless, new information can be gained from the gravity data. Sensible modelling can be realised much more easily when the parametrization corresponds to the model ideas of the user. Practical experience helps in translating the a priori information into numbers and equations. The results are generally not very sensitive to error estimates (which may be difficult to find numerically), but users must learn to "honestly" quantify the uncertainty of the a priori information without being too keen or too lax.

The correlation matrix and the resolution matrix, as well as their eigen-values and eigen-vectors have proven useful for evaluating the modelling and planning its continuation. But with increasing numbers of variables, it becomes more and more complex to derive the important information. This stresses the importance of a compact parametrization, even if the problem then becomes more strongly non-linear. The presentation, graphical or otherwise, of the eigen-vectors should be made as easy interpretable as possible. Relations of even very heterogeneous variables, also without a natural succession, should be clearly recognizable. It has proven difficult for the search of new measuring methods or of new kinds of information, that several poorly resolved eigen-vectors can be put together in arbitrary linear combinations, such that a possibly useful method could become obvious only by suitably "rotating" the base of eigen-vectors.

Flexibility of a program in its capacity to incorporate every kind of a priori information is usually in conflict with its speed and ease of use. Narrow limiting the wanted minimum speeds up the optimization, but requires complex a priori information to be handled, which is often computing time-intensive. Furthermore, a parametrization fitted to a given problem must be programmed by the user or be provided with the program. A practicable compromise between the two possibilities is always an aim.

Inversion should not be looked at as a magic tool solving all interpretation problems unerringly and automatically approaching the truth, while intuition would be generally misleading. Learning greatly gains from mistakes, doubtful methods and errors. In particular, the researcher still has much influence on the results, for example, by setting the error bounds. which may be very wide. The two-sigma rule (Sect. 7.2.2.3.2) then permits an estimate of realistic a priori errors. The second example demonstrates this. Inversion thus suggests experimentation with various forms of a priori information. It can then guide the search for plausible or acceptable models where independent detailed knowledge from other sources is missing. During experimenting the open-minded researcher will go through a learning process. In the end, hopefully a better, more convincing model will be found, but also a better, more creative researcher.

The program INVERT (PS97) is an example of an efficient gravity inversion routine which incorporates the aspects and devices described in this chapter. INVERT was developed as a research tool in applications to many problems and is today extensively used in industrial exploration applications. An executable version, *INVERT.exe*, and a handbook, *INVERT.doc*, are attached to this book on a CD.

References

Arfken, G.: Mathematical Methods for Physicists, 3rd ed. *Academic Press, San Diego, CA*, 1200pp., 1985

Baarda, W.: S-transformations and criterion matrices. *Publications on Geodesy*, 5, 1168pp., *Netherlands Geodetic Commission, Delft*, 1973 (2nd ed. 1981)

Backus, G.E., Gilbert, F.J.: The resolving power of gross earth data. *Geophys. J. R. Astron. Soc.*, 16, 169–205, 1968

Darbyshire, F.A., White, R.S., Priestley, K.F.: Structure of the crust and uppermost mantle of Iceland from a combined seismic and gravity study. *Earth Planet. Sci. Lett.* 181, 409–428, 2000

Deal, N., Nolet, G.: Nullspace shuttles. *Geophys. J. Int.*, 124, 372–380, 1996

Fedorova, T., Jacoby, W.R., Wallner, H.: Crust-mantle transition and Moho model for Iceland and surroundings from seismic, topography, and gravity data. *Tectonophysics*, 396, 119–140, 2005

Fogel, D.B.: Evolutionary Computation: Toward a New Philosophy of Machine Intelligence, 3rd ed. *IEEE Press, Piscataway, NJ.*, 296pp., 2006

Frazer, A., Burnell, D.: Computer Models in Genetics. *New York, McGraw-Hill*, 1970

Goldberg, D.E.: Genetic algorithms in Search, Optimization and Machine Learning. *Addison-Wesley Longman Publ. Co., Boston*, 372pp., 1989

Gollub, H.G., van Loan, C.F.: Matrix Computations. *Johns Hopkins Univ. Press, Baltimore*, 1989.

Hadamard, J.: Lectures on Cauchy's Problem in Linear Partial Differential Equations. *Yale University Press, New Haven, 1923; Reprint: Dover Publ., New York*, iv+316pp., 1953

Hansen, P.C.: Regularisation tools, a Matlab package for analysis and solution of discrete ill-posed problems. *Report UNIC-92-03, as Postscript via Netlib* (netlib@research.att.com) *from library NUMERALGO*, 1993

Hawthorne, J.B.: Model of a kimberlite pipe. *Phys. and Chem. of the Earth*, 9, 1–15, 1975

Henk, A.: Zur Geologie und Geophysik des Meerfelder Maares und seiner Umgebung/Westeifel. *Diplomarbeit (Master's thesis), Mainz*, 1984

Holland, J.H.: Genetic algorithms. *Scientific American*, 267, 17–30, 1992

Hunsche, H.: Geomagnetische, geoelektrische und magnetotellurische Messungen im Rahmen einer Exkursion zum Mosenberg und Meerfelder Maar in der Westeifel, *Gamma, 22, Braunschweig*, 1973

Jackson, D.D.: Interpretation of inaccurate, insufficient and inconsistent data. Geophys. *J. R. Astron. Soc.*, 28, 97–109, 1972

Jackson, D.D.: Marginal solutions to quasi-linear inverse problems in geophysics: The edgehog method. *Geophys. J. R. Astron. Soc.*, 35, 121–136, 1973

Jacoby, W.: Schweremessungen auf Helgol and Auswertung mit Ausgleichsverfahren Z. Geophysik, 32, 340–351, 1966.

Kaban, M., Flóvenz, O. and Pálmason, G.: Nature of the crust-mantle transition zone and the thermal state of the upper mantle under Iceland from gravity modelling, *Geophys. J. Int.* 149, 281–299, 2002

Kodaira, S., Mjelde, K., Gunnarsson, H., Shiobara, H., Shimamura, H.: Structure of the Jan Mayen microcontinent and implications for its evolution. *Geophys. J. Int.* 132, 383–400, 1998

Lanczos, C.: Linear and Differential Operators. *Van Nostrand, London*, 564pp., 1961

Levenberg, K.: A method for the solution of certain non-linear problems in least squares. *Quart. Appl. Math.* 2, 164–168, 1944

Lorentz, V.: On the growth of maars and diatremes. *Bull. Volcanol.*, 48, 265–274, 1986

Lorentz, V., Büchel, G.: Zur Vulkanologie der Maare und Schlackenkegal der Westeifel Mitteilungen Polichia, 68, 28–100, 1980.

Marquardt, D.W.: An algorithm for least squares solutions of nonlinear parameters. *J. SIAM*, 11, 431–441, 1963

Menke, W.: Geophysical Data Analysis. *Academic Press, Orlando*, 289pp., 1984

Menke, W.: Crustal isostasy indicates anomalous densities beneath Iceland. *Geophys. Res. Lett.* 26, 1215–1218, 1999

Mertes, H.: Aufbau und Genese des westeifeler Vulkanfeldes. *Bochumer geol. Geotech. Arb.*, 415pp., 1983

Meyer, W.: Geologie der Eifel. *Schweitzerbart, Stuttgart*, 615pp., 1988

Mitchell, M.: An Introduction to Genetic Algorithms. *MIT Press, Canbridge MA*, 221pp., 1998

Nelder, J.A., Mead, R.: A simplex method for function minimization. *Computer J.*, 7, 308–313, 1965

Nolet, G.: Seismic wave propagation and seismic tomography. In "Seismic Tomography, Riedel", ed., G. Nolet, 1–23, *Norwell, Mass.*, 1987

Paige, C.C., Saunders, M.A.: LSQR, an algorithm for sparse linear equations and sparse least squares. *ACM Trans. Math. Software*, 8, 43–71, 1982

Penrose, R.C.: Computerdenken: die Debatte um künstliche Intelligenz, Bewußtsein und die Gesetze der Physik (Original title: The Emperor's New Mind. Concerning Computers, Minds, and the Laws of Physics. *Oxford. U.P., New York*, 1989). *Spektrum d. Wissenschaft, Heidelberg*, 454pp., 1991

Press, W.H., Teukolsky, S.A., Vetterling, W.T., Flannery, B.P.: Numerical recipes in FORTRAN, 2nd ed. *Cambridge University Press, New York*, xxvi + 1007pp., 1992

PS97: see Smilde, 1997, 1998

Rao, C.R.: Linear Statistical Inference and its Applications, 3rd ed. *Wiley, New York*, 656pp., 1973

Sabatier, P.C.: On geophysical inverse problems and constraints. *J. Geophys.*, 43, 115–137, 1977

Sebazungu, E.: Investigations of maar diatreme volcanoes by inversion of magnetic and gravity data from the Eifel area in Germany. *PhD thesis [unpubl.], Mainz*, 2005

Smilde, P.: GRAVI, unpublished computer program for calculation of relative gravity values from field measurements. *Available on request from author*, 1995

Smilde, P.L.: Verwendung und Bewertung von a priori Information bei potentiell singulären Inversionsproblemen am Beispiel der gravimetrischen Bestimmung von Dichteverteilungen. *PhD thesis [unpubl.], Mainz*, 1997 *[referred to as PS97]*

Smilde, P.L.: Verwendung und Bewertung von a priori Information bei potentiell singulären Inversionsproblemen am Beispiel der gravimetrischen Bestimmung von Dichteverteilungen. *Deutsche Geodät. Komm., C 490, München*, 1998 *[also referred to as PS97]*

Stachel, T., Büchel, G.: Das Döttinger Maar: Fallstudie eines großen tertiären (?) Tuffschlotes im Vulkanfeld der Hocheifel. Z. dt. *Geol. Ges*, 140, 35–51, 1989

Tarantola, A.: Inverse Problem Theory and Model Parameter Estimation, 2nd ed. *SIAM Publ., Philadelphia*, xvi+613pp., 2005

Tikhonov, A.N., Arsenin, V.Y.: Solutions of Ill-posed Problems. *Winston & Sons, Washington, D.C. and Wiley, New York*, XIII+258pp., 1977

Appendix: Analytical Geometry

M1

Definition of a plane in x, y, z by three non-identical and non-collinear points $P_i = (x_i, y_i, z_i)$, where $i = 1, 2, 3$. Normal vector $\boldsymbol{p}$ from O (0, 0, 0) to plane

The following expressions describe a plane in x, y, z:

$$z = a + bx + cy \tag{M1.1}$$

$$Ax + By + Cz + D = 0 \tag{M1.2}$$

$$x/x_o + y/y_o + z/z_o = 1 \tag{M1.3}$$

$$xp_x + yp_y + zp_z - p = 0, \text{ with } p = (p_x, p_y, p_z) \tag{M1.4}$$

where x_o, y_o and z_o are the axis intercepts of the plane and $\boldsymbol{p}$ is the plane-normal vector from O (0, 0, 0) to the plane and $p = |\boldsymbol{p}|$).

Some of the relations between the various parameters are:

$$\left.\begin{array}{l} a = -D/C,\ b = -A/C,\ c = -B/C; \\ x_o = -D/A = -a/b, y_o = -D/B = -a/c,\ z_o = -D/C = a; \\ p_x = p^2/x_o; p_y = p^2/y_o; p_z = p^2/z_o, p^2 = (p_x{}^2 + p_y{}^2 + p_z{}^2) \end{array}\right\} \tag{M1.5}$$

Inserting the three-point coordinates leads to three linear equations for the unknowns, a, b, c:

$$\underline{A}v = w \tag{M1.6}$$

with

$$\underline{A} = \left\{\begin{array}{ccc} 1 & x_1 & y_1 \\ 1 & x_2 & y_2 \\ 1 & x_3 & y_3 \end{array}\right\}; \quad v = (a, b, c); \quad w = (z_1, z_2, z_3) \tag{M1.7}$$

The solution is obtained by applying Cramer's rule with the determinants $|\boldsymbol{A}|$, $|\boldsymbol{A_a}|$, $|\boldsymbol{A_b}|$, $|\boldsymbol{A_c}|$, where in $|\boldsymbol{A_a}|$ the first column of $|\boldsymbol{A}|$ is replaced by $\boldsymbol{b}^T$, and in $|\boldsymbol{A_b}|$ the second and in $|\boldsymbol{A_c}|$ the third, accordingly:

$$\begin{aligned}
|\boldsymbol{A}| &= x_1 y_2 - x_2 y_1 + x_2 y_3 - x_3 y_2 + x_3 y_1 - x_1 y_3 \\
|\boldsymbol{A_a}| &= x_1(y_2 z_3 - y_3 z_2) + x_2(y_3 z_1 - y_1 z_3) + x_3(y_1 z_2 - y_2 z_1) \\
|\boldsymbol{A_b}| &= -[y_1 z_2 - y_2 z_1 + y_2 z_3 - y_3 z_2 + y_3 z_1 - y_1 z_3] \\
|\boldsymbol{A_c}| &= x_1 z_2 - x_2 z_1 + x_2 z_3 - x_3 z_2 + x_3 z_1 - x_1 z_3
\end{aligned} \tag{M1.8}$$

Then

$$a = |\boldsymbol{A_a}|/|\boldsymbol{A}|, \quad b = |\boldsymbol{A_b}|/|\boldsymbol{A}|, \quad c = |\boldsymbol{A_c}|/|\boldsymbol{A}| \tag{M1.9}$$

From a, b, c any of the other parameters can be calculated by applying the appropriate relations of (M1.5), especially the distance

$$p = a/(1 + b^2 + c^2)^{1/2} \tag{M1.10}$$

M2

Normal vector $\boldsymbol{p}$ from O $(0.0,0)$ to straight line $\boldsymbol{s}_{12}$ in x, y, z through two points P_1, P_2 or point vectors $\boldsymbol{r}_1$, $\boldsymbol{r}_2$

The line $\boldsymbol{s}_{ij}$ in x, y, z is defined by

$$\boldsymbol{s}_{12} = \boldsymbol{r}_2 - \boldsymbol{r}_1 = (x_2 - x_1, y_2 - y_1, z_2 - z_1) = (\Delta x, \Delta y, \Delta z)$$

The scalar products $\boldsymbol{r}_{ij}\boldsymbol{p} = p^2$ $(i = 1, 2)$ and $\boldsymbol{s}_{12}\boldsymbol{p} = 0$ render three linear equations for p_x, p_y, p_z which can be written as

$$\underline{\boldsymbol{C}}\boldsymbol{p} = \boldsymbol{q}$$

with

$$\underline{\boldsymbol{C}} = \left\{ \begin{matrix} x_1 & y_1 & z_1 \\ x2 & y_2 & z_2 \\ \Delta x & \Delta y & \Delta z \end{matrix} \right\}; \quad \boldsymbol{p} = (p_x, p_y, p_z); \quad \boldsymbol{q} = p^2(1, 1, 0) \tag{M2.1}$$

The solution is obtained the same way as in Appendix M1 with the determinants $|\boldsymbol{C}|$, $|\boldsymbol{C_x}|$,

$$\begin{aligned}
&|\boldsymbol{C_y}|, |\boldsymbol{C_z}|, \text{accordingly;} \\
|\boldsymbol{C}| &= \Delta x(y_1 z_2 - y_2 z_1) + \Delta y(x_2 z_1 - x_1 z_2) + \Delta z(x_1 y_2 - x_2 y_1 \\
|\boldsymbol{C_x}| &= p^2[(y_2 - y_1)\Delta z + (z_1 - z_2)\Delta y] = p^2 A \\
|\boldsymbol{C_y}| &= p^2[(z_2 - z_1)\Delta x + (x_1 - x_2)\Delta z] = p^2 B \\
|\boldsymbol{C_z}| &= p^2[(x_2 - x_1)\Delta y + (y_1 - y_2)\Delta x] = p^2 C
\end{aligned} \tag{M2.2}$$

In this case, p is obtained in the same way as in M1:

$$\left.\begin{aligned}
p &= D/(A^2+B^2+D^2)^{1/2} \text{ and}\\
p_x &= AD/(A^2+B^2+D^2)\\
p_y &= BD/(A^2+B^2+D^2)\\
p_z &= CD/(A^2+B^2+D^2)
\end{aligned}\right\} \qquad \text{(M2.3)}$$

A1

The cylindrical cycloid (Fig. A.1)

Many geological bodies are outlined on maps by closed curves. Near the surface, gravity has the best resolution. The geometrically simplest approximation for such geometries is the cylinder. A circular cylinder is described in 3D by 8 variables: 2 axis coordinates, 1 radius, 1 surface depth or height, 1 bottom depth, 1 axial dip and its corresponding azimuth and 1 density. For vertical or horizontal cylinders the latter two angles can be dropped.

For more complicated outlines than circles, cycloids are appropriate (Fig. A.1). In polar coordinates the radius r_o of a circle is modulated as a function of the angle α by a sine wave of a certain frequency, n, and phase, α_n. The frequency must be an integer to get a closed curve. The curve is described by a periphery point of a circle of radius r_n which rotates with the frequency n while its centre rotates with frequency 1 about the main circle in the same direction making an epicycloid or in the opposite direction making a hypocycloid; n determines how many maxima and minima the radius length has, r_n determines the greatest deviation from r_o and α_n gives the orientation of the maxima and minima. The hypocycloid of frequency 2 is an ellipse with the axes $r_o + r_n$ and $r_o - r_n$. Several frequencies can be linearly superimposed, each with its frequency, amplitude and phase.

The N regularly distributed points of the periphery of the cycloid are calculated for $k = 0, N$ by

$$\begin{aligned}
x_k &= x_c + |r_o|\cos[2\pi(r_o/|r_o|)(k-1/2)/N)]\\
&\quad + \sum_i r_i \cos[2\pi((n_i - r_i/|r_i|)(r_o/|r_o|)(k-1/2)/N) - n_i\alpha_i]\\
y_k &= y_c + |r_o|\sin[2\pi(r_o/|r_o|)(k-1/2)/N)]\\
&\quad + \sum_i r_i \sin[2\pi((n_i - r_i/|r_i|)(r_o/|r_o|)(k-1/2)/N) - n_i\alpha_i]
\end{aligned} \qquad \text{(A.1)}$$

where x_c and y_c are the centre coordinates, i counts the superimposed frequencies, n_i are their integer frequencies, α_i are their phases and r_i are their amplitudes which make hypocycloids if positive and epicycloids if negative. A cylindrical cycloid of frequency n requires $2n + 8$ variables, and only $2n + 6$ if the axial orientation and density are fixed.

While such an outline cannot approximate any arbitrary form, it can describe with few parameters rather complex bodies, and the parameters are imaginable so that a priori information can be easily translated into the parameter values and their scatter. The transformation into cycloid parametrization can be applied to reparametrization of triangulated surfaces (polyhedra), plates and 2D Talwani beams.

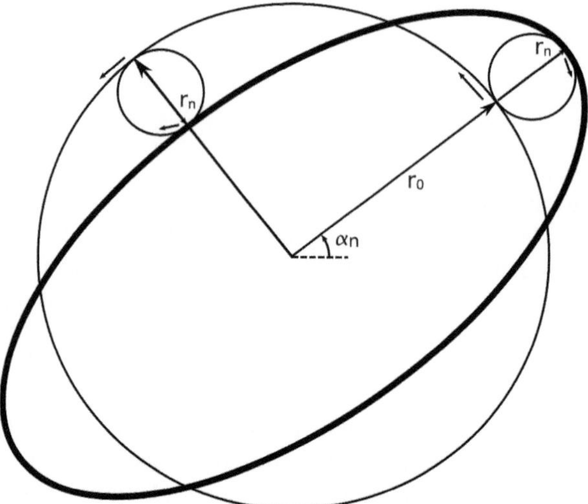

Fig. A.1: Cylindrical cycloid. The example shows a hypocycloid of frequency $n = 2$ describing an ellipse

Index